The Neutron and its Applications, 1982

Sir James Chadwick (1891–1974)

The Neutron and its Applications, 1982

Plenary and invited papers from the Conference to mark the 50th Anniversary of the Discovery of the Neutron held at Cambridge, 13–17 September 1982

Edited by P Schofield

Conference Series Number 64

The Institute of Physics
Bristol and London

Copyright © 1983 by The Institute of Physics and individual contributors. All rights reserved. Multiple copying of the contents or parts thereof without permission is in breach of copyright but permission is hereby given to copy titles and abstracts of papers and names of authors. Permission is usually given upon written application to the Institute to copy illustrations and short extracts from the text of individual contributions, provided that the source (and, where appropriate, the copyright) is acknowledged. The code at the bottom of the first page of an article in this book indicates that copies of the article may be made in the USA for personal or internal use, on condition that the copier pays the stated per-copy fee to the Copyright Clearance Center, Inc, for copying beyond that permitted by Section 107 or 108 of the US Copyright Law.

CODEN IPHSAC 64 1–522

British Library Cataloguing in Publication Data
The Neutron and its Applications, 1982. – (Conference series/Institute of Physics, ISSN 0305-2346; no. 64)
    1. Neutrons–Congresses
    I. Schofield, Peter, 1983   II. Institute of Physics
    I. Title
    539.7'213   QC793.5.N462
    ISBN 0-85498-155-1

Organising Committee
    P Schofield (Chairman), J P Baldwin, A D Boardman, M G Brereton, M D Cohler, J E Enderby, D Greene, M R Hawkesworth, M T Hutchings, J M Irvine, R J Stewart, D S Whitmell, B T M Willis

International Advisory Panel
    A Abragam, A Bohr, G E Bacon, B N Brockhouse, W E Burcham, W J L Buyers, G Caglioti, A H Cook, D Cribier, T Ericson, J F Fowler, M Goldhaber, Y Ishikawa, P K Iyengar, J Janik, N Kroo, A M Lane, K E Larsson, W B Lewis, A R Mackintosh, W Marshall, M Oliphant, R E Peierls, N F Ramsey, T Riste, C G Shull, T Springer, L Van Hove, G H Vineyard, S Weinberg, D H Wilkinson, M K Wilkinson

Special Plenary Sessions
    D H Wilkinson, E R Dobbs, L Cohen, S Dugdale, P Schofield

Honorary Editor
    P Schofield

The following groups and sub-committees contributed to the organisation of the conference:
    Institute of Physics:— Nuclear Physics Sub-Committee, Solid State Physics Sub-Committee, Atomic Collisions in Solids Group, Computational Physics Group, Magnetism Group, Materials and Testing Group, Neutron Scattering Group†, Nuclear Interactions Group, Crystallography Group, Polymer Physics Group†, Carbon Group†. The Hospital Physicists' Association
    † Joint Groups of The Institute of Physics and The Royal Society of Chemistry

Sponsors
    The conference and exhibition to mark the 50th anniversary of the discovery of the neutron was organised by The Institute of Physics, in collaboration with Harwell and sponsored by the European Physical Society and the UK Atomic Energy Authority with support from The Royal Society, Institut Laue-Langevin, European Research Office of US Army and National Science Foundation

Published by The Institute of Physics, Techno House, Redcliffe Way, Bristol BS1 6NX and 47 Belgrave Square, London SW1X 8QX, England.

Printed in Great Britain by J W Arrowsmith Ltd, Bristol

These Proceedings are dedicated to the memory of Sir James Chadwick

Preface

Of all the great scientific discoveries between the two World Wars, it is undoubtedly that of the neutron by Chadwick in 1932 that has had the most profound consequences. It was therefore appropriate that a special scientific conference should be held to mark the 50th anniversary of the discovery and that it should be held around the site of the old Cavendish Laboratory in Cambridge, where the discovery was made.

The conference, with accompanying exhibitions, was organised by The Institute of Physics in collaboration with Harwell and took place from 13th to 17th September, 1982. It was attended by some 500 delegates, about half from overseas; the attendance would have been greater were it not for the economic blight which has hit our universities since planning of the conference started in 1978. In addition to the 12 plenary and 65 invited papers published here, there were some 230 contributed papers presented orally in parallel sessions or as posters, many of which will appear in The Institute of Physics journals.

The scientific programme of the conference had three main themes, reflected in the ordering of the papers in these Proceedings. 'Neutron Physics' covered the properties of the neutron and its role in nuclear physics and in astrophysics. 'Neutron Scattering' explored the ever-expanding role of the neutron as a probe of the molecular structure and dynamics of matter, which has led to major advances in many branches of science. 'Technology and Applications' dealt with the wide variety of uses of neutrons through activation analysis, radiography and in medicine, as well as in present and possible future methods of power generation.

There were two special sessions. The conference opened with a Commemorative Session. In his opening address Lord Sherfield recalled his association with Chadwick in Washington during the Manhattan Project and offered some personal reflections on 50 years of advances in physical sciences. The discovery of the neutron was a classic example of the value of pure basic research, unencumbered by the demands of customers, be they government departments or industrial managers. Citing C P Snow's two cultures, he pointed out that the serious communication gap was not between scientists and those trained in the arts but between science and government, and he stressed the importance of finding a solution to the problem of bringing the scientist and the engineer into the policy and decision making process. There followed reminiscences and reflections of some of those who were working in the Cavendish in 1932 and other pioneers of neutron research. A summary is found on pp 1–4. (A recording of this session on cassette is obtainable from The Institute of Physics, 47 Belgrave Square, London SW1X 8QX.)

The meeting would have been incomplete without attention to the two crucial public issues of Nuclear Weapons and Nuclear Power. These were the subjects of a special plenary session on Wednesday, 15th September, under the Chairmanship of Lord Zuckerman. The principal speakers were Richard Garwin (IBM) and Sir Walter Marshall (CEGB) respectively, whose papers are printed here. They were followed, in turn, by panel discussions. On weapons, Sir William Cook, Sir Sam Edwards and Sir Harry Tuzo contributed to a sober and informative assessment of current realities. The Nuclear Power discussion was enlivened by Lord Bowden's impassioned advocacy of the Canadian 'CANDU' system; other speakers were Gareth Price (Shell) and Terry Price (The Uranium Institute) who contributed an overview of 'Policy Issues Relating to Nuclear Power'.

I wish to express my gratitude to the majority of the invited speakers who submitted their papers on time and to the restrictive requirements on length. They may rest assured that their contributions gained in clarity what was sacrificed in detail. Others proved more adept at special pleading. A few authors took the opportunity of the occasion to indulge in philosophical speculation: Bowden, in the Commemorative Session, on 'the basic improbability of nuclear physics', D Clayton, in pointing out the sensitivity of our present model of the universe to the mass, lifetime and other properties of the neutron, Armbruster in relating his current activities to those of the alchemists, and Brockhouse in illustrating his view of the nature of Physics through the history of slow neutron spectroscopy; the Editor, not wishing to act as censor, has been indulgent in his turn.

Many people contributed to the success of the meeting. First, I should record our gratitude to those bodies, listed on p. iv, who provided direct financial support. An exhibition of books and scientific equipment, organised by Miss Anne Bellion, also contributed to the income and we are grateful to Miss Bellion and her exhibitors for their support. A fascinating exhibition on the history of neutron science was assembled by J W White (Oxford) and C G Windsor (Harwell) and thanks are due to the many individuals and institutes who provided documents, apparatus and models. An exhibition illustrating the theme of the conference was coordinated by F J Stubbs (Harwell) with contributions from the Institut Laue-Langevin, the Rutherford Appleton Laboratory, the Hammersmith Hospital as well as UKAEA laboratories. These two exhibitions were presented by the Harwell Design Studio under P Mulford.

The suggestions of the International Advisory Panel regarding programme and speakers were invaluable and of great help to the Organising Committee in the attempt to achieve a reasonable balance to the programme, the success of which may be judged by the contents of this volume.

Finally, a great debt of gratitude is due to Miss Delia Mayston and staff of the Meetings Department of The Institute of Physics for the smooth running of the Conference. Perhaps only those associated with the organisation can appreciate the extraordinary complexity involved in arranging such an event in Cambridge!

The success of a conference depends ultimately on the participants. It was gratifying that so many entered into the multi-disciplinary spirit of the occasion to explore fields other than their own. It is hoped that the Proceedings will have a similar appeal.

P Schofield
October, 1982

Contents

Preface

Section 1: Introduction

1–4 The commemorative session
   J Hendry

Section 2: Nuclear physics and astrophysics

5–14 Particle properties of the neutron (Plenary Lecture)
   N F Ramsey

15–20 The neutron lifetime
   J Byrne

21–27 The structure of the neutron (Plenary Lecture)
   C H Llewellyn-Smith

29–31 The question of baryon conservation
   M Goldhaber

33–43 The neutron and the universe (Plenary Lecture)
   D D Clayton

45–50 Neutron stars: the first fifty years
   G Baym

51–56 Neutron superfluidity
   M Ruderman

57–60 Grand unified theories and cosmology
   J Ellis

61–64 Neutrons in the early universe
   R J Tayler

65–70 Cosmic neutrinos
   G Steigman

71–74 Solar neutrinos
   J N Bahcall

75–81 Reactor neutrinos
   F Reines

83–88 The role of neutrons in nucleosynthesis in supernovae
   W A Fowler

89–94 Light element nucleosynthesis
J Audouze

95–100 The role of the neutron in heavy element nucleosynthesis
J W Truran

101–104 The transuranium elements
G T Seaborg

105–117 Heavy ion fusion and exotic nuclei studies at SHIP
P Armbruster

119–124 Neutron-induced fission and the structure of the fission barrier
J E Lynn

125–135 Nuclear reactions caused by neutrons (Plenary Lecture)
A M Lane

137–142 The neutron optical model
P E Hodgson

143–148 Neutron single particle states in nuclei
E Friedman

149–156 Isotope shifts
F Träger

Section 3: Neutron optics

157–168 Wave properties of the neutron (Plenary Lecture)
C G Shull

169–176 The present state of neutron interferometry
H Rauch

177–180 Production and properties of ultracold neutrons
A Steyerl

181–186 Neutron spin echo
F Mezei

Section 4: Neutron scattering

(a) Condensed matter physics

187–192 Neutron diffraction
G E Bacon

193–198 Slow neutron spectroscopy: an historical account over the years 1950–1977
B N Brockhouse

199–214 Neutron scattering and magnetism (Plenary Lecture)
A R Mackintosh

215–220 Neutron scattering review – magnetic studies
W C Koehler

221–226 Linear and non-linear magnetic excitations in perfect and disordered crystals
W J L Buyers

227–232 Neutron scattering studies of itinerant electron magnets
Y Ishikawa

233–238 Magnetism in the actinides: the role of neutron scattering
G H Lander

239–244 Elementary excitations in one-dimensional magnets
J Villain

245–250 Neutron scattering and phase transitions
R A Cowley

251–254 Phonons and their interactions
R M Nicklow

255–260 Incommensurate structures
J D Axe

261–266 Liquid helium
E C Svensson

(b) Molecular physics

267–270 Neutron studies of liquids and dense gases
P A Egelstaff

271–276 The structure and dynamics of ionic solutions
J E Enderby

277–282 Disorder and correlations in molecular and liquid crystals
A J Leadbetter

283–288 Neutron scattering in chemistry (Scattering from layer lattices and their intercalation compounds – an illustration)
J W White

289–294 Diffraction measurements on physisorbed films
M Nielsen, J Bohr, K Kjaer and J P McTague

295–300 Molecular reorientation and tunnelling motions
S Clough

(c) Materials science

301–304 Neutron small angle and diffuse scattering
W Schmatz

305–310 Neutron diffraction studies of amorphous solids
A C Wright

(d) Polymers

311–315 Contribution of neutron scattering experiments to the understanding of conformation and dynamics of polymer molecules
J S Higgins

317–319 Application of neutron scattering to crystalline polymers
A Keller

321–328 Neutron scattering of flowing polymers
R Oberthür

329–333 Problems involving H and D polymer mixtures
S F Edwards

(e) Biology

335–346 The contribution of neutron scattering to molecular biology (Plenary Lecture)
H B Stuhrmann

347–350 Label triangulation
R P May

351–355 Neutron scattering studies of virus structure
S Cusack

357–359 Neutron studies of connective tissue
A Miller

361–364 Neutrons and proteins
B P Schoenborn and V Raghaven

Section 5: Reactor physics

365–370 Physics design methods for pressurized water reactors
R L Hellens

371–376 50 years of neutron transport
N R Corngold

377–382 Hilbert space method for the numerical solution of reactor physics problems
R T Ackroyd

383–388 Neutron spectra
R J Brissenden

389–394 Nuclear data for fission reactors
A Michaudon

395–402 The role of the neutron in controlled thermonuclear fusion
V S Crocker, T D Beynon and L J Baker

Contents

Section 6: Radiation effects

403–410 Radiation damage (Plenary Lecture)
M W Thompson

411–416 Atomic processes occurring during irradiation damage by neutrons
R S Nelson

417–422 Radiation damage effects in fast reactor design and performance
J F W Bishop

423–428 Radiation effects in fuel materials for fission reactors
Hj Matzke

429–433 The use of sink strengths in the theory of radiation damage
R Bullough

Section 7: Applied neutron physics

435–446 Uses of neutrons in engineering and technology (Plenary Lecture)
J Walker

447–450 Neutron radiography – accomplishments and potential
J P Barton

451–456 Neutron applications in the applied earth sciences
C G Clayton

457–462 Laboratory neutrons – a breakthrough in non-nuclear disciplines
R E Jervis

463–468 Neutron techniques in safeguards
M S Zucker

Section 8: Medical

469–478 Neutrons in medicine (Plenary Lecture)
J F Fowler

479–484 Neutron radiobiology: dependence of the relative biological effectiveness on neutron energy, dose and biological response
G W Barendsen

485–490 In vivo activation analysis in medicine
K Boddy

Section 9: Sources

491–496 Neutron sources, future plans and possibilities
G A Bartholomew

497–502 The IBR-2 reactor as a pulsed neutron source for scientific research
I M Frank et al

Section 10: Nuclear power

503–512 Neutrons and nuclear power (Plenary Lecture)
W Marshall

Section 11: Nuclear weapons

513–520 The neutron and nuclear weapons (Plenary Lecture)
R L Garwin

521–522 **Author Index**

The commemorative session

John Hendry

Historians' Office, UKAEA, 11 Charles II Street, London SW1Y 4QP

The Commemorative Session was introduced by Sir Denys Wilkinson, who first recalled one of his own encounters with Sir James Chadwick, the discoverer of the neutron, and then sketched the origins of the neutron concept during the 1920s. Comparing the present day concept of the neutron with that of fifty years ago he noted that on neither view was the neutron an elementary particle. In 1932, physicists were unanimous in interpreting the newly discovered neutron as a bound state of the two established elementary particles, a proton and an electron. Soon after, the concept of an elementary neutron had been proposed and accepted, but in the ensuing decades the neutron had grown to be a far more complex entity than at first envisaged. Its components to date included three quarks, two kinds of vacuum, and a variety of instantons, gluons, and other ingredients.

The first recollections of 1932 were given by John Ratcliffe, who worked at the Cavendish Laboratory in Cambridge from 1924 to 1960 teaching and researching on radio waves and ionosphere. Ratcliffe showed first a portrait of Lord Rutherford, director of the Laboratory in 1932, a "big man with a big voice", and a "good straight-forward simple-minded sort of person"; and then a picture of the Laboratory buildings, "dingy, dirty and dismal". He next drew attention to that important part of the work of the Laboratory that had not been concerned with nuclear physics, including that of C.T.R. Wilson on thunderstorms and atmospheric conditions, of G.I. Taylor on fluid motion and turbulence, of Kapitza on low temperature physics and strong magnetic fields, and of the group in which he himself worked, originally under Appleton. He then emphasised and illustrated, through anecdotes and photographs, the fundamental theme of "simplicity" which ran through Rutherford's research and that of the Laboratory in general, and which was manifest both in the experiments performed and in the theoretical thinking behind them.

The next speaker was Maurice Goldhaber, who had left Germany while still a student in 1933 and had been accepted by Rutherford for research at the Cavendish. Goldhaber first described something of Chadwick's background before 1932, including his discovery of the continuous beta-spectrum and his time spent as an internee in Germany during the First World War, studying radioactivity and investigating a German toothpaste advertised as radioactive. He then described his own arrival in Cambridge and his admission, in those days without any formalities, at Magdalene College. He explained that he started out as a theoretician and had therefore been assigned to R.H. Fowler, but that he had then switched over to Chadwick to work on the nuclear photoelectric

that had affected him in particular was the increasing presence of
theoretical physics in and around the laboratory, resulting in part
from the good sense of Rutherford's daughter in marrying Fowler, who
had himself been originally a pure mathematician, but who in due course
took a room in the Cavendish and began a school of theoretical physics.

The second part of the session was opened by Lord Bowden, who spoke
of the basic improbability of nuclear physics, illustrating his thesis
through three examples. The first improbability, he said, concerned
the scintillation screens used by Rutherford to detect alpha rays. The
crystal used, zinc sulphide with about one part in ten thousand of
copper, was uniquely and unexpectedly sensitive to the impact of alpha
rays; not only was a full quarter of their energy transferred to radiation
but this radiation was also in the yellow-green range of the spectrum
to which the eye is most sensitive. Rutherford's discovery of the atomic
nucleus, Cockroft and Walton's demonstration of artificial disintegration,
and much of nuclear physics depended upon improbable coincidence that
this crystal should exist and have been available just when Rutherford
first needed it. A second improbability concerned C.T.R. Wilson's
cloud chamber, also essential to the development on nuclear physics
in the 1920s and 1930s, which, said Lord Bowden, owed its existence to
Wilson's love of mountain-climbing and his resulting interest in the
condensation phenomena he observed. The third improbability concerned
Wynn-Williams's amplifier, again a crucial ingredient of nuclear physics
research during the early 1930s. The construction of amplifiers for
particle detection had been fraught with difficulties, and Wynn-Williams's
success rested on the under-running of a special type of valve not
generally available but made by Marconi for use in RF amplifiers.
Without these valves the amplifiers could not have been made when they
were, and they were in fact non-reproducible for some years. Lord
Bowden concluded by noting that the improbabilities were in one sense
relative, in that if Rutherford had not made the discoveries he did,
he would almost certainly have done something else equally earth-shaking:
the whole world would have been very different, and physics might have
developed very differently.

Eduardo Amaldi recalled how the discovery of artificial radioactivity
by the Joliots in 1934 had provided the occasion for Enrico Fermi to
switch over from theoretical to experimental work, and described the
experimental neutron research conducted by Fermi's group in Rome, of
which he was a member, from 1934 to 1936. He paid particular attention
to their study of the behaviour of thorium and uranium under neutron
bombardment in 1934-1935, and to their conclusions, supported by Hahn
and Meitner in Berlin, on the existence and nature of transuranic elements.
He also discussed the way in which the developing theoretical ideas
about nuclear structure both depended on and illuminated the experimental
work on neutron cross sections.

Finally, Sir Rudolf Peierls offered some reflections on the development of our theoretical understanding of neutron physics. He discussed
the reasons why the neutron was at first interpreted as a bound state of
proton and electron despite the known problems arising from the assumed
existence of electrons in the nucleus, and he traced the steps by which
it became a fundamental particle in its own right, following the
discovery of the positron and Fermi's theory of beta decay. Sir Rudolf
then followed the history of the concept and growth of neutron physics
through the changing subject categories used in the scientific journals

effect. He continued by describing this work, which included an accurate determination of the neutron mass, thus ending a three-way debate between Chadwick, the Joliots and Ernest Lawrence. He recalled that their realisation of the importance of slow neutrons had preceded Fermi's famous work on the subject, but had been thought in the Cavendish to be over-speculative.

In conclusion, Goldhaber paid tribute to the years of preparatory work put in by Chadwick before the discovery of the neutron, which had made possible both that discovery and the rapid succession of results that had followed. He also mentioned the help given over many years by Chadwick's assistant, Nutt.

The Chairman of the session, W.E. Burcham, then read some reminiscences of Chadwick's discovery of the neutron by P.I. Dee who was unable to be present himself at the conference. Dee was working with C.T.R. Wilson in Cambridge at the time that Chadwick was examining the properties of the "Beryllium radiation". He recalled his suggestion that measurements of the range of electrons recoiling from this radiation would very clearly distinguish between photons and neutrons because of the great difference in recoil energy to be expected. The recoil electrons were sought for in a cloud chamber but not found in these first experiments because of the smallness of the collision cross section; Dee commented on the reactions of Rutherford and of Chadwick, each of whom confidently believed in the neutron, to this negative result.

Ernest Walton was the next to speak, describing the background to his famous work on the artificial disintegration of lithium with Cockroft, which reached fruition at about the same time as the neutron was discovered. Walton described his arrival as a student at the Cavendish in 1927 and his original proposal to work on a form of betatron. As he explained, this was premature. Nothing was known about orbital stability and, besides, the Cavendish had nothing like the equipment he would have needed. He had then turned his attention to a linear accelerator, but this again suffered from a shortage of suitable equipment. Following a visit by Gamow, however, the possibility of penetration of the potential barrier of a light nucleus by protons of relatively low energy was realised and he and Cockroft immediately began to assemble the necessary apparatus. He concluded by describing the lithium experiment and Rutherford's pleasure in its success.

Closing the first part of the session, Sir Harrie Massey, who had also been a research student at the Cavendish in 1932, recorded his impressions from the "Garage". In this large room work related to old style atomic physics rather than to nuclear physics was conducted. The room, he recalled, was often graced by the presence of J.J. Thomson, who came in around lunchtime to work on electrodeless discharges and what we should now term plasma waves. Massey himself was working on the scattering of slow electrons by metallic vapours, and he mentioned some of the changes, especially in equipment and instrumentation, that were taking place at the time. One very important change, he recalled, was the replacement of soda glass by pyrex; this made vacuum work very much easier, and avoided difficulty with the "highly conservationist" stores keeper. Other significant introductions were the new counting devices being built, with Chadwick's encouragement, by Bowden, Lewis and Wynn-Williams, despite Rutherford's lack of enthusiasm for such things. Another very important change that had only recently taken place and

and the numbers of papers within these categories. He concluded by discussing the impact of Bohr's 1936 liquid drop model of the nucleus and the way in which the authority and success of Bohr's ideas led to the alternative shell model being neglected for a good many years.

Particle properties of the neutron

Norman F Ramsey

Mount Holyoke College and Harvard University

> Abstract. Particle properties of the neutron and experiments which have determined these properties are reviewed. Properties discussed include mass, electric charge, magnetic monopole, electric and magnetic dipole moments, neutron mean life for beta decay, ratio of axial to vector coupling constants for the weak interaction, electric and magnetic polarizability, period for $n\bar{n}$ oscillations and various quantized properties such as spin, statistics and isospin.

1. Introduction

When I was a first year undergraduate at Columbia University, the two most exciting events during my first physics course were the announcements by Chadwick of the discovery of the neutron and by Cockcroft and Walton of the first nuclear transmutions by accelerated particles. These two events probably influenced my decision to shift from mathematics to physics and they certainly influenced my decision to study at the Cavendish Laboratory where these and other great events in physics had occurred. It is therefore to me a special pleasure and honor to return from Cambridge, Massachusetts, to Cambridge, England, to help celebrate the Fiftieth Anniversary of the Neutron.

2. Neutron Mass

Neutron mass measurements go back to the very beginning: Chadwick's (1932) original discovery of the neutron was essentially a mass measurement. Chadwick studied the recoils of H, Li, Be, B, C and N atoms from the objects that were produced when α-particles bombarded Be and he showed that his observations were mutually compatible only if the new neutral object producing the recoils had a mass approximately equal to that of the proton. Chadwick (1932, 1933) also showed that a similar mass was obtained by applying the conservation laws to the production processes.

Chadwick and Goldhaber (1934, 1935) soon discovered the photodisintegration of the deuteron which permitted an accurate determination of the binding energy of the deuteron from the energy of the gamma rays and the energy of the photoprotons. From the binding energy of the neutron and the difference in mass spectroscopic measurements on H_2^+ and D^+, the neutron mass could be accurately determined. Subsequently there have been many improvements in the measurements, particularly by measuring the energy of the photons emitted when slow neutrons are captured by hydrogen. A particularly effective experiment of the latter nature was that of Knowles (1962). The present official value for the mass of the neutron comes from a least squares combination both of various measurements on the deuteron (Mattauch 1965) and on the fundamental constants affecting all mass scales (Cohen, et al 1973) and Particle Data Group 1982). The interaction between different varieties of measurements in the determination of most fundamental constants is now so great that the values are set by agreement in an international committee (Cohen et al 1973) and ordinarily retain those values until the next readjustment of the constants, which is scheduled to take place within another year. The present official value (Particle Data Group 1982) for the rest mass of the neutron in different units are

$$m_n = 1.008665012(37) \text{ u} \quad (0.011 \text{ ppm})$$
$$= 1.6749543(86) \times 10^{-27} \text{ kg} \quad (5.1 \text{ ppm}) \quad (1)$$
$$= 939.5731(26) \text{ MeV} \quad (2.8 \text{ ppm})$$

where the first numbers in parentheses are the standard deviation uncertainties in the last digits of the quoted value, u is the atomic mass unit and ppm is parts per million. The marked variation in uncertainties from 0.011 ppm on the atomic mass scale to 5.1 ppm on the kg scale show that the limiting error in determining the mass of the neutron in kg is not the nuclear physics determination of the deuteron binding energy but the value for the Avogadro constant.

Although the present official values for the neutron mass are those given in Eq. (1), there have been improvements in such measurements since these numbers were officially set. The improvements have come in part from new measurements of the γ-rays emitted when neutrons are captured by γ-rays (Greenwood, Helmer, Gehrke and Chrein 1979 and Vylor et al 1978) and in part from a new γ-ray energy scale determined by Kessler, Deslattes, Henins and Sanders (1978) and Deslattes, et al (1980). A new evaluation of the fundamental constants is anticipated early in 1983 and it is now anticipated (Taylor 1982) that with the new data the atomic mass of the neutron will probably shift to 1.008664904(14) (0.014 ppm) and with corresponding changes occurring in the other neutron mass values. However, until the new evaluation is published I recommend the continued use of the official values in Eq. (1), unless the differences are really significant for the purpose at hand; in that case, however, the new value must be used with great care to be certain that the quantities with which it is compared are consistently determined.

3. Neutron Electric Charge

There have been a series of limits set on the electric charge of the free neutron by Shapiro and Estulin (1956), by Shull, Billman and Wedgewood (1967) and most recently by Göhler, Kalus and Mampe (1982) at the Institut Laue-Langevin (ILL) at Grenoble. The experiment of Göhler et al used a focussed beam of 200 m/s neutrons which passed for 10 meters through an electric field of 5.9 kV/mm. The experimenters found that the charge q_n of the neutron was

$$q_n = -(1.5 \pm 2.2) \times 10^{-20} \text{ e} \quad (2)$$

where e is the charge of the proton.

This experiment was more sensitive than any of the previous neutron experiments. However, there are experiments reviewed by Dylla and King

(1973) which measure a related quantity -- the charge on a neutral molecule such as SF_6 and show

$$q_{SF_6} = (0 \pm 4.3) \times 10^{-21} \text{ cm.}$$

4. Neutron Magnetic Monopole

With the recent announcement of the possible discovery of a free magnetic monopole (Cabrera 1982) there is some reason to discuss the limit on the magnitude of a magnetic monopole that might be associated with the neutron. No experiment has yet been reported which is primarily a search for a magnetic monopole charge on the neutron. However, past neutron beam experiments can be analyzed for the limits that they place on such a magnetic monopole. In the theories of Dirac (1948) and others, magnetic monopoles -- if they exist at all -- are ordinarily multiples of $e/2\alpha$ in Gaussian units.

The author, has done experiments (Cohen, Corngold and Ramsey 1956) in which neutrons at $v = 1000$ m/sec were deflected less than $\delta = 0.3$ cm on passing through a magnet of length $\ell = 100$ cm at $B = 7.5$ kG. From these values the magnetic pole q_m is limited by

$$q_m < \frac{4\alpha M_n v^2 \delta}{e B \ell^2} = 1.8 \times 10^{-15} \text{ } (e/2\alpha) \quad (3)$$

It is of interest to note that, if the experiment of Göhler, Kalus and Mampe (1982) on the neutron electric charge were carried out with equal sensitivity in the presence of a 500G field, it could improve the above magnetic monopole limit by a factor of 10^7.

5. Neutron Electric Dipole Moment

Although the electric dipole moment of the neutron must vanish for a theory which is symmetric under either parity (P) or time reversal symmetry (T), a non-zero value is predicted by most theories which account for the known T violation in the decay of K_L^o. As a result there is considerable theoretical interest in the experimental limits on the neutron electric dipole because of the constraints these limits places on theories that are not symmetric under T.

Most early measurements of the neutron electric dipole moment were magnetic resonance experiments with polarized neutron beams in which the electric dipole moment was measured by the change in the neutron precession frequency when an electric field was shifted from parallel to antiparallel the magnetic field (Dress et al 1977). Scientists at Leningrad (Altarev et al 1980) and Grenoble (Harvard-Sussex-Rutherford-ILL collaboration 1975) are currently making measurements on the neutron electric dipole moment and both groups are using bottled ultracold neutrons with storage times greater than 5s for which the resonance is more than 1000 times narrower than in the earlier beam experiments. The lowest reported limit at present is that of Altarev et al (1981) who find for the neutron electric dipole moment d that

$$d < 6 \times 10^{-25} \text{ e cm.} \quad (4)$$

6. Neutron Magnetic Dipole Moment

The most accurate value for the neutron magnetic moment is the one recently obtained at the ILL by Greene, et al (1979). They used the neutron beam magnetic resonance apparatus previously used to set a limit on the neutron electric dipole moment. They attained high precision by successively passing water and neutrons through the same tube and by calibrating the magnetic field with the separated oscillatory field proton resonance when water passed through the tubes. Their value for μ_n, the neutron magnetic moment, when expressed in various relevant units is

$$\mu_n = -1.04187564(26) \times 10^{-3} \mu_B \quad (0.25 \text{ ppm}) \quad (5)$$

$$= 1.04066884(26) \times 10^{-3} \mu_e \quad (0.25 \text{ ppm}) \quad (6)$$

$$= -0.68497935(17) \mu_p \quad (0.25 \text{ ppm}) \quad (7)$$

$$= -1.91304308(54) \mu_N \quad (0.28 \text{ ppm}) \quad (8)$$

where μ_B is the Bohr magneton, μ_e the magnetic moment of the free electron, μ_p is the magnetic moment of the proton and μ_N is the nuclear magneton.

7. Neutron Decay

The early mass measurements of Chadwick and Goldhaber (1935) showed that the neutron was heavier than the proton and electron combined, with the implication that the free neutron itself was unstable against beta decay; this prediction was confirmed experimentally (Snell and Miller 1948) when high flux neutron beams from nuclear reactors became available. The neutron decay is of particular interest because there are no nuclear structure effects to be taken into account in analyzing the data obtained from the study of the neutron decay into a proton, an electron and an antineutrino. This decay is consequently an important source of information on the weak interaction.

Although the lifetime of the neutron is of great importance and interest, it has been a difficult quantity to measure accurately and many of the measurements disagree well beyond the estimated experimental error. Most of the experiments involve measurements on a neutron beam with consequent difficulties in determining accurately the appropriate normalization and the effective solid angle for which the decay products are intercepted. These problems have been overcome in experiments with bottled ultra cold neutrons. In one case the neutrons were retained by inhomogeneous magnetic fields acting on the neutron magnetic moment (Kuegler, Paul and Trinks 1978). In the other case (Kosvintsev, Kushnir, Morozov and Terekhov 1980) the ultra cold neutrons were retained by total reflection at the walls of an aluminum container, with the wall absorption being measured by the insertions of aluminum vanes in the container. Neither of these methods, however, has so far claimed an accuracy as good as the best beam measurements.

The measurements that are consistent with each other and have the smallest estimated errors are the neutron beam experiments of Christensen et al (1972) and of Byrne et al (1980). In the first of these the decay electrons were detected in approximately 4π solid angle. In the second method, if the neutron decayed within a determined volume, the decay protons were electrically trapped and later counted. The results of these two experiments are in close agreement. The customary weighted averaging of the neutron measurements (Particle Data Group 1982) gives for the mean life τ

$$\tau = t_{\frac{1}{2}}/\ln 2 = (925 \pm 11) \text{ sec.} \qquad (9)$$

Although the full array of the studies of the beta ray spectrum of the neutron clearly go beyond the scope of a report on the particle properties of the neutron, it is appropriate to report at least briefly, measurements of the weak interaction coupling constants of the neutron. The ratios of the axial-vector (or Gamow-Teller) coupling constant g_A to the vector (or Fermi) coupling constant g_V can be found by at least three methods: (a) from the observed neutron lifetime in conbination with calculations of corrected values for ft and f where f is the phase space factor for the electron in neutron decay (Christensen et al 1972, Kropf and Paul 1974, Byrne et al 1980 and 1982 and Wilkinson 1981), (b) from the correlation between the spin of polarized neutrons and the momentum of the emitted electrons (Krohn and Ringo 1975 and Erozolinskii, et al 1979) and (c) from the shape of the proton recoil spectrum in free neutron decay (Stratowa, Dobrozemsky and Weinzierl 1978). The results are (Particle Data Group 1982)

$$g_A/g_V = |g_A/g_V| e^{i\phi} \qquad (10)$$

$$|g_A/g_V| = 1.255 \pm 0.006 \qquad (11)$$

$$\phi = (180.11 \pm 0.17)^\circ \qquad (12)$$

or

$$g_A/g_V = -1.255 \pm 0.006 \qquad (13)$$

8. Neutron Electric and Magnetic Polarizabilities

So far there have been no accurate measurements of either the electric or the magnetic polarizability of the neutron. There have been numerous theoretical calculations of the neutron electric and magnetic polarizabilities as discussed by Schroeder (1980). The electric and magnetic polarizabilities of the proton have been measured by Baranov et al 1974 and others but there are no measurements of the magnetic polarizability of the neutron and the few measurements that do exist on the electric polarizability of the neutron are mutually contradictory. Thus Aleksandrov, Samosvat, Sereeter and Sor (1966) found by studying the angular distribution of neutrons scattered by lead that the neutron electric polarizability α_n is experimentally limited by

$$|\alpha_n| < 6.1 \times 10^{-3} \text{ fm}^3 \qquad (14)$$

whereas Anikin and Kotukhov (1972) from the scattering of neutrons by uranium conclude

$$\alpha_n = (0.26 \pm 0.10) \text{ fm}^3 \qquad (15)$$

Not only are these values mutually contradictory but the last value appears unreasonable theoretically since a typical theoretical estimate by Schroeder (1980) is $\alpha_n = 8.5 \times 10^{-4} \text{ fm}^3$. A good and reliable measurement of both the electric and magnetic polarizabilities would be a most welcome addition to the properties of the neutron.

9. Neutron Oscillations

Recently, serious consideration has been given to the possibility that the free neutron might be a superposition of neutron (n) and antineutron ($\bar{n}$) states similar to the known superposition of K^0 and $\overline{K^0}$ for the free kaon. These considerations give rise to another measurable property of the free neutron: the period $\tau_{n\bar{n}}$ of the oscillation between the n and $\bar{n}$ states. An approximate limit of $\tau_{n\bar{n}} > 10^5$ sec can be set from observations on the stability of nucleons in matter (Learned et al 1979, Glashow 1979, Kuzmin 1970, Chetrykin et al 1980, Marshak and Mohapatra 1980 and Green 1982) but such a limit is subject to considerable uncertainty due to nuclear effects. Several experiments are now in progress to measure $\tau_{n\bar{n}}$ for the free neutron but only one of these experiments has so far given a result. The CERN-ILL-Padua-Rutherford-Sussex experiment at the ILL in Grenoble, France (Green 1982) has set the limit for the free neutron

$$\tau_{n\bar{n}} > 1.2 \times 10^5 \text{ secs.} \qquad (16)$$

It is expected that with improvements to this and the other experiments the limit or $\tau_{n\bar{n}}$ during the next few years will be progressively raised to 10^7 or 10^8 secs.

10. Quantized Properties of the Neutron

A number of particle properties are quantized, so an extended series of measurements with increasing accuracy is not appropriate.

A quantized property with a long history is the neutron spin. As soon as the neutron was discovered it was apparent that if all nuclei were to consist of neutrons plus protons, the neutron would need to have a half integral spin and that simple nuclei could most easily be explained in terms of the value $J = \frac{1}{2}$ for the neutron spin. Schwinger (1937) showed that the almost complete interference between singlet and triplet scattering in parahydrogen provides a direct proof that the spin of the neutron (its angular momentum in units of $\hbar$) is given by

$$J = \frac{1}{2} \qquad (17)$$

A closely related discrete property is the quantum statistics followed by the neutron. Even before the neutron was discovered it had been concluded from the rotational spectra of homonucleur diatomic molecules (Heitler and Herzberg 1929 and Rasetti 1930) that nuclei with an odd mass number A obey Fermi-Dirac statistics while nuclei with even A obey Bose-Einstein statistics. This conclusion was a major difficult for theories which had nuclei consisting of protons plus electrons but these difficulties immediately disappeared when nuclei were assumed to consist of protons plus neutrons if, and only if, the neutrons were assumed to be fermions.

Other quantized properties of the neutron have been set for the neutron either by convention or by definition and have originated from the need to distinguish the neutron from other particles, most of which were discovered after the neutron. For this reason I shall merely list these properties without further discussion. They include:

Intrinsic parity	$= P = +1$
Isospin	$= I = \frac{1}{2}$
Component of isospin	$= I_3 = -\frac{1}{2}$
Baryon number	$= B = 1$
Lepton number	$= L = L_e = L_\mu = 0$
Strangeness	$= S = 0$
[Hypercharge	$= Y = 1$]
Charm	$= c = 0$
Bottomness	$= b = 0$
Topness	$= t = 0$

11. Other Neutron Properties

There is considerable arbitrariness as to what should be called a particle property of the neutron and what should be considered neutron structure or particle scattering -- both of which topics are covered in other reports at this meeting. Therefore, I shall merely list without discussion some properties which from one point of view could be considered particle properties of the neutron but which more appropriately fit in to other reports. These properties include the neutron-electron interaction, the charge distribution within the neutron and in particular the mean squared charged radius $= \langle \sum_i e_i r_i^2 \rangle$, the form factors and structure functions of the neutron as determined by electron-deuteron scattering experiments, and the neutron-proton interaction as determined from the ground state of the deuteron and from neutron proton scattering experiments.

12. Conclusions

In conclusion I have assembled in Table I the particle properties of the neutron which I have discussed. It is apparent that much has been learned about the neutron in the fifty years since its discovery. What more will be learned in the next fifty years?

Table I. Particle Properties of the Neutron

m_n = 1.008665012(37) u (0.011 ppm)
= 1.6749543(86) × 10^{-27} kg (5.1 ppm)
= 939.5731(26) MeV (2.8 ppm)

q_n = − (1.5 ± 2.2) × 10^{-20} e

q_m < 1.8 × 10^{-15} e/2α

d < 6 × 10^{-25} e cm

μ_n = − 1.04187564(26) × 10^{-3} μ_B (0.25 ppm)
= 1.04066884(26) × 10^{-3} μ_e (0.25 ppm)
= − 0.68497935(17) μ_p (0.25 ppm)
= − 1.91304308(54) μ_N (0.28 ppm)

$\tau = t_{1/2}/\ln 2$ = 925 ± 11 secs

g_A/g_V = − 1.255 ± 0.006

$|\alpha_n|$ ⩽ 6.1 × 10^{-3} fm^3

$\tau_{n\bar{n}}$ > 1.2 × 10^5 secs

J = ½   P = + 1

I = ½ I_3 = − ½

B = 1   L = L_e = L_μ = 0

s = 0   Y = 1

t = 0   b = 0

References

Aleksandrov Y A, Samosvat G S, Sereeter Z and Sor T G 1966 JETP Lett 4 134

Altarev I S et al 1980 Nucl. Phys. A341 269
——————— 1981 Phys. Lett. 102B 13

Altarev I S, Barisov Y V, Borovikova N V, Branden A B, Egorov A I, Porsev G D, Ryabov V L, Serebov A P, Taldaev R R 1980 Nucl. Phys. A341 269
——————— 1981 Phys. Lett. 102B 13

Anikin G V and Kotukhov I I 1972 Sov. Journ. of Nucl. Phys. 14 152

Baranov R, et al 1974 Phys. Lett. 52B 122

Byrne J 1982 Reports Progress Physics 45 115

Byrne J, Morse J, Smith K F, Shaikh F, Green K, and Greene G L 1980 Phys. Lett. 92B 274

Cabrera B 1982 Phys. Rev. Lett. 47 1738

Chadwick J 1932 Nature 129 312 and Proc. Roy. Soc. Lond. A136 692
——————— 1933 Proc. Roy Soc. Lond. A142 1

Chadwick J and Goldhaber M 1934 Nature 134 237
——————— 1935 Proc. Roy. Soc. London A151 479

Chetrykin K G, Kozamonsky M V, Kuzmin V A and Shaposhnikov 1981 Phys. Lett. 99B 358

Christensen C J, Nielsen A, Bafasen A, Brown W K and Rustad B M 1972 Phys. Rev. D5 1628

Cohen E R and Taylor B N 1973 Phys. Chem. Ref. Data 2 663 (CODATA Report)

Deslattes R D, Kessler E G, Sauder W C and Heinins A 1980 Annals of Phys. 129 378

Dirac P A M 1948 Phys. Rev. 74 817

Dylla H F and King J G 1973 Phys. Rev. Letters A7 1224

Erozolimskii B G, Franck A I, Mostovoi Y A, Arzumanov A S, Vortski L R 1979 Yad. Fiz. (USSR) 30 692 [Sov. J. Nucl. Phys. (USA)]

Gähler R, Kalus J and Mampe W 1982 Phys. Rev. D25 2887

Glashow S 1979 Harvard Preprints HUTP79/A029, 79/A040 and 79/A059

Green K 1982 Rutherford Laboratory Preprint SP 82-60 (Invited talk at Rencontre de Moriond, Les Aris March 1982)

Greene G L, Ramsey N F, Mampe W, Pendlebury J M, Smith K, Dress W B, Miller P D and Perrin P 1979 Phys. Rev. D20 2139

Greenwood R C, Helmer R G, Gehrk R J and Chrien R E 1979 Atomic Masses and Fundamental Constants 6 219

Harvard-Sussex-Rutherford-ILL Collaboration (Baker C, Byrne J, Golub R, Green K, Heckel B, Kilvington A, Mampe W, Morse J, Pendlebury J M, Ramsey N F, Smith K and Sumner T) 1975

Heitler W and Herzberg G 1929 Naturwiss 17 673

Kessler E G, Deslates R D, Henins A and Sauder W C 1978 Phys. Rev. Lett. 40 171

Knowles J W 1962 Can. Journ. Phys. 40 257

Kosvintsev Y Y, Kushnir Y A, Morozov V I and Terekhov G I 1980 Pis' ma Zh. Eksp. Teor. Fiz. 31 257 [JETP Lett. 31 237 (1980)]

Krohn V E and Ringo G R 1975 Phys. Lett. 55B 175

Kropf A and Paul H 1974 Z. Physik 267 129

Keugler K J, Paul and W Tinks U 1978 Phys. Lett. 72B 422

Kuzmin V A 1970 Pis'ma zh. Eksp. Teor. Fiz. 13 335

Learned J, Reines F and Soni A 1979 Phys. Rev. Lett. 43 907

Marshak R E and Mohapatra R N 1980 Phys. Rev. Lett. 44 1316 and Phys. Lett. 94B 193

Mattauch J H E, Thiele W E and Wapstra J H Nuclear Phys. 67 1, 32 and 73

Particle Data Group 1982 Phys. Letters 111B i (Roos M, et al)

Rasetti F 1930 Z. Physik 61 598

Schroeder V E 1980 Nucl. Phys. B166 103
Schwinger J S 1937 Phys. Rev. 52 1250
Shapiro I S and Estulin I V 1956 Zh. Eksp. Teor. Fiz. 30 579 [Sov. Phys. JETP 3 626 (1957)]
Shull C G, Billman K W and Wedgewood F A 1967 Phys. Rev. 153 1414
Snell A H and Miller L C 1948 Phys. Rev. 74 1217
Stratowa C, Dobrozemsky R and Weinzierl P 1978 Phys. Rev. D18 3970
Taylor B N 1982 (private communication July 19, 1982)
Vylor T, Gromov K Y, Ivanov A I, Osipenko B P, Frolov E A, Chumin V G, Schus A F and Yudin M F 1978 Sov. J. Nucl. Phys. 28 585
Wilkinson D 1981 Progr. in Part. and Nuclear Phys. 6 325

The neutron lifetime

J Byrne, School of Mathematical and Physical Sciences, University of Sussex, Brighton, Sussex, BN1 9QH.

1. Introduction

When Chadwick and Goldhaber made the first precise determination of the neutron mass they predicted that the neutron would be β-active, decaying spontaneously into a proton, electron and neutrino (Chadwick and Goldhaber 1935). The process was first detected by Snell and Miller who observed positive ions extracted electrically from within an intense neutron beam and subsequent work confirmed that the production rate was consistent with neutron β-decay over a half-life in the range 10-30 minutes (Snell and Miller 1948, Snell et al 1950). At about the same time Robson verified that the positive ions were indeed protons and estimated a half-life of 9-25 minutes (Robson 1950). The first accurate determination of the neutron lifetime was also carried out by Robson, to a precision of about 20% and in the same experiment he measured the electron spectrum (Robson 1951).

The end-point of the electron kinetic energy spectrum is now known to have the value E_o = 782.318 ± 0.017 keV as determined from the neutron-hydrogen atom mass difference (Wilkinson 1982). This energy is about typical for a β-active nucleus. The corresponding end-point for the spectrum of recoil protons is 755 eV which is too low to be recorded easily in conventional charged particle detectors.

2. Significance of the Neutron Lifetime

Before considering the experimental problem of measuring the neutron lifetime, still the least accurately determined lifetime of those elementary particles which are known to be unstable while remaining stable against strong interactions, we should consider briefly why this particular fundamental constant is so important. Of course the neutron lifetime is crucial to astrophysics where it is central to the problem of helium production in the early universe (Tayler 1968, 1979, 1980). However, its importance to elementary particle physics derives from the fact that, in combination with $0^+ \to 0^+$ ft-values, it yields values for the weak coupling constants, g_V and g_A in a manner which is largely free of uncertainties associated with nuclear structure. Essentially what the neutron lifetime provides is a measure of the combination $|g_V|^2 + 3|g_A|^2$. There is however another route whereby these coupling constants can be determined, a route relying on neutron β-decay alone assuming that this process proceeds through a time-reversal invariant pure (V-A) coupling. Under these assumptions we can derive expressions for the parity conserving electron-neutrino angular correlation coefficient 'a' and the parity violating

0305-2346/83/0064-0015 $02.25 © 1983 The Institute of Physics

electron-neutron spin correlation coefficient 'A'

$$a = (1 - \alpha^2)/(1 + 3\alpha^2), \quad A = -2(\alpha^2 - \alpha)/(1 + 3\alpha^2)$$

where $\alpha = |g_A/g_V|$. Other correlations may also play a role but, viewed from an experimental angle the two noted above are the most important. The number α also provides a link to the strong interactions since it is connected to the pion-nucleon coupling constant through the Goldberger-Treiman relation. In any case a knowledge of α and the neutron lifetime is sufficient to determine the weak coupling constants uniquely.

3. Measurements

The results of all measurements which have yielded precise numbers are shown in Fig 1 and here the most striking feature is the way the measured values of the neutron half-life have consistently fallen in time. One should observe that with the exception of the measurements from Kugler et al (1978) and Kostvintsev et al (1980), which made use of bottled ultra cold neutrons, all of the results shown in Fig 1 were obtained by beam methods. Furthermore the experiments carried out by Spivak et al (1955), Sosnovski et al (1959) and Bondarenko et al (1978) used essentially the same apparatus with some development. The two values quoted by Christensen et al (1967, 1972) were based on the same experimental data.

The first serious attempt to perform an accurate measurement of the neutron half-life was made by Robson three decades ago. In his apparatus, decay protons were deflected out of a neutron beam onto the first stage of an electron multiplier which acted as a detector. The β-particles were focused by a magnetic spectrometer onto an anthracene scintillator. Thus, since both decay particles were recorded, the background could be eliminated by accepting only coincidence events from the two detectors. The effective volume of the source was determined from an elaborate computation based on trajectories of the charged particles as determined from a mechanical model, a calculation made even more difficult by the necessity of taking into account the spatial variation of the efficiency for detecting coincidences.

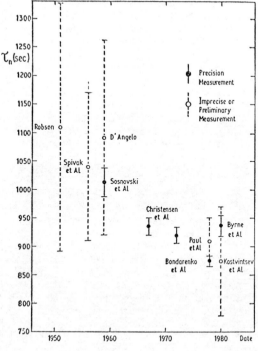

Figure 1 : Measured values of neutron lifetime.

The value of the neutron half-life obtained by Robson was $t = \tau \ln 2 =$

12.8 ± 2.5 min and the 20 per cent error quoted reflects to a great extent
the difficulty of determining with any precision the exact position of
the decaying neutron. This source of uncertainty was largely eliminated
in an experiment carried out by d'Angelo, who photographed individual
decay events in a diffusion cloud chamber, however, the result of
12.7 ± 2.5 min showed no improvement in accuracy over Robson's determination. This was mainly because of poor counting statistics and the
difficulty of eliminating the background of spurious events originating
from nuclear reactions within the chamber gas. An interesting feature
of this experiment was the use of a magnetic mirror to deflect the neutron
beam into the chamber and thereby reduce the γ-ray background, a forerunner perhaps of the neutron guide.

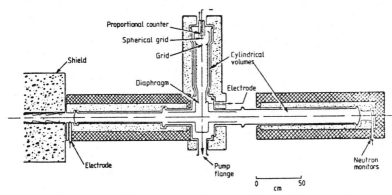

Figure 2 Apparatus for determining the neutron half-life (Sosnovski *et al* 1959, Bondarenko *et al* 1973).

The first precision measurement was carried out by Sosnovski and collaborators using the apparatus illustrated in Fig 2. These experiments were
begun about 1950 and a preliminary result was presented at the first
Conference on the Peaceful Uses of Atomic Energy held at Geneva in 1955.
A collimated beam of slow neutrons passes down the length of a hollow
electrode maintained at a potential of 20 kV and some of the decay protons
leaving the beam travel in a field free region and are focused onto the
window of a ball proportional counter maintained at ground potential.
Since the protons travel in straight lines and are then detected with an
efficiency approaching 100 per cent, determination of the effective source
volume is a straightforward computational problem. However, between the
source and the sensitive volume of the counter, the protons have to pass
through no fewer than four grids and the calculation of the corresponding
transmission factors is not a simple matter. The value obtained for the
half-life was 11.7 ± 0.3 min, a result which has recently been abandoned
in favour of the much lower value of 10.13 ± 0.09 min, (Bondarenko et al
1978). The discrepancy between the old and new results was accounted for
as arising from proton loss by charge exchange collisions due to
excessive background pressure in the earlier work.

In these experiments only a small fraction of the decays occurring in
the source volume was recorded directly, since the solid angle for
collection was small and it was therefore necessary to arrive at the true
source volume by scaling up the results. In the determination of the

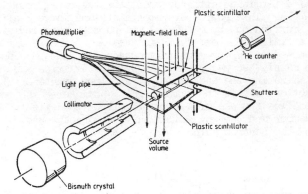

Figure 3 Apparatus for determining the neutron half-life (Christensen *et al* 1967, 1972).

neutron half-life, carried out by Christensen and his colleagues at the Danish nuclear laboratory at Risö, an 8 kG-magnetic field established transverse to the beam direction allows all the decays taking place in a given volume to be recorded. The apparatus is shown in Fig 3. The presence of the magnetic field forces decay electrons leaving the beam to spiral about the field lines so that they must eventually hit a large plastic scintillator. The resultant light pulses are fed through a light pipe to a photomultiplier located outside the magnetic field. If an electron is backscattered from the scintillator it cannot fail to reach a second scintillator located at the opposite side of the beam and in every case the event is recorded. This device removes much of the difficulty associated with the effective source volume determination. It might be noted that in this experiment a ^{3}He proportional counter, based on the ^{3}He (n,p) ^{3}H reaction, replaced the conventional thin foil as a neutron detector, although ^{3}He is, of course, a $1/v$ absorber for slow neutrons.

The final result of the measurement was that $t = 10.61 \pm 0.16$ min, a result which differs from the conclusions of the Russian group, both in their earlier and later versions, by between three and four standard deviations.

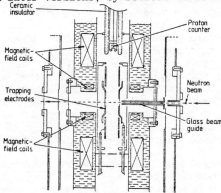

Figure 4 Apparatus for determining the neutron half-life (Byrne *et al* 1980).

The most recent measurement of the neutron half-life based on a beam method is that performed at the ILL Grenoble using the apparatus shown in Fig 4 (Byrne et al 1980). The operation of the system may be summarised as follows. The beam of very cold ($\sim 10^{-4}$ eV) neutrons emerged from the reactor normal to a magnetic field of the order of 12 kG produced by a superconducting magnet; in such a field both electrons and protons move in tight spiral orbits of maximum radius $\lesssim$ 3.5 mm. The beam passed transverse to a hollow cylindrical electrode aligned parallel to the magnetic axis and maintained at ground potential. On either side of this central electrode two coaxial electrodes were maintained at potentials of 1 kV, the whole combination of electric and magnetic fields acting as a potential well for the low energy ($\leq$ 0.75 keV) protons from neutron decay. Thus the protons could be trapped for significant periods of time (10^{-3} - 10 s) before being released by suitable pulses and detection in a silicon surface barrier maintained at - 30 kV.

The essential point of the device described above is that, if protons are stored for a time T_s and the output spectrum is sampled for a period $T_c \ll T_s$ following the application of a release pulse, the background is reduced by a factor T_c/T_s; in practice values of the order of 10^{-4} were achieved for this ratio and the proton spectrum was observed essentially free of background. The value obtained for the neutron half-life was t = 10.82 ± 0.21 min.

4. Results and conclusions

Examination of existing experimental data on the correlation coefficients for neutron decay reveals one value of 'a' (Stratowa et al 1978) and two values of 'A' (Krohn and Ringo 1975, Erozolimskii et al 1979) of comparable accuracy. From these data we can derive a value for the ratio of the weak coupling constants

$$\alpha = 1.257 \pm 0.009,$$

which, in combination with the ft-values for $0^+ \to 0^+$ transitions permits us to compute a value for the neutron half-life

$$t = 10.45 \pm 0.13 \text{ min}$$

This estimate is to be compared with the three measured values of t of comparable precision (cf Fig 1) : t = 10.61 ± 0.16 min (Christensen et al 1972), t = 10.13 ± 0.09 min (Bondarenko et al 1978) and t = 10.82 ± 0.21 min (Byrne et al 1980). These numbers are not consistent within their quoted errors and the result t = 10.13 ± 0.09 min is out of line. We might therefore adopt the view of Wilkinson (1981) and reject this result; averaging the remaining pair of values which are consistent with each other we find a best estimate for the direct measurement

$$t = 10.69 \pm 0.13 \text{ min}$$

The calculated and measured values of t given above are consistent within one standard deviation but only just. It therefore seems reasonable to adopt the weighted mean as the true value of t and we find

$$t = 10.57 \pm 0.10 \text{ min}$$

Wilkinson (1982) has recently carried out an analysis of all the data and recommends the value

$$t = 10.39 \pm 0.10 \text{ min}$$

which is somewhat smaller than the result given above since it includes
the contribution of the experimental measurement of Bondarenko et al
(1978).

An alternative approach would be to calculate the value of α by combining
the data on each separate measurement of the neutron half-life with the
ft-value for the $0^+ \rightarrow 0^+$ transitions and compare the results with the
measured values of α derived from the correlation data. The results
of such an analysis again show that the result of Bondarenko et al is out
of line. Rejecting this result the remaining neutron decay data are in
reasonable agreement.

The simple truth is that a substantial degree of uncertainty exists as
to the true value for the neutron half-life although it is likely that
assigning a value t = 10.6 ± 0.2 min (in essense the Christensen result)
will probably cover most eventualities. This situation is very unsatis-
factory and is likely to remain unresolved until a new generation of
experiments, probably but not necessarily using bottled neutrons, gets
properly under way.

References

Bondarenko, L N Kurguzov, V V, Prokofiev Yu A, Rogov, E Y and Spivak P E,
1978 JETP Lett 28 203.
Byrne J, Morse, J, Smith, K F, Shaikh, F, Green, K, and Greene, G L
1980, Phys Lett 92B 274.
Chadwick, J and Goldhaber, M, 1935, Proc Roy Soc A151 479.
Christensen, C J, Nielsen, A, Bahnsen, A, Brown, W K and Rustad, B M,
1967, Phys Lett 26B, 11; 1972 Phys Rev D5 1628.
D'Angelo, N 1959, Phys Rev 114, 285.
Erozolimskii, B G, Frank, A I, Mostovoi, Yu, A Arzumanov, S S and
Voitzik, L R, 1979, Sov J Nucl Phys 30 356.
Kostvintsev, Yu Yu, Kushnir, Yu A, Morozov, V I and Terekhov, G I, 1980,
JETP Lett 31 236.
Krohn, V E and Ringo, G R, 1975, Phys Lett 55B, 175.
Kugler, K J, Paul, W and Trinks, U, 1978, Phys Lett 72B, 421.
Robson, J M, 1950, Phys Rev 78, 311; 1951 Phys Rev 83, 349.
Snell, A H and Miller, L C, 1948, Phys Rev 74, 1217.
Snell, A H, Pleasonton, F and McCord, R V, 1950, Phys Rev. 78, 310.
Sosnovski, A N, Spivak, P E, Prokofiev, Yu A, Kutikov, I E and Dobrinin,
Yu P, 1959, Nucl Phys 10, 395.
Spivak, P E, Sosnovski, A N, Prokofiev, A Y, and Sokolov, V S, 1956, Proc
Int Conf on Peaceful Uses of Atomic Energy, Geneva 1955, 2, 33 (New York:
UN).
Stratowa, C, Dobrozemsky, R and Weinzierl, P, 1978, Phys Rev D18, 3970.
Tayler, R J, 1968, Nature 217, 433; 1979 Nature 282, 559; 1980, Rep Prog
Phys 43, 253.
Wilkinson, D H, 1981, Prog Particle Nucl Phys 6, 325; 1982 Nucl Phys
A377, 474.

The structure of the neutron

C H Llewellyn Smith

Department of Theoretical Physics, 1 Keble Rd, Oxford OX1 3NP

The neutron is made of three quarks (q) held together by the exchange of vector (spin 1) bosons, called "gluons", with a small admixture of quark-antiquark ($q\bar{q}$) pairs. The three "valence" quarks consist of two isospin down or d quarks, with electric charge $Q = -1/3$ (in units in which the proton's charge is +1), and an isospin up or u quark, with $Q = 2/3$. The quarks have a partially hidden three valued degree of freedom called colour, the three valence quarks having different colours - say red, green and blue. We now know that the gluons couple to the colour degree of freedom or, in other words, that colour plays the role of a strong charge. The theory which describes the interactions of quarks and gluons is known as Quantum Chromodynamics or QCD. I shall describe this picture of the neutron in more detail in the rest of this talk.

The first success of the quark model is that it gives an excellent description of the spectroscopy of baryons as qqq states and mesons as $q\bar{q}$ states. As an example, consider neutral baryons made of ddu, of which the neutron is the ground state. If there is no internal orbital angular momentum, the total angular momentum (J) will be equal to the total spin (S) of the quarks which can be $1/2$ or $3/2$. The lowest radial states are identified with known particles thus

$J = 1/2$, $I = 1/2 \rightarrow n(940$ MeV$)$

$J = 3/2$, $I = 3/2 \rightarrow \Delta^{\circ}(1232$ MeV$)$

In the $J = 3/2$ state the two d quarks are coupled symmetrically to $S = 1$ in apparent conflict with the Pauli principle. This paradox is resolved by the introduction of colour. We will discuss the significance of colour later. For the moment we shall take it as a rule of thumb that the quarks are antisymmetric in their colour variables and must therefore be symmetric in spin and space variables. According to this rule, the totally symmetric $S = 3/2$ state can accomodate uuu and ddu as well as ddu and uud and therefore has $I = 3/2$, as observed for the Δ. The $S = 1/2$ state which has mixes spin symmetry can only accomodate uud or ddu and has $I = 1/2$.

Next consider negative parity baryons with one unit of internal angular momentum, whose orbital wave function therefore has mixed symmetry. In the (symmetric) radial ground state, the $I = 3/2$ combinations must have $S = 1/2$, with mixed spin symmetry, which can couple with the orbital angular momentum to give $J = 1/2$ or $3/2$. For $I = 1/2$, however, both $S = 1/2$ giving $J = 1/2$ or $3/2$ and $S = 3/2$ giving $J = 1/2$, $3/2$ or $5/2$ are allowed. Negative parity baryons with these quantum numbers are indeed seen in the 1520 - 1700 MeV range. The ordering of these states is easy to understand in the quark model, as we shall see below in the case of the "hyperfine splitting" between the Δ^0 and the neutron.

We have seen that the two d quarks in the neutron are coupled to $S = 1$. Adding a u quark with the appropriate Clebsch - Gordon coefficients to make an $S = 1/2$ state, we obtain

$$n^\uparrow = \frac{\sqrt{2}}{3} d^\uparrow d^\uparrow u^\downarrow - \frac{1}{3}\sqrt{}(\frac{d^\uparrow d^\downarrow + d^\downarrow d^\uparrow}{\sqrt{2}}) u^\uparrow$$

for the wave function of a spin up neutron. We can then immediately calculate the magnetic moment moment of the neutron in the approximation that it is the sum of the contributions of the u and d quarks:

$$\mu_n = \mu_u(-\frac{2}{3} + \frac{1}{3}) + \frac{4}{3} \mu_d$$

The magnetic moment of the proton is obtained by interchanging u and d so that

$$\frac{\mu_n}{\mu_p} = \frac{-\mu_u + 4\mu_d}{-\mu_d + 4\mu_u}$$

The natural assumption that $\mu_u/\mu_d = Q_u/Q_d = -2$ then gives $\frac{\mu_n}{\mu_p} = -\frac{2}{3}$, remarkably close to the experimental value of -0.685. It is equally easy to calculate the matrix element for the M1 quark spin flip transition $\Delta^0 \to n\gamma$. Adjusting $\mu_{u,d}$ to obtain the correct value of μ_p, the prediction turns out to be 0.85 ± 0.05 times the observed value, which is not bad considering that recoil effects were neglected.

Turning from purely magnetic properties, we can ask whether the quark model casts light on the old mystery of why $M_n > M_p$. The electrostatic interaction between the quarks contributes more to M_p than M_n so we are forced to conclude that $M_d < M_u$ and one mystery is replaced by another. However, the quark model does give the successful mass difference relation

$$\Xi^0 - \Xi^- + P-N = \Sigma^+ - \Sigma^-$$
$$-6.4 \quad\quad -1.3 \quad\quad -7.98 \pm 0.08$$

where the symbols stand for particle masses and the experimental values of each term are given in MeV.

The origin and pattern of quark masses are not really understood at present. The standard electroweak unified gauge theory can certainly accomodate the data as there is a free parameter for every mass. However, a more satisfactory scheme would explain the mass spectrum, including the fact that although $M_d > M_u$ the other $Q = -1/3$ quarks are lighter than their $Q = 2/3$ partners ($M_s < M_c$, $M_b < M_t$). If $M_d - M_u$ has the same origin as $M_c - M_s$ it seems surprising that isospin symmetry is so good. In fact it is possible to isolate the values that $M_{u,d}$ would have in the absence of strong interactions, which are $M_u^o \simeq 6$ MeV and $M_d^o \simeq 10$ MeV. However, the strong interactions generate an additional mass of order 300 MeV which is the same for u and d (fermion mass renormalizations are generally proportional to M^o but in this case a phenomenon known as spontaneous chiral symmetry breaking, for which there is good evidence, allows a piece which is finite even for $M^o \to 0$). Thus although $M_d^o/M_u^o \simeq 1.8$, so that isospin appears to be badly broken in the Hamiltonian, the effective masses are almost equal and isospin works. The mystery is then why $M_{d,u}^o \ll 300$ MeV while M_t, M_b, M_c, $M_s \gtrsim 300$ MeV. The origin and relation of these and other mass scales, such as those associated with electroweak and grand unification, is one of the most interesting open questions in particle physics.

Returning to static properties of the neutron, the quark model allows us to calculate the ratio of axial and vector couplings (g_A/g_V) for β decay. We assume that $g_A/g_V \simeq -1$ for quarks. A value very close to -1 is required by highly inelastic neutrino scattering and the successes of current algebra. Furthermore -1 is necessary for the construction of a unified electroweak gauge theory in which quarks are treated as fundamental constituents. A simple calculation using the wave function above and the analagous proton wave function gives

$$\left(\frac{g_A}{g_V}\right)_N = -\frac{5}{3}(1-\delta)$$

where δ is the probability of a quark being in the lower Dirac component (to first approximation such relativistic effects do not alter the magnetic moment ratios discussed above). The experimental value of -1.25 requires $\delta = 0.25$ and values of this order are obtained in models in which Dirac quarks are bound in an infinite potential.

The quark model accounts for many other static properties of baryons and mesons. Turning to dynamic properties, the structure of nucleons is most clearly revealed by large momentum transfer inelastic lepton scattering experiments. These processes can be described approximately by an impulse approximation known as the parton model in which we add incoherently the contributions of different quarks, treated as point-like particles. For example, the main contribution to $\nu n \to \mu^- x$ is given by

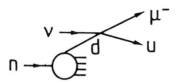

By using different beams ($\nu, \bar{\nu}$ and e or μ) and targets (p,n) we can check the consistency of this picture and study the properties of quarks. In particular
1. The angular distributions show that the struck constitutents have spin 1/2.
2. Comparison of electromagnetic and weak scattering verifies the assignment of charges to the quarks.
3. Combining ν and $\bar{\nu}$ data it can be shown experimentally that the difference of the number of fermions and antifermions in the nucleon

is three.

These lepton scattering experiments actually allow us to measure how the momentum of a rapidly moving neutron is distributed among its constituents. In experiments at momentum transfers of order $(10\,\text{GeV})^2$ it is found that roughly 40% of the momentum is carried by the valence quarks ddu and 10% by $q\bar{q}$ pairs. The remaining 50% is attributed to neutral gluons. These momentum fractions are not independent of the momentum transfer/energy at which they are measured. As the energy increases there is more phase space for gluon bremsstrahlung and the creation of $q\bar{q}$ pairs, so that a quark observed at one energy might be resolved into a quark plus gluons and $q\bar{q}$ pairs at higher energies. Thus the ratio of the momentum carried by $\bar{q}$'s and q's is expected to tend to one for $E \to \infty$. This ratio is indeed observed to increase very slowly with energy. Going down in energy the fraction of momentum in $q\bar{q}$ pairs and gluons decreases. The parton picture ceases to make sense at very low energy but it seems that the fraction of momentum in the glue falls rapidly and the high energy results are not incompatible with the picutre of a static neutron as just ddu.

We now consider the properties of the gluons, whose existence is revealed by inelastic lepton scattering experiments, and the nature of the strong force. We have seen that baryon spectroscopy requires that quarks have three colours so that, for example, we can build the Δ^{++} resonance of $u^\uparrow u^\uparrow u^\uparrow$ in s waves. There is other ample evidence for the existence of three colours e.g. from calculations of $\sigma(\bar{e}e\to\text{hadrons})/\sigma(\bar{e}e\to\bar{\mu}\mu)$ which is proportional to the number of colours (N_c) and $\Gamma(\pi^\circ\to\gamma\gamma)\Gamma(\pi\to\mu\nu)$ which is proportional to N_c^2. We can construct nine different hues of π meson from tricoloured quarks. Since only one is seen, it follows that the gluons must carry and be coupled to colour. In other words colour is the strong charge.

The simplest possibility is that there is an SU(3) colour symmetry. With the gluons belonging to an octet i.e. having the colour quantum numbers of $\bar{q}q$ with the colour singlet removed (an SO(3) symmetry is also possible but turns out not to work). The exchange of these coloured gluons then generates a colour dependent force. For example, the simplest colour exchange diagram is

Forces of this sort would certainly separate coloured and non-coloured states. It is easy to imagine dynamics which might explain why coloured states have not been observed, by making them heavy or removing them from the physical spectrum. This would then explain why only states with the quantum numbers of qqq, $q\bar{q}$ or combinations thereof have been observed since all colour singlets have this structure.

There is convincing evidence that gluons have spin one from the success of relations based on the assumption of spontaneously broken chiral symmetry. I will not explain this assumption but particle physicists may recall that it leads to $M_\pi^2 \ll M_\rho^2$, the Goldberger-Treiman relation and successful calculations of $I=1/2$ and $3/2$ πN scattering lengths, the slope parameter in $K \to \pi e \nu$ and $\Gamma(\pi^\circ \to \gamma\gamma)$. In the framework of the quark model chiral symmetry requires vector gluons.

Vector forces have many desirable features. They are attractive for $q\bar{q}$ system, as we would expect by analogy with $e\bar{e}$ in QED. However, they are also attractive for qqq in a colour singlet, in contrast to eee in QED. If we combine a green quark with a red and blue quark to form a singlet, the red blue pair must couple to form a state with anti green colour quantum numbers which attracts the green quark. Vector forces generate a spin dependent strong hyperfine splitting force which is repulsive for q and $\bar{q}$ in the S = 1 state. This explains why $M_\rho > M_\pi$, in analogy with the hyperfine splitting between ortho and para positronium. However, whereas in QED vector forces have the opposite sign for particle-antiparticle systems and systems made only of particles, the signs are the same for $q\bar{q}$ and qqq in a colour singlet for the reason discussed above. Vector exchange therefore explains why $M_\Delta > M_N$, and also $M_\Sigma > M_\Lambda$ if we make the natural assumption that the "chromomagnetic moment" of the s quark is less than that of the lighter d quark. Furthermore the hyperfine force between the two d quarks in the neutron is repulsive because they are coupled to S = 1. They therefore tend to be on the outside of the neutron which explains why its charge radius is negative.

It turns out that in order to make a mathematically satisfactory theory of coloured gluons coupled to coloured quarks it is necessary to suppose that there is a gluon self interaction. This is not surprising as gluons couple to colour and carry colour themselves. Similarly the graviton must have a self interaction as it couples to energy and momentum which it carries. When the necessary self interaction is included, the quark gluon theory sketched above is (the non-Abelian gauge theory) known as Quantum Chromodynamics or QCD. QCD has a remarkable property known as asymptotic freedom: if we measure the strong charge of a quark it gets less as the quark is approached! The idea that charges depend on distance is familiar from classical electromagnetism. If we place a test charge in a dielectric the charge will decrease from Q when it is measured at much less that the molecular spacing to Q/ϵ at large distances because of the screening effect of polarization. Unfortunately there is no simple analogy for the "anti screening" which results from vacuum polarization in QCD and is responsible for asymptotic freedom.

The successes of the parton model, in which interactions are neglected, shows that the strong force is indeed weak at the short distances involved in high momentum transfer processes. Asymptotic freedom therefore justifies the parton picture. It also allows us to calculate perturbatively how the momentum fractions discussed above change with energy due to gluon bremsstrahlung and $q\bar{q}$ pair creation. Such calculations work reasonably well and provide a semi-quantitative description of this and many other phenomena.

Asymptotic freedom implies that the strong force grows as we move out from short distances but just how strong it becomes requires a non-perturbative calculation. The data suggest that it becomes so strong that quarks can never be pulled apart. It is widely supposed that the colour flux lines

linking, say, a quark and antiquark tend to tie themselves together due to
the gluon self interaction and form a narrow "flux tube" when the q and $\bar{q}$
are far apart, so that the interaction energy grows linearly with the
separation. An analogy is often drawn between the "colour electric" flux
lines in this case and the magnetic flux lines joining a magnetic monopole
and an antimonopole placed in a superconductor, which would form a narrow
non-superconducting tube because of the Meissner effect. In fact the true
vacuum of QCD is probably as different from the "bare" vacuum as the
superconducting state is from the normal state of a metal. It may be
useful to think of hadrons as relatively normal bubbles in the QCD vacuum
occupied by quarks and gluons. Even if the flux tube picture is correct
the energy will not grow indefinitely as the $q\bar{q}$ pair are separated.
Eventually it will become energetically favourable for the tube to break
and form two mesons by the creation of a $\bar{q}$-q pair at the junction. Thus
attempting to separate a quark by "stretching" a meson would be like
trying to isolate a magnetic monopole by cutting an ordinary bar magnet.

In the recent past there has been exciting progress in studying the
properties of QCD in an approximation in which quarks and gluons are only
defined on discrete points on a lattice in space-time. In this
approximation there is only a finite number of field variables and the
theory can be studied numerically. In calculations performed up to now
the lattice spacing is not as small nor the overall size as large as might
be wished and virtual $q\bar{q}$ pairs have been neglected. Nevertheless
the results are very encouraging. They provide fairly convincing evidence
that QCD does confine quarks and gluons into colour singlet states and
that the flux tube picture discussed above has some validity. Furthermore
a qualitative description of the spectrum of light hadrons and very
reasonable values for the magnetic moments of the neutron and proton are
obtained. It seems possible that in a few years we may be able to
calculate the properties of the neutron and other hadrons quite accurately
from the first principles using lattice QCD.

I have only discussed the strong force. Asymptotic freedom allows the
possibility that as the strong coupling decreases at short distances it
merges ultimately with the weak and electromagnetic couplings. Even if
this idea of "grand unification" is correct and fundamentally the
different forces are the same, it is of course a good approximation to
neglect electromagnetic and gravitational forces when discussing the
structure of the neutron. However, because of its unique properties, the
neutron can play a vital role in studying the properties of these forces
and unified models e.g. in searches for a neutron electric dipole moment,
neutron decays such as $n \rightarrow \pi^- e^+$ and neutron-antineutron oscillations, all of
which are discussed by other speakers at this conference.

In conclusion there is very convincing evidence that the neutron is made
of quarks and gluons whose interactions are controlled by QCD. Is this
the end of the story or do the quarks and gluons themselves have
substructure? We have seen that quarks have $\mu_q \propto Q_q$ and $g_A = -g_V$, to a
reasonable approximation at least. The example of the neutron shows that
neither of these relations is true for a composite system in general.
Furthermore the success of the parton model in describing inelastic lepton
scattering and e^+e^- annihilation puts quite stringent limits on the size
of quarks. Thus if quarks are composite the binding must be very tight.
Nevertheless various arguments have been advanced for further structure,
the most obvious being historical analogy coupled with unease at the
number of different quarks, which might be explained in terms of a smaller

number of subquarks. A rather different possibility, which is not incompatible with substructure, is suggested by models that incorporates the elegant idea of supersymmetry which relates fermions and bosons. In these models there must be scalar and pseudoscalar partners of the usual quarks and spinor partners of the vector gluons. If they are relatively heavy they would have escaped detection up to now but there is reason to think that they might be discovered at the next generation of accelerators such as LEP. Finally the most promising supergravity models which attempt to unify all forces do not seem to be able to accomodate all the known "elementary" particles directly and many must therefore be treated as composites.

Whatever the fate of such speculations, it seems certain that the neutron will continue to play an important role in unravelling the fundamental laws of nature.

References

For a general introduction to quarks and partons see Close (1979). For an overview of QCD see Llewellyn Smith (1982). For a recent review of lattice QCD see Rebbi (1982). For a discussion of quark masses see Gasser and Leutwyler (1982). For a review of grand and super unification see Ellis (1981). For a review of composite models of quarks see Peskin (1981).

Close F E 1979 An Introduction to Quarks and Partons (London: Academic Press)
Ellis J R 1981 CERN TH-3174 to be published in Proc. 1981 Les Houches Summer School
Gasser J and Leutwyler H 1982 Physics Reports (in press)
Llewellyn Smith C H 1982 Phil. Trans. Roy. Soc. Lond. A. $\underline{304}$ 5
Peskin M E 1981 Proc. 1981 International Symposium on Lepton and Photon Interactions at High Energy ed W Pfeil (University of Bonn) pp880-907
Rebbi C 1982 to be published in Proc. XXI International Conference on High Energy Physics

The question of baryon conservation

M. Goldhaber

Brookhaven National Laboratory, Associated Universities Inc., Upton, New York, U.S.A.

What has developed into the law of baryon conservation was first postulated by Herman Weyl when he talked of proton stability more than half a century ago. It was elaborated later by Stückelberg and Wigner. (For a general review see e.g., Goldhaber, Langacre and Slansky 1980 and Goldhaber and Sulak 1981.) A modern version of the law of baryon conservation might read: The net number of baryons ($\Sigma B - \Sigma \bar{B}$) does not change spontaneously or in any known interactions (inside a nucleus or in high energy collisions). For a long time it was widely believed that protons are absolutely stable, and though we have learned that a free neutron decays by beta emission with a half life of about 11 min, neutrons sufficiently strongly bound by nuclei were also considered absolutely stable. In order not to have to qualify whether we are dealing with free or bound neutrons, I shall only talk of proton stability implying tacitly that bound neutrons are included in most statements.

The above postulate is however just an ad hoc postulate. As our astronomical friends warn us: absence of evidence is not evidence of absence. It seemed therefore worthwhile to ask the experimental question: "How stable is the proton really?" A limit could be estimated from the absence of spontaneous fission of ^{232}Th. But deliberate experiments can go much further. The first of these experiments was done by Reines, Cowan and myself (Reines et al 1954). We could give limits in a very simple experiment, $>10^{21}$ years for free protons and $>10^{22}$ years for bound nucleons. Over the years, largely due to the efforts of Reines and his collaborators, the limit for bound nucleons was extended to $\sim 10^{30}$ years. Then a few years ago the grand unified theories (GUTS) were proposed in which strong, weak and electromagnetic interactions are combined, leading to the possibility that protons decay. Their lifetime is predictable in some of these theories. In particular, the simplest of the grand unified theories, SU5, predicts a lifetime for the proton of about $10^{29\pm 2}$ years. Since this is at most a little longer than the previous experimental limit, it seemed feasible to devise experiments which could test these predictions. About a dozen such experiments are now in various states of completion. Some of these experiments have already preliminary data, some have just started taking data, some are being prepared, some are still in the design stage, and some are only in the planning stage.

An experiment in the Kolar Gold Fields in India (Indian-Japanese collaboration) has now been taking data for more than a year and has obtained a number of events which are candidates for proton decay. The detector (~ 160 tons) consists of iron plates and proportional counters. A similar experiment in the Mont Blanc tunnel (CERN-Frascati-Milano-Turino collaboration) has started working a few months ago and has observed one candidate for proton decay. There was a detailed discussion of these events at the recent International Conference on High Energy Physics at Paris where Perkins (1982) summed up the situation thus: One cannot at present be sure whether these candidates represent proton decays or neutrino induced events. It would therefore be desirable to record the neutrino background and to have, if possible, more specific signatures characteristic of proton decay. It is hoped to achieve this with water Cerenkov counters which can be easily made quite large and can detect directionality of decay particles. Two water Cerenkov counters have recently started taking data: the Harvard-Purdue-Wisconsin experiment in a silver mine in Utah, and the Irvine-Michigan-Brookhaven experiment, with which I am connected.

Here is the list of our collaborators:

Irvine-Michigan-Brookhaven Collaboration:

R.M. Bionta[2], G. Blewitt[4], C.B. Bratton[5], B.G. Cortez[2,a], S. Errede[2], G.W. Foster[2,a], W. Gajewski[1], M. Goldhaber[3], J. Greenberg[2], T.W. Jones[2,7], W.R. Kropp[1], J. Learned[6], E. Lehmann[4], J.M. LoSecco[4], P.V. Ramana Murthy[1,2,b], H.S. Park[2], F. Reines[1], J. Schultz[1], E. Shumard[2], D. Sinclair[2], D.W. Smith[1], H. Sobel[1], J.L. Stone[2], R. Svoboda[6], L.R. Sulak[2], J.C. van der Velde[2], and C. Wuest[1].

(1) The University of California at Irvine
Irvine, California 92717

(2) The University of Michigan
Ann Arbor, Michigan 48109

(3) Brookhaven National Laboratory
Upton, New York 11973

(4) California Institute of Technology
Pasadena, California 91125

(5) Cleveland State University
Cleveland, Ohio 44115

(6) The University of Hawaii
Honolulu, Hawaii 96822

(7) University College
London, U.K.

(a) Also at Harvard University

(b) Permanent address: Tata Institute of Fundamental Research, Bombay, India.

Our detector is shown schematically in fig. 1. It is filled with water and has 2,048 five-inch photo-tubes on its sides. The fiducial volume contains more than 4 ktons. It is located at a depth of $\sim$1600m water equivalent. We expect to identify signals characteristic of proton decay, e.g., $p \rightarrow \pi^o + e^+$. Though the water was clear enough immediately after filling, it is being purified further and the background is being

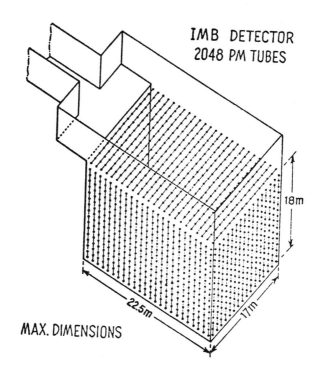

Figure 1

studied. Simple events can be clearly recognised, and delayed electrons from μ-decay have been recorded, showing that our reconstruction scheme from recorded Cerenkov light to track works.

Special attention is being given to events with upward moving particles, for which the muon background is unimportant.

We have no definitive results yet, so I am not even tempted to give you what may be called "symposiogenic results".

References

Goldhaber M., Langacker P., and Slansky R. 1980 Science 210 pp 851-860
Goldhaber M. and Sulak L.R. 1981 Comments Nucl.Part Phys. 10 No. 5 pp 215-225
Marciano W. 1982, Proc. of the Int. Conf. on High Energy Physics (to be published)
Perkins D. 1982, Proc. of the Int. Conf. on High Energy Physics (to be published)
Reines F., Cowan Jr. C.L., and Goldhaber M. 1954 Phys. Rev. 96 pp 1157-8.

The neutron and the universe

Donald D. Clayton

Rice University, Houston and Max-Planck-Institut f. Kernphysik, Heidelberg

Abstract. The rich astrophysical legacy of the neutron derives from its basic properties (spin, mass, charge, decay rate) as well as the strength of its interactions with nuclei (neutron separation energies, capture cross sections, nuclear fluid). Astrophysical phenomena that depend strongly on those quantities include helium production in the big bang, hydrogen burning power in stars, solar neutrino experiment, heavy element nucleosynthesis, supernova core collapse, and neutron stars. We consider how our universe would have differed if the neutron's properties had been different.

1. Introduction

The neutron has a rich astrophysical legacy. Many features of the universe depend upon its properties. To an astrophysicist this is a gratifying thing — that macroscopic events require for their understanding the microscopic properties of Chadwick's baryon. Rather than attempting to list these, I will prefer the more spirited approach of imagining how our world would have been different if the neutron had been different. This is no small task, so interwoven are physical and astrophysical theory, so I ask understanding for whatever oversights I may make in this attempt.

For sake of discussion, let us consider these properties of the neutron: its spin, its mass, its lifetime against beta decay, and its interaction with other nucleons and nuclei. How would our physical universe and its contents differ if the known properties had been different? Among a host of changes, some of great sweep, a few especially noteworthy ones can be singled out.

2. Spin

The half-integral spin of the neutron imposes Fermi-Dirac statistics upon it. We could, perhaps, imagine a universe in which the neutron (and proton) were instead Bosons. Nuclei would then not be forced into a shell-model structure by the exclusion principle, so that the natural abundances of the elements would have evolved to quite different results via processes of nucleosynthesis. For example, the A = 5 nuclei would then probably have been stable, so that nucleosynthesis in the big bang (Wagoner et al 1967) would not have effectively stopped at ^{4}He. Likewise, hydrogen-burning stars would not have to wait for the denser and hotter helium plasma in order to bridge the A = 5 gap via the complicated 3α process (Salpeter 1952), but could instead build directly from ^{4}He, a single mass unit per step. So sweeping would the changes be in the evolution of the chemical composition of the universe that we will hastily close this door to further consideration. Suffice it to observe that very neutron-rich isotopes would have been possible, because each neutron could enter the ground state. Creation of elements of higher proton number would have ceased at low Z owing to the dominance of Coulomb energy, which would render beta decays impossible on energetic grounds. But it would be a major research project to consider how different the matter of our world would have been without need of Fermi-Dirac statistics. Of course, we might not be here to consider the problem had that been the case, because the dominantly large abundances of carbon and oxygen so important to our existence were dictated by the instability of all A = 5 isobars, forcing fusion in a ^{4}He gas.

The neutron star bears the distinction of being the only astronomical object predicted before discovery (Baade and Zwicky 1934; Oppenheimer and Volkoff 1939). The very existence of these exciting objects, virtual cosmic laboratories for nuclear physics, depends on the neutron's adherence to Fermi-Dirac statistics. In a gas of free fermions, the density of continuum states having momentum $p < p_F$ is simply $N(p < p_F) = 4\pi p_F^3/3h^3$. Because only two fermions (with opposite spins) can occupy each kinematic state, any increase in the density of fermions forces them into higher momentum states, even at zero temperature. It raises the Fermi momentum p_F, which, in turn, increases the pressure of the gas. For a free neutron gas at typical densities in a neutron star, $P \propto n_n^{5/3}$ for this reason. Such an equation of state was used by Oppenheimer and Volkoff in predicting 0.7 $M_\odot$ for the maximum mass of a neutron star. But the spin effects are even more pervasive than that. The strong atractive tensor force in the $^3S_1$ state that binds the deuteron is not allowed in the scattering of two neutrons because of the required antisymmetry. The repulsive part of the nn interaction stiffens the equation of state so that neutron stars up to about 2.5 $M_\odot$ are possible (Cameron 1970). Some of astronomy's most exciting measurements are of objects that could not exist if the neutron did not in fact have s = 1/2. So let us hastily restore that fundamental property and proceed to others.

3. Mass

If the neutron mass were different, the physical universe would be different, either subtly or dramatically, depending on the severity of the change. Our stars, galaxies, and interstellar media are dominated by hydrogen — ions, atoms, or molecules. Had the neutron been less massive than a hydrogen atom, that would, of course, not be possible. Free hydrogen would then have decayed into neutrons, yielding a universe dominated by free neutrons. It would have been a much less visible universe! Equally dramatic changes would persist as that interstellar neutron gas collapsed to form stars. What kind of stars? The pressure support of stars comes overwhelmingly from the electron gas, virtually absent if $M_n < M_H$. Without such support, the interstellar medium would have collapsed directly to neutron stars and blackholes! We would not, in any case have come into existence to view that dark and violent world, where X- and γ-ray luminosities would have far exceeded optical powers.

Perhaps we should consider changes less extreme. Suppose that we retain the inequality $M_n > M_H$, but with a mass difference altered from the observed value 0.78 MeV. Where better shall we begin than with the beginning of our universe. Uncertain though that is, we may adopt the standard model (e.g., Harrison 1973; Wagoner et al 1967) of the hot big bang. In the lepton era, $e^{\pm}$ pairs dominate the particle density, so that $n_+ \simeq n_-$. The neutron-proton ratio is determined by the weak captures

$$n + e^+ \to p + \bar{\nu}, \qquad p + e^- \to n + \nu \qquad (1)$$

because they are both much faster at those temperatures and densities than is the free decay of the neutron. But the second reaction is the slower one because it utilizes only those electrons with energies above threshold. The equilibrium ratio is

$$n_n/n_p = \exp(-Q/kT) \qquad (2)$$

where $Q = (m_n - m_p)c^2$. This ratio is unity at high temperature, and falls as the temperature falls toward values $kT \simeq Q$. Reactions (1) and their inverses remain in equilibrium, leading to (2), only for temperatures high enough that the weak rates proceed faster than does the cooling due to the rate of expansion. That equilibrium (2) breaks down when the temperature has fallen to $T \simeq 1.0 \times 10^{10}$ °K for the standard expansion rate of a hot big bang. From then on the nucleons are rapidly freed from weak captures. The value of the ratio (2) at $T = 1.0 \times 10^{10}$ °K is numerically

$$n_n/n_p = \exp -1.5(1 + \frac{\Delta Q}{1.29 \text{ MeV}}) = 0.22 \exp -1.5(\Delta Q/1.29 \text{ MeV}) \qquad (3)$$

where $\Delta Q \equiv (m_n - m_p)c^2 - 1.29$ MeV is the imagined alteration of the neutron-proton mass difference from its observed value of 1.29 MeV. Equation (3) results, after allowing about 200 sec for free neutron decay, in about 30% ^{4}He by mass emerging from the big bang for $\Delta Q = 0$, which agrees rather well with estimates for the He density of our universe. But if $m_n - m_p$ were twice as great, for example, giving $\Delta Q/1.29$ MeV = 1, the He-concentration would have been only 1/5 as great. Although this change, or others in the same spirit, would lead to interesting differences in some aspects of stellar evolution and nucleosynthesis, no great changes in the basic workings of the universe would result. Of greatest concern to us here on Earth would be the difference in the luminosity of our sun owing to its greater internal number of free pressure-generating particles per gram. One can reliably estimate that, for the initially homogeneous sun, reducing He from 30% to 6% would have cut the initial solar constant by a factor of three, resulting in an undoubtedly chilly Earth.

Another astrophysically interesting aspect of the neutron-proton mass difference occurs in the equation of state of a neutron gas, such as in a neutron star or in a collapsing supernova core. We must first understand how a neutron gas can exist, considering that the free decay $n \to p + e^- + \bar{\nu}$ transforms the neutron in a vacuum to a proton plus leptons. The same decay occurs in a gas of free neutrons, but only up to a point. The decay creates an electron gas that is capable of filling the final electron states normally created in neutron decay. Then the decay is blocked by the exclusion principle acting on the electrons. If the thermal energy kT is small in comparison with the electron energy required by the exclusion principle, then each electron state will be occupied up to a Fermi momentum $p_F(e) = h(3n_e/8\pi)^{1/3}$. When the electron density n_e is great enough, the Fermi energy associated with $p_F(e)$ becomes greater than $(m_n - m_p - m_e)c^2$, so that further neutron decay is blocked owing to the occupancy of the low-energy states of the electron that could otherwise be created by the neutron decay. In essence, the neutron has become stable. This degeneracy of the electrons is a very big difference from the situation in the big bang. Now it will be clear that the greater is $m_n - m_p$, the greater will be the associated density of electrons (and protons) n_e required to fill the available phase space. The numbers work out about like this. Neutron decay is first inhibited near a nucleon density $\rho_n \simeq 10^8$ g cm^{-3}, where some neutrons are forced to remain in the gas. At a density some twenty times greater (still 10^5 less than nuclear density) the neutrons must outnumber protons and electrons by about 100:1. So it is clear that if $m_n - m_p$ were doubled, for example, the value of the electron Fermi energy would be correspondingly increased, which would require a full order of magnitude increase in the electron density n_e. The proton density must equal the electron density. Thus, the contribution of electrons to the pressure and the free-proton density (which can actually assemble into heavy nuclei within the neutron gas) depend on the neutron-proton mass difference. In fact, because the electrons are relativistic, $P_e \propto p_F(e)^4 \propto (m_n - m_p - m_e)^4$. The associated free protons, on the other hand, reduce the pressure by combining with free neutrons into heavy neutron-rich nuclei, thereby decreasing the number of independent free particles in the gas. In short, the physics of a neutron gas would result in different neutron stars, and hence different pulsars, if $m_n - m_p$ were different.

Because the neutron is only slightly more massive than the proton, nucleosynthesis in a hydrogen gas can begin with the weak-decay formation of a deuteron from two protons. But if the neutron were only 0.3% more massive than the proton, the deuteron would instead decay into two protons! Stars would have then been hugely altered by their inability to initiate fusion in this manner. Nor could the problem then be circumvented by the CN cycle, because any process that fuses He from H must be accompanied by two low-energy $e^+\nu$ emissions that change protons to neutrons, now a prohibitively endothermic change. Even the alpha particle would have been instable if $m_n - m_p$ were as great as a few percent. The very face of our universe depends upon the neutron having a mass nearly degenerate with that of the proton!

We need not confine attention to the $m_n - m_p$ difference. Suppose that difference were preserved, but the absolute value of the neutron mass changed. For example, the neutron could have had half its present mass with the proton correspondingly reduced to preserve $m_n - m_p$. How then would the universe differ? Certainly nuclear binding energies would change, requiring all of stellar fusion and nucleosynthesis to differ. But suppose the nuclear force could be adjusted in such a way that all nuclear binding energies were preserved. How then would our universe be altered? How does our understanding depend upon the absolute mass of the nucleon?

A very great difference would occur in the structure of stars, whose pressure support comes from both electrons and nucleons, but whose weight results almost totally from nucleons. If the nucleon mass were halved, the number of free particles per gram would double. The sun (with the same number of nucleons) would be only 10^{-2} as luminous as it is, leaving our Earth totally frozen. A nucleon twice as heavy, by contrast, would have produced a sun of such short life and brightness that our own lives could not have arisen, much less had the time to evolve to our level of curiosity about such matters. The end products of stellar evolution would be especially strongly affected. The maximum mass that can be supported by electron-degeneracy pressure is $M_{WD} < 5.8/\mu_e^2$ $M_\odot$, where μ_e is the number of mass units per electron (Chandrasekhar 1935). Normally, $\mu_e \simeq 2$ for hydrogen-exhausted stellar cores, leading to $M_{WD} < 1.4$ $M_\odot$; but if the nucleon mass is altered by a factor 2, say, the M_{WD} limit is changed by a factor of four. Considering that white dwarfs are the second most abundant star in the sky (next to main-sequence stars), the sky would be much changed. For a neutron star, the effects are even more profound. A less massive neutron coud make the formation of neutron stars much more comfortable, whereas a more massive neutron could not resist the crush toward a black hole.

But the nature of stars might be more academic than this exercise indicates. Both galaxy formation and star formation are such difficult processes to bring about that they are not well understood. If the nucleon were only half as massive, it is conceivable that galaxies and stars could never have achieved the gravitational instability required in the expanding chaos of the big bang. Conversely, if the nucleon had been twice as massive, star formation would be so efficient, but stellar lifetimes so short that now, 10^{10} years after the big bang, a dark and violent stellar graveyard might be all that would remain!

It is not clear if we are justified in going as far as the "anthropic principle" (Carter 1974). Its simple application in this case might take the form of the following: the mass of the neutron is as it is because if it were much different we would not be here to measure it. Certainly, the idea remains endlessly fascinating, even if philosophically uncertain.

4. Halflife

If the lifetime of the neutron against its free decay $n \rightarrow p + e^- + \bar{\nu}$ were much different, our astrophysical universe would also differ. Its rate (with masses fixed as known) measures the weak coupling constants $G_V^2 + 3G_A^2$. Because of the strong dependence on the axial-vector coupling constant, the neutron lifetime is a major datum for determining G_A^2 (with G_V^2 determined from $0^+ \rightarrow 0^+$ decays like ^{14}O).

The primary reaction for initiating thermonuclear power in hydrogen burning stars is (Bethe and Critchfield 1938; Bahcall and May 1968)

$$p + p \rightarrow d + e^+ + \nu \qquad (4)$$

during proton scattering. This rate is proportional to G_A^2, and hence depends strongly on the rate of the decay of the neutron. If the neutron decayed more slowly, for example, G_A^2 would be smaller (at known G_V^2), as would therefore the average rate of the pp reaction in a hydrogen thermonuclear plasma. Maintenance of the solar luminosity would in that case require a higher temperature inside the sun. The measurability of solar neutrinos would be much affected for the following reason. Those neutrinos having sufficient energy to produce the allowed absorption by ^{37}Cl are the highest energy neutrinos emitted by the sun. Those come from the ^8B decay, whose production is strongly dependent upon the solar central temperature. It turns out that the expected flux of ^8B neutrinos scales inversely as the 2.5 power of the pp reaction cross section (Bahcall et al 1969). In short, if the neutron lifetime were much different, so would be the expectations for the solar neutrino experiment based on ^{37}Cl absorption (Bahcall and Davis 1982).

In a broader sense, all of that part of astrophysics that depends upon weak decay rates would be altered if the neutron decay were altered. A sweeping example comes from the big bang itself. If the weak rates were much slower, the reactions (1) in the early lepton phase of the big bang would cease to be effective at a higher freeze-out temperature. Then the equilibrium equation (2) would have a value near unity instead of near 0.2. Were that ratio unity, the nucleons from the big bang would have fused into ^{4}He nuclei during the expansion, with no leftover hydrogen. In a universe without hydrogen there would be no hydrocarbons, no water, and no people (at least with the known chemistry of life). Once again we think of the anthropic principle.

Another example comes from supernova core collapse. A typical current question (e.g., Bethe 1982) is whether the absorption of neutrinos by neutrons ($\nu + n \rightarrow p + e^-$) can replenish electrons fast enough behind the accretion generated shock to make up for the pressure deficit that will otherwise be there because of electron capture during the shock. That pressure deficit makes a rarefaction that damps the propagation of the shock wave. Whether that shock can eject matter (a supernova) depends on the hydrodynamic maintenance of its strength through such subtle equation-of-state questions. This particular reaction for generating electrons from the dominant neutrons and neutrinos has a cross section directly proportional to the decay rate of the neutron. Standing as it does on the forefront of research, it is not yet known if the rate of this effect is hydrodynamically important; indeed, it is not yet known whether core-collapse models of supernovae even eject mass.

For many problems in stellar nuclear astrophysics the weak decay rates are unimportant. Hydrostatic burning in stars, for example, occurs over such long times that weak decays, when they are energetically allowed, happen relatively instantaneously. For such applications the nuclear burning would proceed almost unchanged even if the neutron halflife were different. There are important exceptions, however. One of the most dramatic occurs in the so-called r-process of heavy element nucleosynthesis. It is characterized by a very intense burst of neutrons during a stellar explosion. The intense flux of neutrons lasts for time of order seconds, whereas the heavy element lifetimes against capture of a neutron are of order milliseconds. Such circumstances are therefore capable of assembling very neutron-rich isotopes of each element. The process is so infrequent, however, that neutron-rich isotopes would be very rare in nature if they depended upon a rare process converting neutron-poor isotopes of an element to neutron-rich isotopes of the same element. Nor would the abundance distribution correctly account for the natural abundance distribution correctly account for the natural abundance peaks near $A = 130$ and near $A = 195$ among the neutron-rich nuclei. These are very evident in Figure 1. It is instead necessary that some abundant seed nucleus (iron) be converted to these nuclei in order for them to be as abundant as observed. But this requires that beta decays occur during the short time of the explosion. In any one element neutron captures proceed only so far (perhaps 20 mass units) before further neutrons are effectively unbound in the face of an intense thermal gamma-ray flux. To make neutron-rich isotopes of a heavy element (Pt, say) from a much lighter parent (Fe, say) requires that the otherwise rapid neutron captures patiently wait for beta decay to move the more abundant elements to higher Z (e.g., Seeger et al 1965; Hillebrandt 1978). The limiting factor becomes the beta decay rate of neutron-rich isotopes of each element. Can the matter move to higher Z during the second available during the explosion? Clearly if the neutron free decay were much altered in its strength, so would be also the decay rate of neutron-rich nuclei. At present it seems that beta decay rates are just barely fast enough to allow this build-up.

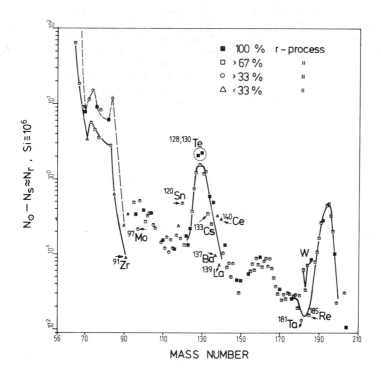

Fig. 1. Abundances per 10^6 Si atoms of the neutron-rich unshielded nuclei resulting from the r-process. Solid points are exclusively of r-process origin, whereas open points have had a portion of their abundances removed owing to their calculable production by the s-process captures on the path of beta stability. From Käppeler et al (1982).

5. Interaction with Nucleons

Another neutron property important for our observed universe is its interaction with other nucleons — the so-called nuclear force. The statement is in part obvious. All of nuclear physics would be changed if the nuclear force had a different strength. So therefore would stellar power and nucleosynthesis. Perhaps one should let it go at that. But it is appropriate to remember a few subtle and beautiful astrophysical consequences of the exact strength of this force.

The deuteron is bound by only about 0.2% of the neutron's mass. If this triplet spatially-symmetric interaction were slightly weaker, it would not (as in the cases of the singlet diproton, dineutron, or singlet deuteron) be bound at all. The cancellations are very close! Were the deuteron not bound, the reaction (4) for initiating hydrogen burning in the sun would not be possible. One might then suppose that the sun would simply have to burn hotter so as to use the CN cycles instead of the pp-chains (e.g., Clayton 1968). However, another problem is immediately evident when one thinks of the big bang itself. Early in that expansion, the n/p ratio freezes out near 10^{10} °K, as in eqs. (1) and (2). But nuclei do not form at that time. Indeed, it is about 200 sec later, when the ambient temperature has fallen to 10^9 °K that the deuteron is first stable in the photon field, which dissociates it according to $\gamma + d \rightarrow n + p$. Only then can the build-up to ^{4}He follow. But suppose the deuteron were not bound, or were bound so weakly that all neutrons would decay to protons before the temperature had fallen to a value low enough for it to survive photodissociation. In this case the universe would have emerged (at least in the standard hot-big-bang model of it) as 100% hydrogen. The problem of initiating fusion in a pure hydrogen gas would therefore not be circumvented.

Suppose instead that same triplet strength had been stronger, so that the deuteron would be more tightly bound. In this case the big-bang products would be little altered, because the stability of the deuteron at higher temperature would simply cause the free neutrons to assemble into ^{4}He at a somewhat higher temperature earlier in the expansion. But the sun would have been redder and less luminous! The reason is that reaction (4), which provides solar power, would have a larger cross section. Using the $\lambda_B \propto W^5$ approximation for the sake of discussion, the low-energy cross section would scale as $\sigma(pp) \propto E_B^5$, where E_B is the binding energy of the deuteron. Were the cross section larger, the sun would burn at lower temperature, hence, by the virial theorem, more distended, and hence <u>less</u> luminous (rather than more luminous, as one might at first intuitively suppose).

If the nucleon-nucleon force were generally stronger, particles now unstable would be stable, with ineluctable alterations to our universe. Three examples will suffice to illustrate our delicate balance: (1) the singlet diproton might have been stable, in which case hydrogen would fuse via $p(p,\gamma)^2$He $(\beta^+\nu)^2$H in both big bang and stars; (2) ^{5}Li might have been stable, in which case the fusion of hydrogen would not be limited to build-up to A = 4; (3) barely unbound ^{8}Be might have been stable, in which case the relatively complicated 3α process (e.g., Clayton 1968) would not have been required for the burning of helium. Both the evolution of stars and the relative abundances of light elements (which strongly influence the chemical basis of our existence) would have been greatly altered by the small energy shifts needed to throw these near cancellations onto the other side of the fence.

Existing objects with observable properties dependent upon n-n scattering are found as neutron stars. The basic problem in understanding them is that of calculating the pressure of a nuclear fluid at densities comparable to and greater than nuclear density $\rho_0 = 2.8 \times 10^{14}$ gm cm^{-3}. This is a neutron fluid, because the electron Fermi energy of a small admixture of protons and electrons easily prevents neutron decay. This one-sidedness is quite unlike symmetric nuclear matter, where $N \approx Z$. And in the neutron star the inner 90% is squeezed by gravity to densities ranging from barely subnuclear to several times ρ_0. To calculate the state of such matter various workers (e.g., Baym and Pethick 1979) have employed either the Brueckner-Bethe-Goldstone nuclear-matter theory, which in lowest order sums contributions from two-body scattering processes, or variational techniques based on trial wave functions. The problems then are the choice of the two-body interaction, the role of the internally excited nucleon Δ (1236 MeV), and the inclusion of tensor correlation effects. The severity of the problem is emphasized by the difficulty of understanding the symmetric (N = Z) state at ordinary nuclear density ρ_0. Very exotic states may be relevant to the structure of the neutron stars (Baym and Pethick 1979). They include: (1) non-zero expectation value for the pion field above $\rho \simeq 2\rho_0$; (2) an "abnormal state" in which neutrons become massless; and (3) a quark fluid when the neutrons are so compressed as to spatially overlap and dissolve. Inclusion of only the first these leads to sufficiently different estimates of the equation of state that calculations of a neutron-star structure having a chosen central density ($\rho_c = 3\rho_0$, say) yield a neutron star mass anywhere between 0.3 $M_\odot$ and 2.8 $M_\odot$, depending upon the nuclear theory used. It will be a source of continuing satisfaction to the student of neutron matter that nature has provided hundreds of such astronomical laboratories of varying masses. Their moments of inertia can be derived from the relation between pulsar power and the slowing of the spin rate.

In a quite different vein it is worthwhile to remember how neutron interactions are believed to determine the abundances of the heavy elements and their isotopic compositions. Return to this end to the concept of the intense burst of neutrons that produces the neutron-rich (β^- unshielded) isotopes of the heavy elements. With neutron densities $n_n \gtrsim 10^{20}$ cm^{-3}, the neutron captures race towards heavier isotopes. The capture flow is quickly hindered by (γ,n) reactions when the neutron separation energy S_n falls to $S_n <$ (10-20) kT $\simeq$ 1-2 MeV (Seeger et al 1965). The (γ,n) rates rise so rapidly with falling S_n that the capture flow is halted, and beta decay must occur before the captures can continue. Given a free neutron density and temperature, the identity (i.e., mass) of these "waiting-point nuclei" depends primarily upon how S_n decreases with increasing neutron number. This depends directly on the nature of neutron interactions in very neutron-rich nuclei. Nor is the question without observable consequences, because the stable abundances in nature show pronounced peaks near A = 130 and A = 195, as shown in Figure 1. So pronounced are these peaks that they are usually attributed (e.g., Seeger et al 1965) to slower-than-average beta decay rates by those special nuclei piling up at waiting points having magic-shell neutron closures at N = 82 and N = 126, respectively. These neutron-rich waiting-point abundances subsequently decay to the unshielded isobars shown in Figure 1. This requires that the astrophysical circumstances (n_n,T) conspire with the energetics of neutron separation in such a way that the sequence of waiting points define a specific path in the Z,N plane. For example, if the mass peak A = 128-130 is to correspond to waiting points having N = 82, the elements involved must be Z = 46-48 (e.g., ^{130}Cd). Clearly the strength of the neutron's binding to such neutron-rich nuclei is a sensitive astrophysical property of the neutron, because it determines how many of those nuclei exist.

Abundances along the path of beta stability, on the other hand, are determined by an entirely different aspect of neutron interactions — the (n,γ) reaction cross sections. The situation envisioned here is a weak neutron flux operating over long times in the interiors of stars. The successive captures lead to the next heavier isotope if it is stable, whereas, if it is not stable, a relatively quick beta decay moves the capture path to the same isobar in the next element. Abundant iron is the basic seed from which heavy nuclei are made in this way. It is the property of such a chain that the product $\sigma_A(n,\gamma)N_A$ along the chain is a monotonically decreasing function of atomic weight if the amount of iron exposed to different fluences is a monotonically decreasing function of the fluence. Basically the system attempts to minimize differences in successive $\sigma_A N_A$ products, so that nuclei with large capture cross section $\sigma_A(n,\gamma)$ have compensatingly small abundances N_A (Clayton et al 1961; Clayton 1968). That this is the case can be seen by looking at the "shielded nuclei," whose abundances are not augmented also by isobaric beta decay chains.

These products are shown in Figure 2. The sources of the (n,γ) cross sections and the abundances used are detailed by Käppeler et al (1982). These $\sigma_A N_A$ products closely follow the calculated curve, which assumes that the amount of iron irradiated is an exponentially decreasing function of the fluence. There can be no doubting the essential correctness of this theory, for those nuclei whose abundances are not shielded on the capture path have $\sigma_A N_A$ products that scatter about the same diagram (e.g., Clayton et al 1961; Clayton 1968). This understanding offers one of the best proofs that nuclear abundances are as they are because of the stellar reactions that have created them (rather than by whim of the Creator, say). The interactions of neutrons with nuclei provides the key to understanding. It simultaneously provides a challenge to the neutron physicist to measure and calculate the appropriate interactions. For nature, and Chadwick, have presented us with a sensitive indicator of the nuclear universe.

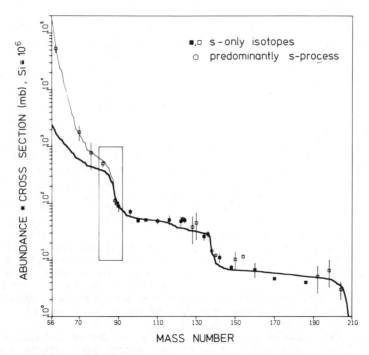

Fig. 2. Product of abundance of shielded (s-process) nuclei times their (n,γ) cross sections. The solid lines are calculations for an exponential distribution of neutron fluences on iron seed. From Käppeler et al (1982).

6. Acknowledgments. Work on this paper was supported in part by the Alexander von Humboldt Stiftung, with the hospitality of the Max Planck Institute für Kernphysik, Heidelberg, and in part by NASA grant NSG-7361 to Rice University.

7. References

Baade W and Zwicky F 1934 Phys. Rev. 45 138
Bahcall J N, Bahcall N A and Ulrich R K 1969 Astrophys. J. 156 559
Bahcall J N and Davis R Jr 1982 Essays in Nuclear Astrophysics eds C A Barnes, D D Clayton and D N Schramm (Cambridge: Cambridge University Press)
Bahcall J N and May R M 1968 Astrophys. J. (Letters) 152 L17
Baym G and Pethick C 1979 Ann. Rev. Astron. Astrophys. 17 415
Bethe H A 1982 Supernovae eds W D Arnett, J M Lattimer, M J Rees and R Stone (Dordrecht: D. Reidel)
Bethe H A and Critchfield C L 1938 Phys. Rev. 54 248
Cameron A G W 1970 Ann. Rev. Astron. Astrophys. 8 179
Carter B 1974 Confrontation of Cosmological Theories with Observation ed M S Longar (Dordrecht: D. Reidel)
Chandrasekhar S 1935 Mon. Not. R. Astron. Soc. 95 207
Clayton D D 1968 Principles of Stellar Evolution and Nucleosynthesis (New York: McGraw-Hill)
Clayton D D, Fowler W A, Hull T E and Zimmerman B A 1961 Ann. Phys. 12 331
Harrison E R 1973 Ann. Rev. Astron. Astrophys. 11 155
Hillebrandt W 1978 Spa. Sci. Rev. 21 639
Käppeler F, Beer H, Wisshak K, Clayton D D, Macklin R L and Ward R A 1982 Astrophys. J. 257 821
Oppenheimer J R and Volkoff G M 1939 Phys. Rev. 55 374
Salpeter E E 1952 Astrophys. J. 115 326
Seeger P A, Fowler W A and Clayton D D 1965 Astrophys. J. Suppl. 11 121
Wagoner R V, Fowler W A and Hoyle F 1967 Astrophys. J. 148 3

Neutron stars: the first fifty years

Gordon Baym

Loomis Laboratory of Physics, University of Illinois at Urbana-Champaign, Urbana, IL

1. Early History

The discovery of the neutron opened up, almost at once, a completely unexpected chapter in astrophysics. In a picture evolved over the period 1932-39, the neutron became the fundamental ingredient of a new class of stellar objects -- highly compact neutron stars, with masses of the order that of the sun but only $\sim$ 10 km. in radius -- that would be looked to in hope of resolving fundamental questions of the end states of stellar evolution, the source of the energy released in supernovae, and even the energy source of ordinary stars.

The story begins, as Léon Rosenfeld recounted at the 1973 Solvay Conference, at the time of the identification of the neutron. When Chadwick's letter of 24 February 1932 to Bohr telling of this new particle arrived in Copenhagen [see Fig. 1] he (Rosenfeld), Bohr and Landau "had a lively discussion on the same evening about the prospects opened by this discovery. In the course of it Landau improvised the concept of neutron stars -- 'unheimliche Sterne,' weird stars, which would be invisible and unknown to us unless by colliding with visible stars they would originate explosions...." (Rosenfeld, 1974) Landau had been concerned with understanding the final states of stars and had the previous month sent for publication (Landau, 1932) an intuitive and independent derivation of the Chandrasekhar mass limit, $M_{Ch} \sim 1.4\ M_\odot$, the maximum mass supportable against gravity by relativistic particle degeneracy pressure (Chandrasekhar, 1931). Although Landau would not write on neutrons in stars until nearly six years later, he was confronting at this time as a theoretical question the collapse of stars of mass greater than M_{Ch} to higher and higher densities, and was toying with ideas of heavier stars having cores in which even "the laws of ordinary quantum mechanics break down ... when the density of matter is so great that atomic nuclei come in close contact, forming one gigantic nucleus."

A second line, rooted more in observation, began the following year, when Baade and Zwicky at Caltech, in carrying out their original studies of supernovae, formulated independently the idea of neutron stars to explain the enormous energy release of these explosions. Reporting their results at the 15-16 December 1933 meeting of the American Physical Society at Stanford, they wrote with remarkable foresight, "with all reserve we advance the view that supernovae represent the transition from ordinary stars into <u>neutron stars</u>, which in their final stages consist of closely packed neutrons" (Baade and Zwicky, 1934a). They amplified this indeed accurate

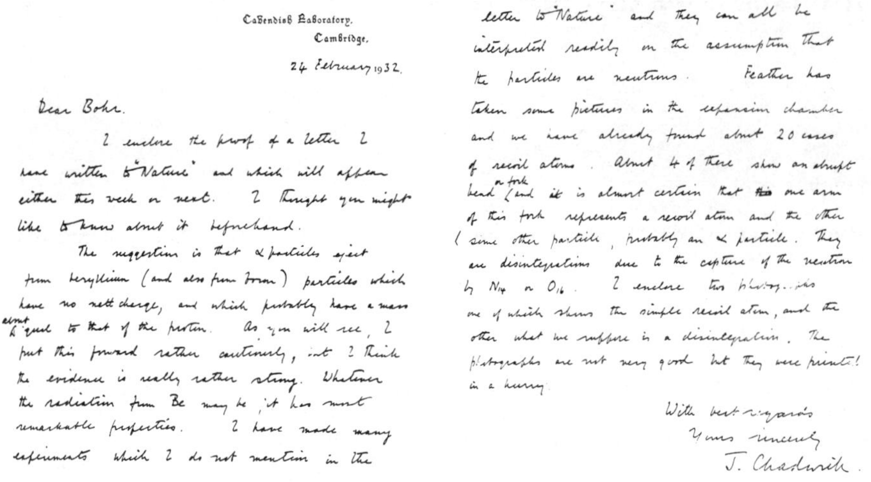

Cavendish Laboratory.
Cambridge.

24 February 1932.

Dear Bohr.

I enclose the proof of a letter I have written to "Nature" and which will appear either this week or next. I thought you might like to know about it beforehand.

The suggestion is that α particles eject from beryllium (and also from boron) particles which have no nett charge, and which probably have a mass about equal to that of the proton. As you will see, I put this forward rather cautiously, but I think the evidence is really rather strong. Whatever the radiation from Be may be, it has most remarkable properties. I have made many experiments which I do not mention in the letter to "Nature" and they can all be interpreted readily on the assumption that the particles are neutrons. Feather has taken some pictures in the expansion chamber and we have already found about 20 cases of recoil atoms. About 4 of these show an abrupt bend or fork (and it is almost certain that one arm of this fork represents a recoil atom and the other (some other particle, probably an α particle. They are disintegrations due to the capture of the neutron by N_{14} or O_{16}. I enclose two photographs one of which shows the simple recoil atom, and the other what we suppose is a disintegration. The photographs are not very good but they were printed in a hurry.

With best regards
Yours sincerely
J. Chadwick.

Fig. 1. Chadwick's letter to Bohr on the discovery of the neutron. [Reproduced with the kind permission of Lady Chadwick, and the American Institute of Physics.]

picture in the second of their papers (1934b) of the following March which proposed cosmic-ray production in supernovae, "Such a star may possess a very small radius and an extremely high density. As neutrons can be packed much more closely than ordinary nuclei and electrons, the 'gravitational packing' energy in a cold neutron star may become very large, and, under certain circumstances, may far exceed the ordinary nuclear packing fractions. A neutron star would therefore represent the most stable configuration of matter as such." Only through such close packing could they understand the energy release in supernovae which they estimated to be equivalent to the annihilation of the order of several tenths of a solar mass. The basic physical mechanism assumed was that, as a consequence of its being neutral, a neutron should not experience radiation (or any nuclear) pressure; hence neutrons made on the surface of an ordinary star will "rain down towards the center," (Baade and Zwicky, 1934c) rapidly transforming the star into a neutron star.

The ultimate fate of stars more massive than M_{Ch} remained a theoretical mystery. One suggestion was that they developed cores of highly compressed matter (see e.g., Milne 1932). While Chandrasekhar in late 1934 at Trinity College had given serious arguments against this possibility, suggesting mass loss or supernovae phenomena instead (Chandrasekhar, 1935), Gamow in 1937, reviewing Chandrasekhar's (1931) and Landau's (1932) earlier arguments, wrote that more massive stars "are subject to the formation of matter in the nuclear state in their interior at some period of their existence." (Gamow, 1937) This state would be formed by inverse beta capture of electrons by nuclei, changing protons to neutrons. [Although Gamow did not refer to it, the first microscopic descriptions of the equation of state of nuclear matter in beta equilibrium were actually given by Hund (1936) a year earlier.] Such a state would not only provide, through gravitational release, energy "enough to secure the life of the star for a very long period of time," but eruptive processes at the surface of the dense nuclear core could give rise to spewing out of "nuclear substance" and formation of various nuclei, thus helping with problems of nucleosynthesis.

Quite independently, Landau, in a paper to Nature (Landau, 1938) submitted from Moscow via Bohr in Copenhagen in late 1937, proposed that in ordinary stars of masses greater than a critical mass $\sim 0.001~M_\odot$, a neutron core, "where all the nuclei and electrons have combined to form neutrons," would be energetically favored over a normal core. Hund's earlier work is referred to here, but only as an afterthought at the suggestion of Bohr and Møller (Bohr, 1937). Landau stressed, as did Gamow, that a neutron core would "give an immediate answer to the question of the sources of stellar energy." It would be half a year until Bethe would present a satisfactory picture of stellar energy generation based on nuclear burning processes.

The stage was now set for the next phase in the development, once again in California. By August 1938, Zwicky, with Tolman, had estimated the maximum binding energy that a neutron star of mass M could have. The result, $0.42~Mc^2$, allowed for more than enough available energy in supernovae. He also made the first attempts to understand spectral shifts observed in supernovae in terms of gravitational redshifts in the strong fields of the neutron star remnants the supernovae would produce, and gave the first estimates of neutron star surface temperatures ($\sim 10^{6-7}~°K$) (Zwicky 1938). Then Oppenheimer and Serber (1938), within the following month, worrying that nuclear mixing might possibly not account for energy release in very bright stars (their example, Capella, was in fact a giant), reexamined the Landau idea of a condensed neutron core, and revised his lower critical mass for neutron star formation to $\sim M_\odot/6$, possibly lowered to $\sim 0.1~M_\odot$ by nuclear forces. They concluded that such a core mass, with its surrounding envelope, would be too large for a star like the sun, but neutron cores in larger stars could not be ruled out.

At this point Oppenheimer and Volkoff began their pioneering numerical general relativistic calculations of neutron star configurations, using the fully general relativistic equation of hydrostatic balance (now called the Tolman-Oppenheimer-Volkoff, or TOV equation) with a free neutron Fermi gas equation of state. Their results, which included calculations of the allowed range of masses of neutron stars, were published in early 1939 (Oppenheimer and Volkoff 1939, also Volkoff 1939) accompanied by a paper by Tolman (1939) submitted simultaneously, giving a number of exact solutions of the TOV equation. [Interestingly, von Neumann and Chandrasekhar had derived the same equation of hydrostatic balance some four years earlier in 1934, in order to study highly collapsed stars (Chandrasekhar 1977); this also took place at Trinity College, just after Chandrasekhar worked out Newtonian stellar configurations with degenerate matter. It is also said (Harrison et al. 1965) that von Neumann integrated exactly the TOV equation for the special limiting case of pressure equal to one third the density. Unfortunately this work was not published.]

Although the promised Tolman-Zwicky paper never appeared, Zwicky wrote a lengthy article (Zwicky 1939) on neutron stars, shortly after the Oppenheimer-Volkoff paper, that discussed them in the framework of general relativity and presented a detailed account of their role in supernovae.

With the Oppenheimer-Volkoff paper the basic framework of the theory of neutron stars was in place; detailed ingredients would not begin to be developed fully for several more decades. And with Baade and Zwicky's work, the astrophysical role of neutron stars in supernovae was laid out. [Curiously, there is no evidence of discussions between Oppenheimer and Zwicky, and virtually no references to the others' work, even though Oppenheimer spent considerable time in this period at Caltech; rather, Tolman appears to be the main link between the two.]

With the coming of the Second World War, interest in neutron stars became dormant, not to awaken for two decades. Important steps in the revival included the development of a more realistic high density equation of state by Harrison, Wheeler and Wakano (1958), in connection with studying final states in gravitational collapse; and studies of cooling of neutron stars (e.g., Tsuruta and Cameron, 1966), inspired by but not supportive of the possibility that X-ray stars such as Sco X-1 discovered in the early 1960's might be thermally radiating neutron stars. Up until this time the lack of any observational information on neutron stars, let alone any firm evidence for their existence, left neutron stars primarily as appealing toys for theorists. As Wheeler wrote in 1966, "a 'cool' superdense star ... of 10^{6}°K, ... is fainter than the 19th magnitude and therefore hardly likely to be seen. The rapidity of cooling makes detection even more difficult" -- a note of pessimism tempered by mention of the possibility of observing radiations (neutrino, X-ray, gravitational and electromagnetic) produced by a neutron star newly made in a supernova (Wheeler, 1966).

No one was prepared for the singular way that neutron stars would make their first appearance only a year and a half later, in the form of pulsars, discovered by Bell, Hewish and coworkers (Hewish et al. 1968) at the Mullard Radio Observatory at Cambridge, some 35 years after the discovery of the neutron itself at Cambridge. With the identification the following Spring by Gold of pulsars as rapidly rotating neutron stars (Gold 1968), neutron star research entered its modern period.

2. Neutron Stars at Present

In the fifteen years since the discovery of neutron stars, an impressive and still growing body of information on them has been gathered by a variety of observational means. Not only do neutron stars exist in abundance but they form rather unusual and active systems, powered by the enormous gravitational energy they make available. At latest count over 320 pulsars have been observed in neighborhood of the galaxy (Taylor and Manchester 1982). Furthermore, neutron stars in close binary orbits with more ordinary stars, from which they accrete matter, are observable as pulsating compact X-ray sources (Lamb and Pines 1979), as X-ray burst sources (Lewin and Joss 1981) and also as γ-ray burst sources (Klebesadel et al. 1982, Lamb 1982).

Optical and X-ray observations of pulsating binary X-ray sources allow one to impose limits on the masses of the neutron stars in these objects, which appear to be quite consistent with present theories of neutron star structure and formation in supernovae. X-ray satellites, such as the Einstein, Rösat, and Exosat, can measure surface luminosities, which combined with theories of neutron star cooling, yield in principle considerable information on the properties of matter in neutron star interiors. Surface magnetic fields can also be inferred from models of pulsars and pulsating X-ray sources, or more directly from features in X-ray spectra, to be generally

$\sim 10^{12}$ G. Information on moments of inertia and radii may be obtained from observations of the secular rates of change of neutron star spin periods, interpreted in the framework of theoretical models of these changes. Also, short term variations of pulse arrival times, for both pulsars and pulsating X-ray sources, offer a wealth of information about the possible internal workings of neutron stars. [For review see Lamb 1981.]

The basic structure of neutron stars, illustrated in Fig. 2 (and reviewed in Baym and Pethick, 1975, 1979) is reasonably well understood. Both the density and electron Fermi energy increase with depth, and through electron capture processes, $e^- + p \rightarrow n + \nu$, the matter becomes more and more neutron rich with increasing depth. Beneath a solid metallic crust, typically ~ 1 km thick, and so neutron rich in its deeper layers that it is permeated there by a sea of free neutrons, is a liquid interior -- a gigantic nucleus with A $\sim 10^{57}$ -- beginning at a density $\sim 3 \times 10^{14}$ g/cm^3, that of nuclear matter. Several possible phases (Fig. 2) of the deep interior have been considered, but the properties of this region, and consequently the allowed range of neutron star masses, remain somewhat uncertain.

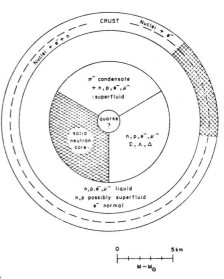

Fig. 2

While Baade and Zwicky provided the fundamental clue that neutron stars would be made in supernovae, the detailed mechanism by which this occurs has only recently (as a result of extensive computer simulations combined with analytic studies) begun to become clear. In a supernova, a more massive star evolves to the point where it no longer burns fuel in its interior and can then no longer support itself against gravitational collapse. The basic problem is to understand how the gravitational energy released in the collapse is transferred to the outer layers, expelling them, and leaving behind enough mass to make a neutron star (but not so much that it will itself collapse to a black hole). In the presently most promising picture, one originally proposed by Colgate and Johnson (1960), a hydrodynamic shock, generated by the collapse and subsequent rebound of the central core, propels the mantle and envelope of the star outwards. As one now understands, the shock forms outside a core of $\sim 0.9 M_\odot$, thus leaving a condensed embryonic neutron star behind. Whether however the shock is sufficiently energetic to produce a supernova explosion is still under active investigation. [For recent overviews see Brown et al. (1982) and Rees and Stoneham (1982).]

In summary, we can only guess at the pleasure that the early originators of the concept of neutron stars, Baade, Landau, Zwicky and others, would take in seeing the present fruits of their insights into this remarkable endstate of stellar evolution.

Research supported in part by NSF Grant DMR-81-17182.

References

Baade W and Zwicky F 1934a Phys. Rev. **45** 138
Baade W and Zwicky F 1934b Proc. N.A.S. **20** 254, 259
Baade W and Zwicky F 1934c Phys. Rev. **46** 76
Baym G and Pethick C 1975 Ann. Rev. Nucl Sci. **25** 27
Baym G and Pethick C 1979 Ann. Rev. Astron. and Astrophys. **17** 415
Bohr N 1937 letter to L. Landau, 6 Dec. 1937 [Bohr Letter Collection, Copenhagen]
Brown G E, Bethe, H A, and Baym G 1982 Nucl. Phys. **A375** 481
Chandrasekhar S 1931 Astrophys. J. **74** 81
Chandrasekhar S 1935 Mon. Not. R.A.S. **95** 207, 227
Chandrasekhar S 1977 Interview by S. Weart [A.I.P., N.Y.]
Colgate S A and Johnson M H 1960 Phys. Rev. Lett. **5** 235
Gamow G 1937 *Structure of Atomic Nuclei and Nuclear Transformations* (Oxford) 234-8
Gold T 1968 Nature **218** 731
Harrison B K, Wakano M and Wheeler J A 1958 La Structure et l'Evolution de l'Univers (11th Solvay Conference), (Stoops, Brussels)
Harrison B K, Thorne K S, Wakano M and Wheeler J A 1965 *Gravitation Theory and Gravitational Collapse* (U. of Chicago) 16
Hewish A, Bell S J, Pilkington JDH, Scott D F and Collins R A 1968 Nature **217** 709
Hund F 1936 Ergebn. exakt. Naturwiss. **15** 189
Klebesadel R et al. 1982 Astrophys. J. **259** L51
Lamb D Q 1982 in *Gamma-Ray Transients and Related Astrophys. Phenomena* (AIP Conf. Proc. 77, N.Y.) 249
Lamb F K and Pines D, eds. 1979 *Compact Galactic X-ray Sources* (UIUC Phys. Dept., Urbana)
Lamb F K 1981 in *Pulsars* [IAU Symp. 95, Riedel, Dordrecht] 303
Landau L 1932 Phys. Z. Sowjet. **1** 285
Landau L 1938 Nature **141** 333
Lewin WHG and Joss P C 1981 Sp. Sci. Rev. **28** 3
Milne E A 1932 Mon. Not. R.A.S. **92** 610
Oppenheimer J R and Serber R 1938 Phys. Rev. **54** 540
Oppenheimer J R and Volkoff G M 1939 Phys. Rev. **55** 374
Rees M and Stoneham R, eds. 1982 *Supernovae*, Proc. Nato Adv. Study Inst., Cambridge (Riedel, Dordrecht)
Rosenfeld L 1979 in *Astrophysics and Gravitation*, Proc. Solvay Conf. on Physics, 16th. [Ed. Univ. Bruxelles, Brussels] 174
Taylor J and Manchester R 1982 Astrophys. J. (in press)
Tolman R C 1939 Phys. Rev. **55** 364
Tsuruta S and Cameron AGW 1966 Can. J. Phys. **44** 1863
Volkoff G M 1939 Phys. Rev. **55** 413, 421
Wheeler J 1966 Ann. Rev. Astron. and Astrophys. **4** 393
Zwicky F 1938 Astrophys. J. **88** 522
Zwicky F 1939 Phys. Rev. **55** 726

Inst. Phys. Conf. Ser. No. 64: Section 2
Paper presented at Conf. on Neutron and its Applications, Cambridge, 1982

Neutron superfluidity

Malvin Ruderman

Department of Physics, Columbia University, New York, N.Y.

1. Introduction

About 10^{-2} of the identified nucleon content of our universe consists of the neutron seas within neutron stars. That these neutrons would most certainly be superfluid (Migdal 1959, Ginzburg and Kirzhnits 1964, Ruderman 1967) was realized almost a decade before the discovery of pulsars but this seemed of rather academic interest then and only several papers developed this notion. But since 1968 neutron superfluidity and its consequences has become a major part of neutron star research.

The neutron sea which gives the neutron star its name does not extend to the stellar surface. The 300 meters just below the surface consists of relatively conventional, although extremely dense, metallic matter: nuclei (rather neutron rich) and electrons with the nuclei arranged into a (bbc) crystalline lattice. (For a review and references see Baym and Pethick 1975.) Below this depth, where the stellar density exceeds $4.3 \times 10^{11} \text{gcm}^{-3}$, a degenerate neutron superfluid fills the space between nuclei: neutrons flow around and through nuclei which, as far as the neutrons are concerned are simply regions where the neutron density is greater than it is in the interstitial regions. A km or so below the stellar surface where the density is near $2 \times 10^{14} \text{g cm}^{-3}$ the crust ends and the "core" begins: the lattice of nuclei disappears and the several percent of protons (and electrons) which coexist with the neutrons are, together with the neutrons, now uniform density seas. The core neutrons are still superfluid, the protons form a charged superfluid, i.e. a superconductor, and the electrons a normal degenerate Fermi sea (except that its Fermi energy is many tens of Mev). It is not yet entirely clear what form of matter will exist in the central part of the core of those particular neutron stars which are abundantly formed in the Galaxy (and often observed as pulsars). If the central density is sufficiently large ($\rho \gtrsim 5.10^{14} \text{g cm}^{-3}$?) a coherent superfluid type condensation is expected for a net negatively charged π-meson field in which the nucleons are no longer distinguishable as protons or neutrons but are rather a coherent mixture of both. These nucleons are, however, in most models for this π-condenstate phase "near neutrons" in that the mixture of proton in each nucleon is only 1-10% (Brown and Weise 1976). Whether these "near" neutrons" form the same kind superfluid as do the outer core neutrons is not known with certainty since there is a competition in achieving the ground state of π-condenstate matter between spatial ordering which favors quasi crystalline structure and the momentum pairing needed to make Cooper pairs in the BCS description of a Fermion superfluid (Tamagaki 1979).

Finally the central part of the core may be best described simply as a degenerate sea of up, down, and strange quarks since the density may be sufficiently high that the quarks which compose the nucleons move easily through the matter despite "confinement". But here also residual attraction between quarks is expected to result in quark superfluids if the uniform density quark matter description is adequate (Bailin and Love 1982).

2. Criteria for Superfluidity

The description of the neutron (and "near nutron") fermion seas as superfluids follows the BCS model of electron superfluidity but with certain important quantitative differences. (For a review of neutron superfluidity and exhaustive references see Shaham 1981) The BCS theory shows how an energy gap Δ for particle states at the top of a degenerate fermi sea and a superfluid transition at temperature $T_c \sim \Delta/k$ will result as long as there is some attraction - no matter how weak - between fermion pairs at the top of the sea. For the nearest laboratory analogue to the neutron superfluid, He^3, $T_c \sim 10^{-3}$°K and $\Delta \sim 10^{-7}$ eV which is very much smaller than the He^3 fermi energy which exceeds 10^{-4} eV. For laboratory electron superconductors Δ/E_f is similarly small and this makes calculations of the magnitude of Δ very parameter and model sensitive. But the neutron-neutron attraction is sufficiently strong (almost enough to bind the $^1S_0$ di-neutron) that calculations of Δ and T_c are expected to be qualitatively reliable through most of the crust and core. For densities characteristic of internuclear crustal neutrons the $^1S_0$ n-n attraction gives rise to a Δ near an MeV and a corresponding $T_c \sim 10^{10}$ °K. The superfluid energy gap Δ is a maximum at a density of several times 10^{13} g cm^{-3}. At higher densities it drops because the effective s-wave attraction between $^1S_0$ neutrons drops and ultimately becomes repulsive with increasing relative momentum between those neutrons at the top of the sea with opposite momenta which form Cooper pairs. (This occurs because of the growing dominance of the strong short range n-n repulsion). But as the superfluidity from $^1S_0$ pairing disappears the growing $^3P_2$ attraction between neutrons replaces it as a cause for superfluid condensation. Coincidentally the transition from $^1S_0$ to $^3P_2$ neutron superfluidity occurs close to where the crust disappears so that the former is characteristic of essentially all crustal neutrons, the latter of core neutrons, perhaps including the "near neutrons" of a π-condensate phase (but not the quarks of a conceivable quark phase). Mainly because of efficient neutrino (ν) and antineutrino ($\bar\nu$) emission processes even young pulsars have internal temperatures which are not more than about 10^{-2} the 10^{10} °K neutron superfluid T_c. Therefore, except possibly near the base of the crust where the switch from $^1S_0$ to $^3P_2$ pairing occurs and Δ is small, one can be quite confident that the crust and core neutrons are superfluid and probably also that core protons condense into a $^1S_0$ BCS superconductor. This neutron superfluidity has two distinct consequences: (I) There is a suppression of thermally excited particle states in the superfluid ($T \ll T_c$) so that the neutron sea, even though it may have a high temperature by normal astrophysical standards, behaves in this respect as if its T=0. (II) There are important restrictions in the allowed hydrodynamic motions of the superfluid which may be especially important in considerations involving its spin-up or spin-down. We turn first to consequences of (I).

3. Thermal Consequences of neutron superfluidity

Because of the suppression of particle excitations the stellar neutron heat capacity is similarly depressed. (Phonon excitation is not important be- of the relative incompressibility of neutron star matter where the sound speed $\sim 10^{10}$ cm s^{-1}). However the expected neutron superfluidity does not greatly increase the cooling rate of a neutron star relative to that which would obtain if the neutrons were "normal" for two reasons. First, there are other contributors to the stellar heat capacity in addition to core and crust neutron excitations. Most important are the electrons, the crustal nuclei, and possibly the "near neutrons" of a π-consensade central core if they are not also superfluid. Second, when neutron excitations are suppressed so are certain of the neutrino emission mechanisms which control the cooling rate of the neutron star. The so-called modified URCA emission consists of two closely related weak interaction reactions whose net effect is the conversion of thermal energy into $\nu + \bar{\nu}$: $n + n \rightarrow n + p + e^- + \bar{\nu}$ and $e^- + n + p \rightarrow n + n + \nu$. But each of these will be suppressed if $kT \ll \Delta$ since the Pauli principle will then not allow accessible final states for the superfluid n and p. But direct $\nu\bar{\nu}$ emission by accelerated electrons (e.g. plasma neutrino emission and neutrino Bremstrahlung) can still take place so that neutrino cooling and the stellar heat capacity are both diminished but not extinguished by neutron superfluidity. As a result there is not a great enough difference between calculated neutron star surface temperatures as a function of the stellar age with and without neutron superfluidity for such measurements to be a crucial test even if they were available (Nomoto and Tsurita 1982). In part this is because the stellar luminosity is equally or even more sensitive to the value of the stellar magnetic field and in part because nonthermal contributions to the x-ray luminosity can be important. (For the crab pulsar, for example, the latter is certainly much the greater.) Consequently only a few upper bounds for pulsar surface temperatures can, at present, be inferred with confidence from x-ray observations. There is, however, a possibility that present measurements for the x-ray luminosity of the Vela pulsar may already be sufficient to tell us something important about the state of matter in that neutron star core. The surface opacity of a pulsar is dependent upon the strength of the surface magnetic field (presumably at least several 10^{12} G) and the angle between energy transfer and the direction of $\underset{\sim}{B}$. A rotating neutron star may therefore have a considerable modulation at the pulsar period of its thermal emission toward a fixed observer. But no such modulated x-ray luminosity has been detected from Vela down to a luminosity about two orders of magnitude less than that calculated for thermal emission from neutron star models with and without neutron superfluid cores. If this is an indication of comparably low thermal emission from the surface of the Vela pulsar an additional strong source of neutrino emission is implied. This may be from the scattering of "near neutrons" by a π-condensate in which they would be imbedded which, in turn, can support both $e^- - \nu$ emission and $e^- \rightarrow \nu$ scattering. Or "normal" quark matter which can be a much more effective emitter of neutrinos that "normal" neutrons may be the needed strong neutrino emission source. But it would seem that the "near neutrons" or the quarks would need to be non-superfluid if they are to play such a dominant role in maintaining strong stellar neutrino emission at low interior temperatures.

4. Superfluid Hydrodynamics in Rotating Neutron Stars

It is well known in laboratory experiments that rotating superfluid He^4 may in some circumstances, be difficult to spin up or down to different rotational states. This suggests that similar phenomena may be important in the rotating containers of neutron superfluid which constitutes neutron stars and perhaps account for various departures from regularity in the observed periods of spinning down pulsars. The hydrodynamic equations of motion of a superfluid are those of an ideal fluid without viscosity with special constraints on the allowed motions: (1) $\nabla \times \underset{\sim}{v} = 0$; the local velocity of the superfluid is irrotational. This does not, of course, mean that the fluid cannot rotate since circulation about a vortex line can satisfy this condition everywhere except on the vortex line itself. But for such vortexes we have a second constraint. (2)$\oint \underline{v} \cdot d\underline{s} = \pm \pi \hbar/m_n$. Here m_n is the neutron mass (and we consider only the lowest vortex excitation which is expected to be sufficient for neutron star fluid motion). Since the vortex circulation in an ideal fluid once established is preserved, the condition (2) does not constrain the subsequent dynamics of vortex lines in a neutron superfluid; that can be calculated from classical hydrodynamics. Conditions (1) and (2) specify that except for essentially trivial parts of the fluid motion, the superfluid velocity is specified by the positions of quantized vortex lines. In the absence of forces on a vortex line it moves with the local fluid velocity. A superfluid in a rotating cylinder with angular velocity $\underline{\Omega}$ mimics the uniform rotation of a classical fluid by establishing an array of vortex lines parallel to $\underline{\Omega}$. In between the vortices $\underline{\nabla} \times \underline{v} = 0$; on them $\underline{\nabla} \times \underline{v}$ diverges. But for averaged values which include the vortex cores $<\underline{\nabla} \times \underline{v}> = 2\underline{\Omega}$ and $<\underline{v}> = \underline{\Omega} \times \underline{r}$. The number density of such vortices $n_v = m_n (\pi \hbar)^{-1} \sim 10^5$ cm^{-2} for the rapidly spinning ($\Omega \sim 200$ s^{-1}) Crab pulsar. The rotational state of a neutron superfluid can be altered only by changing the radial positions of these vortex lines and this can be difficult in certain circumstances. In the laboratory a rotating porous ceramic filled with H_e^4 superfluid will contain such vortex structures. When at very low temperatures the container rotation is reduced or even stopped the vortex lines generally cannot readjust their positions and the superfluid rotation therefore continues indefinately. This is because the vortex lines are "pinned" on protuberant surface irregularities from which they will not move under the relatively negligible thermal perturbations. (A vortex line has an effective tension about 10^8 dynes in a neutron star, which makes it energetically difficult to move if such motion would lengthen the vortex.) Another indication of the possible slowness of response of an essentially zero temperature rotating superfluid to changes in angular speed of its container occurs also with a relatively smooth container. A superfluid can have a quasi-turbulent behavior where vortex lines mix and tangle on a small scale. At maximum effective turbulence the superfluid could then act as if it had a viscosity $\nu \sim \hbar/m_n$. A tea cup of superfluid could spin up or down by the establishment of a "turbulent" Ekman boundary layer of thickness $(\nu/\Omega)^{\frac{1}{2}}$ just as in a spun up cup of tea and a time scale for changing its rotation in response to that of the tea cup given by $t \sim R(\nu \Omega)^{-\frac{1}{2}}$ with R the cup radius (Tsakadze 1975, Alpar 1978). The rotating container with neutron superfluid which is the neutron star differs from the cup of superfluid tea in two important ways. First we observe only the tea cup (the stellar surface) not the tea. Second the neutron star container is much more complicated than the tea cup. It consists of an inner crust porous to the $^1S_0$ neutron superfluid which fills the internuclear region and all the other charged components which are effectively tied to the crust by the very strong magnetic field which pervades the star. Thus the "container" includes the core protons and electrons as well as all components of any possible π-condensate or quark phase. If we express the container - core coupling time in terms of the time needed for the lighter container to adjust its rotation speed to that of the heavier core neutron superfluid, Γ(Ekman) $\sim 10^6$ sec but could be longer if the boundary layer is not fully turbulent. There are many other ways in which the core neutron superfluid can change its rotational motion in response to changes in the rotation of its "container". These involve additional vortex motion from forces pushing directly on vortex line cores. Such a force per unit length $\underline{F}$ will cause a vortex line to move relative to the local fluid velocity with a $\underline{v}_v = \hat{\underline{k}} \times \underline{F}(2m_n/\rho \hbar)$ where $\hat{\underline{k}}$ is the vortex line direction. The real vortex

core is not a line with singular circulation velocity but rather has a
finite radius $\xi \sim$ (Fermi Energy/Δ) x (interparticle spacing)$\sim 10^{-12}$cm. The
total volume of vortex line core in the rotating stellar neutron superfluid
is only 10^{-1}cm^3 compared to the total superfluid volume of about 10^{18} cm^3.
Nevertheless it is the core electrons scattering on that small volume of
vortex core which is the main way in which the spinning down stellar crust,
magnetically tied to the core electrons, causes the core neutron superfluid
to spin down. The main electron - neutron vortex core interactions so far
estimated are 1) Electron scattering on thermally excited neutrons in the
vortex core where the superfluid energy gap Δ approaches zero. This gives
a very temperature sensitive n-core crust coupling and a crust relaxation
$\hat{\Gamma}$ of order 10^2 sec at $T \sim 10^8$ °K and 10^7 sec at $T \sim 10^7$°K (Feibelman 1971).
2) Electron scattering on the magnetic field from partial alignment at the
vortex core of the J=2 $^3P_2$ neutron pairs in the direction of the vorticity.
The neutron magnetic moments then contribute a $B \sim 10^{11}$ G over a radius
$\xi \sim 10^{12}$ cm which can scatter electrons. This gives a $\hat{\Gamma} \sim 10^7$ sec (Sauls,
Stein, and Serene 1982). 3) Electron scattering on the magnetic field from
the proton cloud which moves with the circulating neutrons of a vortex.
Of course electrons generally accompany the protons so that no net current
flow of this sort will occur unless the proton orbits are smaller than the
electron screening radius $r_{sc} \sim c(4\pi n e^2 c^2/E_f(e))^{-\frac{1}{2}} \sim 10^{-12}$ cm. Therefore
quite fortuitously the current is unscreened over a radius quite comparble
to ξ. The resulting $\hat{\Gamma} \sim 10^2$ sec is by far the shortest of the three
(Alpar, Langer, and Sauls 1982). Therefore on this and all longer time
scales the core is effectively well coupled to all the charged components
of the star which therefore will spin down (or up) together.

5. Spin Down of the Crustal Neutron Superfluid

There are two related period anomolies in the young Crab and Vela pulsars
as well as in some older ones. An occasional sudden speed up with
$\Delta\Omega \sim 10^{-6}\Omega$ in Vela and $10^{-9}\Omega$ in the Crab pulsar is followed by a very much
greater increase in $|\Delta \dot{\Omega}/\dot{\Omega}|$ which lasts for 150 days in Vela and about a
week in the Crab. Because of the strong and crust - neutron core coupling
of mechanism (3) above, it is hard to understand this in terms of core neutrons but tempting to try to understand it in terms of the response of the
$^1S_0$ neutron superfluid of the inner crust to a sudden increase of angular
momentum of that crust. There will be a tendency of vortex line cores to
be strongly pinned to crustal lattice nuclei where the internuclear neutron
density is in the approximate range $\rho \sim 2-7 \times 10^{13}$g cm^{-3}. In this regime
the minimum neutron energy is achieved when the neutrons in the nucleus
where Δ is smaller than it is in most of the rest of the superfluid coincides with the vortex core were Δ is also reduced. (Outside of it the
nucleus neutron Δ exceeds that of the internuclear neutrons or both are
relatively small.) Because of the strong pinning in this region of the
crust a substantial difference will develop between the angular speed of
the spinning down crust which carries with it the pinned vortex lines and
the inclosed neutron superfluid which will not spin down until its pinned
vortex lines can finally move radially outward. As the superfluid streams
by its own vortex lines with relative velocity v_R a strong Magnus (Bernoulli)
force $\sim (\rho \hbar/m_n) \underline{v}_R \times \hat{k}$ pushes on the vortex core and thus on the lattice which
pins them until finally the vortices unpin. A gradual outward "creep" of
vortex lines of this sort is very sensitive to both kT and the magnitude
of the difference in angular velocity between crust and superfluid when a
steady state is finally achieved. In a series of papers Alpar, Anderson,
Pines, and Shaham (1981) argue that the difference between pre- and post-
glitch $\dot{\Omega}$ is caused by the temporary temination of the creep after the sud-

den small jump in Ω at the glitch. As the pre-glitch Ω is ultimately reestablished because of crust spin down from the external torque, the approximate pre-glitch steady state of vortex creep will be approached again. From Vela timing data they conclude that such a model fits for an internal stellar temperature $kT \sim 1$ Kev and about 10^{-2} of the neutron superfluid involved. Sudden collective relaxation of the pinned vortices may also explain the occurrence of glitches themselves. This may be a hydrodynamic phenomenon in which some vortex unpinnings trigger others especially where vortices pile up where they move radially out into the beginning of the strong pinning region (Alpar et al 1981). Or the crustal lattice may simply break because of the great strains in it which develop from the Magnus forces on all the pinned vortices (Ruderman 1976). It is difficult to estimate the onset of the latter because of uncertainty in the strength of a real as compared to an idealized crystal.

References

Alpar, A. 1978 Journ. Low Temp. Phys. 31
Alpar, M., Anderson, P., Pines, D., and Shaham, J. 1981 Ap.J.(Lett) 249 L29 and references therein.
Alpar, A., Langer, and Sauls, J. 1982 preprint
Bailin, D., and Love, A. 1982 preprint
Baym, G., and Pethick, D. 1975 Ann. Rev. Nucl. Sci. 25 27 and references therein
Brown, G., and Weise, W. 1976 Phys. Reports 27c, 1
Feibelman, P. 1971 Phys. Rev. D4 1589
Ginsburg, V., and Kirzhnits, D. 1964 Zh Exp. Teor. Fiz 47 2006
Migdal, A. 1959 Zh. Exp. Teor. Fiz 37 249
Nomoto, D., and Tsuruta, S. 1982 preprint
Ruderman, M. 1976 Proc. Fifth Eastern Theor. Phys. Conf., D. Feldman, ed. (New York: Benjamin)
Ruderman, M. 1976 Ap.J. 203 213
Sauls, J., Stein, D., and Serene, J. 1982 Phys. Rev. D. 25 967
Shaham, J. 1981 Jour: De Physique Colloq. C2 supple. 3 Tome 41 and references therein
Tamagaki, R. 1979 Nucl. Phys. A328 352
Tsakadze, J., and S. 1975 Sov. Phys.- Usp. 18 242

Grand unified theories and cosmology

John Ellis

CERN, Geneva, Switzerland

Abstract. After a brief introduction to grand unification and to the standard Big Bang cosmology, the possible use of grand unified theories to explain the matter-antimatter asymmetry in the Universe is explained. A possible connection with the electric dipole moment of the neutron is mentioned, and speculations made about the future of the Universe.

1. Introduction

After the successes of the standard model of elementary particle interactions comprising Quantum Chromodynamics (QCD) and the Glashow-Weinberg-Salam model, there is now great interest in the possibility of a Grand Unified Theory (GUT) embracing all elementary particle interactions. Such GUTs explain some low energy parameters of the standard model and also suggest that baryons should decay. These same baryon-number violating interactions enable one to realize a mechanism first suggested by Sakharov (1967) for baryosynthesis in the very early hot Big Bang, via the out-of-equilibrium decays of superheavy particles when the temperature of the Universe was about 10^{28} °K corresponding to particle energies of order 10^{15} GeV. Section 2 of this paper reviews the motivations and structure of simple GUTs (Georgi and Glashow 1974, Georgi et al. 1974, Buras et al. 1978, Ellis 1981). Section 3 reminds you of salient features of the hot Big Bang theory (Weinberg 1972). Section 4 introduces the problem of Big Bang baryosynthesis, reviews the criteria of Sakharov (1967) for baryon creation, and outlines the proposed GUT mechanism (Yoshimura 1978, Ellis et al. 1979, Weinberg 1979). Section 5 describes the possible connection with the neutron electric dipole moment (Ellis et al. 1981a,b) while section 6 contains speculations about the future.

2. Introduction to Grand Unification

The fundamental interactions of leptons and the quark constituents of hadrons such as the neutron are well described by gauge theories. The strong interactions are described by QCD, a gauge theory based on the non-Abelian group SU(3), while the weak and electromagnetic interactions are described by an SU(2) x U(1) gauge theory (Glashow 1961, Weinberg 1967, Salam 1968). QCD successfully fits the qualitative features of all strong interaction data, while the SU(2) x U(1) model fits all weak interaction data with one parameter called $\sin^2\theta_W$.

The philosophy of grand unification is to replace this ramshackle group-theoretical structure with three independent gauge couplings and at least

20 parameters by a more economical theory based on a single semi-simple group with just one gauge coupling. The strong coupling is much larger than the weak couplings at present energies but is expected to decrease logarithmically (asymptotic freedom). If no new physics intervenes (the desert hypothesis) equality of the couplings occurs, and grand unification becomes possible, at an energy scale of order 10^{15} GeV. The logarithmic variation of the coupling means that $m_X/m_{neutron} = \exp(O(1)/\alpha_{em})$: to get $m_X > 10^{14}$ GeV as required by proton and bound neutron stability, and $m_X < 10^{19}$ GeV so that we can neglect gravity we need the fine structure constant to lie in the range $\frac{1}{120} > \alpha_{em} > \frac{1}{170}$ (Ellis and Nanopoulos 1981).

The simplest GUT group is SU(5) which has 24 gauge bosons (Georgi and Glashow 1974). Twelve of these are the familiar γ, $W^{\pm}$, Z^0 and gluons, while there are 12 new superheavy bosons X,Y of mass $O(10^{15}$ GeV. Quarks and leptons are assigned to at least three generations of reducible $\underline{5} + \underline{10}$ representations of SU(5). The charges of quarks and leptons are now related in such a way that $|Q_e|/|Q_p| = 1$. One can calculate the weak mixing parameter $\sin^2\theta_w$ and gets 0.215 ± 0.002 to be compared with the experimental value of 0.215 ± 0.012 (Georgi et al. 1974, Buras et al. 1978, Ellis 1981). It is also possible to calculate the bottom quark mass in terms of the τ lepton mass, getting $m_b \approx 5$ GeV if there are only 3 generations of quarks and leptons (Buras et al. 1978). It is nice to have these circumstantial pieces of evidence for GUTs, but the crucial test will be whether baryon decay occurs with the expected lifetime of 10^{27} to 10^{31} years. The present lower limit on the lifetime is about 2×10^{30} years, but there are two ongoing experiments reporting candidates for baryon decay corresponding to a lifetime of a few times 10^{30} years (Krishnaswamy et al. 1981, 1982). Many other experiments now starting up should be able to confirm or exclude these tentative indications.

3. Hitch-hiker's Guide to the Big Bang

There are three main reasons for believing in the standard Hot Big Bang cosmology (Weinberg 1972). One is the present Hubble expansion: the Universe appears fairly homogeneous and isotropic on a large distance scale, with everybody receding from everyone else at a rate of about 100 km per second per Mpc of separation. Naive extrapolation and sophisticated theorems suggest there was a hot singularity in our past. A confirmation of this is the 3°K microwave background radiation with a density of about 400 photons per c.c., which is generally believed to be the relic of an earlier epoch when matter and radiation were in equilibrium with a temperature 10^3 higher and the cosmic scale factor 10^3 smaller than today. A third reason for believing in the Big Bang is the success of cosmological nucleosynthesis calculations of the primordial abundances of light elements, especially ^{4}He. These suggest that the Universe was once 10^9 times hotter and smaller than it is today.

We will extrapolate back another 18 orders of magnitude using the standard Robertson-Walker-Friedmann metric. In the early Universe when radiation dominated the energy density, we estimate that the age in seconds is $O(T^{-2})$ with the temperature T expressed in MeV. Nucleosynthesis took place at $T = O(0.1)$ MeV i.e. an age of $O(100)$ seconds: we will be interested in $T \approx 10^{15}$ GeV corresponding to an age of $O(10^{-36})$ seconds. It will be important for baryosynthesis to know whether GUT reactions were in thermal equilibrium, i.e. whether the collision rate was faster than the expansion rate at that epoch.

4. Baryon Generation

The problem (Steigman 1976) to be solved is that the Universe apparently contains no large concentrations of antimatter, while it does contain matter. From one point of view there is very little matter, probably $O(1/10)$ baryon per cubic metre implying a baryon-to-photon ratio $n_B/n_\gamma = O(10^{-10}$ to $10^{-9})$. From another point of view there is a surprisingly large amount of matter, since if the Universe was matter-antimatter symmetric at the epoch of baryon-antibaryon annihilations we would only expect a small statistical fluctuation $n_B/n_\gamma = O(10^{-20})$ today. This suggests that the Universe was not baryon-antibaryon (or quark-antiquark) asymmetric at that epoch, and our job is to understand why and how.

Sakharov (1967) pointed out the three criteria that must be satisfied if one is to generate an asymmetry from an initially asymmetry Universe. One must have interactions which violate baryon number — these are present in GUTs. These B-violating interactions must discriminate between matter and antimatter by violating charge conjugation C and its combination CP with parity. Since the weak interactions violate C and CP, and since GUTs contain the weak interactions, we also expect GUTs to violate C and CP. Finally, there must be a departure from thermal equilibrium in the interactions violating B, otherwise Boltzmann's H-theorem would guarantee the absence of any matter-antimatter asymmetry. Detailed calculations (Ellis et al. 1979) suggest that the GUT reactions probably departed from thermal equilibrium as the temperature fell below $O(10^{14})$ GeV, thereby enabling a matter-antimatter asymmetry to be generated.

A favoured mechanism for this is the out-of-equilibrium decays of heavy particles such as X bosons or Higgs bosons (Yoshimura 1978, Ellis et al. 1979, Weinberg 1979). While the CPT theorem guarantees that the total decay rates for X and $\bar{X}$ particles must be the same, their partial decay rates may differ if C and CP are violated. A net quark asymmetry could be generated in X decay if the branching ratios $B(X \to qq)$, $B(\bar{X} \to \bar{q}\bar{q})$ were different.

In simple models the largest CP violation occurs in Higgs decays, but it is not enough to generate the observed matter-antimatter asymmetry. This means that one must appeal to a larger GUT, with corresponding ambiguities. There are also uncertainties due to the possible effects of 2-2 scattering cross-sections and possible complications in the evolution of the Universe. For this reason the GUT mechanism for baryosynthesis is only qualitative, not quantitative.

5. The Neutron Electric Dipole Moment

Of particular interest to participants in the neutron conference is a possible connection between baryosynthesis and the neutron electric dipole moment d_n (Ellis et al. 1981a,b). There are strong similarities between the diagrams contributing to n_B/n_γ and some of those contributing to d_n via renormalization of the CP-violating QCD vacuum parameter θ. From these we infer an order-of-magnitude "lower bound"

$$d_n \gtrsim 2 \times 10^{-18} (n_B/n_\gamma) \text{ e-cm} \gtrsim 3 \times 10^{-28} \text{ e-cm}$$

to be compared with the present experimental upper limit of about

4×10^{-25} e-cm. We wish good luck to the two experimental groups now trying to improve this upper limit.

6. The Future

The Universe will expand forever if its density ρ is less than the critical one, corresponding to a neutrino mass of order 30 eV. If the Universe does expand forever then we expect baryons to decay in about 10^{30+n} years' time, and the only activity thereafter will be the occasional decay and reformation of black holes. If the Universe eventually collapses, then we may meet again in recycled form in about 10^{11} years' time, at another meeting 50 years after the rediscovery of the neutron.

References

Buras A J, Ellis J, Gaillard M K and Nanopoulos D V 1978 Nucl. Phys. B135 66.
Ellis J 1981 CERN preprint TH 3174.
Ellis J, Gaillard M K and Nanopoulos D V 1979 Phys. Lett. 80B 360.
Ellis J, Gaillard M K, Nanopoulos D V and Rudaz S 1981a Phys. Lett. 99B 101; 1981b Nature 293 41.
Ellis J and Nanopoulos D V 1981 Nature 292 436.
Georgi H and Glashow S L 1974 Phys. Rev. Lett. 32 438.
Georgi H, Quinn H R and Weinberg S 1974 Phys. Rev. Lett. 33 451.
Glashow S L 1961 Nucl. Phys. 22 579.
Krishnaswamy M R et al. 1981 Phys. Lett. 106B 339; 1982 Phys. Lett. 115B 349.
Sakharov A D 1967 Zh. Eksp. Teor. Fiz. Pis'ma Red. 5 32.
Salam A 1968 Proc. 8th Nobel Symposium ed N Svartholm (Stockholm: Almqvist and Wiksells) p 367.
Steigman G 1976 Am. Rev. Astron. and Astrophys. 14 339.
Weinberg S 1967 Phys. Rev. Lett. 19 1264.
Weinberg S 1972 Gravitation and Cosmology (New York: Wiley).
Weinberg S 1979 Phys. Rev. Lett. 42 850.
Yoshimura M 1978 Phys. Rev. Lett. 41 381.

Neutrons in the early universe

R. J. Tayler

Astronomy Centre, University of Sussex, Falmer, Brighton, BN1 9QH

Abstract. Several cosmological theories have been proposed in which the neutron plays a crucial role. An account is first given of theories which are now known to be incapable of accounting for the observed properties of the Universe. A discussion follows of the significance of the neutron half-life in the hot big bang and of the possible role of phase transitions involving neutrons in a cold big bang.

1. Introduction

Two main concerns of cosmology are with content and structure: why is the Universe made up of particular chemical elements and elementary particles and why does it have its observed structure both in terms of galaxies and clusters of galaxies and of near homogeneity and isotropy of its average properties? This article is concerned with the role of the neutron in these topics. I shall concentrate mainly on content but will finish with some remarks about structure.

Early studies of the chemical composition of objects in the Universe indicated that they might all have essentially the same composition once allowance had been made for the loss of volatile light elements from the less massive members of the solar system. It seemed that the chemical elements might have been produced in a single event to be understood in terms of the established laws of physics. Several theories were proposed in each of which there is a crucial involvement of neutrons. We shall discuss them in the next section and we shall see that they are not capable of explaining the observations. This is fortunate as it is now known that there is a wide variation of heavy element abundances from star to star, although it seems that some light elements and isotopes were produced cosmologically. The production of light elements in the hot big bang depends on the half-life of the neutron, as will be discussed in Section 3. Finally in Section 4 we discuss how the neutron might have influenced the development of structure in the Universe, if it had a cold origin.

2. Early Cosmological Theories

Almost as soon as the neutron had been discovered there was interest in the equilibrium theory of the origin of the elements (Sterne, 1933). The basis of this theory is: given a number of neutrons, protons and electrons at a particular density and temperature and sufficient time for nuclear reactions to reach equilibrium, what element abundances will result? This e-process was studied by various authors including particularly Klein, Beskow and Treffenberg (1946). Beskow and Treffenberg (1947a,b) discussed the structure of objects in which the e-process might have occurred but they found it impossible to reproduce the whole range of observed abundances. At 10^{10}K some approximation to the abundances of the light elements was obtained but the heavy elements were underabundant by many orders of magnitude. Other authors showed that reasonable relative abundances of iron and its neighbours could be obtained at a few times 10^9K but then there was a serious deficit of light elements.

The neutron played a more prominent role in the polyneutron theory of Mayer and Teller (1949). They argued that the heavy elements must have been formed in a neutron rich environment. They gave two reasons for this belief. The first was the higher abundances of neutron-rich isotopes than of lighter isotopes. The second was the impossibility of building up the heaviest elements by charged particle reactions because of photodisintegration of nuclei at high temperatures. Mayer and Teller proposed that the heavy elements were formed by the disintegration or fission of a cold nuclear fluid of objects called polyneutrons. These had such a mass that self-gravitation was less important than nuclear forces. A related but earlier idea was that of Lemaître who supposed that initially the entire Universe was a Primeval Atom or heavy isotope of the neutron (see Lemaître 1950).

Mayer and Teller showed that a polyneutron would become unstable and form surface droplets from which a process of β-decay, evaporation of neutrons and fission formed the observed nuclei, with something like the observed relative abundances of the heavy elements and isotopes. They did not however discuss how the polyneutrons formed in the actual expanding Universe. Peierls, Singwi and Wroe (1952) discussed the evolution of a cold big bang cosmological theory. They supposed that a Universe composed mainly of neutrons expanded from an initial singularity. They showed that, as matter passed through normal nuclear density at $t \sim 10^{-4}$s, it entered a state of tension and that it then became energetically favourable for it to break up into fragments at about nuclear density with cavities between them. Peierls et al. pointed out that these fragments were the polyneutrons postulated by Mayer and Teller so that, with the chosen initial conditions polyneutrons would form and disintegrate automatically. Unfortunately, as they showed, the theory had a fatal defect. Essentially all of the matter in the Universe went into polyneutrons and although some light elements were produced in their decay the amount of matter in heavy elements was comparable with that in light elements instead of being orders of magnitude less.

The neutron was central to the theory of Alpher, Bethe and Gamow (1948)(αβγ theory), which has developed into the current hot big bang theory. Gamow suggested that initially the Universe was very hot and dense and was composed of neutrons and photons alone. As the Universe expanded, neutrons decayed and a sequence of reactions starting with $n+p \rightarrow d+\gamma$ was supposed to produce the elements heavier than hydrogen. In fact the absence of stable nuclei of mass number 5 and 8 together with the very rapid fall in the density of the Universe led to only very minimal production of heavy elements. Thus the αβγ theory was also incapable of explaining universal abundances.

The neutron also entered into a version of the steady state theory discussed by Gold and Hoyle (1959). They suggested that matter was continuously created as neutrons and that neutron decay produced an intergalactic gas at $T \sim 10^9$K. They showed that thermal instabilities in the gas could produce self-gravitating condensations and hence the new galaxies which were required in the steady state theory. In addition high energy particles could be produced providing a universal origin for cosmic rays. The theory was short-lived because it soon became clear that insufficient bremsstrahlung radiation from the hot gas was being received at Earth.

3. The Hot Big Bang Theory

It has become clear since 1950 that the element abundances are not universal. Although the relative amounts of hydrogen and helium do not vary very much, the quantity of heavy elements relative to hydrogen varies by a factor of at least a thousand from the most metal-deficient to the most metal-rich star. It is now believed that the heavy elements have been produced by nuclear reactions in stars, so that young stars contain elements produced in earlier stars. The discovery of the cosmic microwave radiation gave strong support to the hot big bang theory and it is generally believed that the light elements were produced in the manner described by Gamow.

Hayashi (1950) pointed out a defect in the original $\alpha\beta\gamma$ theory. The Universe could not consist of neutrons and photons alone because of creation of particle/antiparticle pairs from photons. At high temperatures the neutron/proton ratio would satisfy

$$n_n/n_p = \exp(-2.5 m_e c^2/kT) \qquad (1)$$

and it would be kept at that value by the reactions

$$n+e^+ \rightleftarrows p+\bar{\nu}_e, \qquad p+e^- \rightleftarrows n+\nu, \qquad n \rightarrow p+e^-+\bar{\nu}_e. \qquad (2)$$

As the Universe expands and the temperature drops, the weak interactions cease to be capable of keeping n_n/n_p at the value (1). This happens when the weak interaction timescale t_{WK} becomes larger than the age of the Universe t. These satisfy

$$t \approx 10^{20} T^{-2}, \qquad t_{WK} \approx 10^{50} T^{-5}, \qquad (3)$$

so that the neutron/proton ratio is frozen when $T \approx 10^{10}$K. After this it only changes very slightly until neutrons and protons react to produce deuterium and other light isotopes. This happens when the temperature is low enough for photo-disintegration of deuterium to cease. At this stage the remaining neutrons and an equal number of protons form ^{4}He with a very small admixture of d, ^{3}He and ^{7}Li.

The light element abundances depend on the precise value of t_{WK} which in turn is directly proportional to the neutron half-life. The larger is the half-life, the more neutrons survive and the greater is the abundance of ^{4}He. As a result a good value of the half-life is essential for a detailed comparison of theory and observation. A reduction of neutron half-life by 0.5 min. reduces the predicted helium content by 4 per cent which is just about large enough to be important. Gamow originally believed the half-life to be $\sim$ 14 min. Subsequent measurements have given the values 11.7 min. (Sosnovsky et al. 1959), 10.6 min. (Christensen et al. 1972), 10.1 min. (Bondarenko et al. 1978) and 10.8 min. (Byrne et al. 1980). It is likely that the neutron half-life is now known to an accuracy of less than 0.5 min. but the main current interest of cosmologists in the neutron is in a definitive value for the half-life. The present experimental position is described in this volume by J. Byrne.

4. The Cold Big Bang and the Origin of Structure

Although the hot big bang theory explains the light element abundances, it has not given a clear explanation of the observed structure in the Universe. It was hoped that galaxies would form by gravitational instability in a homogeneous isotropic Universe but this only appears to work if some structure is put in as an initial condition. Many workers hope that this will be solved by an understanding of elementary particle physics in the extremely early (t $\sim 10^{-35}$s) Universe but Hogan (1982) has suggested its origin in a cold big bang. This theory has obvious disadvantages in that it provides no natural explanation of the light element abundances and the microwave radiation, which have to be produced by pregalactic stars, but I will not discuss these problems.

Hogan suggests that a cold homogeneous isotropic Universe expanding from very high density spontaneously shatters into fragments as a result of metastability during a phase transition. The phase transition near to nuclear density is supposed to occur as a result of pion condensation or of neutron solidification or liquefaction. This is very similar to the discussion of Peierls et al. but according to Hogan his fragments do not behave like polyneutrons. Instead his calculations suggest that the fragments provide the essential 'seed' perturbations which grow by gravitational instability to produce pregalactic stars. These in turn produce helium, heavy elements, the microwave radiation and drive further instabilities leading to the formation of galaxies. If his theory is correct neutrons may influence the observed structure in the Universe.

References

Alpher R A, Bethe H A and Gamow G 1948 Phys. Rev. 73 803
Beskow G and Treffenberg L 1947a Ark. Mat. Astr. Fys. 34A No 13
Beskow G and Treffenberg L 1947b Ark. Mat. Astr. Fys. 34A No 14
Bondarenko L N, Kurguzov V V, Prokof'ev YuA, Rogov E V and Spivak P E 1978
    Soviet Phys. JETP Lett. 28 303
Byrne J, Morse J, Smith K F, Shaikh F, Green K and Greene G L 1980
    Phys. Lett. 92B 274
Christensen C J, Nielsen A, Bahnsen A, Brown W K and Rustad B M 1972
    Phys. Rev. D 5 1628
Gold T and Hoyle F 1959 9th Symp. Int. Astr. Un. (Stanford: U.P.) p 583
Hayashi C 1950 Progr. Theor. Phys. Japan 5 224
Hogan C J 1982 Astrophys. J. 252 418
Klein O, Beskow G and Treffenberg L 1946 Ark. Mat. Astr. Fys. 33B No 1
Lemaître G 1950 The Primeval Atom (New York: Van Nostrand)
Mayer M G and Teller E 1949 Phys. Rev. 76 1226
Peierls R E, Singwi K S and Wroe D 1952 Phys. Rev. 87 46
Sosnovsky A N, Spivak P E, Prokoviev YuA, Kutikov I E and Dobrinin YuP 1959
    Nucl. Phys. 10 395
Sterne T E 1933 Mon. Not. R. astr. Soc. 93 736

(A fuller version of this paper is being submitted for publication in Quarterly Journal of the Royal Astronomical Society)

Inst. Phys. Conf. Ser. No. 64: Section 2
Paper presented at Conf. on Neutron and its Applications, Cambridge, 1982

Cosmic neutrinos

Gary Steigman

Bartol Research Foundation, University of Delaware, Newark, DE 19711 USA

Abstract. The early Universe was hot and dense ensuring that high energy collisions were frequent. The early Universe was the Ultimate Accelerator - a hot, dense soup of all elementary particles. Neutrinos are produced copiously and survive in abundance to influence the subsequent evolution of the Universe. The effect of cosmic neutrinos on primordial nucleosynthesis is discussed and the role that massive relic neutrinos may play in the recent evolution of the Universe is outlined. Constraints on the number of kinds of neutrinos and on the limits to their masses are presented.

1. Introduction

As the Universe expands, it cools and becomes more dilute. The early Universe was hot and dense. Since high density and high temperature correspond to high flux and high center of mass energy in collisions, the early Universe was a Cosmic Accelerator - the Ultimate Accelerator. In comparing the reaction rate with the universal expansion rate we find there was an epoch during the early evolution of the Universe when collisions were frequent, producing all elementary particles - those known and those not yet discovered because they are too heavy and/or too weakly interacting. Neutrinos in particular are produced copiously and survive in abundance to influence the subsequent evolution of the Universe.

2. Survival of Relic Neutrinos

The early Universe is "Radiation Dominated" - the energy density is dominated by the contribution from extremely relativistic (ER) particles.

$$\rho_{ER} = \sum_B \rho_B + \sum_F \rho_F = (g/2)\rho_\gamma, \tag{1a}$$

$$g(T) = \sum_B g_B (T_B/T_\gamma)^4 + 7/8 \sum_F g_F(T_F/T_\gamma)^4. \tag{1b}$$

In (1) the sum runs over all ER bosons (B) and fermions (F); ρ_γ is the photon energy density; $g(T)$ is the effective number of relativistic degrees of freedom; g_B (g_F) the helicity of each boson (fermion) species; T_B, T_F, T_γ are the corresponding temperatures. For the early evolution, the expansion rate varies as $t^{-1} \propto \rho^{1/2} \propto g^{1/2}T^2$; the age of the Universe is: $t(sec) \approx 2.4 g^{-1/2} T^{-2}_{MeV}$. The Neutral Current Weak Interactions (NCWI) - $e^+ e^- \leftrightarrow \nu_i + \bar{\nu}_i$ ($i=e,\mu,\tau,...$) - govern the production and survival of relic neutrinos. At high temperatures ($T>M_W$), the reaction rate varies as $\Gamma_{\nu\bar\nu} = n\sigma_{\nu\bar\nu} \propto T$, so that the number of collisions in an expression time increases with decreasing T (as the Universe expands) as $\Gamma_{\nu\bar{\nu}} t \propto T^{-1}$. For $T \lesssim T_{GUT}$, the cosmic neutrinos are in equilibrium. As the Universe cools further ($T < M_W$), the reaction rate falls more rapidly: $\Gamma_{\nu\bar{\nu}} \propto T^5$ and the number of collisions in an expansion time now decreases with decreasing T: $\Gamma_{\nu\bar{\nu}} t \propto T^3$. When the temperature drops below a few MeV, equilibrium can no longer be maintained. Since neutrino pairs no longer annihilate nor are they produced, their number in a comoving volume is preserved. We note that massive neutrinos with $m_\nu \gg$ few MeV would have decoupled earlier at a higher temperature. All light neutrinos ($m_\nu \ll$ few MeV) remain in equilibrium down to decoupling at a few MeV as a result of the NCWI - μ-neutrinos (and all other light neutrinos) are as abundant as e-neutrinos. We first concentrate on these light neutrinos.

In equilibrium, $T_\gamma = T_e = T_\nu$ and the ratio of neutrinos to photons (by number and by energy density) is

$$n_\nu/n_\gamma = 3/4(g_\nu/2); \quad \rho_\nu/\rho_\gamma = 7/8(g_\nu/2). \tag{2}$$

Neutrinos have decoupled somewhat before primordial nucleosynthesis begins in earnest. However, at nucleosynthesis $T_\nu \sim T_\gamma$ and the relative contribution of light neutrinos to the total energy density is significant.

$$(\rho_\nu/\rho_\gamma)_{NUC} = (7/8)N_\nu; \quad N_\nu = \sum_i (1/2) g_{\nu i}. \tag{3}$$

Today, the relic (microwave!) neutrinos are cooler than the microwave photons because the photons were heated when the $e^\pm$ pairs annihilated when $T < m_e$ at which time the neutrinos were already decoupled. It is easy to estimate the ratio of temperatures by considerations of entropy conservation (see, for example, Steigman 1979).

$$(T_\nu/T_\gamma)_0^3 = g_{after}/g_{before} = 2(11/2)^{-1} = 4/11. \tag{4}$$

At present, then, the ratio of cosmic neutrinos to cosmic photons is $(n_\nu/n_\gamma)_0 = (3/11)N_\nu$. Primordial nucleosynthesis - as we shall see shortly - limits the nucleon to photon ratio to $\eta = n_N/n_\gamma < 10^{-9}$ so that the present ratio of cosmic neutrinos to nucleons is large.

$$n_\nu/n_N = 3N_\nu/11\eta \gtrsim 10^9. \tag{5}$$

If neutrinos have a finite rest mass then the universal ratio of mass in neutrinos to that in nucleons is

$$M_\nu/M_N = m_\nu n_\nu/\rho_N \sim 3m_\nu(eV) \eta_{10}^{-1} > m_\nu/3eV \tag{6}$$

If $m_\nu > 3eV$, we live in a neutrino dominated Universe.

3. The Role of Neutrinos In Primordial Nucleosynthesis

The abundance of primordial ^{4}He synthesized in the hot big bang is controlled by the ratio of neutrons to protons at the time of nucleosynthesis. During early epochs, the Charged Current Weak Interactions (CCWI): $p+e^- \leftrightarrow n+\nu_e$, $n+e^+ \leftrightarrow p+\bar{\nu}_e$; $n \leftrightarrow p+e^-+\bar{\nu}_e$ interconvert neutrons and protons. The competition between the universal expansion rate (t^{-1}) and the CCWI rate determines the neutron abundance. It is hardly necessary to remind this audience that the CCWI rate is normalized by the value of the neutron half-life ($\tau_{1/2}$). The universal expansion rate depends on the number of species of light particles. For more neutrino flavors ($N_\nu\uparrow$), the density at a fixed temperature increases ($\rho_{ER}\uparrow$) and the Universe expands faster ($t^{-1}\uparrow$). The faster expansion means that the CCWI will depart from equilibrium earlier, at a higher temperature, leaving behind more neutrons. More neutrons are then available to produce more ^{4}He. Too many species of neutrino, then, could lead to the overproduction of ^{4}He (Shvartsman 1969; Steigman, Schramm and Gunn 1977).

Deuterium and Helium-4 are the only elements whose observed abundances indicate a big bang origin. Since D is easily destroyed, the primordial abundance should exceed that observed today (D/H $\gtrsim 1\times10^{-5}$; for a review of the data on the abundances of the light elements see Yang et al. 1982 — hereafter YTS^{2}O). This requirement constrains the nucleon abundance to be small: $\eta \leq 1\times10^{-9}$ (YTS^{2}O). In very low (nucleon) density models, the primordial abundance of D is very large. In the course of galactic evolution it may not be difficult to burn the primordial D but, there is potential problem. D is burned to ^{3}He which is more difficult to destroy; indeed, it is likely that stellar production in the course of galactic evolution enhances the primordial abundance of ^{3}He (Rood, Steigman and Tinsley 1976). To avoid overproducing D+^{3}He today requires that the nucleon abundance be constrained to be no smaller than $\eta \geq 2\times10^{-10}$ (YTS^{2}O). Armed with these limits for η ($2 \leq \eta_{10} \leq 10$), we turn to ^{4}He for constraints on the number of neutrino species.

The abundance of ^{4}He is the most accurately known of all the light elements. It is, however, difficult to derive — from observations at present — a primordial abundance with sufficient accuracy to truly constrain the cosmology and/or the particle physics. The fairest estimate from the current data is that the primordial abundance (by mass) is in the range $0.22 \leq Y_p \leq 0.26$ (YTS^{2}O). In the standard (particle physics) model the τ-neutrino is expected to be a light neutrino so that $N_\nu \geq 3$. For the neutron half-life in the range $10.4 \leq \tau_{1/2}(MIN.) \leq 10.8$, $N_\nu=3$ is consistent with our estimates of Y_p for the range of η allowed by D and ^{3}He (YTS^{2}O). A fourth family of leptons, $N_\nu=4$, would only be consistent if $Y_p \geq 0.25$; just barely allowed by the data.

Should future observations lead to a reduction in our estimate of the upper limit to the primordial abundance of ^{4}He, more severe constraints on N_ν and η will be forthcoming. For example, if we were led to $Y_p \leq 0.25$, then $N_\nu \leq 3$; for $N_\nu=3$, $\eta \leq 8\times10^{-10}$. If a further reduction were indicated so that $Y_p < 0.23$, then the "standard" model ($N_\nu=3$) would be excluded. The constraint $N_\nu < 3$ could, of course, only be satisfied if ν_τ is a heavy neutrino; accelerator studies only yield the limit $m(\nu_\tau) \leq 250$ MeV (Bacino et al 1979). The standard (cosmological) model would have to be rejected (or, at least, modified) for $Y_p \leq 0.22$ for which $N_\nu < 2$.

To recapitulate, the standard model ($N_\nu=3$, $10.4 \leq \tau_{1/2} \leq 10.6$ min, $2 \leq \eta_{10} \leq 10$) is in excellent agreement with all current data (the abundances of D, ^{3}He and ^{4}He; the recent work of Spite and Spite (1982) suggests that we may also add ^{7}Li to this list). The data constrain the number of (two component) neutrino species to be no greater than four ($N_\nu \leq 4$). It is worth noting that this constraint, $N_\nu \leq 4$, actually applies, in addition, to particles other than neutrinos such as Axions, Photinos, Gravitinos, etc. which may be ER at nucleosynthesis. In fact, all light particles contribute to an effective N_ν,

$$N_\nu = \sum_F{}'(g_F/2)(T_F/T_\nu)_0^4 + (8/7)\sum_B{}'(g_B/2)(T_B/T_\nu)_0^4.$$

The primes in the summations in equation (7) indicate that for fermions (bosons), electrons (photons) have been omitted.

Finally, what of neutrinos with $m_\nu \approx O(MeV)$? This case was considered recently by Kolb and Scherrer (1982) who found that light neutrinos are truly light ($\Delta N_\nu = 1$) for $m_\nu \lesssim 0.1$ MeV. Neutrinos with $m_\nu \gtrsim 25$ MeV are too heavy to contribute ($\Delta N_\nu \approx 0$). Neutrinos with $0.1 \lesssim m_\nu(MeV) \lesssim 10$ significantly influence the universal expansion rate at nucleosynthesis ($1 \lesssim \Delta N \lesssim 2$); those with $m_\nu \gtrsim 10$ MeV, less so ($\Delta N_\nu \lesssim 1$).

4. Why Massive Neutrinos

A comparison of the predictions of primordial nucleosynthesis with the observed abundances of the light elements leads to the conclusion that ours is a low (nucleon) density Universe (YTSSO). The density Ω_N is the ratio of the nucleon density ρ_N to the critical density ρ_C separating low density universes ($\rho<\rho_C$) which expand forever from high density universes ($\rho>\rho_C$) whose expansion will stop to be followed by collapse. For $\eta \lesssim 10^{-9}$, corresponding to $D/H \gtrsim 1\times 10^{-5}$, $\Omega_N \lesssim 0.2$; nucleons alone fail to close the Universe by at least a factor of five. There are – perhaps – some problems with such a low density universe (Schramm and Steigman 1981a,b). For example, some studies of the dynamically determined mass on large scales (Davis et al. 1978; Peebles 1979; Davis et al. 1980) suggest that $\Omega_{Dyn} \gtrsim 0.2$, in apparent conflict with the upper limit to the nucleon density. This suggests that the "dark mass" may be non-nucleonic (Schramm and Steigman 1981a,b). Furthermore, galaxy formation is inhibited in a low density universe. Perturbations can't grow during the radiation dominated epoch which lasts longer in a low density universe and perturbations stop growing when the universe becomes curvature dominated which occurs earlier in a low density universe. Larger initial perturbations are thus required. But these would have left temperature fluctuations in the microwave background larger than observed: $\delta T/T \gtrsim 0.2 n_{10}^{-2} \gtrsim 0.002$. Finally, it should be noted that the dark mass in the Universe seems to be less dissipative than the luminous mass; the mass to light ratio increases as the scale increases. For these reasons, massive relic neutrinos have proved to be a popular panacea (Cowsik and McClelland 1973; Szalay and Marx 1976; Schramm and Steigman 1981a). As we have seen, relic neutrinos are abundant and, if $m_\nu \gtrsim 3eV$, would dominate the universal mass density: $\Omega_\nu / \Omega_N \approx 3m_\nu(eV) n_{10}^{-1} \gtrsim 0.3 m_\nu(eV)$. For massive (but light: $m_\nu \ll 1$ Mev) relic neutrinos the contribution to the overall mass density is

$$\Omega_\nu = \rho_\nu/\rho_C \gtrsim m_\nu/100 \text{ eV}. \tag{8}$$

There are some interesting cosmological constraints on the masses of relic neutrinos.

5. Cosmological Constraints On Neutrino Masses

Massive neutrinos are excellent candidates for the dark mass in the Universe since they decoupled early in the evolution of the Universe and are dissipationless. For these reasons it has been suggested that such neutrinos may dominate the universal mass density in general and the mass in rich clusters of galaxies in particular (Cowsik and McClelland 1973; Szalay and Marx 1976; Schramm and Steigman 1981a). Massive neutrinos could, however, prove to be too much of a good thing. If they are too heavy they would provide too much dark mass. From data on rich clusters we are led to the constraint,

$$0.3 m_\nu (eV) \lesssim \Omega_\nu / \Omega_N \lesssim M_{DARK}/M_{LUM} \lesssim 10\text{-}30 \Rightarrow m_\nu (eV) \lesssim 30\text{-}100. \qquad (9)$$

A similar constraint follows from different considerations. The higher the universal mass density, the faster the Universe expands and the younger it is today (Gershtein and Zeldovich 1966). That the Universe be at least as old as the minimum estimates for the ages of the oldest stars ($t_0 \gtrsim 10\text{-}13$ billion years) limits the sum of the neutrino masses to: $m_\nu \lesssim 25\text{-}75$. Finally, if indeed ours is a neutrino dominated universe ($m_\nu \gtrsim 3eV$), then light but massive neutrinos play an important role in the evolution of perturbations which lead to the emergence of structure in the Universe. Free-streaming (non-interacting) neutrinos would mix regions of high and low density, damping any perturbations on scales they traversed (Bond et al. 1980; Wasserman 1981). The lighter the neutrinos (provided that $m_\nu > 3eV$), the faster they move and the larger the scale on which they damp any initial perturbations. The damping mass scale is (Bond et al. 1980; Wasserman 1981).

$$M_{D\nu} \simeq 4 \times 10^{18} m_\nu^{-2} M_\odot \qquad (10)$$

If we require that perturbations survive as scales no larger than superclusters ($M \lesssim 10^{16} M_\odot$), the $m_\nu \gtrsim 20$ eV.

6. Heavy Relic Neutrinos

Neutrinos with $m_\nu \gg 1$ MeV will become nonrelativistic before they decouple (Lee and Weinberg 1977; Dicus et al 1977). When they do finally decouple, their abundance is small compared to that of their relativistic cousins.

$$n_\nu / n_\gamma \simeq 6 \times 10^{-8} m_\nu^{-3}. \qquad (11)$$

In (11) the neutrino mass is in GeV. Comparing the universal mass density in "heavy" neutrinos with that in nucleons we find

$$\Omega_\nu / \Omega_N \sim 68 m_\nu^{-2}(10^{-9}/n) \gtrsim 68 m_\nu^{-2}. \qquad (12)$$

The same data which constrains the masses of light neutrinos may be employed to constrain the masses of heavy neutrinos. Heavy relic neutrinos will cluster in galaxies where there is much less dark mass than on scales of clusters. For galaxies $M_{DARK} \lesssim (1-3) M_{LUM}$ so that (from equation (12)) $m_\nu \gtrsim (5-8)$ GeV. A somewhat weaker constraint follows from the age of the Universe: $t_0 \gtrsim 10 \times 10^9 yr \Rightarrow m_\nu \gtrsim 2$ GeV.

7. Summary

The last few years have witnessed the establishment of a symbiotic relationship between elementary particle physics and astrophysics. Cosmic neutrinos provide a clear example of the benefits of such an interdisciplinary approach to fundamental problems in physics. From the successes of big bang nucleosynthesis in accounting for the observed abundances of the light elements we learn that at most one new family of leptons is allowed: $N_\nu \leq 4$. We are certain that $N_\nu \geq 2$ (ν_e, ν_μ) but, is $N_\nu \geq 3$ (is $m(\nu_\tau) \ll 1$ MeV)? We have seen that a nucleon dominated universe is a low density universe ($\Omega_N \leq 0.2$). For neutrino masses in the range $3\text{eV} \leq m_\nu \leq 8$ GeV, ours would be a neutrino dominated Universe. However, too much mass would be provided – at present – by cosmic neutrinos with masses in the range $100\text{eV} \leq m_\nu \leq 2$ GeV; such neutrinos must decay before the present epoch. For stable, massive, relic neutrinos four regimes are possible: for $20 \leq m_\nu(\text{eV}) \leq 100$ or $2 \leq m_\nu(\text{GeV}) \leq 8$ the Universe is neutrino dominated and cosmic neutrinos play a considerable cosmological role; for $m_\nu \leq 3\text{eV}$ or $m_\nu \geq 8$ Gev, relic neutrinos are incidental players on the cosmic stage.

8. References

Bacino W et al. 1979 Phys. Rev. Lett. **42** 749
Bond J R, Efstathiou and Silk J 1980 Phys. Rev. Lett. **45** 1980
Cowsik R and McClelland J 1973 Ap. J. **180** 7
Davis M, Geller M J and Huchra J 1978 Ap. J. **221** 1
Davis M, Tonry J, Huchra J and Latham D W 1980 Ap. J. (Lett.) **238** L113
Dicus D A, Kolb E W and Teplitz V L 1977 Phys. Rev. Lett. **39** 168
Gershtein S S and Zeldovich Ya B 1966 JETP Lett. **4** 120
Lee B W and Weinberg S 1977 Phys. Rev. Lett. **39** 165
Peebles P J E 1979 A.J. **84** 730
Rood R T, Steigman G and Tinsley B M 1976 Ap. J. (Lett.) **207** L57
Schramm D N and Steigman G 1981a GRG **13** 101
Schramm D N and Steigman G 1981b Ap. J. **243** 1
Shvartsman V F 1969 JETP Lett. **9** 184
Spite M and Spite F 1982 Nature **297** 483
Steigman G, Schramm D N and Gunn J E 1977 Phys. Lett. **66B** 202
Steigman G 1979 Ann. Rev. Nucl. Part. Sci. **29** 313
Szalay A S and Marx G 1976 Astron. Astrophys. **49** 437
Wasserman I 1981 Ap. J. **248** 1
Yang J, Turner M S, Steigman G, Schramm D N and Olive K A 1982 In Preparation

Solar neutrinos

John N. Bahcall

Institute for Advanced Study, Princeton, New Jersey 08540, U.S.A.

Abstract. The main features of the solar neutrino problem are reviewed. Special emphasis is given to the comparison between theory and observation for the ^{37}Cl experiment. The present status of the proposed ^{71}Ga experiment is also summarized.

1. Introduction

In this talk, I will describe the neutrino sources that are important for the solar neutrino problem, review the basis for expecting the calculated capture rates in different experiments, summarize the results of the Brookhaven observations (Davis 1964, 1978) using ^{37}Cl, and describe the ongoing ^{71}Ga experiment.

Before we begin discussing the details, I want to give you an overview of the solar neutrino problem. The standard theory of stellar evolution combined with conventional weak interaction theory yields (Bahcall 1964, Bahcall et al. 1982) a predicted capture rate of 7.6 ± 3.3 SNU (solar neutrino units). The error is intended to be an effective 3-σ uncertainty (defined later). The observations yield a capture rate in the Brookhaven tank of 2 ± 0.3 SNU, where the quoted experimental uncertainty corresponds to 1-σ. There is no accepted solution for this discrepancy, although many have been proposed.

The conflict between theory and observation is of interest to both astronomers and physicists. The solar neutrino experiment is a well defined test of the theory of nuclear energy generation in stars and of stellar evolution. We know more about the Sun than we do about any other stars and it is also in the simplest stage of stellar evolution - sitting quietly on the main sequence in a (presumably) quiescent state. If we can't predict the correct results for the Sun, how can we have confidence in our calculations for other cases? This question is particularly troublesome to astronomers because the theory of stellar evolution is widely used in astronomy to interpret many kinds of observations, e.g., in dating stars and galaxies, in inferring stellar histories from measured surface parameters of stars, and in making cosmological inferences from spectroscopically determined chemical compositions.

Physicists are especially interested because solar neutrino experiments allow one to test the stability of the electron's neutrino over proper times that are much longer than those that are accessible in the laboratory. This aspect of solar neutrino astronomy has become of greater topical concern as a result of recent work on Grand Unified Theories, in some of which lepton non-conservation and neutrino oscillations are a natural result.

About 15 experiments have been proposed in an effort to clarify the solar neutrino problem (nearly all of these are reviewed briefly in Bahcall, 1981). In short, the ^{37}Cl experiment is on-going, the ^{71}Ga experiment is going, and other experiments are necessary and feasible for a full understanding of the problem. (I have been encouraged on my visit here to learn about the major effort being made by Norman Booth and his associates to carry out an electronic experiment using ^{115}In as a target.)

2. Stellar Evolution, Neutrino Sources, and the Predicted Capture Rates

I will begin with a three minute course in stellar evolution and then list for you the main neutrino sources that are important in solar neutrino experiments.

Only the most basic principles of the theory of energy generation in stars and of stellar evolution are required to calculate the expected capture rates for solar neutrino experiments. The principles that are most important for our purposes are: hydrostatic equilibrium in a spherical Sun, energy generation by nuclear fusion among light elements, energy transport by radiation and convection, and a primordial uniform composition. Hydrostatic equilibrium is obviously achieved or the Sun would collapse in a half-hour. The Sun is also known to be spherical to high accuracy from optical observations of the surface. Energy transport in the solar interior is primarily by radiation; hence one must calculate accurately the opacity of the material under the solar interior conditions. Putting all of these ingredients together, stellar evolution occurs as a result of the burning of hydrogen to make helium - the reason the Sun shines.

The bottom line of this brief course is that only the ^{37}Cl solar neutrino experiment is inconsistent with the standard theory of stellar evolution. The theory is widely applied in interpreting astronomical observations and is generally successful in accounting for the (less direct) observations made of photons emitted by stellar surfaces.

The major neutrino sources are listed in Table 1, together with their contributions (Bahcall et al. 1982) to the ^{37}Cl experiment. Of special importance are the p-p neutrinos, which are a signature of the basic reaction that initiates all of nuclear burning, and the ^{8}B higher-energy neutrinos which dominate the expected capture rate for the ^{37}Cl experiment. Nearly all of the neutrinos that are predicted to reach us from the Sun are from the basic p-p reaction. These neutrinos are below the energy threshold for detection in the ^{37}Cl experiment of Davis but are the most important neutrino source for the ^{71}Ga experiment. The ^{8}B neutrinos are extremely rare (0.01% the total flux of solar neutrinos) and are relatively sensitive to astrophysical uncertainties.

In order to interpret the observed capture rate in a solar neutrino experiment, one must know the predicted capture rate and the uncertainties in the prediction. I define an effective 3-σ uncertainty such that if the actual rate lies outside this range then someone has made a career-damaging blunder (experimental or theoretical). With this definition as an implicit guide, the most important recognized uncertainties in the various solar neutrino experiments have been computed (Bahcall et al. 1982). For the ^{37}Cl experiment, the major areas of uncertainty are (with their estimated 3-σ uncertainties): nuclear parameters (3 SNU), chemical composition (1 SNU), stellar opacity (0.5 SNU), neutrino absorption cross sections (0.8 SNU). The incoherent sum of these error estimates is the previously quoted value of 3.3 SNU. The uncertainty due to the neutron lifetime is about 1 SNU.

Table 1. Neutrino Sources

Reactions	Energy (MeV)	Flux Standard 10^{10} cm^{-2} sec^{-1}	SNU: Model (^{37}Cl)
$p+p \rightarrow D+e^{+}+\nu_e$	0-0.4	6.1	0
$p+e^{-}+p \rightarrow D+\nu_e$	1.4	0.015	0.23
$^7Be+e^{-} \rightarrow {}^7Li+\nu_e$	0.86(90%) 0.34(10%)	0.43	1.02
$^8B \rightarrow {}^8Be^{*}+e^{+}+\nu_e$	0-14	0.00056	6.05
$^{13}N \rightarrow {}^{13}C+e^{+}+\nu_e$	0-1.2	0.05	0.08
$^{15}O \rightarrow {}^{15}N+e^{+}+\nu_e$	0-1.7	0.04	0.26
			7.6

In quoting the above uncertainties, I have not taken account of the Munster (Krawinkel et al. 1982) value for the cross section factor of the ^{3}He$(\alpha,\gamma)^4$He reaction, the predicted capture rate is 5.0 ± 2.1 SNU. The Munster result is inconsistent with the recent CalTech (Osborne et al. 1982) and Los Alamos (Robertson et al. 1982) measurements.

The ^{71}Ga experiment is much less subject to uncertainties since it is primarily sensitive to the neutrinos from the basic p-p reaction. In a sense, the ^{71}Ga experiment may be thought of as a (very inefficient) neutrino calorimeter - an expensive detector that measures the present rate of nuclear energy generation in the interior of the Sun. The predicted capture rate for this experiment is 106 ± 10 SNU. The difference between the CalTech and Munster measurements gives rise to an uncertainty of less than 5% for the ^{71}Ga experiment.

3. The Experiments

The ^{37}Cl experiment of Davis (1964, 1978) is inconsistent with the predicted capture rate. This experiment has shown an approximately constant capture rate of 2 SNU for the past 15 years. Either something is seriously wrong with the standard solar models or the neutrinos do not reach the earth in the form in which they are produced at the Sun. In order to decide which of these possibilities is causing the problem, we need a new experiment that detects the astronomically secure p-p neutrinos.

A ^{71}Ga solar neutrino experiment is underway (Hampel 1981, Bahcall et al. 1978). It is an international collaboration between Brookhaven National Laboratory, the Max Planck Institutes, the Weizmann Institute, and the Institute for Advanced Study (Princeton). It is planned to test in 1983 the modular detection system using an intense radioactive source of neutrinos (^{51}Cr) produced in a high flux reactor.

Other experiments are needed and are feasible. They can be combined with the results of the ^{37}Cl and ^{71}Ga experiments to yield detailed information about the rates of individual nuclear reactions in the solar interior, about the spectrum of neutrinos that are emitted, and possibly even the time-dependence (over periods of millions of years) of the neutrino fluxes (see, for example, the papers in the conference proceedings edited by Friedlander 1978, also Bahcall 1978, 1981, Cowan and Haxton 1982, Raghavan 1976, Scott 1976, and references quoted therein).

Acknowledgment

This work was supported in part by the National Science Foundation under grant #PHY79-19884.

References

Bahcall J N 1964 Phys. Rev. Lett. 12 300
Bahcall J N 1978 Rev. Mod. Phys. 50 pp 881-904
Bahcall J N 1981 in Neutrino 81 ed. by R J Cence, E Ma and A Roberts (University of Hawaii High Energy Physics Group) Vol. 2 253
Bahcall J N, Cleveland B T, Davis R Jr., Dostrovsky I, Evans J C Jr., Frati W, Friedlander G, Lande K, Rowley J K, Stoenner W and Weneser J 1978 Phys. Rev. Lett. 40 1351
Bahcall J N, Huebner W F, Lubow S H, Parker P D and Ulrich R K 1982 Rev. Mod. Phys. 54 pp 767-799
Cowan G A and Haxton W C 1982 Science 216 (4541) 51
Davis R Jr. 1964 Phys. Rev. Lett. 12 303
Davis R Jr. 1978 in Proceedings of Informal Conference on Status and Future of Solar Neutrino Research edited by G Friedlander (Brookhaven National Laboratory Report No. BNL 50879) Vol. 1 1
Hampel W 1981 in Neutrino 81 ed. by R J Cence, E Ma and A Roberts (University of Hawaii High Energy Physics Group) Vol. 1 6
Krawinkel H, Becker H W, Buchmann L, Gorres J, Kettner K U, Kieser W E, Santo R, Schmalbrock P, Trautvetter H P, Vlieks A, Rolfs C, Hammer J W, Azuma R E and Rodney W S 1982 Z. Phys. A 304 307
Osborne J L, Barnes C A, Kavanagh R W, Kremer R M, Mathews G J, Zyskind J L, Parker P D and Howard A J 1982 Phys. Rev. Lett. 48 1664
Raghavan R S 1976 Phys. Rev. Lett. 37 259
Robertson R G H, Dyer P, Bowles T J, Brown R E, Jarmie N, Maggiore C J and Austin S M 1982 submitted to Physical Review C
Scott R D 1976 Nature 264 729

Reactor neutrinos

Frederick Reines

Department of Physics, University of California, Irvine, California 92717.

1. Introduction

The subject of fission reactor neutrinos owes a multiple debt to the neutron: first in the production of the neutrino source by the neutron propagated fission chain reaction and second in the inverse beta decay reaction ($\bar{\nu}_e + p \to n + e^+$) used in the detection of the neutrino. In this report we discuss the status and future prospects for neutrino research of reactors indicating its unique role despite the small number of reaction types available for study.

2. Reactions Studied at Nuclear Reactors

The number of $\bar{\nu}_e$ reactions which have been detected and studied is quite limited:

$$\bar{\nu}_e + p \to n + e^+ \tag{1}$$

$$\bar{\nu}_e + d \to n + n + e^+ \tag{2}$$
$$\searrow n + p + \bar{\nu}_e \tag{3}$$

$$\bar{\nu}_e + e^- \to \bar{\nu}_e + e^- \tag{4}$$

In addition to the above observed reactions,

$$\bar{\nu}_e + {}^{37}Cl \to {}^{37}Ar + e^- \tag{5}$$

though sought, has not been seen to occur.

It is characteristic of this list that each entry is related to one or more fundamental characteristics of the weak interaction.

(1) (a) Used to establish existence of the free antineutrino ($\bar{\nu}_e$)

     (b) Cross-section reflects a factor of two in changed weak current.

     (c) e^+ spectrum (as a function of distance from reactor or detector) measures neutrino stability.

* Supported in part by United States Department of Energy

(2) Provides independent check of charged weak current. Used in conjunction with (3) to test neutrino stability.

(3) (a) Makes possible a clean cut measure of neutral weak current. Involves only isovector-axial vector coupling, no "weak angle" (i.e. Weinberg, Salam angle).

(b) By comparison with ν_μ reaction at accelerators provides a test of $\bar{\nu}_e$, $\bar{\nu}_\mu$ universality.

(4) Unique (so far) measure of diagonal interaction predicted in this case to be due to a mixture of intermediate Bosons ($\omega^\pm, Z^0$) i.e. charged and neutral weak currents.

(5) Negative result shows either $\bar{\nu}_e \neq \nu_e$ due to the chirality of the interaction which produces the neutrino but $\bar{\nu}_e \neq \nu_e$ (i.e. Majorana) or $\bar{\nu}_e \neq \nu_e$ and neutrino a Dirac particle*. We discuss these reactions in turn indicating the experimental methods used and the current status.

3. $\underline{\nu_e + p \rightarrow n + e^+}$

This reaction was first employed (Reines 1982) to detect the free neutrino because of its highly distinctive character which enabled the use of a delayed coincidence between the product neutron and the positron. In addition, the simplicity of the target and the consequent behaviour of the interaction cross-section $\sigma_{\bar{\nu}_e, p}$ for reactor neutrinos provided the opportunity to study the absolute dependence of this cross-section on $\bar{\nu}_e$ energy. A series of experiments employed organic scintillators in various configurations near fission reactors to exploit different aspects of this reaction (Reines 1982, Reines and Cowan 1959). The first try to detect $\bar{\nu}_e$ via reaction (1) was made at the Hanford reactor in 1953. In this attempt a 300 litre cylindrical liquid scintillation counter with 90, 2 inch photomultiplier tubes viewing a scintillator consisting of toluene, αNPO and a neutron capturer, cadmium propionate (7.5%) was employed. White paint aided in light collection. A reactor-associated delayed coincidence was sought in a heavily shielded detector between the product e^+ and n capture gammas. This experiment was of limited accuracy, primarily because the background produced by cosmic ray neutrons. These neutrons recoiled from protons in the scintillator producing a "first" pulse and then slowing down and capturing, mocking up a neutrino signal. Despite this cosmic ray background a hint of a reactor-associated signal, 0.41±0.20/min was seen, corresponding to an interaction cross-section $\sigma_{\bar{\nu}_e} \sim 10^{-43}$ cm^2/proton.

i.e. reactor on (10,000 sec) 2.55±0.15/min
reactor off (6,000 sec) 2.14±0.13/min

At the time of the experiment the predicted rate was 0.2/min. In view of the various uncertainties involved e.g. statistical, and $\bar{\nu}_e$ spectrum, the result was taken to be suggestive and served primarily as a stimulus to perform an improved experiment.

* The Majorana nature of the ν is best tested by a double beta decay experiment in which the ν emitted is absorbed and the total energy is shared by the two decay electrons.

4. Definitive Identification - Savanna River Plant, 1956

This experiment (located 11.2m from a 700MW reactor) employed a delayed coincidence between prompt e^+ annihilation radiation (2,0.5MeV γ's) and ≥ 2 neutron capture gammas. The e^+ kinetic energy was not detected. The detector scheme is shown in figure 1. The result of this experiment was the observation of a reactor-associated signal in which the first pulse was shown to be due to an e^+, the second pulse to a neutron. The reactor-associated accidental background was $\leq 1/25$ the reactor correlated signal; the ratio of reactor-associated signal to reactor independent correlated background was 5/1 and signal to accidental reactor independent background was 4/1. A bulk shielding test ruled out reactor neutrons and gamma rays. The free neutrino existence was demonstrated but the integral interaction cross-section was measured only approximately

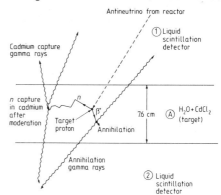

$$\sigma_{\bar{\nu}_e,p} = \frac{R}{3600 f N \varepsilon_{e^+} \varepsilon_n} = (1.2^{+0.7}_{-0.4}) \times 10^{-43} \text{cm}^2/\text{fission } \bar{\nu}_e$$

where: $R = 3.0 \pm 0.1 \text{ h}^{-1}$

$\varepsilon_n = 0.17 \pm 0.06$ neutron detection efficiency

$\varepsilon_{n^+} = 0.15 \pm 0.02$ position detection efficiency

$N = 2.2 \times 10^{28}$ target protons

$f = 1.2 \times 10^{13} \bar{\nu}_e/\text{cm}^2 \text{sec}$

This is in fair agreement with the theoretically expected value of 10^{-43}cm^2. The detectors used in the identification experiment were used in a modified form (Reines and Cowan 1959) reminiscent of the earlier attempt at Hanford. Unlike that effort the improved cosmic ray shielding and increased reactor power made it possible to see the reaction and obtain a hint of the directly determined $\bar{\nu}_e$ spectrum.

The observed cross-section was compared with that expected for two component neutrinos:

$$\frac{\sigma_{\text{expt'l}}}{\bar{\sigma}_{\text{th}}} = 1.1 \pm 0.4$$

This early result tended to favor the two component theory* but was far from definitive.

An improved experiment at Savannah River (1965) in which the kinetic energy of the positron as well as the annihilation radiation was measured in delayed coincidence with the neutron (Nezrick et al 1966). An improved comparison with theory yielded the ratio

$$\frac{\sigma_{expt'l}}{\bar{\sigma}_{th}} = 0.88 \pm 0.13$$

supporting the two component theory.

5. Detection of ν_e+d charged current branch - Savannah River Plant, 1969

This reaction was first sought in 1956 incidental to detection of $\bar{\nu}_e$. It was observed in 1969 (Jenkins et al 1969) using a deuterated scintillator enclosed in a transparent bag which was in turn suspended in a hydrogenous non-deuterated scintillator. Location of the target deuterons well inside the large non-deuterated scintillator provided nearly total absorption of the positron annihilation and neutron capture gamma rays (from captures in gadolinium). In addition the outer region provided sheilding against neutrons produced by the capture of muons originating external to the detector. The time correlated reactor-associated signal was characteristic of two successive neutron captures and disappeared when the deuterons were removed. The observed cross-section was $(2.9 \pm 1.5) \times 10^{-45} cm^2 / fission \bar{\nu}_e$ a number to be compared with an expected value of $(2.4 \pm 0.4) \times 10^{-45} cm^2$. These results were judged too crude to allow a detailed comparison to be made with theory.

6. Detection of changed and neutral branch in ν_e+d (1979)

An initial attempt to detect the neutral branch (Munsee and Reines 1969) by means of a delayed coincidence between the proton and neutron pulses was unsuccessful because of the background in the proton gate. A second attempt (Pasierb et al 1979) succeeded in seeing both branches via the capture of single and double neutrons in a 268 kg D_2O target in which ^{3}He neutron counters were immersed (Fig. 2). Backgrounds from the reactor

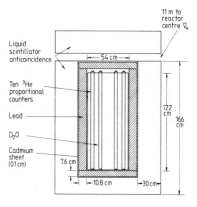

* If because of parity non-conservation beta decay of the neutrino produces only one final neutrino state instead of the two previously expected when it was believed that parity was conserved: then the beta decay coupling constant must be increased by a factor of two to account for the observed neutron decay rate. But this is the same coupling constant that enters into the rate for reaction (1).

and cosmic rays were reduced by surrounding the D_2O with a large liquid scintillation anticoincidence counter. Reactor-associated neutron capture signals were observed and tested as due to $\bar{\nu}_e$ interactions with deuterons. The resultant charged and neutron current cross-sections for fission $\bar{\nu}_e$ were:

$$\bar{\sigma}_{ccd} = (1.5 \pm 0.4) \times 10^{-45} cm^2$$

$$\bar{\sigma}_{ncd} = (3.8 \pm 0.9) \times 10^{-45} cm^2, \text{ values in fair agreement with expectations}$$

$$\bar{\sigma}_{ccd}^{th} = 2.1 \times 10^{-45} cm^2$$

$$\bar{\sigma}_{ncd}^{th} = S \times 10^{-45} cm^2$$

It is interesting that this low energy experiment tests the existence of weak neutral currents for $\bar{\nu}_e$ but, unlike the high energy muon neutrino experiments does not depend on the weak mixing angle.

7. Elastic Scattering of $\bar{\nu}_e$ on e^- - Savanna River Plant, 1976

Despite the fact that this reaction does not appear to involve a changing of charge it nevertheless is associated with the charged and neutral currents as seen by the relevant Feynman diagrams (Fig. 3).

FIG. 3 Collision of ν_e with e^-

The detection of this process (Reines et al 1976) was made difficult by the tiny cross-section $\sim 10^{-45} cm^2/e$ and the nondescript nature of the process - an electron struck by a gamma ray had to be distinguished from an electron nudged by a $\bar{\nu}_e$. In addition such processes as electrons from the beta decay of the ubiquitous ^{214}Bi and e^+ from the $\bar{\nu}_e + p$ reactions in the detector had to be ruled out. A key feature in the detection was the recognition that, unlike a $\bar{\nu}_e$ which collided with at most one electron, a gamma ray in the relevant energy range (>1MeV) would in a low Z medium such as an organic scintillator lose its energy preferentially by multiple Compton collisions. As a result it was possible to employ a spatial anticoincidence against the gamma ray background. This idea plus the development (from 1960-70) of large (300kg) composite NaI crystals was incorporated into the detection scheme shown in Fig. 4. The time scale of the experiment was in large part dictated by the counting rates with which we dealt. At first it was a reactor-associated signal of a few counts per week. The rate rose to several counts/day as the target was increased and reduced background enabled use of a lower detection threshhold. The signal was a pulse in only a single plastic target i.e. unaccompanied by pulses in the surrounding NaI or liquid scintillator. Pulse shape discrimination enabled separation of pulses in the plastic and NaI located between the plastic and photomultiplier tubes.

The accepted reactor-associated rates were $5.9 \pm 0.25/d$ and $1.2 \pm 0.25/d$ in the energy ranges $1.5 \to 3meV$ and $3 \to 4.5meV$, respectively. These data were obtained from 64.40 (live) with reactor on the 60.7d with reactor off.

The requisite instrumental
stability was obtained by
reference to the internal
beta emitting contaminant
^{214}Bi which was identified
by a delayed coincidence
with the α emitted from
the daughter ^{214}Po.

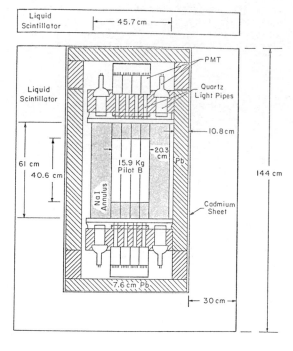

The experimental results
were analysed in terms of
the vector and axial
vector (C_V and C_A) current
strengths and the Weinberg-
Salam angle $\sin^2\theta_w$ based on
the general expression for
the differential scattering
cross-section $\frac{d\sigma}{dE}$. Folding
this differential cross-
section with the reactor
$\bar{\nu}_e$ spectrum and the various
experimental uncertainties
the result, though not
clearly incompatible with
V-A theory, favored the
Glashow, Salam, Weinberg
theory with
$\sin^2\theta_w =0.29\pm0.05$. The
observed cross-section for
the production by fission $\bar{\nu}_e$ of recoil electrons in the 1.5→4.5MeV range
is:

$$\bar{\sigma}_{expt} = (7.5\pm1.5) \times 10^{-46} cm^2/fission\ \nu_e$$

An improvement in the precision with which this interesting cross-section
is known remains an experimental challenge.

The elastic scattering process has occasionally been proposed as a means
to detect solar ^{8}B neutrinos using the angular correlation of the struck
electron to relate the signal to the sum. At best it is an experiment
of great dificulty because of the low energies (<14MeV) and large masses
$\sim 10^3$ tons involved.

8. <u>Radiochemical ν_e Experiment at reactor - Savanna River Plant, 1956</u>

The Pontecorvo-Alvarez-Davis radiochemical ^{37}Cl approach was used by
R. Davis to measure the production of ^{37}Ar via the reaction
$\bar{\nu}_e+^{37}Cl \rightarrow ^{37}Ar+e^-$. The idea was to expose some kilograms of CCl_4 to reactor
$\bar{\nu}_e$ and remove any product ^{37}Ar which would be counted via K-capture in a
small low background detector. The result was negative as mentioned
earlier in this report. As described by Bahcall in the present conference
this method has since been adapted by Davis to search for solar neutrinos
giving results at variance with expectation from the standard solar model.

9. Neutrino oscillations and reactor $\bar{\nu}_e$

An interesting suggestion made by Pontecorvo and by Maki et al (see Rosen and Pakvasa, 1981) was that, in analogy with the $K°$ system, the observed neutrinos can be considered as a superposition of base states which changes character as it travels from source to detector. Such a possibility would be a natural consequence of massive neutrinos and has attracted considerable attention because of its possibly profound effect on cosmology and particle physics. It should be emphasised that neutrino masses and oscillations are not currently mandated by theory nor does the present experimental evidence require oscillation. On the other hand the results do not rule out oscillations in any absolute sense and the idea is so intriguing that many experiments are underway or planned at reactors and accelerators to explore the range of parameters involved.

References

Jenkins, T.L., Kinard, F.E. and Reines, F., 1969 Phys. Rev. 185, 1599.
Munsee, J.H., and Reines, F., 1969 Phys. Rev. 177, 2002.
Nezrick, F.A. and Reines, F., 1966 Phys. Rev. 142, 852.
Pasierb, E., Gurr, H.S., Lathrop, J., Reines, F. and Sobel, H.W., 1979 Phys. Rev. Lett. 37, 315.
Reines, F. The detailed development of the detection schemes developed by C.L. Cowan and the author may be found in the contribution to the International Colloquium on the History of Particle Physics. Paris, July 1982.
Reines, F., and Cowan, C.L. Jr., 1959 Phys. Rev. 113, 273.
Reines, F., Gurr, H.S., Sobel, H.W. 1976 Phys. Rev. Lett. 37, 315.
Rosen, S.P., and Pakavasa, P., 1981, Physics Today.

The role of neutrons in nucleosynthesis in supernovae

William A. Fowler

W. K. Kellogg Radiation Laboratory, Caltech, Pasadena, CA

1. Introduction

This week we celebrate the 50th anniversary of the discovery of the neutron by James Chadwick here in Cambridge in 1932. The title of my talk has been chosen because 1982 is also the 25th anniversary of an early paper on the subject of my title by Burbidge, Burbidge, Fowler & Hoyle (1957), which had its beginnings in Cambridge. It is appropriate because neutrons play an important role in nucleosynthesis in supernovae. However, I am not interested in looking back either 25 or 50 years ago. I follow the wise advice of one of America's baseball immortals, Satchel Paige, 1906-1982, who said "Don't look back. Something might be gaining on you." I don't have to look back. A horde of supernova buffs are working the core collapse-mantle explosion problem and the consequent nucleosynthesis. See "Essays in Nuclear Astrophysics" (1982) and "Supernovae: A Survey of Current Research" (1982) for copious references. In spite of the global aspects of the attack on the problem this discussion will be centered on what we have been doing recently on the experimental and theoretical aspects of the nuclear physics involved in nucleosynthesis in supernovae in the Kellogg Radiation Laboratory at Caltech in Pasadena. Kellogg was built in 1931 for research by Charles C. Lauritsen on X-rays and their application in cancer therapy using funds obtained by Robert A. Millikan from Will Keith Kellogg, the corn flakes king. However, 1932, the golden year of classical nuclear physics, started the laboratory on its role in nuclear physics and eventually nuclear astrophysics. Remember 1932? Urey discovered deuterium, Anderson discovered the positron and Chadwick discovered the neutron. In a practical fashion these discoveries were overshadowed by Cockcroft & Walton's demonstration, also in Cambridge, that nuclei could be disintegrated by protons raised to relatively low energies in accelerators. Before 1932 was over Lauritsen and his student, H. Richard Crane, had converted their alternating voltage X-ray tube into a positive ion accelerator and in Crane et al (1933) were the first to publish evidence for the production of neutrons in an accelerator using the $^9Be(\alpha,n)^{12}C$ and $^9Be(d,n)^{10}B$ reactions.

2. Reaction Network

This discussion is restricted to carbon, neon, oxygen and silicon burning under quasistatic conditions during the presupernova evolution of massive stars ($\gtrsim 8$ M$_\odot$) and under explosive conditions during the supernova event. Even so, in addition to $^{12}C + ^{12}C$ and $^{16}O + ^{16}O$ fusion, strong reactions between n, p, α and γ and hundreds of intermediate mass nuclei from A = 12 to A = 80 are involved. The γ-rays are not only produced in capture reactions and in cascade transitions but also counterbalance capture reactions by photodisintegration. In addition, thermalization by excited states is mainly promoted by excitation and de-excitation by γ-rays. Cujec & Fowler (1980) have shown that interactions with D, T, ^{3}He and heavier nuclei can be ignored in advanced stellar evolution with the exception of the fusion reactions already noted. Weak interactions involving the same nuclei take place: electron capture and electron and positron emission. Neutrino emission and transfer become important. In the advanced stages of quasistatic evolution and during the supernova stage the temperatures become high enough $T_9 \equiv T/10^9 K \sim 20$ ($kT \sim 2$ MeV) that the low lying states of nuclei become populated. For example, the famous Mössbauer state in ^{57}Fe at .0144 MeV with $J = 3/2$ becomes more important *a priori* than the ground state with $J = 1/2$ once kT exceeds $.0144/\ell n$ 2 MeV or $T_9 > .24$. The rate of population and depopulation of excited states, especially long-lived isomers via higher short-lived states, is an important problem which has been treated in some detail by Ward & Fowler (1980). For example, when the isomeric state of $^{26}Al^m$ with $J = 0$ at .228 MeV and $\bar{\tau} = 9.2$ s becomes fully populated relative to the ground state with $J = 5$ and $\bar{\tau} = 10^6$ yr the effective positron emission lifetime changes to $11 \times 9.2 \approx 100$ s, a decrease by a factor of 3×10^{11}. The equilibration times can be substantial, for example ~ 100 s at $T_9 = 0.3$.

3. The Strong Interactions

Experimental cross sections with n, p and α can of course be measured only for stable nuclei and a few long-lived radioactive nuclei with isomers produced in reactors or accelerators. Experiments using accelerated short-lived radioactive nuclei may be possible in the future (Boyd 1981). Current practice is to fit measurements made in a number of laboratories, notably Caltech, Colorado, Kentucky, Lucas Heights, Melbourne, Münster, Oak Ridge, Ohio State, Toronto, and others, to a Hauser-Feshbach calculation of average cross sections using a minimum number of global parameters such as nuclear radii, optical model potentials, known low energy strength functions particularly for neutrons, amount of isospin mixing along with width fluctuation corrections. By "global" we mean that a few parameters are chosen not to fit each measured reaction but to give the best fit to all measurements. The known excited states of nuclei are introduced and extended by level density expressions so that rates for the ensemble of thermally populated states of a nucleus are calculated. The reaction rates results to date are made available in convenient analytic expressions with tabulated numerical coefficients in Holmes et al (1976) and Woosley et al (1978). The ground state rates agree with those calculated directly from measured cross sections within $\pm 50\%$ with notable exceptions. There is much work to be done especially in sorting out the degree of isospin mixing in intermediate mass nuclei and in learning how to handle marked shell structure effects near shell closures.

4. A Special Neutron Role

It is well known that (n,γ) reactions are important in nucleosynthesis especially in the s- and r-processes and in the cosmochronologies based on these processes. These processes will not be discussed here. For recent developments see Käppeler et al (1982), Winters & Macklin (1982) and McEllistrem and Hershberger (1982). However, (n,γ) reactions and also (n,α) and (n,p) reactions are important in our reaction network but reactions producing neutrons play a very special role in the network. It is well known that the Hauser-Feshbach theory is the sum over terms consisting of the product of the transmission functions for the incoming and outgoing channels divided by the sum over transmission functions for all allowed channels. Bohr's idea of competition in the decay of the compound nucleus is thus incorporated. In many cases (p,n) and (α,n) reactions are endoergic and play no role at low reaction energies and low temperatures. However, above the threshold for these reactions the neutron transmission functions rapidly increase over an energy range of several hundred keV, resulting in deep competition cusps or decreases by a factor of ~ 10 in the competing (p,γ) and (α,γ) reactions. This effect has been noted in the number of reactions and the matter is treated in some detail for $^{51}V(p,\gamma)^{52}Cr$ vs $^{51}V(p,n)^{51}Cr$ in Zyskind et al (1980). $^{51}V(p,\gamma)$ synthesizes a new stable nuclear species, ^{52}Cr while $^{51}V(p,n)$ synthesizes ^{51}Cr which for time scales longer than $\bar{\tau} > 40$ d decays back to ^{51}V. It is true that the neutron released in $^{51}V(p,n)$ will eventually be captured, but not necessarily on ^{51}Cr but rather on lighter nuclei with greater abundance or on heavier nuclei with greater capture cross sections. In other words, $^{51}V(p,n)^{51}Cr(n,\gamma)^{52}Cr$ cannot be simply treated as $^{51}V(p,\gamma)^{52}Cr$ in the network — the bookkeeping must be meticulous. The rapid rise above threshold for endoergic reactions producing neutrons and their marked competition effects on the corresponding capture reactions must both be taken into account.

5. Weak Interactions

The drift along the reaction network toward more neutron-rich nuclei can only be accomplished through the weak interactions. Electron capture plays the major role, especially in the gravitational collapse of the supernova core once the iron-group nuclei are reached and no further nuclear energy is available for stabilization. The collapse is initiated by photodisintegration of the heavier nuclei which requires energy and reduces the pressure which opposes gravity. Eventually in the collapse heavy nuclei once again become the predominant constituent. Collapse proceeds as electrons which make the main contribution to the pressure are captured by protons, free or in nuclei, to form neutrons. The collapsing core is neutronized and eventually becomes a neutron star and most probably a pulsar. Fuller et al (1980, 1982a, 1982b) and Fuller (1982) have made systematic calculations of the rates of electron and positron emission and of <u>continuum</u> electron and positron capture rates as well as the associated neutrino energy loss rates for free nucleons and 226 nuclei with $21 \le A \le 60$. Measured nuclear level information and matrix elements for discrete to discrete state transitions were taken from Lederer & Shirley (1978). The total of known matrix elements for $21 \le A \le 38$ was greatly enhanced by measurements in Kellogg by Wilson et al (1980). Gamow-Teller matrix elements where allowed but unmeasured were assigned an average value corresponding to log ft = 5, $|M_{GT}|^2 = .04$ after exhaustive study of measured values in the mass range of interest. Extensive use was made of information available in "mirror" transitions. Simple shell arguments were used to estimate Gamow-Teller sum rules and collective state resonance energies. The discrete state contribution to the rates, dominated by experimental information and Fermi transitions involving excited as well as ground states, determines the nuclear rates in the regions of temperature and density characteristic of the quasistatic phases of presupernova stellar evolution. At the higher temperatures and densities characteristic of the core collapse phase, the rates are dominated by the Fermi and the Gamow-Teller collective resonance contributions. Of great importance is the effect of neutron shell closure in blocking electron capture on neutron-rich nuclei. Excitation of excited states and forbidden transitions become important. For example, the rate for $^{48}Ca(e,\nu)^{48}K$ at $\rho Y_e = 10^{11}$ g cm^{-3} (Y_e = electrons per nucleon), $T_9 = 1$ is dex - 18.203 compared to dex + 3.663 for $^{48}Se(e,\nu)^{48}Ca$ but is dex + 4.479 at $T_9 = 100$ compared to dex + 5.463. The ground state of ^{48}Ca has a closed shell of 28 neutrons so that allowed transitions are forbidden until the shell is opened by neutron excitations at high temperature. A good part of the effect at low temperature is the fact that the threshold is 12.0 MeV for electron capture on ^{48}Ca and only .28 MeV on ^{48}Sc. This follows from the closed shells for both neutrons and protons in ^{48}Ca. Fuller (1982) has shown that the closed neutron shell blocking strongly decreases electron capture on heavy nuclei in supernova core collapse. This leads to a slowing down of neutronization and to the formation of a more massive homologous core from which the gravitational energy of collapse is extracted. On the transformation of the gravitational energy the outgoing shock energy is substantially increased. The riddle whether spherically symmetric models can lead to explosive ejection of the outer mantle is still unsolved as of this writing but the Fuller effect certainly brightens the prospects. However, it is probably true that red giant mass loss, rotation, magnetic fields and Rayleigh-Taylor instabilities in the shock front will have to be considered to obtain successful models of supernova explosions.

6. Abundances

Given a model of presupernova evolution and a presupernova explosion with arbitrary shock energy sufficient to eject the mantle, nucleosynthesis calculations can be made using the strong and weak interaction rates previously discussed. The most recent attempts have been made by Weaver et al (1978), Weaver & Woosley (1980) and Woosley et al (1980) for stars with M = 10, 15 and 25 $M_{\odot}$. Comparisons with solar system abundances are made and fair agreement is obtained but it is clear that additional masses must be studied and an average over the appropriate mass range in the Salpeter mass formation distribution function obtained. There are numerous puzzles. For the 15 $M_{\odot}$ ejecta the relative abundances of ^{40,42,43,44,46}Ca are close to the solar values but no ^{48}Ca is produced. See, however, the nβ-process.

7. The nβ-Process

Much of the current study of nucleosynthesis in supernovae is motivated by recent discoveries of isotopic anomalies in the Allende meteorite, a carbonaceous chondrite thought to constitute the most primitive of solar system materials. See Begemann (1980) for a recent review of the subject. The discovery in certain Allende inclusions by Lee et al (1976) of large excesses of ^{26}Mg due to the decay of ^{26}Al ($\bar{\tau} = 10^6$ y) probably *in situ* has revived the old idea that the formation of the solar system was triggered by a nearby supernova explosion. The subject of short-lived nuclides in the early solar system is reviewed by Wasserburg & Papanastassiou (1982). Small anomalies in the isotopes of a number of elements have been found which are not due to radioactive decay. These anomalies indicate that the solar nebula was not homogeneous in isotopic composition and that nucleosynthetic contributions from late sources had not been completely mixed. Of special interest are the small anomalies of the order of 10^{-3} to 10^{-2} relative to solar system abundances found by Wasserburg and his colleagues in the Lunatic Asylum of the Division of Geological and Planetary Sciences at Caltech. Originally designed to investigate lunar samples, the high resolution and high precision mass spectrographs of the asylum have revealed small isotopic anomalies as noted with a precision of a few parts in 10^4. In order to seek an explanation of the anomalies in the Ca and Ti isotopes observed in an Allende inclusion, catalogued as EK-1-4-1, Sandler et al (1982) have studied a neutron-capture/beta-decay (nβ) process. This process includes both neutron-capture rates, λ_n, and beta-decay rates, λ_β, and in the appropriate extremes can represent either the canonical s-process ($\lambda_\beta > \lambda_n$) or the r-process ($\lambda_n > \lambda_\beta$). In these studies the seed nuclei were taken to consist of all isotopes of the elements from Si to Cr with normal solar system abundances. Hauser-Feshbach values for λ_n were taken from Woosley et al (1978) and observed values for λ_β from Lederer & Shirley (1978). It was found that when the (nβ) process operates at neutron densities ≈ 10^{-7} mole cm^{-3}, temperatures ≈ 30 keV and exposure times ≈ 10^3 s, small admixtures ($\lesssim 10^{-4}$) of the exotic material produced are sufficient to account for most of the Ca and Ti isotopic anomalies found in EK-1-4-1. SKF (1982) go so far as to suggest that ^{48}Ca which constitutes .2% of Ca may have been produced entirely in the nβ-process since none was produced in the mainline synthesis of Weaver et al (1978). Agreement for ^{46}Ca and ^{49}Ti was obtained only by increasing the Hauser-Feshbach rates for ^{46}K(n,γ)^{47}K and ^{49}Ca(n,γ)^{50}Ca by at least a factor of 10. ^{46}K ultimately decays to ^{46}Ca as its major progenitor and ^{49}Ca decays to ^{49}Ti. Justification for these increases through thermal resonances just above threshold for neutron capture was provided in detail using current shell models by SKF (1982). Hauser-Feshbach theory averages over resonances. In intermediate mass nuclei strong particle-hole resonances can enhance rates at certain energies and temperatures. Fluctuations in level densities and strength functions can also raise or lower rates over the Hauser-Feshbach values. Experimental measurements are the last resort. SKF (1982) suggest studies of ^{48}Ca(d,^{3}He)^{47}K*, ^{48}Ca(t,α)^{47}K* and ^{48}Ca(t,p)^{50}Ca* to establish the existence of the necessary excited states in ^{47}K and ^{50}Ca. Since Wasserburg and his colleagues are now attempting to detect anomalies in the Cr isotopes in EK-1-4-1, SKF (1982) extended their calculations through chromium and predict rather smaller anomalies in Cr than in Ca and Ti with roughly equal anomalies in ^{53}Cr and ^{54}Cr relative to ^{50}Cr and ^{52}Cr as standards for the chromium studies. The ^{53}Cr anomaly depends critically on the cross section for ^{43}Ti(n,γ)^{44}Ti since ^{43}Ti is the main progenitor for ^{53}Cr. The fact that the optimum exposure time was found to be 10^3 s points to ^{13}N ($\bar{\tau}$ = 862 s) as the source of the neutrons via the chain of reactions ^{12}C(p,γ)^{13}N(β+)^{13}C(α,n). This implies mixing of hydrogen into ^{12}C-rich regions during supernova explosions. It is for this and other reasons that rotation induced mixing and Rayleigh-Taylor instabilities in shock fronts must be studied in supernova models. The existence of different anomalies in various Allende inclusions, in other carbonaceous chondrites and even in iron meteorites indicates that numerous materials of exotic composition were injected into the solar nebula in small proportions and not completely mixed. It is not at all unreasonable that the solar system and other stellar systems received material produced in "rare" processes not revealed to B^{2}FH (1957) or even to Cameron (1957).

8. Observational Evidence for Nucleosynthesis in Supernovae

The launch on 13 November 1978 of the Einstein (HEAO-2) Observatory introduced the use of focusing high-resolution optics to X-ray astronomy, Giacconi & Tananbaum (1980). This sensitive space instrument was used by Becker et al (1980) to obtain spectra in the range .8 to 4.2 keV from the remnants of Tycho Brahe's 1572 Supernova and other supernovae. The Tycho spectrum showed resolved peaks well above the continuum near 1.84, 2.47, 3.20 and 4.04 keV which are just the K-shell energies of Si, S, Ar and Ca respectively. For one who has been interested in nucleosynthesis in supernovae for 25 years this is the most beautiful and convincing evidence for such synthesis observed to date. Of course, everything depends on the abundances derived from these observations. Relative to solar system abundances, Shull (1982) found Ne = .37, Mg = 2.0, Si = 7.6, S = 6.5, Ar = 3.2, Ca = 2.6 and Fe = 2.2 with CNO assumed to be solar. Tycho's supernova will eventually add to the abundance of Mg to Fe relative to CNO in the interstellar medium. No two supernovae are alike. Others will produce different abundance ratios. A grand average over certain types of supernovae or over a certain mass range produced the solar system abundances. There was a main-line synthesis and there were rare, minor processes. Neutrons played a role constructively, destructively and competitively throughout. Chadwick created the character for that role.

9. References

Becker R H, Holt S S, Smith B W, White N E, Boldt E A, Mushatzky R F & Serlemitsos P J 1980 Ap. J. Lett. $\underline{235}$ L5
Begemann F 1980 Rep. Prog. Phys. $\underline{43}$ 1309
Boyd R N 1981 Editor: Proc. Workshop Radioactive Ion Beams & Small Cross Section Measurements Aug. 31-Sept. 4, The Ohio State University
Burbidge, E M, Burbidge G R, Fowler W A & Hoyle F 1957 Rev. Mod. Phys. $\underline{29}$ 547 B^{2}FH (1957) hereinafter
Cameron A G W 1957 Pub. A.S.P. $\underline{69}$ 201
Crane H R, Lauritsen C C & Soltan A 1933 Phys. Rev. $\underline{44}$ 514, 692
Cujec B & Fowler W A 1980 Ap. J. $\underline{236}$ 658
"Essays in Nuclear Astrophysics" 1982 eds C A Barnes, D D Clayton & D N Schramm (New York: Cambridge University Press)
Fuller G M 1982 Ap. J. $\underline{252}$ 741
Fuller G M, Fowler W A & Newman M J 1980 Ap. J. Suppl. $\underline{42}$ 447; 1982a Ap. J. $\underline{252}$ 715; 1982b Ap. J. Suppl. $\underline{48}$ 279
Giacconi R & Tananbaum H 1980 Science $\underline{209}$ 865
Holmes J A, Woosley S E, Fowler W A & Zimmerman B A 1976 At. Data and Nucl. Data Tables $\underline{18}$ 305
Käppeler F, Beer H, Wissnak K, Clayton D, Machlin R & Ward R 1982 Ap. J. $\underline{257}$ 821
Lederer C M & Shirley V A 1978 "Table of Isotopes" 7th edition (New York: Wiley & Sons)
Lee T, Papanastassiou D A & Wasserburg G J 1976 Geophys. Res. Lett. $\underline{3}$ 109
McEllistrem M & Hershberger R 1982 Proc. Workshop on Neutrons and Astrophysics, April 15, Denison University
Sandler D G, Koonin S E & Fowler W A 1982 Ap. J. $\underline{259}$ August 15 issue SKF (1982) hereinafter
Shull J M 1982 Ap. J. November 1 issue
"Supernovae: A Survey of Current Research" 1982 eds M J Rees & R J Stoneham (Dordrecht: Reidel Publ. Co.)
Ward R A & Fowler W A 1980 Ap. J. $\underline{238}$ 266
Wasserburg G J & Papanastassiou D A 1982 "Essays in Nuclear Astrophysics" (for editors etc. see above)
Weaver T A & Woosley S E 1980 Ann. New York Acad. Sci. $\underline{336}$ 335
Weaver T A, Zimmerman G B & Woosley S E 1978 Ap. J. $\underline{225}$ 1021
Wilson H S, Kavanagh R W & Mann F M 1980 Phys. Rev. $\underline{C22}$ 1696
Winters R & Macklin R 1982 Phys. Rev. $\underline{C25}$ 208
Woosley S E, Fowler W A, Holmes J A & Zimmerman B A 1978 At. Data and Nucl. Data Tables $\underline{22}$ 371
Woosley S E, Weaver T A & Taam R E 1980 "Type I Supernovae" ed J Craig Wheeler (Austin: University of Texas Press) p. 96
Zyskind J L, Barnes C A, Davidson J M, Fowler W A, Marrs R E & Shapiro M H 1980 Nucl. Phys. $\underline{A343}$ 295

Acknowledgment

This work was supported in part by the National Science Foundation [PHY79-23638].

Light element nucleosynthesis

Jean Audouze

Institut d'Astrophysique du CNRS, 98 bis Bd Arago, F-75014 Paris, France

1. Introduction

The nucleosynthetic processes responsible for the formation of the light elements from D to ^{11}B play a very important role in many major fields of astrophysics : cosmology, evolution of galaxies, cosmic ray physics... The purpose of this presentation is first to review the main characteristics of these processes and then to point out the principal consequences on these problems such as some aspects of the particle physics.

With the exception of ^{3}He, ^{4}He and ^{7}Li, the light elements cannot be produced by the standard stellar nucleosynthesis which is responsible for the formation of the heavier elements ^{12}C and up. In the case of ^{6}Li, ^{9}Be, ^{10}B, ^{11}B and especially of D they are indeed destroyed by thermonuclear reactions in the stellar interiors. Even in the case of ^{3}He, ^{4}He and ^{7}Li stars play a fairly minor role in their synthesis. Two processes are invoked to form these elements (i) the nucleosynthesis triggered by the primordial explosion or Big Bang does explain the formation of D, ^{3}He, ^{4}He and ^{7}Li (ii) for the other elements the interaction between the galactic cosmic rays and the interstellar medium is able to account for the observed abundances of ^{6}Li, ^{9}Be, ^{10}B and ^{11}B. Due to the lack of space available for this paper, the reader is referred to Audouze 1981a and 1982 for an analysis of the relevant observed abundances which are summarized in Table 1.

TABLE 1 - THE "RELEVANT" ABUNDANCES OF THE LIGHT ELEMENTS
(see details in Audouze 1981a and 1982)

Primordial value	Present interstellar value
D/H $\approx (2 \pm 0.5)10^{-5}$	^{6}Li/H $\approx 9 \times 10^{-11}$
^{3}He/H $\approx (1 - 2)10^{-5}$	^{9}Be/H $\approx 1.5 \times 10^{-11}$
^{4}He/H $\approx 0.24 \pm 0.01$ (by mass)	^{10}B/H $\approx 5 \times 10^{-11}$
^{7}Li/H $\approx 10^{-10}$ or 10^{-9} (see text)	^{11}B/H $\approx 1.5 \times 10^{-10}$

2. The main features of the primordial nucleosynthesis

The primordial nucleosynthesis which can form D, ^{3}He, ^{4}He and ^{7}Li occurs when the temperature has a fairly definite value of about 10^9 K. In the "standard" models of Big Bang this is the temperature of the Universe at a time of one to three minutes after the explosion. The standard models of Big Bang assume that (i) the Universe was homogeneous and isotropic (ii) the expansion of the Universe is well described by the general relativity theory (iii) the amount of antimatter is negligible compared to the amount of matter and (iv) the number of leptons (including neutrinos) is much less than that of photons.

The temperature of 10^9 K should be low enough for deuterium which is formed by the $p + n \rightarrow D + \gamma$ reaction not to be destroyed by photodesintegration reactions. The key reaction of the set of nuclear reactions (see Wagoner 1980 for a complete presentation of this set) which occur during this primordial phase is the neutron absorption reaction. The rate of this reaction depends obviously on the relative densities of neutrons X_n and protons X_p. They are given by :

$$\frac{X_n}{X_p} = \exp \frac{-(M_n - M_p)}{KT_f}$$

where $M_n - M_p$ are the difference between the masses of the proton and the neutron and T_f is the freezing temperature at which the equilibrium between protons and neutrons governed by the following weak interactions $n + \nu_e \leftrightarrow e^- + p$ and $n + e^+ \leftrightarrow p + \nu_e$ stops (in other words T_f is the temperature below which the primordial neutrons start to decay). T_f cannot be much higher than a few 10^9 K that is why the temperature range in which the primordial nucleosynthesis can take place is fairly restricted. One should notice here how important it is to determine precisely the mass of the neutron on which this conference is devoted.

The net result of the primordial nucleosynthesis depends basically on two parameters : the average baryon density of the Universe when the primordial nucleosynthesis takes place. With the assumption that the explosion is adiabatic one can relate this baryon density to the present density of the Universe. Fig. 1 provides the quite classical representation of the resulting abundances of the elements produced during this early phase relative to the present density of the Universe obtained by Wagoner 1973. From the current discussions of the abundances of D, ^{3}He and ^{7}Li, this representation would favour a present baryon density $\rho_B \simeq 3 \ 10^{-31}$ g cm^{-3}. This baryon density is significantly lower than the so-called critical density which marks the border between the cosmological models for which the expansion (open universes with low densities) and undergo a succession of expansions and

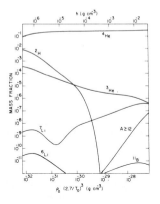

Figure 1 - Resulting abundances of the light elements at the end of the primordial nucleosynthesis with respect to the present density of the Universe (in the frame of the standard model of Big Bang): these abundances agree with the observations for $\rho \simeq 3 \ 10^{-31}$ g cm^{-3} corresponding to an open Universe (after Wagoner 1973).

of the Universe lasts for ever and those for which the Universe contractions (closed universes with higher densities). The present deuterium abundance is considered as the best argument in favour of an open Universe. Quite recently two French astronomers F. and M. Spite (1982) determined the Li abundance of several old halo stars and found $^7Li/H \approx 10^{-10}$. It might be still premature to state that the primordial $^7Li/H$ abundance is 10^{-10}. However the arguments presented by these two workers concerning the absence of variation of $^7Li/H$ with the stellar effective temperature are already fairly convincing. If this low value is supported by further observations this would reinforce the conclusion on the open character of the Universe.

While the D, ^3He and ^7Li nucleosynthesis depends mainly on this baryon density the production of ^4He is related to the expansion rate of the Universe. If the expansion is very fast the freezing temperature at which the neutron starts to decay is large and then the ^4He abundance is high. With a slow expansion the resulting ^4He abundance is much lower as shown in Fig. 2. According to the General Relativity theory which is now very well accepted the expansion rate depends on the total density of the Universe including not only the baryons but also leptons and especially neutrinos, the number of which being smaller but still comparable with that of photons. Very

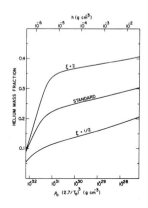

Figure 2 - Abundance (in mass) of helium with respect to the present density of the Universe calculated for three expansion rates : $\xi = 1/2$ corresponds to an expansion slower than the standard deduced from the general relativity by a factor 2 and $\xi = 2$ to an expansion twice more rapid (Wagoner 1973).

recently as reviewed for instance in Audouze 1981b two types of progress have been made i) in view of a better knowledge of the primordial helium abundance (see e.g. Kunth 1981) deduced from the "lazy" blue compact galaxies ii) in view of a proper analysis of the relations between this abundance, the baryon/photon ratio, the neutron life-time and the maximum number of different types of neutrinos (one assumes now that there are at least three different families: the electronic neutrino, the muonic neutrino and the tau neutrino). According to the thorough study of Olive et al 1981, with $Y_p \le 0.25$ (where Y_p is the primordial ^4He abundance by mass), with the density of the Universe of $\approx 2\ 10^{-30}$ g cm^{-3} deduced from the dynamics of small group of galaxies and a neutron life-time of 10.6 minutes there should not be any room for a new neutrino (and therefore lepton) family (Fig. 3).

To end this too quick survey of the main features of the primordial nucleosynthesis, this process accounts very satisfactorily for the formation of the observed D, ^3He, ^4He and ^7Li.

One deduces from it strong arguments in favour of an open Universe and a restricted number of different types of neutrinos (and therefore leptons).

3. The interaction between the galactic cosmic rays and the interstellar medium

It has been proposed in the early 70's that the light elements Li,Be,B could be produced by the interaction between the galactic cosmic rays and the interstellar medium (see Audouze and Reeves 1982 for a recent review). By bombarding the interstellar atoms during their journey in the galactic disk between their sources and the point where they stop after being slowed down by their inelastic interaction with the interstellar electrons, the galactic cosmic rays induce spallation reactions which produce not only the light elements but also some other rare species like for example the odd nuclei between Si and Fe. By the same process, one is able to explain the so-called "cosmic" ^{6}LiBeB abundances and the LiBeB/CNO ratio observed in the galactic cosmic rays which is as high as 0.23 (instead of 10^{-5} in the "cosmic" abundances). Table 2 is extracted from the work of Meneguzzi et al 1971 who used the so-called "leaky box" model to evaluate the effect of the cosmic ray interstellar medium interaction provides the resulting LiBeB abundances. The "leaky box" model assumes an homogeneous distribution of the galactic cosmic ray sources in the galactic disk. It assumes also that there is an equilibrium between the effects which diminish the cosmic ray flux at a given energy (slowing down by electrons-destruction by spallation reactions-escape out from the galactic disk) and those which increase it (production either by spallation or by the sources). This model not only reproduces the observed LiBeB abundances in the cosmic rays and in the interstellar medium ; it

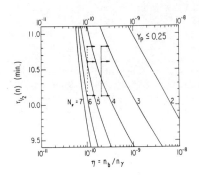

Figure 3 - The maximum number of neutrino families for $Y_p \leq 0.25$ with respect to η the ratio between baryons and photons and the half-life of the neutron. The two sets of arrows corresponds to a cosmological factor $\rho_{present}/\rho_{critical} = 0.04$ (solid line). The dashed line is the lower limit of $\rho_{present}$ compatible with the D and ^{3}He nucleosynthesis. One can notice that for the most probable present density the number of different families of neutrinos depend on the neutron life-time (after Olive et al 1981).

provides also an estimate of the average amount of interstellar matter encountered by the cosmic rays which is $\simeq 6$ g cm^{-2}. As a result one has a relation between the transport life-time of the cosmic rays inside the galactic disk and the average interstellar density. From recent data on the cosmic ray radioactive ^{10}Be this life-time should $\simeq 10^7$ years and the average interstellar density $\simeq 0.1$ particle cm^{-3}. Moreover by transmitting their energy to the interstellar electrons, the cosmic rays account for the heating and the partial ionization of this medium.

To come back to the by-products of this cosmic ray nucleosynthesis one can realize from Table 2 that the resulting ^{7}Li/^{6}Li ratio is 1.5 instead of 10 (or higher) and the resulting ^{11}B/^{10}B ratio is 2.8 instead of 4.

Primordial nucleosynthesis is one way to explain the present ^{7}Li/^{6}Li ratio if the conclusion of Spite and Spite 1982 is found to be untenable. Other possible mechanisms occurring in some stars might account for the missing interstellar ^{7}Li. Cameron and Fowler (1971) have shown that ^{7}Li can be produced by the ^{3}He + ^{4}He reaction which could occur in flashes undergone by red giants inside which the helium rich and the hydrogen rich zones can

TABLE 2 - FORMATION RATES OF LIGHT ELEMENTS DUE TO THE INTERACTION BETWEEN
THE GALACTIC COSMIC RAYS AND THE INTERSTELLAR MATTER
(from Meneguzzi et al 1971)

Light elements	Galactic cosmic ray production rate	Abundance due to observed galactic cosmic rays (with respect to H)
^{6}Li	$1.1 \ 10^{-4}$	$8.8 \ 10^{-11}$
^{7}Li	$1.7 \ 10^{-4}$	$1.2 \ 10^{-10}$
^{9}Be	$2.8 \ 10^{-5}$	$2 \ \ \ 10^{-11}$
^{10}B	$1.2 \ 10^{-4}$	$8.7 \ 10^{-11}$
^{11}B	$2.8 \ 10^{-4}$	$2 \ \ \ 10^{-10}$

mix. This process is actually supported by Li abundance determinations for some red giants. Starrfield et al 1978 have shown that ^{7}Li could be also produced during nova outbursts. Therefore if Spite and Spite 1982 are found to be right the interstellar ^{7}Li comes mainly from stellar sources either red giants or novae (or why not supermassive stars, see e.g. Norgaard and Fricke 1976).

To transform the resulting ^{11}B/^{10}B ratio from 2.8 to the observed value of 4 the easiest way first suggested by Meneguzzi et al 1971 and reanalysed by Reeves and Meyer (1978) is to assume the existence of low energy cosmic ray fluxes which cannot be observed inside the solar cavity because of the solar modulation but which could account for the missing ^{11}B. This is mainly because the ^{14}N (p,α) ^{11}C(β^+) ^{11}B spallation reaction has a very low threshold (of about 3 MeV). With realistic low energy cosmic ray fluxes one could easily account for the simultaneous production of ^{6}Li, ^{9}Be, ^{10}B and ^{11}B.

4. Summary : the evolution of the light element abundances during the history of the Galaxy

The nucleosynthesis of the light elements from D to ^{11}B can be considered as a well understood problem in the frame of the current views on cosmology, particle and nuclear physics. From the study of these mechanisms one has some pleasing (if not satisfactory) answers to many problems such as the overall dynamical behaviour of the Universe, some constraints on the physics of the lepton particles, the interaction between the cosmic rays and the interstellar medium, the existence of low energy cosmic rays unvisible on earth, etc... These processes account quite nicely for the observed abundances with quite a restricted number of ad-hoc hypotheses.

These nucleosynthetic processes are at least in part dependent on some of the physical parameters of the neutron like its mass and its beta decay life-time. To end up with the point of view of an astrophysicist, these mechanisms can be put in the perspective of the evolution of the Galaxy as it has been done a few years ago by Audouze and Tinsley 1974 and more recently by Reeves and Meyer 1978 : when the Galaxy evolves the interstellar matter density decreases such as the rate of high mass star explosions. At first sight one should expect a significant decrease of D during the galactic evolution since it is destroyed by all the nuclear processes occurring during this evolution. On the other hand one should expect a small but measurable enrichment of ^{4}He which is mainly produced during the Big Bang but is also produced in the normal course of the stellar evolution.

The models of chemical evolution of galaxies which are presently available show for instance that D is at most destroyed by a factor of about 2 during the whole galactic history. Most of the light elements should have about the same abundance now as they had $4.5\ 10^9$ years ago at the birth of the Solar System. From the comparison between ^{7}Li and ^{11}B (taking a primordial ^{7}Li value of 10^{-9}) Reeves and Meyer 1978 argued that some infall of external matter onto the galactic disk occurring at a rate of 1 - 2 M_0 per year for the whole Galaxy during its history would provide a coherent agreement for all the observed light element abundances. Audouze, Boulade, Malinie and Poilane are currently investigating models of chemical evolution with a primordial ^{7}Li/H abundance as low as 10^{-10} (as proposed by Spite and Spite). The calculations performed so far do not support (while they do not exclude) anymore the need for some infall of external gas. The observed rate of nova outbursts seem to be sufficient to build up the ^{7}Li/H abundance from its primordial value of 10^{-10} up to the interstellar value of $\simeq 10^{-9}$.

To sum up the study of the light element nucleosynthesis appears as one of the best occasions to tackle at the same time cosmological, nuclear and particle physics and in general overall astrophysics. This is may be why such a subject related to the physics of the neutron is so exciting.

5. References

Audouze J 1981a in Nuclear Astrophysics Ed. D Wilkinson (Oxford: Pergamon Press) pp 125-157
Audouze J 1981b in Cosmology and Particles Ed J Audouze (Bures: Editions Frontières) pp 231-240
Audouze J 1982 in Astrophysical Cosmology Ed H A Brück et al (Rome: Pontificiae Academiae Scientiarum Varia-48 pp 395-425
Audouze J Reeves H 1982 in Essays in Nuclear Astrophysics Ed C A Barnes et al. Cambridge: University of Cambridge Press)
Audouze J and Tinsley B M 1974 Astrophys. J. 192 487
Cameron A G W and Fowler W A 1971 Astrophys. J. 164 11
Kunth D 1981 Thèse de Doctorat d'Etat ès Sciences Physiques Université Paris 7
Meneguzzi M, Audouze J and Reeves H 1971 Astron. Astrophys. 15 33
Norgaard H and Fricke K J 1976 Astron. Astrophys. 49 337
Olive K A Schramm D N Steigman G Turner M S and Yang J 1981 Astrophys. J. 246 557
Reeves H and Meyer J P 1978 Astrophys.J. 226 613
Spite M and Spite F 1982 Nature 297 483
Starrfield S Truran J W Sparks W M and Arnould M 1978 Astrophys.J. 222 600
Wagoner R V 1973 Astrophys.J. 179 343
Wagoner R V 1980 in Physical Cosmology Ed. R Balian et al (Amsterdam: North Holland) pp398-442

The role of the neutron in heavy element nucleosynthesis

James W. Truran

Department of Astronomy, University of Illinois

1. Introduction

Recognition of the fact that neutrons play a critical and pervasive role in the processes involved in the formation of the heavy elements in nature dates back now more than thirty years. Guided by early compilations of the abundances of the elements in solar system matter (Goldschmidt 1937) and by laboratory studies of neutron activation cross sections for a large number of elements (Hughes 1946), Alpher, Bethe and Gamow (1948) noted that there exists an approximate inverse relationship between neutron capture cross section and relative abundance. This behavior has since been confirmed by refined analyses of element abundances (Suess and Urey 1956; Cameron 1973).

Alpher, Bethe and Gamow sought an explanation for this correlation within the framework of cosmology. They recognized that free neutrons would constitute a substantial fraction of matter at high densities and temperatures. They argued that, during the early stages of expansion of the universe following the cosmological big bang, the interaction of these neutrons with protons would lead via successive neutron captures to the production of progressively heavier elements. We have since learned that universal big-bang nucleosynthesis could not have been responsible for the formation of the bulk of the nuclear species heavier than helium. Revised analyses of the weak interaction processes proceeding at early epochs indicate that the neutron to proton ratio characterizing the onset of the nucleosynthesis era is of order 1/7, leading to the production of a ^{4}He mass fraction of order 25 percent but only trace amounts of heavy elements (Peebles 1966; Wagoner 1973).

An alternative to cosmological synthesis is clearly required. Here once again neutron processes played an historically important role in the formulation of nucleosynthesis theory. The recognition that nucleosynthesis is a continuing process in stellar interiors followed the detection by Merrill (1952) of the presence of the element technetium in the atmospheres of red giant stars. As technetium has no stable isotopes, and the lifetime of the longest lived isotope is less than the ages of the stars in which it has been observed, its presence confirms that thermonuclear processes involving heavy nuclei have occurred in the interior.

The processes by which heavy elements are synthesized in stellar and supernova environments fall broadly into two categories, distinguished by whether the associated nuclear transformations are dominated by charged-particle reactions or by neutron-capture reactions. The formation of most of the nuclear species in the mass range $A \lesssim 60$ is believed to be associated with charged-particle induced reactions occurring either during relatively stable phases of nuclear burning in stellar interiors and in supernovae. The manner in which neutrons influence particularly the explosive thermonuclear burning phases will be reviewed in the following section. The defining role played by neutron capture processes in the synthesis of nuclei beyond the iron abundance peak (A $\gtrsim$ 60) is then discussed and recent studies of both s-process and r-process neutron-capture nucleosynthesis are reviewed.

2. Explosive Nucleosynthesis

While the supernova phase constitutes only an extremely small fraction of the lifetimes of a small fraction of all stars, theoretical studies nevertheless suggest that supernova environments represent the likely site of the synthesis of most of the elements in the range of mass number $20 \lesssim A \lesssim 60$ observed in nature. From the point of view of nucleosynthesis, one has the advantage that constructive nuclear transformations accompany the shock-induced ejection of both core and envelope matter, ensuring that the abundance patterns thus achieved will not be distorted by subsequent evolution. Successive exoergic stages of burning of hydrogen, helium, carbon, oxygen and silicon fuels define the presupernova evolution (Arnett 1978). When the ashes of these burning epochs are subsequently subjected to high temperatures and densities associated with their ejection in supernova events, further thermonuclear processing yields elemental and isotopic abundance patterns which mimic very closely those of solar system matter (Arnett 1973; Truran 1973).

Neutrons play a critical role in defining the character of these thermonuclear burning phases. Neutron-induced reactions, including particularly (n,γ), (n,p) and (n,α) reactions, will of course proceed under these conditions. More critical, however, is the question of the relative numbers of neutrons and protons in the processed stellar matter. We call attention, in particular, to the fact that the most abundant nuclear constituents of the core immediately prior to the supernova event are self-conjugate nuclei (^{4}He, ^{12}C, ^{16}O, ^{20}Ne, ^{24}Mg, etc). Trace amounts of neutron-rich isotopes of various of these elements are also present in the gas, such that the total number of neutrons exceeds that of protons by a factor or order several parts in a thousand. Thermonuclear processing of such matter in the wake of supernova shocks typically occurs on timescales of the order of seconds or less, ensuring that weak interaction processes cannot act to convert any appreciable fraction of protons into neutrons.

It follows that the final products of these explosive burning episodes must lie along or very near to the Z = N line. This remains true even when the buildup proceeds past the last stable alpha-particle nucleus (^{40}Ca): the dominant species in situ include the nuclei ^{44}Ti, ^{48}Cr, ^{52}Fe, ^{56}Ni, and ^{60}Zn. When these nuclei and their immediate neighbors subsequently decay, isotopic abundance patterns compatible with those of solar system matter are realized. We note, for example, that the isotopic composition of chlorine formed in such circumstances is dictated by the nuclear properties of ^{35}Cl and ^{37}Ar: following the decay of the ^{37}Ar formed in situ, the terrestrial ^{35}Cl/^{37}Cl ratio is quite accurately reproduced. This same phenomenon occurs for many other cases: the potassium isotopes ^{39}K and ^{41}K (formed as ^{41}Ca), the chromium isotopes ^{50}Cr, ^{52}Cr and ^{53}Cr (the last two formed as ^{52}Fe and ^{53}Fe), the iron isotopes ^{54}Fe, ^{56}Fe, and ^{57}Fe (the last two formed as ^{56}Ni and ^{57}Ni), and the nickel isotopes ^{58}Ne, ^{60}Ni, ^{61}Ni and ^{62}Ni (the last three formed as ^{60}Zn, ^{61}Zn and ^{62}Zn). Furthermore, the relative abundances of ^{51}V (formed as ^{51}Mn), ^{55}Mn (formed as ^{55}Co) and ^{59}Co (formed as ^{59}Cu) are reproduced very well in these calculations. This general behavior and the fact that the resulting isotopic patterns are so closely in agreement with those of solar system matter strongly support the view that supernovae provide the appropriate nucleosynthesis environment.

3. Neutron Capture Processes Defined

The production of most of the nuclear species more massive than iron proceeds in nature by means of neutron capture processes. Scrutiny of the abundance patterns in the heavy element regime reveals signatures of two distinct neutron fluxes. This has led historically to the definition of two nucleosynthesis processes, which are identified with quite different astrophysical environments. The distinction is made here largely on the basis of relative lifetimes for neutron captures (τ_n) and electron decays (τ_β). The condition that $\tau_n > \tau_\beta$, where τ_β is a characteristic lifetime for beta-unstable nuclei near the valley of beta stability, ensures that as captures proceed the neutron capture path will itself remain close to the valley of beta stability. This defines the astrophysical "s-process" of neutron capture. When $\tau_n < \tau_\beta$, it follows that neutron captures will typically proceed into the neutron-rich regions off the beta stable valley. This "r-process" neutron capture mechanism is expected to operate in an environment characterized by a more violent epoch of generation of neutrons -- heretofor assumed to be associated with supernova explosions. The characteristics of neutron-capture nucleosynthesis for these two cases are reviewed below.

4. s-Process Nucleosynthesis

Early efforts in the development of s-process theory were concentrated upon questions of timescales, analyses of the observed abundance patterns, and the identification of possible neutron sources. The constancy of the product of neutron capture cross section σ and s-process abundance N_s between closed shells and the sharp decreases at the positions of the shell closures were both recognized. Building upon these systematic behaviors and critical experimental determinations of the neutron capture cross sections of nuclei participating in these astrophysical processes (Macklin and Gibbons 1965; Allen et al 1971; Käppeler et al 1982), the relative s-process and r-process contributions have been determined as a function of mass number and the overall characteristics of these two processes have thereby been better defined. It was also recognized quite early that a single neutron exposure could not account for the observed pattern of s-process abundances in solar system matter.

Significant progress toward an understanding of the stellar environments in which s-process nucleosynthesis can proceed has resulted over the past decade (see the reviews by Truran (1980) and Ulrich (1982)). This recent work is built upon the operation of two neutron sources long ago identified by Cameron (1955; 1960): the ^{13}C(α,n)^{16}O and the ^{22}Ne(α,n)^{25}Mg reactions. The ^{13}C and ^{22}Ne source elements both are formed in the normal course of stellar evolution: ^{13}C follows from the reactions ^{12}C(p,γ)^{13}N(e$^+\nu$)^{13}C and ^{22}Ne is produced from ^{14}N, the product of CN cycle hydrogen burning, by the reaction sequence ^{14}N(α,γ)^{18}F(e$^+\nu$)^{18}O(α,γ)^{22}Ne. Some production of heavy elements is believed to accompany helium burning in the cores of massive stars (Couch et al 1974; Lamb et al 1977), but it is the operation of the ^{22}Ne source in the convective helium-burning shells of red giant stars undergoing thermal pulses that provides the most promising environment.

Schwarzschild and Härm (1967) first identified the environment provided by low and intermediate mass stars possessing both hydrogen- and helium-burning shells as a promising site for neutron-capture synthesis. In an examination of the possible role of the ^{13}C neutron source in this environment, Ulrich (1973) was able to show that a succession of thermal pulses naturally provides an exponential distribution of neutron exposures, comparable to that determined by Seeger et al (1965) to be necessary to fit the solar system distribution of s-process abundances. The neutron release associated with the $^{22}Ne(\alpha,n)^{25}Mg$ reaction, which operates during thermal pulses in stars of intermediate mass, has been found to provide an appropriate s-process flux. On this model, the outward progression of the hydrogen-burning shell during the interpulse phase leaves in its wake a substantial concentration of ^{14}N, which is readily transformed into ^{22}Ne when subsequently mixed downward into the helium shell during the next thermal pulse. As has been emphasized by Iben (1975), this neutron source thus arises as a natural consequence of the evolution of intermediate mass stars along the asymptotic giant branch. Truran and Iben (1977) have demonstrated that the operation of the $^{22}Ne(\alpha,n)^{25}Mg$ source in thermally pulsing stars will produce the s-process abundances in the mass range 70 > A > 204 in solar proportions; furthermore, sufficient matter is found to be processed in this manner to explain the abundance level of s-process nuclei in the galaxy (Iben and Truran 1978).

One very important constraint on the operation of the $^{22}Ne(\alpha,n)^{25}Mg$ reaction as a neutron source is that the temperature at the base of the convective burning shell exceed ~ 300-500 million degrees K: only for higher temperatures will the timescale for neutron liberation be less than that of the thermal pulse. This implies that s-process abundance enhancements can be realized in red giant atmospheres only for relatively massive stars of high luminosities. Observational studies (Scalo and Miller 1979) of the peculiar red giants which show s-process abundance enhancements reveal, however, that many of the stars have masses and luminosities, both of which are well below those compatible with the operation of the $^{22}Ne(\alpha,n)^{25}Mg$ neutron source. An alternative neutron capture environment appears to be demanded. It is generally assumed that the neutron source appropriate to these peculiar red giants may be provided by the $^{13}C(\alpha,n)^{16}O$ reaction. It is argued that this can occur as a result of mixing of a small contamination of hydrogen into helium and carbon rich matter, associated either with shell flashes (Scalo and Ulrich 1973) or with helium flashes in the cores of lower mass stars (Paczyński and Tremaine 1977). The episode of convective mixing of carbon into the hydrogen-rich convective envelope recently encountered by Iben and Renzini (1982) in red giant models of small core mass may hold important implications in this regard.

5. r-Process Nucleosynthesis

The high neutron densities demanded for r-process nucleosynthesis would seem to suggest that it is associated with a rather violent event. Promising environments which have recently been examined are associated with supernovae. For example, the expansion and cooling of highly neutronized matter from the regions immediately adjacent to those which become incorporated into neutron star remnants can be accompanied by the production of r-process nuclei. One might typically expect matter characterized by total neutron to proton ratios up to $N/Z \sim 2$-8, established by weak interaction processes at temperatures $T > 10^{10}$ K and densities $\rho > 10^{10}$ g/cc. It has been demonstrated (Cameron et al 1970; Hillebrandt 1978; Schramm 1973) that the subsequent thermonuclear evolution of this matter through expansion and cooling will give rise to the formation of heavy nuclei ($A > 60$) by successive neutron captures. The gross features of solar system r-process abundances can be reasonably reproduced in such calculations: the positions of the r-process peaks, their relative abundances, and the absence of substantial odd-even abundance variations.

Shock heating of the helium layers of supernovae can also provide an environment compatible with r-process nucleosynthesis (Hillebrandt and Thielemann 1977; Truran et al 1978). According to this model, neutrons are liberated by the ^{22}Ne(α,n)^{25}Mg reaction and related (α,n) reactions on a dynamic timescale, following the passage of the shock. Capture of these neutrons on preexisting iron-peak and heavier nuclei can result in the production of r-process heavy elements. The heavy element abundance patterns resulting from these shock-induced helium-shell exposures have been determined to be extremely sensitive to the assumed seed abundance distributions. In fact, the achievement of an r-process abundance pattern resembling that of solar system matter demands that the stellar matter possess an initial distribution of heavy elements which is distinctively non-solar (Cowan et al 1980; Hillebrandt et al 1981).

The presence of ^{13}C in sufficient concentrations to provide a significant neutron source accompanying the explosive burning of helium-rich stellar matter is a further possibility which has been considered. By analogy with the s-process case, such a nucleosynthesis event may be associated with either a helium core flash or a helium shell flash; the difference envisioned here is primarily one of the timescale of the neutron release. While these environments have not previously been expected to provide significant r-process sites, Cowan et al (1982) have identified conditions under which r-process heavy element production may be realized. A series of thermal pulses associated with the helium shells of low mass red giants, in which convective mixing of protons and ^{12}C can provide an appropriate ^{13}C concentration, could provide the necessary environment. Such a nucleosynthesis event might serve at least to explain either some of the peculiar red giants or some of the diverse isotopic anomalies in heavy elements in meteorites (Lee 1979), which are believed to reflect neutron capture exposures.

6. Concluding Remarks

We have seen how neutrons act to help define some crutial characteristics of various processes of nucleosynthesis. Charged particle reactions dominate the sequences of nuclear transformations involved in the formation of nuclei from carbon and oxygen to iron and nickel, but it is the ratio of the total number of neutrons to that of protons in the stellar matter to which the emerging elemental and isotopic abundance patterns are particularly sensitive. Most of the elements heavier than iron are formed as a consequence of neutron-capture reaction mechanisms. It appears that the bulk of the designated s-process nuclei have their origin in luminous asymptotic-branch red-giant stars in which the neutron source is the ^{22}Ne(α,n)^{25}Mg reaction. The precise conditions under which the r-process nuclei in our galaxy have been synthesized have not been established. Supernova explosions would seem to represent a likely site. Since definitive theoretical statements concerning the nature of the supernova events themselves are not yet possible, the extent to which processed material will contribute to the abundances in the galaxy is not known. To place these matters in perspective, we note that only 10^{-6} solar masses of r-process elements need be ejected per supernova event to provide galactic requirements (Truran and Cameron 1971).

7. References

Allen B J, Macklin R L and Gibbons J H 1971 Adv. Nucl. Phys. **4** 205
Alpher R A, Bethe H and Gamow G 1948 Phys. Rev. **73** 803
Arnett W D 1973 Ann. Rev. Astr. Ap. **11** 73
Arnett W D 1978 Ap. J. **219** 1008
Cameron A G W 1955 Ap. J. **121** 144
Cameron A G W 1960 Astron. J. **65** 485
Cameron A G W 1973 Space Sci. Rev. **15** 121
Cameron A G W, Delano M D and Truran J W 1970 "International Conference on the Properties of Nuclei Far from the Region of Beta-Stability" (Geneva: CERN)
Couch R G, Schmiedekamp A B and Arnett W D 1974 Ap. J. **190** 95
Cowan J J, Cameron A G W and Truran J W 1980 Ap. J. **231** 1090
Cowan J J, Cameron A G W and Truran J W 1982 Ap. J. **252** 348
Goldschmidt V M 1937 Skrifter Norske Videnshaps-Acad. Oslo I: Mat. Naturv. Kl. No. 4
Hillebrandt W 1978 Space Sci. Rev. **21** 639
Hillebrandt W, Klapdor H V, Oda T and Thielemann F K 1981 Astr. Ap. **99** 195
Hillebrandt W and Thielemann F K 1977 Mitteilungen der Astron. Gesellschaft **43** 234
Hughes D J 1946 Phys. Rev. **70** 106A
Iben I Jr 1975 Ap. J. **196** 525
Iben I Jr and Renzini A 1982 Ap. J. Letters **259** L79
Iben I Jr and Truran J W 1978 Ap. J. **220** 980
Käppeler F, Beer H, Wisshak K, Clayton D D, Macklin R L and Ward R A 1982 Ap. J. **257** 821
Lamb S A, Howard W M, Truran J W and Iben I Jr 1978 Ap. J. **217** 213
Lee T 1979 Rev. Geophys. Space Phys. **17** 1591
Macklin R L and Gibbons J H 1965 Rev. Mod. Phys. **37** 166
Merrill P W 1952 Science **115** 484
Paczyński B and Tremaine S D 1977 Ap. J. **216** 57
Peebles P J E 1966 Ap. J. **146** 542
Scalo J M and Miller G E 1979 Ap. J. **223** 596
Scalo J M and Ulrich R K 1973 Ap. J. **183** 151
Schramm D N 1973 Ap. J. **185** 293
Schwarzschild M and Härm R 1967 Ap. J. **150** 961
Seeger P A, Fowler W A and Clayton D D 1965 Ap. J. Suppl. **11** 121
Suess H E and Urey H C 1956 Rev. Mod. Phys. **28** 53
Truran J W 1973 Space Sci. Rev. **15** 23
Truran J W 1980 Nukleonika **25** 1463
Truran J W and Cameron A G W 1971 Ap. Space Sci. **14** 179
Truran J W, Cowan J J and Cameron A G W 1978 Ap. J. Letters **222** L63
Truran J W and Iben I Jr 1977 Ap. J. **216** 797
Ulrich R K 1973 "Explosive Nucleosynthesis" eds D N Schramm and W D Arnett (Austin: University of Texas Press)
Ulrich R K 1982 "Essays in Nuclear Astrophysics" eds C A Barnes, D D Clayton and D N Schramm (New York: Cambridge University Press)
Wagoner R V 1973 Ap. J. **179** 343

The transuranium elements

Glenn T. Seaborg

The University of California, Berkeley, CA 94720 USA

Abstract. Following the synthesis and identification of the first transuranium element (neptunium, atomic number 93) in 1940, an additional 14 have been synthesized through neutron and charged particle (including heavy ion) irradiations. The recognition that numbers 93-103 should be placed in the periodic table as an actinide transition series made their identification possible. It should be possible to discover elements beyond 107, perhaps even some in the "superheavy" group.

1. Introduction

The discovery by James Chadwick of the neutron fifty years ago opened the door to the transuranium elements. Two years later Fermi (1934) and co-workers bombarded uranium with neutrons and observed beta particle-emitting radioactivities which they interpreted as transuranium isotopes. The later experiments of Hahn et al (1936) appeared at first to confirm this point of view, and for several years the "transuranium elements" were the subject of much experimental work and discussion. Early in 1939 Hahn and Strassmann(1939), after Meitner had been forced to leave Germany, described experiments that confirmed they had observed radioactive barium isotopes as a result of the bombardment of uranium with neutrons. Subsequent work showed that all of the radioactivities previously ascribed to transuranium elements are actually due to uranium fission products, and hundreds of radioactive fission products of uranium have since been identified.

2. Neptunium, Plutonium and the Actinides

The actual discovery of the first transuranium element resulted from experiments aimed at understanding the fission process. McMillan and Abelson (1940) deduced by chemical means that a non-recoiling product, with half life 2.3 days, formed by neutron capture in uranium, is an isotope with mass number 239 of the element with atomic number 93. This element was given the name neptunium because it is beyond uranium, just as the planet Neptune is beyond Uranus.

Plutonium (atomic number 94) was discovered late in 1940 and early in 1941 by Seaborg et al (1946) as a result of their bombarding uranium with deuterons in the Berkeley 60-inch cyclotron. This particular isotope was later shown to be $^{238}94$. Early in 1941, the isotope of major interest, $^{239}94$, was discovered by Kennedy et al (1946) in uranium bombarded with neutrons produced in the Berkeley cyclotron; it was shown to be fissionable with slow neutrons.

The tracer experiments performed on neptunium and plutonium showed them to have chemical properties like uranium, and not at all like those of rhenium and osmium (as suggested by the periodic table of that time). Such a "uranide" concept led to the erroneous prediction of chemical properties for elements 95 and 96. Then, in 1944, Seaborg (1945) suggested that perhaps all elements heavier than actinium (atomic number 89) had been misplaced in the periodic table and should actually be represented as an "actinide" series analogous to the "lanthanide" elements.

3. Transplutonium Actinide Elements

The actinide concept meant that elements 95 and 96 should have some properties in common with actinium and some in common with their rare earth "sisters," europium and gadolinium. Experiments designed according to this concept led Seaborg et al to the discovery of elements 95 (1945) and 96 (1944) at the wartime Metallurgical Laboratory of the University of Chicago. Not only did this new understanding lead to americium (95) and curium (96) but to the synthesis and identification of berkelium (97) by Thompson et al (1950a), californium (98) by Thompson et al (1950b), einsteinium (99) and fermium (100) by Ghiorso et al (1955a), mendelevium (101) by Ghiorso et al (1955b), and nobelium (102) by Ghiorso et al (1958). It also signaled the end of the actinide series when lawrencium (103) was discovered by Ghiorso et al (1961).

All known elements beyond element 100 were first produced on a "one-atom-at-a-time" basis. Increasing instability leads to a decreasing half life for the heaviest elements. This, together with sharply decreasing yields for the nuclear production reactions, results in great difficulties for the synthesis and identification of still heavier elements. However, there are possibilities for the discovery of more "near transactinide" elements and some prospects for "superheavy" elements (Z=110 to 120) produced by bombardments with heavy ions.

4. Transactinides and Superheavy Elements

Since the actinide series terminates at element 103, all the elements beyond element 103 are referred to as "transactinide" elements. The first three of the transactinide elements--rutherfordium (104) (1969), hahnium (105) (1970) and element 106 (as yet unnamed) (1974)--have been synthesized and identified by Ghiorso and coworkers, at Berkeley using bombardments of isotopes of actinide elements with heavy ions (C, N, and O ions). Flerov and coworkers (1978) at the Joint Institute for Nuclear Research (Dubna, Soviet Union) have made competing claims for the discovery cf elements 104, 105 and 106 that do not seem to satisfy published criteria by Harvey et al (1976) for the discovery of chemical elements.

Element 107 (unnamed as yet), with the highest atomic number of a known element, was discovered at the GSI Laboratory in Darmstadt, Germany, by Münzenberg et al (1981) as the result of the bombardment of bismuth with chromium ions.

Study of chemical properties of rutherfordium (104) by Silva et al (1970) has confirmed that it is indeed homologous to hafnium, as demanded by its position in the periodic table. Studies of the chemical properties of the heavier elements will be very difficult: the longest-lived isotope of rutherfordium (^{261}Rf) has a half life of only about one minute, that of hahnium (^{262}Ha) a half life of about 40 seconds, and that of element 106

(^{263}106) a half life of about one second.

The superheavy elements are predicted to form an "island" of relative stability extending both above and below Z=114 and N=184, and separated from the "peninsula" of known nuclei by a "sea" of instability. Attempts to synthesize and identify superheavy elements, through bombardments of a wide range of heavy nuclides with a wide range of heavy ions, have so far been unsuccessful. A review article by Seaborg et al (1979) describes the various attempts and some future possibilities. The ^{48}Ca plus ^{248}Cm fusion reaction seems to offer the most hope at present.

A great deal of effort has been expended to pedict the chemical properties of the superheavy elements, because this information is needed in order to devise procedures for their chemical identification following their possible production in bombardments of heavy target nuclei with heavy ions or for finding them in natural sources. Keller and Seaborg (1977) have written a review article covering the broad predictions of these chemical properties based on extensive calculations of electron structures using modern high-speed computers and on the judicious use of the periodic table to make extrapolations.

The calculations show that elements 104 through 112 are formed by filling the 6d electron subshell, which makes them, as expected, homologous in chemical properties with the elements hafnium (72) through mercury (80). Elements 113 through 118 result from the filling of the 7p subshell and are expected to be similar to the elements thallium (81) through radon (86). The calculations indicate that the 8s subshell should fill at elements 119 and 120, thus making these an alkali and alkaline earth metal, respectively. Next the calculations point to the filling, after addition of a 7d electron at element 121, of the inner 5g and 6f subshells, 32 places in all, which I have termed the "superactinide" elements and which terminate at element 153. This is followed by the filling of the 7d subshell (elements 154 through 162) and 8p subshell (elements 163 through 168). Unfortunately, the possibilities for the eventual synthesis and identification of superheavy elements appear to be limited, at best, to those at the island of stability located near Z=114 and N=184.

5. Conclusion

Thus the addition of the transuranium elements to mankind's natural heritage of elements has led to an expansion of about 15 percent in the fundamental building blocks of nature. Investigation of these man-made elements beyond uranium has led to a tremendous expansion of our knowledge of atomic and nuclear structure. Each of these elements has a number of known isotopes, all radioactive, thereby leading to an overall total of about 200. Synthetic in origin, they are produced in a variety of transmutation reactions by neutrons or charged particles, including heavy ions. (Neptunium and plutonium are, in addition, present in nature in very small concentrations.)

Many of the transuranium elements are produced and isolated in large quantities through the use of neutrons furnished by nuclear fission reactions: plutonium (94) in ton quantities; neptunium (93), americium (95), and curium (96) in kilogram quantities; berkelium (97) in 100 milligram quantities; californium (98) in gram quantities; and einsteinium (99) in milligram quantities. Of particular interest is the unusual chemistry and impact on the periodic table of these heaviest elements. Their practical impact, particularly that of plutonium, has been extraordinary.

References

Fermi E 1934 Nature <u>133</u> 898
Flerov G N et al; the original papers relating to this work at the Dubna Laboratory are reproduced in the Benchmark Book (1978) Transuranium Elements: Products of Modern Alchemy ed. Seaborg G T, Dowden, Hutchinson and Ross
Ghiorso A, Harvey B G, Choppin G R, Thompson S G and Seaborg G T 1955b Phys. Rev. <u>98</u> 1518
Ghiorso A, Nitschke J M, Alonso J R, Alonso C T, Nurmia M, Seaborg G T, Hulet E K and Lougheed R W 1974 Phys. Rev. Lett. <u>33</u> 1490
Ghiorso A, Nurmia M, Eskola K, Harris J and Eskola P 1970 Phys. Rev. Lett. <u>24</u> 1498
Ghiorso A, Nurmia M, Harris J and Eskola K 1969 Phys. Rev. Lett. <u>22</u> 1317
Ghiorso A, Sikkeland T, Larsh A E and Latimer R M 1961 Phys. Rev. Lett. <u>6</u> 473
Ghiorso A, Sikkeland T, Walton J R and Seaborg G T 1958 Phys. Rev. Lett. <u>1</u> 18
Ghiorso A, Thompson S G, Higgins G H, Seaborg G T, Studier M H, Fields P R, Fried S M, Diamond H, Mech J F, Pyle G I, Huizenga J R, Hirsch A, Browne C I, Smith H L and Spence R W 1955a Phys. Rev. <u>99</u> 1048
Hahn O, Meitner L and Strassmann F 1936 Ber <u>69</u> 905
Hahn O and Strassmann F 1939 Naturwiss. <u>27</u> 11
Harvey B G, Herrmann G, Hoff R W, Hoffman D C, Hyde E K, Katz J J, Keller Jr. O L, Lefort M and Seaborg G T 1976 Science <u>193</u> 1271
Keller Jr. O L and Seaborg G T 1977 Ann. Rev. Nucl. Sci. <u>27</u> 139
Kennedy J W, Seaborg G T, Segrè E and Wahl A C 1946 Phys. Rev. <u>70</u> 555
Münzenberg G, Hofmann S, Hessberger F P, Reisdorf W, Schmidt K H, Schneider J H R, Armbruster P, Sahm C C and Thuma B 1981 Z Phys. <u>A300</u> 107
Seaborg G T 1945 Chem. Eng. News <u>23</u> 2190
Seaborg G T, James R A and Ghiorso A 1944 Met Lab Report CS-2135 Aug 1944 p 15; this paper is reproduced in Benchmark Book (1978) Transuranium Elements: Products of Modern Alchemy ed. Seaborg G T Dowden, Hutchinson and Ross
Seaborg G T, James R A, Morgan L O and Ghiorso A 1945 Met Lab Report CS-2741 Feb 1945 p 3; this paper is reproduced in Benchmark Book (1978) Transuranium Elements: Products of Modern Alchemy ed. Seaborg G T Dowden, Hutchinson and Ross
Seaborg G T, Loveland W and Morrissey D J 1979 Science <u>203</u> 711
Seaborg G T, McMillan E M, Kennedy J W and Wahl A C 1946 Phys. Rev. <u>69</u> 366; Seaborg G T, Wahl A C and Kennedy J W 1946 Phys. Rev. <u>69</u> 367
Silva R, Harris J, Nurmia M, Eskola K and Ghiorso A 1970 Inorg. Nucl. Chem Lett. <u>6</u> 871
Thompson S G, Ghiorso A and Seaborg G T 1950a Phys. Rev. <u>77</u> 838
Thompson S G, Street Jr. K, Ghiorso A and Seaborg G T 1950b Phys. Rev. <u>78</u> 298.

Inst. Phys. Conf. Ser. No. 64: Section 2
Paper presented at Conf. on Neutron and its Applications, Cambridge, 1982

Heavy ion fusion and exotic nuclei studies at SHIP

P. Armbruster
GSI, Darmstadt, FRG

Talking about the latest developments in element production profits from reflection on the roots of this very old branch of science. The problem of element production by transmutation has been defined early by the alchemists. We know, they did not solve the task to transmute any elements into gold. We solved their problem, but forgot their primary human concern, the transmutation of the transmuter. Alchemists believed the insight they gained by successful transmutations will help them to look at the state of the world with a mind which has been changed and enlighted. I wished this to be true. We need enlightment to bring some ghosts we released back into the bottle.

There are many stations from the finding of natural radio-activity by Becquerel (1896), via the finding of the neutron by Chadwick (1932), to the finding of fission isomers by Polikanov (1962), which lead to the speculation of superheavy elements (SHE), until we reach the problems I want to talk of. Some of them are presented not as an ever growing tree, but as an ever turning wheel, Fig. 1. I will talk on recent findings at the last stations: Dynamics of nuclear matter, limits of stability and transmutations. They have been obtained from experiments I had the privilege to take part in, which have been performed at GSI, Darmstadt and at ILL, Grenoble in the years 1978-82 by a collaboration of many physicists. Their names are given in the various references which follow.

Dynamics of Nuclear Matter

The stability of SHE is due to the special stability of spherical closed shell nuclei, a property of the ground state of these nuclei. The production of SHE by nuclear reactions involves large rearrangements of many nucleons, which is a question to nuclear dynamics, a question of the relative strength of nuclear and Coulomb forces. Experimentally we ask, which nuclei will fuse in order to obtain surviving evaporation

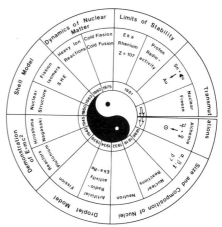

Fig. 1: The wheel of nuclear physics.

0305–2346/83/0064–0105 $02.25 © 1983 The Institute of Physics residues (EVR). Fig. 2 demonstrates a compilation of reactions studied at GSI. We see a well defined border line between successful reactions and failures. The border line is found at a value of $x_{eff} = 0.80$. x_{eff} is defined as the ratio of attractive nuclear forces and repulsive Coulomb forces in a configuration of two touching spheres (Bass 1974, Swiatecki 1931).

Recently in the problem of nuclear dynamics within the frame of validity of the liquid drop model (LDM) a major break through has been achieved. The mathematical problem of energy transfer from a moving wall of the nuclear system and the changing window between the two converging systems to the intrinsic degrees of freedom, that is the coupling from the radial motion via the changing shape to the nucleon system has been solved (Swiatecki 1981, Bjørnholm 1982, Feldmaier 1982). This "New Dynamics" holds for liquid drops characterized by their shape and density, a nuclear surface constant, a Coulomb force constant, and a mean Fermi velocity of the nucleons. The model predicts an increase of the fusion barrier due to the energy losses during the mounting of the barrier, which sets in at a critical value of x_{eff} and depends quadratically on x_{eff}. Fig. 3 shows the measured increase of the fusion barrier for a number of systems. A quadratic dependence as proposed by Swiatecki (1981) is shown for comparison. The increase sets in for $x_{eff} = 0.70$ and becomes marginal for $x_{eff} \sim 0.80$. The "New Dynamics" prohibits fusion above $Z = 120$ for any combination of projectiles and targets. Beams up to Ti ($Z = 22$) may be used to produce SHE from the dynamics' point of view. But the stability of SHE is a nuclear structure effect. These are no ingredient of the "New Dynamics" and will come into play only at small excitation energies of the nuclear system. How is the dynamics modified by nuclear structure effects? Here experiments have to convey a message.

Nuclear Dynamics and Nuclear Structure. Cold Fission and Fusion

Are rearrangements of large nuclear systems, as we find in fission and fusion, at all feasible at low excitation energy and how small may the excitation energy be at all? We investigated this question and found that there are reaction channels much colder than has been known previously. Figure 4 shows the intrinsic excitation of the two fission fragments in ^{233}U (n_{th},f) at a yield level of 10^{-5} (Armbruster 1981a). It is as low as 3.8 MeV. The two measured fragments have less excitation energy than the energy gain from capture of the fission inducing neutron. Fig. 5 shows excitation functions for heavy ion fusion (Schmidt 1981, Münzenberg 1981a). 1n-deexcitation channels are observed. EVR's are still seen at excitation energies below 20 MeV.

Moreover, fusion even far below the barrier is seen. Fig. 6 presents the fusion cross section measured from the barrier down to the μb-regime, that is about 10^{-5} of the fusion cross section above the barrier. As was shown by W. Reisdorf et al. (1982) two different nuclear structure effects are of

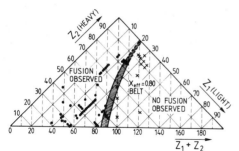

Fig. 2: How element ($Z_1 + Z_2$) may be fused from elements Z_1 and Z_2. Beyond the belt x_{eff} = 0.80 no fusion has been observed.
● EVR's detected at GSI,
× failures in different laboratories.

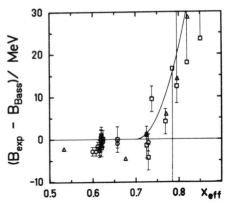

Fig. 3: The extra-push energy according to Bjørnholm et al. 1982 (solid line) and barrier measurements in different GSI-experiments (Sahm 1982).

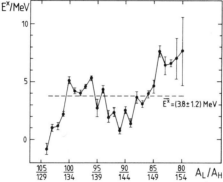

Fig. 4: Excitation energy of ^{234}U-fission fragments (Armbruster et al. (1981a) at a mass yield 10^{-5}. Odd masses are colder than even masses.

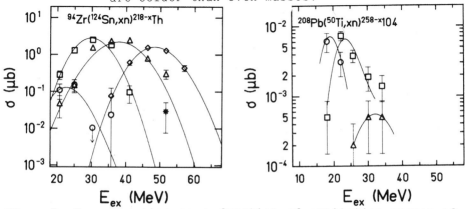

Fig. 5: Cross section as a function of excitation energy of the fused system. Production of Th- (Schmidt et al. 1981) and Element 104 (Münzenberg et al. 1981) isotopes by xn-channels.
○ 1n, □ 2n, △ 3n, ◇ 4n-5n, * 6n-7n

importance. Static deformations of nuclei allow fusion in collisions tip-to-tip even below the barrier as defined for spherical nuclei and a suitable nuclear potential. For fusion reactions involving ^{154}Sm fusion has been observed up to 13 MeV below the barrier. On the other hand zero point vibrations of nuclei allow fusion below the barrier as the barrier transition time of the two nuclei may be smaller than the time constants of the zero point motion. Again prolate deformations, but in this case transient ones, help to fuse below the barrier. It was shown that the shift of fusion to smaller energies correlates with the deformabilities of nuclei. Shifts of 7 MeV have been observed for the (Ar+Sn)-system. Fusion at energies as low as 15 MeV below the barrier seems still to be possible for heavy systems, an encouraging result for the production of systems the stability of which mainly depends on nuclear structure effects.

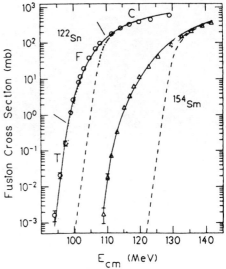

Fig. 6: Subbarrier fusion of the systems ^{40}Ar+^{122}Sn and ^{40}Ar+^{154}Sm (Reisdorf et al. 1982). Comparison of measurements (symbols) with calculations using the same nuclear potential but taking into account barrier fluctuation and static deformation (solid lines) and neglecting the latter (dashed line).

Shell correction energies may stabilize spherical as well as deformed systems. The experiemnts show that shell stabilized deformed systems are more stable to intrinsic excitation than spherical systems. Fig. 7 shows mass distributions of ^{234}U-fission (Schmidt 1981, Croall and Cuninghame 1969). We know since long that the shell corrections leading to asymmetric fission give in to liquid drop fission at an excitation energy of about 40 MeV. The nascent fission fragments in thermal fission are highly deformed. The shell stabilization of spherical systems, e.g. of N = 82-nuclei is of no importance. Recent investigations at ILL using the recoil spectrometer "LOHENGRIN" (Moll et al. 1975) or a mass separating device using detectors (Signarbieux

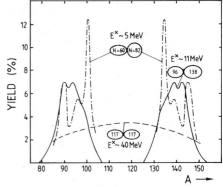

Fig. 7: Mass distributions of ^{234}U-fission at different intrinsic energies of the fission fragments. In cold fragmentations spherical N = 82 fragments are dominant.

1981) showed that the spherical nuclei at N = 82 dominate the fission yield at low excitation energies. Configurations of a deformable nucleus as ^{100}Zr and the spherical nucleus ^{134}Te are the only ones which have the compactness necessary for a breaking, which allows to transfer nearly all the Q-value into kinetic energy of the two fragments. The predominance of the spherical configuration seen for excitation energies at 5 MeV is lost rapidly, if the intrinsic excitation energy increases. Already at about 10 MeV excitation energy the dominance of the shell stabilized spherical nuclei is lost, whereas the shell stabilization of the deformed fragments leading to asymmetric fission is seen up to 40 MeV. Not only cold fission experiments of ^{233}U, ^{235}U, and ^{238}Pu support our hypothesis. Recently the spontaneous fission of ^{258}Fm was investigated (Hoffman 1979). The symmetric breaking into two spherical nuclei with small intrinsic excitation energy has been observed. Already in thermal fission of ^{257}Fm, that is at E* = 6 MeV, the predominance of the shell stabilized spherical Z = 50 nuclei is lost. Again the unexpected sensitivity of shell stabilized spherical nuclei to excitation energy is demonstrated.

Fusion studies on the deformed actinide nuclei (N = 152) and spherical nuclei (N = 126) show the same behaviour. The competition between fission and nucleon emission determines the survival probability of fused systems. The difference (B_n-B_f) for the deformed No-isotopes and the spherical Th-isotopes is about the same, Fig. 8. An evaporation calculation involving the shell corrections reproduces within an order of magnitude the reduced 5n-cross sections (E*- = 45) MeV (Keller 1981). However, it falls by orders of magnitude to reproduce the reduced 4n-cross sections (43 MeV) for Th-isotopes (Vermeulen, 1982). Different entrance channels show the same discrepancies. The exit channel determines the difference between calculated and measured cross sections. The measured cross sections for Th-isotopes are small, as if shell stabilization at E* = 43 MeV would be nonexisting. Even the 1n-cross section observed in (Zr-Sn) reactions at E* = 24 MeV is overestimated by a large factor (Sahm et al. 1982).

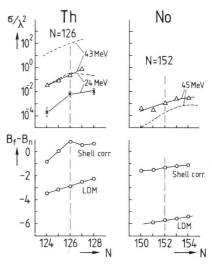

Fig. 8: (B_f-B_n) and reduced cross sections σ/λ^2 for isotopes near the spherical shell (N = 126) and the deformed shell (N = 152). The measurements are compared to evaporation calculations. For N = 126 the measurements cannot be reproduced, unless collective enhancement is introduced (Bjørnholm et al. 1973).

The disappearance of shell effects cannot be explained by their disappearance in the level density formulae entering the evaporation codes. The 1/e-damping energy of shell effects in the level densities ($E_{1/e}$ = 17 MeV), as introduced by Ignatyuk (1975), was reproduced recently by a microscopic level density calculation by K.-H. Schmidt et al. (1982) for spherical Th-isotopes. The reason for the rapid damping of shell effects in spherical nuclei has to be sought in the possibility of coexistence of different shapes in a given nuclear system, as speculated by K.-H. Schmidt already in 1979. Bjørnholm et al. (1974) pointed to the enhancement of level densities for axially and nonaxially deformed systems. This enhancement may lead to a larger level density for deformed shapes of Th-isotopes, which will not be shell-stabilized. Compared to the damping of shell effects in spherical shapes collective enhancement thus may lead to a damping of shell effects in Th-isotopes which is as strong as the value reproducing the experiments ($E_{1/e} \sim$ 8 MeV). Fig. 9 summarizes the experimental findings. Shell stabilized spherical systems have to be synthesized as cold as possible. Unless E* < 15 MeV, the shell effect is of minor importance in the production cross section.

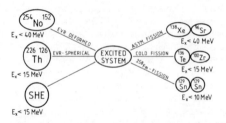

Fig. 9: The sensitivity of spherical shells to excitation energy observed in fission and fusion.

SHE-Production

The shell stabilized spherical SHE have to be produced with low excitation energy, otherwise their stability will be lost. The excitation energy at the barrier in possible fusion reactions, as indicated in Fig. 10, is about 30 MeV. At these energies the fused system may not survive, as collective enhancement in the level densities will reduce the chance of an EVR to survive. There is a good chance to produce SHE with energies below the barrier, as e.g. actinide target nuclei have large static deformations, which enhances subbarrier fusion. The feasibility of SHE production is a delicate game. It depends on the size of the ground state

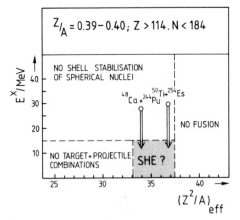

Fig. 10: The traps of SHE-production: no fusion, no isotopes, no shell stabilization. Subbarrier fusion of the given systems is the most promising reaction for SHE-production.

shell correction energy, the size of the damping energy of spherical shell effects due to collective enhancement, and the possibility of cooling the system by subbarrier fusion. Reviewing production reactions for SHE, which have been published (Herrmann 1981) (x in Fig. 2), we see that none having a good sensitivity fulfilled the condition of low excitation energy. Fusion experiments in the range of excitation energies between (15 - 25) MeV using ^{48}Ca-beams and the heaviest actinides targets will be performed still this year by a LBL-GSI collaboration. ^{248}Cm targets thus finally will become available for SHIP-experiments.

SHIP - By New Methods to New Frontiers

All experimental results I reported and still shall report have become possible by a break-through in experimental methods. Recoil spectrometers with μsec-separation times, independence of the efficiency from the chemistry of the separated species combined with high total efficiencies ($\sim$ 20 %) allow to perform experiments of high selectivity for a wanted species (10^{-6} yields) (Armbruster 1976). Various detector systems which identify the separated reaction products unambiguously have been developed. I shall not discuss further the vehicles which carried us to the new boundaries, but we should be aware that thanks to them I can give this report. It is a very short section put in the center of my contribution underlying its central importance. I had a dream - to detect and identify a new nuclear species by one event, a method known from subnuclear physics. It is symbolized by a bouquet of flowers, Fig. 11 (Armbruster 1977), each of which presenting a measurable parameter characterizing the new species or its remnants of decay. The whole bouquet presents all we can learn about the new nucleus, how it was created, how it lived, and how it died.

Fig. 11: The bouquet of parameters to characterize single nuclei of heavy elements. A dream - new techniques fulfill.

The velocity filter SHIP (Separator for Heavy Ion Reaction Products) used at GSI has been described by Münzenberg et al. (1979 and 1981b).

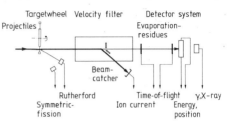

Fig. 12: Experimental set-up to detect EVR-cross sections below 10^{-33} cm^2.

The detection system making use of implantation into Silicon (Schmidt et al. 1979b) registers position, energy, the time of each count occuring in the surface barrier detectors (Hoffman 1979), Fig. 12 presents a scetch of principle of the main components of our system.

Production of Exotic Nuclei at GSI

In the period since 1976 about 90 new isotopes have been produced at GSI using either the on-line separator (Bruske 1981) or the velocity filter SHIP. Fig. 13 shows our hunting grounds. The lines $\Gamma_n = \Gamma_p$ and $\Gamma_n = \Gamma_f$ separate the chart of nuclides into three domains. In the two regions, predominant p-emission or fission as main deexcitation channels, most of our studies have been performed. I will restrict myself to a short and to the subject inadequate description of recent discoveries of heavy elements and of proton radioactivity.

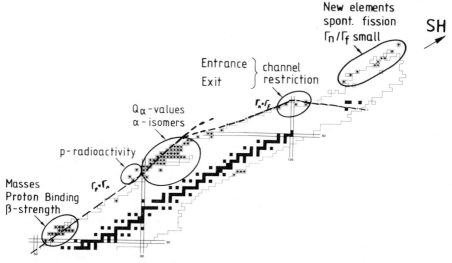

Fig. 13: Hunting grounds of GSI far off stability. The lines $\Gamma_p = \Gamma_n$ and $\Gamma_f = \Gamma_n$ according to Myers (1977). ■ nuclei found at GSI.

Heavy Element Production

In reactions between stable isotopes as targets and projectiles isotopes of nearly all elements beyond Z = 100 have been produced. Fusion reactions with Pb- and Bi-targets have been introduced into the field of heavy element production by Y. Oganessian et al. (1978). We showed that isotopes of most elements can be produced by the cold fusion reactions via 1n- and 2n-channels with excitation energies between (15 - 25) MeV. One example for element 104 is shown in Fig. 5. Especially for isotopes with very small fission barriers the cold way is superior to the production of heavy elements via actinides targets, a method formerly used at LBL to produce the elements above Z = 100 (Ghiorso 1982). The excitation energy of the compound systems in this case reaches values of about 40 MeV. One third of all isotopes known for elements beyond Z = 100 have been seen in the few experiments, when we applied the cold way in bombardments of Pb- and Bi-targets. This demonstrates the power of the new production method and our newly developed experimental techniques (Münzenberg et al. 1982). Fig. 14 shows one decay chain out of 7 detected events in the reaction ^{54}Cr(^{209}Bi,1n)^{262}107 at 4.85 Me/u (Münzenberg et al. 1981c). The implanted EVR ^{262}107, which has a fission barrier of ~ 2 MeV, but profits from the specialisation energy of odd-odd nuclei, decays by several α-decays finally to ^{246}Cf. The α-energies and correlation times have been registered at a well defined detector position following a preceding registered implantation of an EVR. An accidental correlation of an EVR and a following α-decay has a probability of less than 1:3000. Accidental correlations of several decays correspondingly are still less probable. The production cross section of the isotope ^{262}107 ($T_{1/2} = (4.7^{+2.6}_{-1.6})$ ms, $E_\alpha = 10.38$ MeV) is about 2×10^{-34} cm^2. The element 107 has been detected unambigously with a total dose of 1.2×10^{17} ^{54}Cr-ions. It is the real Eka-Rhenium physicists in the thirties were searching for.

During a 2-week irradiation of ^{209}Bi with 6×10^{17} ^{58}Fe-ions at energies (4.95, 0.05, and 5.15) Me/u we caught a very rare fish. There was only one in the net. The event and its preliminary, possible interpretation is presented in Fig. 15. We think we have produced one atom of ^{266}109, the heaviest isotope ever made. The production cross section would be 7×10^{-36}

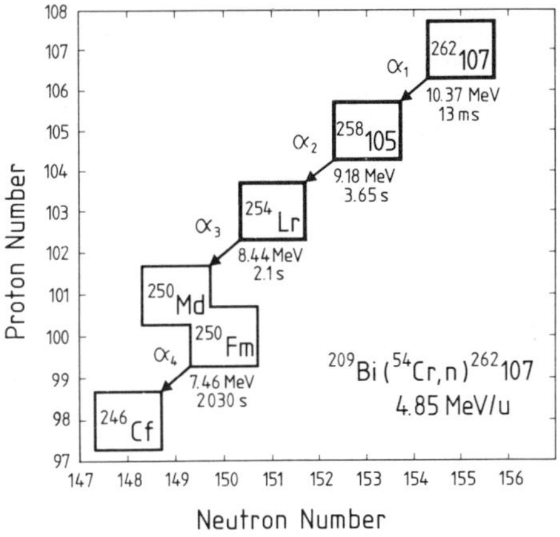

Fig. 14: A sequence of α-decays and corresponding energies proving the existence of ^{262}107 (Münzenberg et al. 1981b).

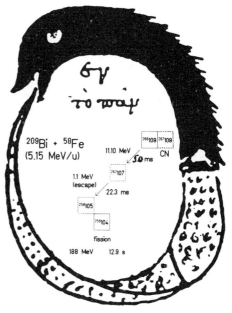

Fig. 15: Possible interpretation of an event observed at 4.10 pm, Aug. 29, 1982. There is only one event, to prove the existence of element 109. Snake: ancient alchemical symbol, εν το παν (One is All).

cm² the α-decay energy 11.10 MeV, the most probable value for the halflife ~4.0 ms. The dream of Fig. 11 has become reality. There was one atom and still we have got a glance of the main properties of the isotope. The atom has been found in a 3-week experiment, the longest experiment performed at GSI until now. ⁵⁸Fe-beams of 10¹² particles/sec, made out of 7 gr enriched material irradiated 500 µgr/cm² Bi-targets rotating at speeds of 40 km/hr continously during 2 weeks. For the single atom of element 109 the alchemist symbol from an ancient hand writing of the 4th century, shown in Fig. 14, saying ευ το παυ (One is All) seems to be an adequate motto.

Proton Radioactivity

In 1913 after α-rays have been shown to be Helium ions speculation about the existence of proton radioactivity went around in some brains of this city and active search for such an activity started. The production of protons from (α,p) reactions was discovered by E. Rutherford in 1919 in the first induced nuclear reaction. Since then it became clear that proton radioactivity will be found, many doubted if at all, for the lightest isotopes of odd elements. The higher the atomic number, the more energetic protons in a given range of halflives are to be expected. Elements above Z = 50 were searched for proton radioactivity systematically. For most of these elements fast α-decays compete seriously with p-radioactivity.

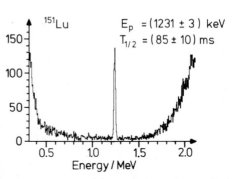

Fig. 16: Energy spectrum showing the 1.23 MeV proton radioactivity of ¹⁵¹Lu (S. Hofmann et al. 1982).

In the reaction ⁵⁸Ni (⁹⁶Ru,-p2n) ¹⁵¹Lu S. Hofmann et al. (1981 and 1982) found a 85 ms activity produced with a cross section of 70 µb, which was shown to be a 1.23 Me proton line. A second emitter was found later in the ⁵⁸Ni (⁹²Mo,p2n) ¹⁴⁷Tm reaction and identified by on-line-mass separation (Klepper et al. 1982), Fig. 16. The halflife and proton energy demand a 11/2⁻-state as parent state in ¹⁵¹Lu. Known N = 80 isotopes of odd elements for Z > 64 have a 11/2⁻-ground state. We assume the emitting state in ¹⁵¹Lu to be its ground state. The populated state in ¹⁵⁰Yb is the ground state, as no coincident γ-rays have been found.

Ground state to ground state decay by proton emission is a fourth mode of radioactive decay besides β-decay, α-decay, and fission. By its discovery we closed a gap in our experimental knowledge. p-radioactivity is the simplest radioactive decay, which seems to pose no problem to a theoretical treatment. Further emitters in the range 56 < Z < 72 should be detected.

Their production cross section may be much smaller and their half lives shorter than in the two cases found. Our search will be continued.

1. Transmutation: Copper and Tin

The transmutation of Cu and Sn into bronze was one of the few successful transmutations of early mankind. It was later used as an argument for further possible transmutations leading to a higher quality nature of the transmuted product. Accidentally the only combination of elements known to ancient alchemists leading to gold is the fusion of a copper (Z = 29) and a tin (Z = 50) nucleus into gold (Z = 79). The elixir to be used goes back to A. Einstein. It is an energy of 166 MeV in the center of mass system. Fig. 17 represents SHIP as a box of transmutation of Cu+Sn into Au. The beam of gold is detected by its characteristic L-x-rays induced in the virgin gold atoms when the beam passes through a thin Sn-foil, see spectrum Fig. 16 (Armbruster et al. 1981b). None of the nuclear properties of the short lived gold isotopes has been used for detection.

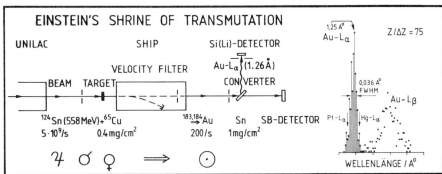

Fig. 17: The 2nd transmutation of the (Cu+Sn) system solving the gold making problem of alchemy. The x-ray spectrum of gold is shown at the left.

2. Transmutation: Nuclear Physicists

The sucessful transmutation of copper and tin into gold is but one of the many possible fusion reactions, it does not solve the budget problems of our governments, nor are we ready to believe that it has changed the minds of the transmuters. We have been made to believe it is one of the innocent games we are allowed to play in our ivory tower having no relation to the players and their problems in society. But the lesson we have to learn is simple. There is no physics in well shielded ivory towers. Whatever we do in our proper white world of fundamental nuclear research, there is participation in the complementary, the dirty black world of the application of nuclear physics to kill each other most effectively. On the other hand all those who work in the black have their part in the white. All nuclear phycisits together carry responsibility for what has been done with our findings, as the circle of Ying and Yang symbolizes in Fig. 1. If nowadays 3 tons of TNT of nuclear explosives are stockpiled per human being, 0.1 tons are sufficient to kill a person, I think it is our common duty to oppose this madness.

An appeal to support "nuclear freeze" from this historical ground to the governments of the world by this group of concerned physicists would not change the state of the world immediately. But it certainly will give some momentum to the spiral of armaments into the right direction.

Armbruster P 1976 Proc. Int. Conf. on Nuclei far from Stability, Cargèse, France (CERN 76-13) 3
Armbruster P 1977 Proc. Int. Conf. on Reactions of Heavy Ions and Synthesis of New Elements, Dubna, SU (GSI-M-2 78)
Armbruster P, Quade U, Rudolph K, Clerc H-G, Mutterer M, Pannicke J, Schmitt C, Theobald J P, Engelhardt W, Gönnenwein F and Schrader H 1981a Proc. Int. Conf. on Nuclei far from Stability, Helsingør, Denmark (CERN 81-09) 675
Armbruster P, Hofmann S and Münzenberg G 1981b (GSI-81-2) 11
Bass R 1974 Nucl. Phys. A231 45
Bjørnholm S, Bohr A and Mottelson E R 1973 Proc. 3rd IAEA Symp. on Physics and Chemistry of Fission, Rochester (IAEA Vienna 1974) Vol. I 367
Bjørnholm S and Swiatecki W 1982 Phys. Rev. to be published
Bruske C, Burkhard K-H, Hüller W, Kirchner R, Klepper O and Roeckl E 1981 Nucl. Instr. and Methods 186 61
Croall I F and Cuninghame J G 1969 Nucl. Phys. A125 402
Feldmaier H 1982 Hirschegg (ISSN-0720-8175)
Ghiorso A 1981 Proc. Int. Conf. on Actinides 81 Asilomar USA (Pergamon Press Ltd. 1982) 23
Herrmann G 1981 Proc. Int. Conf. on Nuclei far from Stability, Helsingør, Denmark (CERN 81-09) 772
Hoffman D C 1979 Proc. Int. Symp. on Physics and Chemistry of Fission Jülich (IAEA Vienna 1980) Vol. II 275
Hofmann S, Faust W, Münzenberg G, Reisdorf W, Armbruster P, Güttner K and Ewald H 1979 Z. Phys. A291 53
Hofmann S, Münzenberg G, Faust W, Heßberger F P, Reisdorf W, Schneider J R H, Armbruster P, Güttner K and Thuma B 1981b Proc. on the Int. Conf. on Nuclei far from Stability Helsingør, Denmark (CERN 81-09) 120
Hofmann S, Reisdorf W, Münzenberg G, Heßberger F P, Schneider J R H and Armbruster P 1982 Z. Phys. A305 111
Ignatyuk A, Smirenkin G N and Tishin A S 1975 Sov. J. Nucl. Phys. 21 255
Keller J 1981 Diplomarbeit (Inst. f. Kernphysik TH Darmstadt)
Klepper O, Batsch T, Hofmann S, Kirchner R, Kurcewicz W, Reisdorf W, Roeckl E, Schardt E and Nymann G Z. Phys. A305 125 1982
Moll E, Schrader H, Siegert G, Asghar M, Bocquet J P, Bailleul G, Gautheron J P, Greif J, Crawford G J, Chauvin C, Ewald H, Wollnik H, Armbruster P, Fiebig G, Lawin H and Sistemich K 1976 Nucl. Instr. and Methods 123 615
Münzenberg G, Faust W, Hofmann S, Armbruster P, Güttner K and Ewald H 1979 Nucl. Instr. and Methods 161 65

Münzenberg G, Hofmann S, Heßberger F P, Reisdorf W, Schmidt K-H, Faust W, Armbruster P, Güttner K, Thuma B, Vermeulen D and Sahm C C 1981 a Proc. on the Int. Conf of Nuclei far from Stability Helsingør, Denmark (CERN 81-09) 755
Münzenberg G, Faust W, Heßberger F P, Hofmann S, Reisdorf W, Schmidt K-H, Schneider W F W, Schött J H, Armbruster P, Güttner K. Thuma B, Ewald H and Vermeulen D 1981b Nucl. Instr. and Methods 186 423
Münzenberg G, Hofmann S, Heßberger F P, Reisdorf W, Schmidt K-H, Schneider J R H, Armbruster P, Sahm C C and Thuma B 1981c Z. Phys. A300 107
Münzenberg G, Armbruster P, Faust W, Güttner K, Heßberger F P, Hofmann S, Reisdorf W, Sahm C C, Schmidt K-H, Schött H J, Thuma B and Vermeulen D 1981 Proc. Int. Conf. on Actinides 81, Asilomar, USA (Pergamon Press Ltd. 1982) 223
Myers W 1977 Droplet Model of Atomic Nuclei (IFI Plenum)
Oganessian Yu TS, Bruchertseifer H, Buklanov G , Chepigin I, Choi al Sek, Eichler B, Gavrilov K A, Gäggeler H, Korotkin YU S, Orlova O A, Reetz T, Seidel W, Ter-Akopian G M, Tretyakova S P and Zvara I 1978 Nucl. Phys. A294 213
Reisdorf W, Heßberger F P, Hildenbrand K D, Hofmann S, Kratz J , Münzenberg G, Schlitt K, Schmidt K-H, Schneider J H R, Schneider W F W, Sümmerer K and Wirth G submitted to Phys. Rev. Lett.
Sahm C C, Clerc H G, Vermeulen D, Schmidt K-H, Armbruster P, Heßberger F P, Keller J, Münzenberg G and Reisdorf W 1982 Hirschegg (ISSN-0720-8715)
Sahm C C 1982 private communication
Schmidt K-H, Armbruster P, Heßberger F P, Münzenberg G, Reisdorf W, Sahm C C, ermeulen D, Clerc H G, Keller J and Schulte H 1981 Z. Phys. A301 21
Schmidt K-H, Delagrange H, Dufour J P and Fleury 1982 to be published in Z. Phys.
Schmidt K-H, Faust W, Münzenberg G, Reisdorf W, Clerc H G, Vermeulen D and Lang W 1979a Proc. Int. Symp on Physics and Chemistry of Fission Jülich (IAEA Vienna 1980) Vol. I 409
Schmidt K-H, Faust W, Münzenberg, Clerc H G, Lang W, Pielenz K, Vermeulen D, Wohlfarth H, Ewald H and Güttner K 1979b Nucl. Phys. A318 253
Schmitt C H 1981 Diplomarbeit (TH Darmstadt)
Quade U 1982 Dissertation (LMU München)
Signarbieux C, Montoya M, Ribrag M, Mazur C, Guet C, Perrin P and Maurel M 1981 Journ. Physique Lettres 42 L-437
Swiatecki W 1981 Phys. Scripta 24 113
Vermeulen D 1982 private communication

Neutron-induced fission and the structure of the fission barrier

J E Lynn

A.E.R.E. Harwell, Didcot, Oxon OX11 ORA

1. Introduction

Fission is the most dramatic of exothermic nuclear reactions. It was discovered in the early period of slow-neutron research, and, because the products of fission include a few neutrons, the technological importance of neutron-induced fission was soon realised; its experimental study has become highly significant on that account alone. But neutron-induced fission is also one of our major tools in the scientific understanding of the nuclear fission process itself. In this paper I summarise the current state of knowledge of fission as revealed mainly by neutron reactions.

The report is divided into two sections. In the first part, the properties of the fission barrier near its saddle point are described. This is the region of the fission barrier that governs cross-sections and fission product angular distributions. In the second section, some of the evidence from mass-yield and kinetic energy distributions of fission products for the nature of the far slopes of the fission barrier between the saddle point and the scission point is discussed.

2. The fission barrier

The current, generally accepted, theory of the fission barrier is that due to Strutinsky (1967), which incorporates shell effects into a liquid drop model of the minimum total energy surface in the deformation space of the nucleus. The importance of the shell effects is that they are of major significance for deformed as well as for spherical nuclei. Thus, for the actinide nuclei in general, the saddle point in the liquid drop energy surface, which is the feature that defines the liquid drop fission barrier, occurs roughly at a deformation corresponding to that of a spheroid with major to minor axis ratio of 2:1; this is also the deformation at which a major shell binding effect occurs in the actinides (for neutron number ~ 144).

The consequence of this juxtaposition of liquid drop maximum and shell minimum is a splitting of the effective fission barrier into two peaks. For the lower-charge actinides ($Z \sim 90$), in which the liquid drop saddle point shape is an extended one, the inner peak of this barrier tends to be considerably lower than the outer. As charge increases, the liquid drop saddle shape shortens towards spherical and the positive shell correction on the low deformation side of the shell minimum tends to reinforce the shrinking liquid drop barrier, while the positive shell term at high deformation is added to a rapidly falling liquid drop term. The net result is that the inner peaks tend to increase from Th and then remain high and relatively constant over the remaining range of the actinides (Larsson and Leander 1974), while the outer peaks, starting higher than the inner ones at thorium fall rapidly from about americium onwards (Möller and Nix 1974). The secondary well contained between the peaks has a depth of 3 to 4 MeV, its base lying some 2 to 3 MeV above the ground state for the lower charge nuclides (Th to Pu) but falling below 2 MeV for Cm upwards.

In detailed calculations using the Strutinsky theory a large region of deformation space has been surveyed including mass asymmetric and axially asymmetric shapes of the elongated nucleus. It has thus been found that the saddle point for the inner peak of the fission barrier is at an axially asymmetric deformation, while that for the outer-peak is mass asymmetric. The secondary well is both axially and mass symmetric.

The consequences of the double-humped fission barrier are striking for all forms of fission proceeding from a compound nucleus with excitation energy near or below the barrier, but appear particularly spectacular in observations of neutron-induced fission, because of the high energy resolution that is often available.

The secondary well in the barrier has the effect of introducing quasi-stationary states into the fission route for decay of the compound nucleus. The nature of these states can be described by writing the Hamiltonian for the compound nucleus in a schematic form that separates the degree of freedom for deformation towards fission from a component that depends only on the remaining (intrinsic) degrees of freedom (at a fixed deformation), including collective motion of other kinds. Thus

$$H = H_{def} + H_{int} + H_{coup} \qquad (1)$$

where H_{coup} is a coupling term that allows for the changes that occur in H_{int} if the deformation variable is changed. Within a potential energy function that contains a secondary well in addition to the primary well centred about normal nuclear deformation, the vibrational type eigen-solutions of H_{def} with eigenvalues lower than the potential maxima separate into two classes, the wave-functions $\Phi^{(1)}$ of class-I solutions having major component of amplitude in the primary well and those of the class-II wavefunctions $\Phi^{(11)}$ having major component in the secondary well. Clearly the class-II vibrational states carry the significant widths for the fission process, and if in neutron absorption the compound nucleus is excited close to a class-II vibrational eigenvalue a major enhancement in the fission cross-section will occur. The height and width of such vibrational resonances yield valuable information on the relative penetrabilities and heights of the peaks in the double-humped barrier.

The coupling term may cause mixing of such simple vibrational states into more complex states of motion associated with the secondary well. These wave-functions of the class-II compound states may be written

$$X_{\lambda_{11}} = \sum_{\nu_{11}\mu} C_{\lambda_{11}(\nu_{11}\mu)} \Phi^{(11)}_{\nu_{11}\mu} X_\mu \qquad (2)$$

the X_μ being the solutions of the intrinsic term in the Hamilton H_{int}; an analogous set of class-I compound states may be defined.

The final compound nucleus states X_λ result from a coupling of the class-II and class-I compound states, but this coupling will be very weak at excitation energies below the inner peak of the barrier. Consequently the spreading of a class-II compound state, which carries an appreciable fission width from its class-II vibrational admixture, into the very much denser final compound states will be very limited, and narrow intermediate structure, the occurrence of considerably separated groups each containing only a few resonances, will be found in slow neutron fission cross-sections. The coupling strength, total fission width and spacing of these groups, each of which corresponds to a class-II compound state, give information on barrier peaks and secondary well depth of the fission barrier.

We thus see that a generalised coupling scheme for fission to the compound nucleus motion will be class-II vibrational state → class-II compound states (moderate density) → class-I compound states (high density). If the spreading of the vibrational state into the class-II compound states is incomplete a damped vibrational resonance, showing sub-structure (of narrow intermediate type), will occur in the cross-section. Alternatively if the coupling between the vibrational state and other class-II states is excessively weak because of low density of the latter (corresponding to low excitation energy relative to the secondary well), the vibrational state may couple directly to the class-I compound states, giving rise to the pure vibrational resonance in the fission cross-section as discussed above.

At excitation energies above the fission barrier the same kind of coupling scheme will operate but the degrees of admixture and spreading will be much greater, so that structure will no longer be apparent in the cross-sections. Nevertheless the secondary well still influences the magnitude of the cross-section by introducing a relatively long-lived intermediate state in the route to fission. This factor must be taken into account in analysing above-barrier fission cross-sections to deduce fission barrier parameters, while correct averaging over intermediate structure must be performed in analysing sub-barrier trends of fission cross-sections for the same purpose (Bjørnholm and Lynn 1980).

In general the barrier peak parameters obtained from these various data substantiate the conclusions from theory to within 1 MeV or so (Bjørnholm and Lynn 1980). But a disturbing discrepancy occurs for the low charge actinides ($Z \approx 90$); the existence of apparently pure, strong vibrational resonances in the cross-sections of these nuclides implies that the inner peaks are of about the same height as the outer peaks and the secondary wells are shallow. Some of the more refined theoretical calculations indicate, however, that in thorium nuclides the outer peak may be further split by a shallow (tertiary) well (Müller and Nix 1974). It has been hypothesized therefore that the gross resonances observed in the Th cross-sections are due to a vibrational state in this third well.

Attempted verification of the triple-humped barrier hypothesis has been centred around studies of neutron-induced fission of the thorium and protoactinium isotopes. The most intensively studied case has been the 710 keV resonance in the neutron fission cross-section of ^{230}Th. On the assumption that the wave-function of the virtual state responsible for this resonance has the simple from $\Phi_\nu \chi_\mu$, with the intrinsic state being a rotational state built on a band-head consisting of a single-particle excitation with a well-defined quantum number K for the projection of its total spin on the cylindrical symmetry axis of the nucleus, it is to be expected that the observed peak in the cross-section is an overlap of individual resonances, each with unique spin J, corresponding to different members of the rotational band. Information on the location of these rotational members can be obtained from the angular distribution of the fission products (Bohr, 1956, Wilets and Chase, 1956).

High resolution measurements on the neutron fission cross-section(Blons et al 1978) have indicated the multi-component nature of the peak(see Fig.1). A full analysis of these data together with angular distribution data from other sources indicates that a fit by a rotational band of unique parity is unlikely (Boldeman et al 1980).

On the other hand the tertiary well deformation, being mass asymmetric, would require the co-existence of a nearly degenerate band of opposite parity. Boldeman et al's analysis of this possibility yields the fit to the cross-section data shown in Fig.1. This fit was guided by the angular distribution data, but owing to the many discrepancies amongst the published data this could not be regarded as definitive. Very recent and careful experimental data (Grainger et al,to be published)are shown in Fig.2 together with the function deduced from Boldeman et al's analysis. The fit is by no means perfect and indicates that reduction of high spin components in the region of 730 keV is required. Scattered values of the angular anisotropy well above the main trend also suggest that the interpretation of a pure vibrational level in a tertiary well is not completely adequate. It is to be noted however that the same barrier penetrability factors are

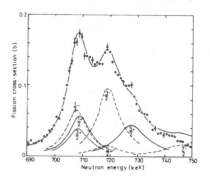

Fig.1 Neutron cross-section measurement of the neutron fission cross-section of ^{230}Th by Blons et al(1980). The fall curve is the fit by Boldeman et al (1980) based on a $K = {}^{1}/_{2}^{+}$ neutron orbital and rotational band with $\hbar^2/2I = 2$ keV (I is the moment of inertia), decoupling parameter $a = -1.3$ and a vibrational level in a tertiary well with reflection asymmetry. The individual rotational resonance components labelled by spin J and parity are also shown.

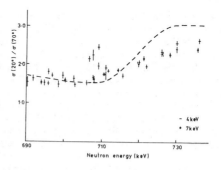

Fig.2 Angular anisotropy of fission fragment emission in neutron-induced fission of ^{230}Th (Grainger, James and Syme, unpublished) for two values of neutron energy resolution. The dashed curve corresponds to the fit of Boldeman et al shown in Fig.1 with resolution broadening of about 7 keV included.

used for both parity bands. Theoretical calculations (Myers,1977) of the
potential energy surface indicate that the potential hill separating the
reflected mass asymmetric shapes is very much lower at the barriers than
at the well; hence the opposite parity band will not be brought down to
near-degeneracy at the barriers, and this will considerably reduce the
widths and strengths of the opposite parities resonances.

The case of neutron-induced fission of ^{232}Th is more complex. Between
neutron energies of about 1.3 and 2 MeV several gross resonance structures
have long been known in the fission cross-section. The angular distribution of fission products at these resonance energies tend to be sideways-
peaked to varying degrees. A recent analysis by (Auchampaugh et al 1981)
suggests quite complicated mixing of cross-section components associated
with $K = {}^{1}/_{2}$, $K = {}^{3}/_{2}$, and $K \geqslant {}^{5}/_{2}$ quantum numbers.

To further substantiate the hypothesis of the tertiary well it is desirable
to consider spectroscopic evidence of this kind, and to consider which
Nilsson single-neutron orbitals at the tertiary well deformation may be
associated with significant fission strength. The coupling between the
neutron orbital and vibrational motion is important here. In eq.(1) the
coupling term must include the change of neutron orbit energy with changing deformation due to the vibration, as well as the residual interaction
of orbitals of the same spin projection and parity. We have done schematic calculations of this kind in a truncated harmonic oscillator well and
determined measures of coupling and fission widths, and hence the fission
strength, from the coefficients and vibrational wave-function amplitude in
an expansion of eq.(2), (the intrinsic wave-function being limited to
single neutron orbitals). Some results for the secondary and tertiary
wells of ^{231}Th and ^{232}Th indicate that the tertiary well model (but not
the secondary well) can reproduce both the simplicity of the ^{231}Th case
and the complexity of ^{233}Th.

Further elucidation of the problem will almost certainly depend on discov-
ering phenomena associated with the secondary well of these nuclides. If
the quantitative theory is correct, narrow and relatively closely-spaced
intermediate structure should be found at excitation energies well below
those of the vibrational resonances i.e. at energies around and below the
inner peak bounding the secondary well. Fission widths of such narrow
intermediate resonances will be very small, but the structure may be
detectable by anomalies in their gamma-ray spectra, those of class-II
states terminating at the intrinsic excitation energy available within the
secondary well.

The latest high resolution data on the ^{232}Th fission cross-section
(Auchampaugh et al,to be published)may contain evidence for the effects of
the secondary well. The cross-section has now been measured to neutron
energies below 100 keV (where its magnitude is of the order of a few µb).
In this low energy range it still shows gross structure but of a magnitude
(too low) and width (too great) to be classed as pure vibrational resonances of the kind that seem apparent at excitation energies more than
1 MeV higher. These could therefore be damped vibrational resonances
associated with the secondary well, and the fluctuations in cross-section
that are apparent about the gross-resonance profile could be due to
class-II compound states.

3. <u>Saddle to scission</u>

In dealing with fission phenomena associated closely with the barrier peak, we have discussed the potential energy surface almost exclusively, the dynamical quantities being contained implicitly in widths and barrier tunnelling parameters. On the slope from saddle to scission the potential energy surface cannot play such a completely dominant role, and we must expect that the dynamical quantities cause major disturbances. Nevertheless it can be supposed that major qualitative features in the fission product mass-yield and kinetic energy distributions are governed by gross variations in the potential energy surface.

Applications of the Strutinsky theory to the region beyond the saddle point indicates that the double-humped (asymmetric), mass-yield of actinide fission is reflected in mass-asymmetric minima in the cross-section across the potential energy curves (Mustafa et al, 1973) while the symmetric mass division of much lighter nuclides like polonium appears to follow from a mass-symmetric potential energy minimum. The major shell effect dominating the mass-yields in fission of the actinides appears to be the spherical shell closure associated with $Z = 50$, $N = 82$ in the incipient fragments; the heavy mass peak in the asymmetric yield from slow neutron-induced fission of the uranium and plutonium isotopes is peaked and centred at $A = 132$, and as the mass of the fissioning nucleus increases the mass centre of the light peak increases correspondingly. This effect is accentuated when the fission fragments have high kinetic energy, but in slow-neutron induced fission of ^{229}Th (Asghar et al, 1982) the principal heavy mass peak jumps to $A = 144$, and the light peak is at $A = 86$, indicating a (spherical) $N = 50$ shell and a (highly deformed) $N = 88$ shell in the incipient fragments.

Symmetric mass yields as well as asymmetric division are found for the lowest charged actinides. Symmetric mass division with some features of asymmetric division are also found for the very heaviest nuclides (Cf and Fm). But the characteristics of those two regions of coincident symmetric and asymmetric mass divisions are otherwise very different. In the Cf and Fm nuclides the average total kinetic energy of the fission fragments is a maximum for symmetric fission (Hoffman et al, 1980) and indeed makes a very sharp increase from about 200 to 240 MeV at ^{258}Fm and ^{259}Fm (Hulet et al, 1980) as the doubly-magic ^{132}Sn symmetric shell effect in the potential energy surface becomes dominant. In the lighter nuclides (^{232}Th and below) symmetric mass division is associated with lower mean kinetic energy, and no shell effect in the potential energy surface appears responsible for this. In fact, the clear existence of a triple-peak in the mass-yield curve of the fission of these nuclides seems to indicate that their symmetric division is a quite separate mode of fission with its own separate saddle-point. This point of view is reinforced by the fission cross-sections of actinium isotopes (Konecny et al, 1973) which show quite different behaviour as a function of excitation energy for symmetric and asymmetric fission, indicating completely different barrier heights and penetrabilities for the two modes.

References

Asghar M et al (1982) Nucl.Phys.A373,225
Auchampaugh G et al (1981) Phys.Rev.C24,503
Boldeman J W et al (1980) Phys.Rev.C22,627
Hoffman D C et al (1980) Phys.Rev.C21,972
Hulet E K et al (1980) Phys.Rev.C21,966
Mustafa M G et al (1973) Phys.Rev.C7,1519
Other references in Bjørnholm S and Lynn J E (1980) Revs.Mod.Phys.52,725

Nuclear reactions caused by neutrons

A M Lane

Theoretical Physics Division, Building 424.4, Atomic Energy Research Establishment, Harwell, Oxon.

1. Introduction

The first landmarks in nuclear reactions were: 1919, the first observed reaction (using radioactivity αs); 1932, the first reaction with an incident beam (of protons); 1933, the first reaction caused by neutrons. In 1935 appeared the first evidence that reaction rates (cross-sections) vary with bombarding energy. Both for neutrons and protons, strong resonance effects were found. Since that time, more and more resonances have been found, and the list is still growing. Until 1950, it was supposed that all reactions were controlled by resonances, but then evidence for a non-resonant or "background" reaction contribution appeared. The implications of the co-existence of the two reaction types are still being worked out, 32 years later.

Until 1970, reaction studies were dominated by proton and neutrons, and most significant advances were made with these particles (the only exception being the finding of the first non-resonant reaction which used deuterons). Since 1970 of course, reactions have been dominated increasingly by heavy-ion studies. They involve quite distinct phenomena, so it does not constrain this survey to exclude them.

It is also no serious constraint to ignore light ions (d, t, h, α), but it would be quite artificial to ignore proton reactions. Neutron and proton reactions have proceeded in close parallel since they both revealed resonances in 1935. There has been an element of competition, for instance in revealing the best set of resonances. We will refer to both, selecting the best illustrative examples irrespective of the particles involved.

There are two obvious sources of asymmetry between neutron and proton studies. The repulsive Coulomb field felt by a proton approaching a nucleus reduces reaction yields and means that no resonances have been reported with protons on targets of $Z > 45$. For $Z \sim 45$, resonances are confined to a small window near 5 MeV incident energy, this being high enough for appreciable yield and low enough for the high experimental resolution needed to observe resonances.

The other difference is that at most neutron resonances, inelastic scattering is not energetically allowed, while many proton resonances do not have this restriction. This apparently minor technical difference turns out to give protons an important practical advantage in that they can reveal certain relative amplitudes of resonances (as we will see below).

2. Resonances

Neutron resonances have been found in almost every element. In certain elements (Zn, Er), the total number of resonances in the various isotopes exceeds 600 (and this excludes many more at higher energies which have not been assigned a unique spin. In certain isotopes, sequences of over 100 s-wave levels have been found (^{64}Zn,(Garg 1981), ^{166}Er (Liou 1972)). Although proton data is confined to Z < 45, it is also very impressive. The studies at TUNL (Bilpuch 1976) have revealed 3600 resonances with assigned spins from Z = 10 to 42, but especially with Z = 18 to 28. The best sequence is one of 125 s-wave states in ^{92}Mo.

What is the cause of these resonances? They arise, of course, from the extension of the bound state spectrum of the compound nucleus into the unbound region. For instance, in the channel ^{207}Pb + n, excited states of the compound nucleus ^{208}Pb are revealed as resonances. In this case, one knows from other sources most of the bound states and one can see that the resonances form an extension of the bound states.

3. Resonance Theory

The theoretical form of a cross-section for a reaction c → c' at an isolated resonance λ is:

$$\sigma_{cc'} = \frac{\Gamma_{\lambda c} \Gamma_{\lambda c'}}{(E_\lambda - E)^2 + \frac{1}{4}\Gamma_\lambda^2}$$

where $\Gamma_{\lambda c}$ is the partial width for channel c, $\Gamma_\lambda = \Sigma_c \Gamma_{\lambda c}$ is the total width, and E_λ is the resonance energy. (We suppress the trivial factors $\pi \lambda_c^2 g$). In appropriate situations, this gives excellent fits to data. (For very low energy neutron resonances, the data may reveal the fact that $\Gamma_{\lambda c}$ contains a factor λ_c^{-1}, such as in the famous case of the resonance in Cd at E_λ = 0.18 eV). The ideal of truly isolated levels is not often attained; if not, then the exact multi-level formula must be used to fit and describe the data:

$$\sigma_{cc'} = |U_{cc'}|^2$$

where:

$$U_{cc'} = \sum_\lambda \frac{g_{\lambda c} g_{\lambda c'}}{E_\lambda - E - \frac{i}{2}\Gamma_\lambda}$$

first given by Kapur and Peierls (1938). In general, $g_{\lambda c}$ are complex and energy-dependent, and only for isolated resonances are they real and satisfy $\Gamma_{\lambda c} = g_{\lambda c}^2$. An alternative exact form with real, constant quantities is due to Wigner and Eisenbud (1947):

$$U_{cc'} = \left[(\underline{1} - i\underline{R})^{-1} \underline{R}\right]$$

where $\underline{R}$ is the real matrix:

$$R_{cc'} = \sum_\lambda \frac{G_{\lambda c} G_{\lambda c'}}{E_\lambda - E}$$

and $G_{\lambda c}$ are real amplitudes. For isolated resonances, $\Gamma_{\lambda c} = G_{\lambda c}^2 = g_{\lambda c}^2$. The energy dependence of $G_{\lambda c}$ is explicitly given by a channel penetration factor P_c : $G_{\lambda c} = \sqrt{2P_c}\, \gamma_{\lambda c}$ with $\gamma_{\lambda c}$ constant. These general formulae contain resonance energies E_λ and channel amplitudes. There are several basic properties of these quantities that need to be settled before one can claim a true comprehension of nuclear reactions:

1. What are the statistical distribution laws for level positions E_λ and spacings $D_\lambda \equiv (E_{\lambda+1} - E_\lambda)$?

2. What is the law for amplitudes $\gamma_{\lambda c}$ of any given channel c for a sequence of levels?

3. What determines the value and behaviour of $\overline{\Gamma}_{\lambda c}$, or equivalently $\overline{\gamma^2_{\lambda c}}$, where the bar signifies average over a level sequence?

4. What connection is there between the set of $\gamma_{\lambda c}$ for two channels?; for instance, are there correlations, as measured by the linear correlation coefficient:

$$\rho(a,b) = \frac{\sum_\lambda (a_\lambda - \bar{a})(b_\lambda - \bar{b})}{\sqrt{\sum_\lambda (a_\lambda - \bar{a})^2 \sum_\lambda (b_\lambda - \bar{b})^2}}$$

where $\bar{a} \equiv (\sum_\lambda a_\lambda)/(\sum_\lambda 1)$? (Presumably since $\gamma_{\lambda c}$ are amplitudes, $\overline{\gamma}_{\lambda c} = 0$).
For a given set of answers to these questions, there remains the technical problem of evaluating $\sigma_{cc'}$, when resonances overlap, as they always do as energy increases and the number of "open" channels increases. Most cross-sections over a few MeV involve overlapping resonances.

4. Distribution of Spacings

This question is separate from questions 2, 3, 4. It appeals to mathematicians, because detailed dynamics are probably irrelevant, and the laws are of very wide applicability. (Of course, these considerations are not limited to unbound states). Dyson and Mehta (1962,3) studied the distribution laws of spacings expected from diagonalisation of a matrix with elements taken at random from a Gaussian distribution. They confirmed Wigner's surmise about the probability distribution:

$$P(D) \propto D e^{-\pi D^2 / 4\bar{D}^2}$$

(which fits the most extensive experimental data). They also pointed out a correlation between spacings which causes local mean spacing to remain closer to the overall mean spacing than would be so if spacings were selected at random. Plotting the accumulating number of levels against energy, the mean square departure from a best straight-line fit is roughly proportional to log n rather than n, where n is the total number of levels. Another view of the same effect is that the correlation of adjacent spacings is -0.27.

5. Channel Amplitudes

Questions 2, 3, 4 depend more on detailed dynamics than question 1, although it is possible that question 2 does not. The first serious model of nuclear reactions was the Bohr "compound nucleus" model (1936), as later worked out by Weisshopf, Peaslee and Feshbach (1947) and Hauser and Feshbach (1952). This model assumes that the strong nuclear interactions invariably capture an incident particle into a long-lived intermediate state, and that the decay of the state is independent of its formation. This picture suggests a random chaotic situation. There are no correlations between channels (answering 4), while the answer to question 2 is obtained from the central limit theorem (Porter and Thomas 1956) which suggests a Gaussian form:

$$P(\gamma) \propto e^{-\gamma^2/2\overline{\gamma^2}}$$

Question 3 is answered by the assumption that, on averaging over energy, the wave-function at the edge of the nucleus is purely ingoing, reflecting the loss of coherence between ingoing and subsequent emerging waves. This gives:

$$(\overline{\gamma^2_{\lambda c}}/\overline{D}_\lambda) \simeq \frac{1}{\pi K a_c}$$

where a_c is the nuclear radius and K a nuclear wave-number. This quantity is the "strength-function" and its value is ~ 0.05.

With these properties, one can calculate the value of $<\sigma_{cc'}>$ in the overlapping region, where the brackets signify average over a region large enough to iron out fluctuations, i.e. $\gg (\overline{\Gamma}_\lambda, \overline{D}_\lambda)$. Early studies (Bethe 1937, Hauser and Feshbach 1952) gave:

$$<\sigma_{cc'}> = \frac{<\sigma_c><\sigma_{c'}>}{\Sigma_c <\sigma_c>}$$

where $<\sigma_c>$ is the absorption cross-section in channel c:

$$<\sigma_c> = <|U_{cc} - <U_{cc}>|^2>$$

This form has been given a rigorous basis by Moldauer (1975), Kawai et al (1973) and Agassi and Weidenmuller (1975). The form of $<\sigma_{cc'}>$ predicts several key properties of cross-sections from the present model:

(1) $<\sigma_{cc'}>$ will decrease sharply with energy since the denominator sum increases as more channels open.

(2) For given incident energy, the dependence on emerging energy in channel c'c' is given by the level density of the final nucleus, approximately exponential.

(3) For two incident channels c,c" forming the same compound system, $(\sigma_{cc'}/\sigma_{c''c'})$ should be independent of energy.

(4) Although we have not mentioned angular distributions, the present model predicts that they should be almost isotropic.

(5) Although not obvious from our discussion, the absorption cross-section above barrier and above the s-wave region, should be πa_c^2. Further, the total cross-section should decrease monotonically with energy, eventually being constant at $2\pi a_c^2$.

Another quantity of interest is the mean square fluctuation in $\sigma_{cc'}$, which for $\overline{\Gamma}_\lambda \gg \overline{D}_\lambda$ is predicted to satisfy (Moldauer 1963)

$$\langle(\sigma_{cc'} - \langle\sigma_{cc'}\rangle)^2\rangle = \langle\sigma_{cc'}\rangle^2$$

Yet another quantity reflecting fluctuations, but in a more interesting way, is the auto-correlation function (Ericson 1963, Moldauer 1975):

$$C(\varepsilon) \equiv \langle\sigma_{cc'}(E + \varepsilon)\sigma_{cc'}(E)\rangle - \langle\sigma_{cc'}(E + \varepsilon)\rangle\langle\sigma_{cc'}(E)\rangle$$

which is predicted to have the form:

$$C(\varepsilon) = \frac{C(0)}{1 + (\varepsilon/\bar{\Gamma}_\lambda)^2}$$

thereby providing a useful method for the determination of the mean total width $\bar{\Gamma}_\lambda$.

This method has been used (especially in the period 1965-75), but is limited by the fact that data usually reveals not just one spin sequence, but many, each with its own value of $\bar{\Gamma}_\lambda$.

6. Inadequacy of the Bohr Compound Nucleus Model

About 1950, several predictions of the model were found to be not in accordance with experiment:

(i) Neutron total cross-sections (Barschall 1952) did not behave as described in (5). At low energies (< 3 MeV) and up to 100 MeV, non-monotonic energy variation was found. At fixed energy, the A dependence was not $\propto A^{2/3}$.

(ii) In conflict with (2), the higher energy end of the emitted particle spectrum did not show an exponential decrease (Gugelot 1954), but often an increase. In conflict with (4), these particles showed a strong forward peak in the angular yield (Rosen and Stewart 1955).

(iii) In conflict with (1), $\sigma_{cc'}$ often did not continue to decrease with energy above a few MeV.

These results showed that the Compound Nucleus Model needed supplementation. It could hardly be said to "fail" since it worked very well in many circumstances. In any case, the resonances at low-energy were inequivocal proof of the existences of the compound mechanism. However the new results made it clear that an incident particle could interact with the target nucleus without necessarily forming a compound nucleus. The mean free path was not negligibly small compared to the radius. This meant that the particle could escape with a refraction without being absorbed. This was described by the phenomenological optical model of Weisshopf, Porter and Feshbach (1954) in which the elastic and total cross-section were obtained from a one-body Schrodinger equation containing a complex potential. Along with this, there was the possibility of a direct process in which for instance the incident particle escaped after a single energy-exchanging collision with a target nucleon. The phenomenological description of this was the DWBA theory (Austern et al 1953, also Butler 1950) in which the initial and final states were generated by the optical potential, and the transition operation was the nucleon-nucleon interaction, thereby treated as a perturbation. (Another direct process arises when the target possesses collective excitations, vibrations or rotations, which are readily excited. In such cases, the DWBA theory may need replacement by CCBA, in which the interaction is treated exactly).

At higher incident energies (say > 15 MeV, and especially near 50 MeV) one can naturally expect that the direct "one-step" process will be supplemented by multi-step processes in which the incident particle escapes after two, three or more collisions, losing energy each time. This observation was made by Griffin (1967) 15 years ago and was found experimentally in the early 1970s, (e.g. Bertrand and Peelle 1973).

7. Implication of Direct Reactions for Resonance Statistics

When the phenomenological theories were firmly established by their success in fitting data, theoretical attention was directed at the problem of fitting them into the exact formal framework of resonance theory described above, in particular to questions 2, 3, 4. First we must try to identify the direct process formally. The natural choice is:

$$\sigma_{cc'}^{(D)} \equiv |<U_{cc'}>|^2$$

which satisfies:

$$<\sigma_{cc'}> = \sigma_{cc'}^{(D)} + \sigma_{cc'}^{(CN)}$$

where:

$$\sigma_{cc'}^{(CN)} \equiv <|U_{cc'} - <U_{cc'}>|^2>$$

is the compound nucleus cross-section.

The direct cross-section amplitude is the average of the scattering matrix element, and the compound nucleus cross-section is the remainder. It is easy to see that the absence of channel correlations in the Bohr Compound Nucleus Model implies $<U_{cc'}> = 0$, so there is no direct cross-section.

The identification of $<U_{cc'}>$ with the direct phenomenological amplitude raises three problems:

(i) can we deduce the phenomenological form, from the exact form of $U_{cc'}$?

(ii) what effect does the presence of the direct amplitude have on the evaluation of $<\sigma_{cc'}^{(CN)}>$?

(iii) what implications does the identification have for resonance statistics, i.e. for questions 2, 3, 4?

8. Deduction of Optical Model and DWBA from the Exact Form of $U_{cc'}$

The first step is to note that $<U_{cc'}(E)> = U_{cc'}(E + i\varepsilon)$, where ε is the averaging interval implied by the brackets. Next, one can show (Brown and de Dominicis 1958) that:

$$<U_{cc'}> - U_{cc'}^o = <\phi_c|\Delta - \Delta(H - E^+)^{-1}\Delta|\phi_{c'}>$$

where ϕ_c are the channel wave-functions that are eigenstates of H_o, $\Delta \equiv (H-H_o)$, and $U_{cc'}^o$ is the scattering matrix of H_o. First let us consider one channel $c = c'$ and demand that H_o is such that $U_{cc'}^o = <U_{cc'}>$.

This means that H_o is the optical model Hamiltonian, and this condition (implying that the right-hand side vanishes) was solved by Feshbach (1958, 1962), who found (ignoring antisymmetry):

$$V + iW = \langle \Phi_c | v - v(QHQ - E^+)^{-1} v | \Phi_c \rangle$$

where Φ_c is the target state, v is the nucleon-nucleus interaction and Q projects on all channels other than c.

For the off-diagonal cases, $c \neq c'$, we may choose H_o to be the optical model choice, then, to first order in $\Delta \equiv (H - H_o) \stackrel{\circ}{=} (v - V - iW)$:

$$\langle U_{cc'} \rangle = \langle \phi_c | \Delta | \phi_{c'} \rangle = \langle \phi_c | v | \phi_{c'} \rangle$$

which is just the DWBA formula. Of course, the same form follows for any other choice of H_o, so the full validation of the DWBA formula requires a demonstration that the neglected term (of 2nd order) is minimised with the optical model choice for H_o. As far as I know this has not been done in any analytic detail, except in a model study by Bloch (1957).

9. **Effect of Direct Reactions on the Expression for $\sigma_{cc'}^{(CN)}$**

To evaluate this was a severe technical problem that was taken on by Kawai et al (1973), Engelbrecht and Weidenmuller (1973) and Moldauer (1975b). In the second work, one transforms from the actual channels to channels between which no direct reactions occur. One can then apply the usual Hauser-Fesbach results to these new channels, then relate back to the original ones. The main practical snag with the results is that no explicit formulae are given except in limiting cases. All open channels experiencing direct coupling affect the results for any pair of channels of interest, i.e. all direct couplings must be specified and used as input in order to evaluate $\sigma_{cc'}^{(CN)}$ for given c,c'.

10. **Implications of Direct Reactions for Resonance Statistics as represented by Questions 2,3,4**

There is no evident affect on question 2, and the assumption has been that the Gaussian form of $P(\gamma)$ continues to apply in the presence of direct reactions.

The answer to question 3 was obtained by Feshbach, Porter and Weisshopf (1954) by applying the last relation in the case $c = c'$. At low-energies, it implies that

$$\overline{\gamma_{\lambda c}^2} / \bar{D}_\lambda \simeq \frac{\frac{1}{2\pi} W \gamma_p^2}{(E_p - E)^2 + \frac{1}{4}W^2}$$

where p is the single-particle level of the optical potential $(V + iW)$ that occurs nearest to the energy range of interest. E_p is its energy and γ_p^2 is its reduced width. One sees the non-monotonic behaviour (or "giant resonance"). W is of order MeVs. It is this behaviour that is ultimately responsible for the non-monotonic total cross-sections. There is a similar behaviour expected in the variation with A rather than energy, and this was confirmed by data on neutron resonances. For s-waves, maxima occur at A $\sim$ 50, 170 corresponding to $p = 3s$, $4s$ single particle states being just unbound in the potential V.

Now we turn to question 4. In the resonance region:

$$\langle U_{cc'}^{(E)} \rangle = \sqrt{4P_c P_{c'}} \left[i\pi s_{cc'}^{(E)} + P. \int \frac{s_{cc'}(E')}{E' - E} dE' \right]$$

so it immediately follows that a non-zero direct amplitude implies a non-zero value of the quantity:

$$s_{cc'}(E) \equiv \overline{\gamma_{\lambda c} \gamma_{\lambda c'}} / \overline{D}_\lambda$$

i.e. that there are linear correlations between channels. One can see that the real and imaginary terms in $\langle U_{cc'} \rangle$ arise from background and local-resonance parts of $U_{cc'}$, the background coming from the net contribution of distant levels. In principle, both of these effects could be detected in experiments. In practice, it has not been easy. The background (from distant levels) will only be visible above the tails of local levels if $\overline{\gamma_{\lambda c} \gamma_{\lambda c'}}$ for distant levels is larger than $|\gamma_{\lambda c} \gamma_{\lambda c'}|$ of local levels. There is only one case where a fairly convincing case for a background viz. ^{37}Cl(n,γ) at thermal energy (Kopecky et al 1974).

The search for the second effect, the linear correlation $\overline{\gamma_{\lambda c} \gamma_{\lambda c'}} \neq 0$, has had a curious history. Until recently there was no data on resonance <u>amplitudes</u> $\gamma_{\lambda c}$ (as opposed to <u>widths</u> $\gamma_{\lambda c}^2$). In lieu of this, one fell back on data on width correlations. In principle these are not necessarily related. However, they will be simply related (Lane 1971) if (i) correlations are linear $\gamma_{\lambda c} = \alpha \gamma_{\lambda c'} + \delta_\lambda$ where α is a constant and δ_λ is random, (ii) the separate probability distributions of $\gamma_{\lambda c}$, $\gamma_{\lambda c'}$ are Gaussian. In this situation:

$$\rho(\gamma_{\lambda c}^2, \gamma_{\lambda c'}^2) = [\rho(\gamma_{\lambda c}, \gamma_{\lambda c'})]^2$$

It was accepted that (i) and (ii) were reasonable conditions. Indeed the best neutron data (Garg 1981) seem to be consistent with the Gaussian assumption.

There are about a dozen cases in resonance neutron capture studies where significant positive correlations between neutron and partial gamma-ray widths are found. The first case was reported in 1968 (Lone et al 1968). (Of course, if a negative correlation were found, that alone would invalidate the above relation between width and amplitude correlations. One negative correlation has been found between partial gamma-ray widths, but it involves only 6 resonances). The best case (Jackson and Toohey 1972) is actually an <u>inverse</u> reaction, viz. ^{91}Zr(γ,n)^{90}Zr where 36 resonance of spin 3/2⁻ have a width correlation $\rho = 0.59$, or (using the above relation) an amplitude correlation $\rho = 0.77$. It would be nice if direct capture theory were accurate enough to predict this value, but it is not. One can only say that, at higher energies (≥ 3 MeV) significant direct capture is observed in many nuclei, so one expects amplitude correlations to be present.

Recently evidence has appeared to check the above formula relating width and amplitude correlations. This is from proton studies (and incidentally is an excellent example of how neutron and proton studies stimulate and supplement each other).

The data is from resonance studies at TUNL (Lane 1978, Wells 1980, Chou 1980, Shriner 1982) in a number of even-even nuclei with masses near 50. Inelastic scattering to the 2+ state is possible. For 3/2- resonances, this mean two inelastic spin channels which may be classified as $p_{3/2}, p_{1/2}$; for 5/2+ resonances, there are three inelastic channels, viz. $s_{1/2}, d_{5/2}, d_{3/2}$. Amplitudes have been measured.

First let us note that in one of the cases, viz 3/2- states with target ^{44}Ca, there is a region where an analogue state occurs and dominates all relevant channels. This is a situation where the correlation ρ is predictable, viz $\rho \sim 100\%$, and indeed this is found (Lane 1978). Thus, in this special case, the specific amplitude correlations implied by theory have been observed.

Returning to the question of the relation between width and amplitude correlation, which is one aspect of the broader issue of the Gaussian nature of nuclear statistics, we find that the proton results do not bear out this picture. Although the relation applies in one case (^{46}Ti, 3/2-, 14 res.), it is usually violated.

In ^{48}Ti, the 3/2- (70 res.) and 5/2 + (45 res.) sequences show that while $|\rho(\gamma,\gamma)| \lesssim 0.05$, $\rho(\gamma^2,\gamma^2)$ can be ~ 0.6 or -0.5, depending on the selection of basis for the two spin states. (Here we ignore the $s_{1/2}$ amplitudes). The value of $\overline{\gamma^4}/(\overline{\gamma^2})^2$, which should be 3, is found to differ by a factor of up to two either way, while $(\overline{\gamma_1^2 \gamma_2^2} - \overline{\gamma_1^2}\,\overline{\gamma_2^2})$, which should be 0.5, is, for most bases, ≤ 0.05. These represent serious departures from Gaussian statistics and mean that, in general, a width correlation cannot be interpreted as an amplitude correlation.

The implications of this discovery have not yet been digested. On the purely theoretical side (Whitehead 1978, Verbaarschot 1979, Grimes 1981), it has been noticed that large shell-model diagonalisations (involving up to 2000 states) show strong non-Gaussian effects in the lowest eigenvectors. The distribution thereby found can be fitted by a sum of Gaussians (up to 4) with that of least width having the largest strength. The Gaussian fitting the largest widths usually accounts for only a small fraction, $10^{-2} - 10^{-1}$. Whitehead (1978) has checked that this is not special to lowest eigenvectors. This value of $\overline{\gamma^4}/(\overline{\gamma^2})^2$ is always > 3, so there is no obvious contact with the data where the value may differ from 3 in both directions.

Finally another result that is not obviously consistent with the spirit of Gaussian statistics should be mentioned, viz. a negative correlation -0.21, between neutron widths of adjacent levels (Liou 1972) in ^{166}Er.

11. Other Subjects in Neutron Reactions

There are three subjects in neutron reactions that should be mentioned in a survey: fission, doorways and effects of target motion.

Fission: This phenomenon was originally discovered in neutron reactions and is still essentially a neutron subject. There was a fascinating development a few years ago (the fission "doorway" states arising from the double-humped barrier) but fission is a self-contained subject outside the mainstream of this survey. (See Bjørnholm 1980).

Doorways: Since the concept of doorway state was recognised twenty years ago, there has been a continuing search, but with little to show.

Theoretically the concept is well-founded. Given the existence of direct inelastic scattering when sufficient energy is available, one can infer that, below threshold, the excitation of the target can occur with the incident particle entering a bound orbit, (rather than a free orbit). This "doorway" state will have an appreciable lifetime (because of the partial transparency of the nucleus implied by the optical potential). This implies that a single particle giant resonance in the strength-function for a given incident angular momentum may contain relatively narrow sub-structures, revealing doorway states. (These doorways may themselves have sub-structures also, and so on, until the ultimate fine-structure resonances terminates the sequence). In practice, the problem of separating single angular momenta has made the search difficult, except for very special categories (analogue states, fission doorway states). One case is an apparent -1^- doorway, common to neutrons and ground-state photons in ^{207}Pb + n p-wave data at 120 keV. (Horen 1978, Raman 1977).

Effects of Target Motion : Recently, effects of the motion of the target nucleus in its molecule or crystal or metallic surroundings have been detected at a neutron resonance (Bowman 1980, Meister 1981), viz. the 6.67 eV resonance in ^{238}U + n for which Γ_n = 1.5 meV, Γ_γ = 27.5 meV and target motion width $\sim$ 80 meV. The first work looked at the difference in resonance shape between targets of UF_6 gas molecules and solid U_3O_8. Differences of order 10% were found, in qualitative agreement with expectation. The second work looked at the differences between different solids, which are of order 1%. There is a peak shift of order 0.1 - 1 meV. When this is related to the expected shift (from different lattice vibrations), there is a residual shift of the same order of magnitude, which is interpreted as an atomic effect. When a neutron joins with a nucleus, there will be a small change in mean square charge radius $\delta<r^2>$, so the atomic energy charges and the resonance energy will shift (relative to a bare nuclear target) by this amount. For two different atomic surroundings, the resonance energy will differ by the difference in these shifts, which $\propto \Delta\rho_{el}^{(o)} \delta<r^2>$ where $\Delta\rho_{el}^{(o)}$ if the difference in electron density at the nucleus. The 5 materials used give differences consistent with $\delta<r^2>$ = $-1.7 \binom{+1.2}{-0.8}$ f^2, suggesting a decrease in radius going from g.s. ^{238}U to excited ^{239}U of about 2%.

References

AgassiD and Weidenmuller H A 1975 Phys. Lett. <u>56B</u> 305
Austern N, Butler S T and McManus H 1953 Phys. Rev. <u>92</u> 350
Barschall H H 1952 Phys.Rev. <u>85</u> 704
Bertrand F E and Peelle R W 1973 Phys. Rev. <u>C8</u> 1045
Bethe H A 1937 Rev. Mod. Phys. <u>9</u> 69
Bilpuch E G, Lane A M, Mitchell G E and Moses J D, 1976 Phys. Rep. <u>28C</u>
    No. 2, 145
Bjørnholm S and Lynn J E, 1980 Rev. Mod. Phys. <u>52</u> 725
Bloch C, 1957 Nucl. Phys. <u>4</u> 503
Bohr N, 1936 Nature <u>137</u> 344
Bowman C D and Schrack R A, 1980 Phys. Rev. <u>C21</u> 58
Brown G E and de Dominicis C, Proc. Phys. Soc. (Lond) 1958 <u>A72</u> 70
Butler S T , 1950 Phys. Rev. <u>80</u> 1095
Chou, Mitchell, Bilpuch and Westerfeldt, 1980 Phys. Rev. Lett. <u>45</u> 1235
Dyson F J, 1962 J. Math. Phys. <u>3</u> 140, 157, 160, 1199; Dyson F J and
    Mehta M L 1963, ibid <u>4</u> 701
Engelbrecht C A and Weidenmuller H A, 1973 Phys. Rev. <u>C8</u> 859

Feshbach H, 1958 Am. Phys. 5 357; ibid 1962 19 287
Garg J B, Tikku V K and Harvey J A, 1981 Phys. Rev. C23 671
Griffin J J, 1967 Phys. Rev. 24B 5
Grimes S M and Bloom S D, 1981 Phys. Rev. C23 1259
Gugelot C, 1954 Phys. Rev. 93 425
Hauser W and Feshbach H, 1952 Phys. Rev. 87 366
Horen D J, Harvey J A, Hill N W, 1978 Phys. Rev. C18 722
Kapur P L and Peierls R E, 1938 Proc. Roy. Soc. (Lond) A166 277
Kawai M, Kerman A K and McVoy K W, 1973 Am. Phys. 75 156
Kopecky J, Spits A M J and Lane A M, 1974 Phys. Lett. 49B 323
Lane A M, 1971 Am. Phys. 63 171
Lane, Dittrich, Mitchell and Bilpuch, 1978 Phys. Rev. Lett. 41, 454
Liou et al. 1972 Phys Rev C5 974
Lone, Chrien, Wasson, Beer, Bhat and Muether, 1968 Phys. Rev. 174 1512
Meister, Pabst, Pikelner and Seidel, 1981 Nucl. Phys. A362 18
Moldauer P A, 1963 Phys. Lett. 8 70
Moldauer P A, 1975a Phys. Rev. C11 426
Moldauer P A, 1975b Phys. Rev. 12 744
Porter C E and Thomas R G, 1956 Phys. Rev. 104 483
Raman S, Mizumioto M and Macklin R L, 1977 Phys. Rev. Lett 39 598
Rosen L and Stewart L, 1955 Phys. Rev. 99 1052
Shriner, Bilpuch, Westerfeldt and Mitchell, Z Phys. 1982 A306 1
Verbaarschot J J M and Brussard P J, 1979 Phys. Lett. 87B 155
Weisshopf V F, Peaslee D C and Feshbach H, 1947 Phys. Rev. 71 145
Weisshopf V F, Porter C E and Feshbach H, 1954 Phys. Rev. 96 448
Wells W K, Bilpuch E G and Mitchell G E, Z Phys. 1980 A297 215
Whitehead, Watt, Kelvin and Conkie, 1978 Phys. Lett. 76B 149
Wigner E P and Eisenbud L, 1947 Phys. Rev. 72 29

Inst. Phys. Conf. Ser. No. 64: Section 2
Paper presented at Conf. on Neutron and its Applications, Cambridge, 1982

The neutron optical model

P.E. Hodgson

Nuclear Physics Laboratory, Oxford, U.K.

Abstract: The use of the neutron optical potential to calculate total, reaction and differential elastic cross-sections, together with the associated polarisations, is briefly reviewed. The effect of coupling between the elastic and inelastic channels is described, together with the use of the isospin term to unify neutron and proton elastic scattering and the (p,n) isobaric analogue state reactions. The methods used to calculate the optical potential from the nucleon-nucleon reaction are described, and the limitations of the concept of an optical potential are discussed.

1. Introduction

The generally smooth variation of the neutron total cross-sections with target nucleus and neutron energy suggested to Feshbach, Porter and Weisskopf (1954) that the interaction of a neutron with a nucleus can be represented by a complex one-body potential that does not depend on the details of nuclear structure, an idea that had already been applied with success to proton scattering by Le Levier and Saxon (1952). In the following years this optical model, as it is called, has been applied with remarkable success to analyse a wide body of neutron scattering data, and several global parameter sets have been found that give generally excellent fits over a range of neutron energies and target nuclei. Among these may be mentioned the non-local potential of Perey and Buck (1962), its local equivalent (Wilmore and Hodgson 1964), and the nucleon potentials of Becchetti and Greenlees (1969).

In the following years the finer details of the neutron optical potential have been examined, and its theoretical foundations established. Lane (1962) introduced the isospin term in the potential and thus unified neutron and proton scattering and the (p,n) reactions to the isobaric analogue state of the target, and the coupled-channels theory unified our understanding of elastic and inelastic scattering. Theoretical studies enabled the optical potential to be calculated from the nucleon-nucleon interaction and the nuclear density distribution, and have helped to unify the phenomenological knowledge of the optical potential over the whole range of energies from bound to scattering states. Most of these studies have been undertaken in order to understand the neutron and its interactions, but in addition the neutron optical model has proved itself of great value for calculating the cross-sections needed in the design of nuclear reactors.

2. The Neutron Optical Potential

The essential idea of the optical model is that all the interactions between an incident neutron and a target nucleus can be represented by a one-body potential $V(r)$. This potential is complex, and as in the analogous optical case the real part refracts and the imaginary part absorbs the incident wave. The imaginary part has the effect of removing flux from the incident channel, and thus takes into account in a global way all the non-elastic processes ocurring in the interaction. It is usual to add a spin orbit potential $V_{so}(r)$ which enables the polarisation of the scattered particles to be calculated.

This optical potential can be defined by the requirement that when inserted into the Schrödinger equation describing the scattering process it gives the measured total and reaction cross-sections, together with the differential elastic cross-sections and polarisations. Such a potential is not necessarily unique; it suffices that it gives wavefunctions having the asymptotic form that fits the experimental data, and there may be many potentials that satisfy this requirement. In principle these potentials depend in a very complicated way on energy and nuclear structure, but provided the experimental data are averaged over energy intervals sufficient to remove the fluctuations due to the compound nuclear states it is found that $V(r)$ can be expressed by quite simple analytical formulae depending only on the radial distance, the neutron energy and the nuclear asymmetry. The utility of the optical potential follows from this remarkable fact.

This neutron optical potential is usually written

$$V(r) = U f(r) + i W g(r) + V_{so}(r) \qquad (2.1)$$

where the Saxon-Woods form factor $f(r)=[1+\exp\{(r-R)/a_u\}]^{-1}$, $R_u=r_u A^{1/3}$ and $g(r)$ has either the volume form $g_V(r)=[1+\exp\{(r-R_V)/a_V\}]^{-1}$, $R_V=r_V A^{1/3}$ or the surface-peaked form $g_S(r)=4a\, g_V'(r)$, where the prime denotes differentiation with respect to r. The spin-orbit potential

$$V_{so}(r) = V_s \left(\frac{\hbar}{m_\pi c}\right)^2 \frac{1}{r} \frac{d\, f_s(r)}{dr} \underline{L}\cdot\underline{\sigma} \qquad (2.2)$$

where $f_S(r) = [1+\exp\{(r-R_S)/a_S\}]^{-1}$, $R_S = r_S A^{1/3}$. The methods used to calculate the measurable quantities from this potential are described by Hodgson (1963).

Many analyses of neutron scattering have been made with this potential, and with suitable values of the parameters it is found able to reproduce quite well the experimental differential cross-sections and polarisations, as well as the associated total and reaction cross-sections. It is found necessary to allow the potentials U and W to depend on energy, but with some exceptions to be discussed later the potentials depend on the target nucleus only through the radius parameter R.

Attempts to calculate the optical potential from the nucleon-nucleon interaction indicate that at least part of it has a non-local form, and this non-locality is partly responsible for the energy dependence of the potential. It is thus possible to fit the data with an energy-independent non-local potential, and this was used by Perey and Buck (1962) in a comprehensive analysis of neutron data. An equivalent local potential has been found by Wilmore and Hodgson (1964) and this has been widely used to calculate total, reaction and differential elastic scattering cross-sections for neutron scattering. Although such potentials fit the data very well on the whole, some systematic departures from their predictions have been found. It is already noticeable in the work of Perey and Buck that the reaction cross-sections for the doubly-magic nuclei are appreciably lower than the calculated values, particularly at lower energies. This is just what would be expected from the lower level density of such nuclei, and can easily be reproduced with a lower value of the imaginary part of the potential. Marked deviations from the global predictions are also found for deformed nuclei, due to the strong coupling between the elastic and inelastic channels. This can be treated theoretically either by altering the parameters of the potential or by developing the optical model so as to include explicitly the coupling to the low-lying nuclear states as described in section 3.

The optical potential cannot be made indefinitely precise by analyses of more accurate experimental data, for two reasons. Firstly very accurate data cannot be fitted within the experimental standard deviations without altering the shape of the potential; this has been shown by Kobos and Mackintosh (1979) for proton scattering and there is no reason to suppose that their conclusions are not also applicable to neutron scattering. Secondly, when the potential is very accurately adjusted in this way to fit a particular interaction, it is unlikely to vary in a smooth way with energy and from nucleus to nucleus. Thus increased accuracy of fitting is paid for by reduced generality. The usefulness of the optical potential is precisely that with similar parameters it can give a good overall fit to a wide range of data. Beyond a certain point it cannot be made more precise without losing this feature.

3. Neutron Scattering from Deformed Nuclei

The problem of accounting for the scattering by deformed nuclei can be tackled either by altering the potential phenomenologically, or by developing the physical model to take explicit account of the effects of the deformation. In the first method, the optical model parameters are varied to optimise the fit, and it is then hoped that the new parameters will in some way take into account the effects of the deformation. The new potentials depend on the magnitude of the deformation, and possibly on other parameters such as the energy as well. This method has been applied to the actinide nuclei, which are characterised by very strong deformations of almost the same magnitude. Madland and Young (1978) have found an optical potential that gives a good overall fit to the total, reaction and elastic differential cross-sections of these nuclei. It is thus possible to take into account the effects of deformation on elastic scattering in a simple phenomenological way.

A more physical analysis of the scattering of neutrons by nuclei is provided by the coupled-channels theory. Deformed nuclei have low-lying excited states that are strongly coupled to the ground state, and this coupling affects the scattering and makes it different from the scattering from spherical nuclei where the coupling between the ground and the excited states is relately weak. This coupling may be included in the calculations by expressing the total wave function as a sum of products of the wavefunctions in the elastic and inelastic channels and the corresponding nuclear state wavefunctions.

If this is inserted into the Schrödinger equation for the whole system one obtains a set of coupled equations for the channel wavefunctions. With appropriate potentials, these equations give the wavefunctions in all the channels and hence the corresponding inelastic scattering cross-sections as well as the total, reaction and differential elastic cross-sections (Buck et al, 1963). In practical calculations only a few inelastic channels are considered explicitly, the remainder being taken into account through the imaginary part of the potential as for the simple optical model. The coupling between the ground and excited states is treated by the appropriate collective model, vibrational or rotational, or in a more detailed microscopic way. The additional parameters introduced are the dynamic or static deformation parameters, and these determine the intensity of the inelastic scattering.

Many analyses of proton elastic and inelastic scattering have shown that it is possible to reproduce the data very well using deformation parameters similar to those expected from knowledge of the structure of the nucleus concerned. The main limitation of the coupled-channels theory is that the compution time increases as the cube of the sum of $(2J+1)$ for all the states included in the calculation, which makes it impracticable to include more than a few states.

After its initial success with proton scattering, the coupled-channels theory has been extensively applied to neutron scattering, with equally satisfactory results. An example of some recent work is the analysis of scattering by actinide nuclei by Haouat (1982).

4. The Isospin Term in the Nucleon Optical Potential

On the basis of microscopic arguments, Lane (1962) introduced the term $4(\underline{t}\cdot\underline{T})V_I(r)$ in the nucleon potential, and showed that this gives terms $\pm\alpha V_I(r)$ in the proton and neutron potentials respectively, where $\alpha = (N-Z)/A$. Providing the form factors are the same, the proton and neutron potentials can then be written

$$U_p = U_0 - \varepsilon E + \alpha U_I + U_C; \qquad U_n = U_0 - \varepsilon E - \alpha U_I \qquad (4.1)$$

where εE gives the energy dependence of the potential and U_C is the term that arises in the proton potential because of the energy dependence of the potential in the presence of the Coulomb field. A simple argument (Perey 1963) gives $U_C = 6e^2 \varepsilon Z/5r_{oA}^{1/3} = 0.4 \, Z/A^{1/3}$.

A global value of the isospin potential may be determined by combining the results of global analyses of neutron and proton elastic scattering. Thus, from (4.1) $U_p - U_n = 2\alpha \, U_I + U_C$. Using the above value of U_C, $(U_p - U_n - U_C)$ can be plotted against α, and the slope gives the global value of U_I.

It is not possible to obtain the isospin potential from analyses of neutron scattering data alone, since the apparent asymmetry dependence of the potential may include a purely geometrical component (Hodgson, 1970).

The isospin potential can be obtained for a particular nucleus in a similar way, and also from analyses of scattering from isotopic sequences. The results of many determinations of the isospin potential made by these methods give the average values $U_I = 22 \pm 2$ MeV, $W_I = 12 \pm 2$ MeV. Since the isospin term is rather a small part of the optical potential, it is not a serious restriction to assume that it has the same radial variation as the remainder of the potential. It is difficult to establish the shape of the isospin potential, and indeed it probably depends rather sensiively on nuclear structure, but if it does differ from the Saxon-Woods form then there still exists an equivalent Saxon-Woods potential, and it is this that is determined in the analyses described here.

Since the isospin potential couples the proton and neutron channels it is also responsible for the (p,n) reaction to the isobaric analogue state (IAS) of the target nucleus. The cross-sections of such reactions may be calculated by solving the coupled equations (the Lane equations) for the wavefunctions in the neutron and proton channels. Since the coupling term is relatively small, it is usually sufficiently accurate to use perturbation theory.

Although at first sight the (p,n) IAS reaction provides a good method of determining the isospin term there are several difficulties: (a) the (p,n) cross-section depends on both the real and the imaginary isospin potentials (b) two-step reaction processes may also be present, and (c) there are smaller contributions like those coming from those parts of the neutron-proton density difference that arise for the Coulomb distortion of the nuclear densities and from the polarizing effect of the excess neutrons on the self conjugate core (Lovas et al, 1981).

5. Energy Dependence of the Nucleon Optical Potential

The real and imaginary parts of the local optical potential are found to be energy dependent, and this is now understood in terms of its non-locality and the coupling to collective modes of excitation. Most of the analyses have been made of proton data, but the results are equally applicable to the neutron potential.

The phenomenological energy dependence of the real part of the potential is of two types. Firstly there is an overall energy dependence that is almost linear from negative to positive energies, with a slope of about -0.3. This is attributed to the non-locality of the nucleon-nucleus interaction; the local equivalent to a non-local potential has an energy dependence related to the range of the non-locality. Secondly, around the Fermi surface there are departures from this smooth behaviour due to the coupling to collective excitations (Bauer et al, 1982). It is convenient to describe this behaviour of the nucelon optical potential in terms of an effective mass m^*, which is the product of the momentum-mass m and the energy-mass $\bar{m}$ associated respectively with the energy dependencies due to the non-locality and to the coupling to the collective modes. The momentum-mass $m=0.64$ corresponding to the smooth overall energy dependence, and the energy mass $\bar{m}$ is peaked at the Fermi surface. The resulting value of the effective mass is approximately unity at the Fermi surface, falling to 0.64 at higher and lower energies.

6. The Theory of the Neutron Optical Potential

Many attempts have been made to calculate the nucleon optical potential from more fundamental information, in particular the phenomenological nucleon-nucleon interaction and the nuclear density distribution. At high energies a first approximation may be obtained by a simple folding of these quantities

$$V(r) \sim \int \rho(\underline{r}') \, \nu(|\underline{r}-\underline{r}'|) d\underline{r}' \qquad (6.1)$$

This expression is quite good above about 500 MeV, but at lower energies it is progressively less accurate due to a number of effects including Pauli blocking, Fermi averaging, dispersion and antisymmetrisation. Qualitatively, the expression (6.1) yields several important features of the optical potential. In particular, the short-range nature of the nucleon-nucleon interaction shows that the real part of the potential has the same radial shape as the nuclear density, namely a uniform variation in the interior and an exponential fall-off in the surface region. These general features are incorporated in the real part of the phenomenological optical potential. This gives quite good estimates of the radius and surface diffuseness of the potential, and its depth can be estimated from the Fermi gas model. The imaginary part of the potential is similarly expected to have approximately the same radial extent as the real part, and its strength can be estimated from the total absorption cross-section by semi-classical arguments. The spin-orbit potential is taken to have the radial derivative form by analogy with the Thomas term in atoms, and it strength obtained from the spin-orbit splitting of the nuclear shell model. Thus quite simple theoretical arguments suffice to give the main features of the optical potential, and more accurate values of the parameters may be obtained by comparison with experimental data.

It is much more difficult to obtain the parameters from the nucleon-nucleon interaction, as the effects already mentioned must each be evaluated. The expression (6.1) may be regarded as the first of a series of terms, or it may be used to define an effective nucleon-nucleon interaction that gives $V(r)$ when inserted in (6.1). The problem is then to calculate this effective interaction.

The method most frequently used is to obtain the effective interaction in nuclear matter by solving the Bethe-Goldstone equation and then to obtain the potential for finite nuclei using the local density approximation. Calculations of varying degrees of complexity have been made by many authors, and among these the work of Jeukenne et al (1976,1977) and of Brieva and Rook (1977,1978) has been particularly successful. Already the calculated potentials are able to give a good overall fit to the experimental data, although not of the same quality as those attained in phenomenological analyses. There is still much work to be done to attain a full understanding of the neutron optical potential.

References

Bauer M, Hernandez-Saldana E, Hodgson P E and Quintanilla J, 1982 J.Phys.G. **8**, 525.
Becchetti F D and Greenlees G W, 1969 Phys.Rev. **182**, 1190.
Brieva F A and Rook J R, 1977 Nucl.Phys. **A291**, 299, 317; 1978 Ibid **A297** 206
Buck B, Stamp A P and Hodgson P E, 1963 Phil.Mag. **8**, 1805.
Feshbach H, Porter C E and Weisskopf V F, 1954 Phys.Rev. **96**, 448.
Glasgow D W and Foster D G, 1971 Phys.Rev. **C3**, 604.
Haouat G, 1982 NEANDC Report 158U.
Hodgson P E, 1963 The Optical Model of Elastic Scattering (Clarendon Press, Oxford); 1970 Nucl.Phys. **A150**, 1.
Jeukenne J P, Lejeune A and Mahaux C, 1976 Phys.Rep. **25C**, 83; 1977 Phys.Rev. **C15**, 10; Ibid **C16**, 80.
Kobos A M and Mackintosh R S, 1979 Ann.Phys.(New York)**123**296; J.Phys.**G5**,97.
Lane A M, 1962 Nucl.Phys. **35**, 676; Phys.Rev.Lett. **8**, 171.
Lovas R G, Brown B A and Hodgson P E, 1981 Nucl.Phys. **357**, 205.
LeLevier R E and Saxon D S, 1952 Phys.Rev. **87**, 40.
Madland D G and Young P G, 1978 Proc.Int.Conference on Neutron Physics and Nuclear Data (OECD Nuclear Energy Agency) 349.
Perey F G and Buck B, 1962 Nucl.Phys. **32**, 353.
Wilmore D and Hodgson P E, 1964 Nucl.Phys. **55**, 673.

Neutron single particle states in nuclei

E. Friedman

The Racah Institute of Physics, The Hebrew University, Jerusalem, Israel

Abstract. The concept of single particle states in nuclei is reviewed with emphasis on the experimental background, as the interest in the model results mainly from its ability to relate in a simple fashion different types of experiments. In recent years significant advances have been made in the variety of experiments which can be described consistently using a single particle approach. These advances have been matched by progress in understanding the basic features of the model.

1. Introduction

Neutron single particle states were observed in nuclei more than 30 years ago and were, in fact, the first single particle states to be identified. The (d,p) reaction and its inverse, the (p,d) reaction which were the easiest to study experimentally, provided ample information on states formed by adding a neutron to a target nucleus or removing a neutron from it. The present review deals with two main periods in the study of single particle states in nuclei: the "classical" period when the bulk of precise experimental results on nucleon transfer reactions with light projectiles was obtained and the concept of single particle states was established throughout the periodic table. The "new" era, covering roughly the last 8 years, is characterised by deeper understanding of the connection between geometrical aspects of single particle states and their spectroscopy and by consistent handling of different types of experiments. Also, new possibilities have been opened up recently by nucleon transfer reactions between heavy ions and by magnetic scattering of electrons which provide, for the first time, information on single neutron states using an electromagnetic probe.

The present review is, obviously, limited in its size and scope, and is aimed primarily at clarifying the notions and presenting a logical development. This does not always coincide with the historical sequence of events and I offer my apology to all authors whose work is not cited.

2. The "Classical" Period

The period which covers roughly the years 1950-1970 can be described as the classical period of the single particle states in nuclei. Initially, such states were identified in (d,p) reactions on closed shell nuclei and the first analysis of Butler (1950) was indeed of the experimental results of Burrows et al (1950) for the (d,p) reaction on ^{16}O target. Within a few years the pioneering measurements of energy spectra by the range curve method were replaced by precision measurements using magnetic spectrographs which were able to locate many excited states throughout the periodic table. The measurements of angular distributions were facilitated by using multi-gap spectrographs (Enge and Buechner 1963) and values of the transferred orbital angular momentum were reliably determined with the help of the Distorted Waves Born Approximation (DWBA).

In nuclei far off closed shells where the "valence" nlj states are seriously fragmented, one is faced with the question of whether the single particle concept is valid at all. An essential ingredient of this concept is a single particle *potential* in which the wavefunction of the transferred nucleon is calculated for use in the DWBA description of the experimental results. Such a potential is also the most attractive feature of the model as it offers the possibility of relating different experiments. Sherr et al (1965) studied (p,d) reactions to ordinary as well as to isobaric analogue states and showed that better fits to the data and better overall consistency were obtained when the single particle wavefunctions were calculated using an effective (common) binding energy (for each shell) rather than the precise separation energy for each state. Therefore, fairly well-defined binding energies of "single-particle" states could be obtained also in cases of strong fragmentation and indeed a rather simple systematics emerged. Figure 1 shows the results of Cohen et al (1963) for neutron single particle states, as obtained from stripping and pickup reactions. Very similar results were obtained by Takeuchi and Moldauer (1969) who calculated single neutron states in the real part of the optical potential obtained from fits to the elastic scattering of low energy neutrons. That clearly demonstrated the capability of the single particle model to relate different experiments with the help of a single particle potential.

Following the general success of the single particle model, attention was focused on finer details in specific regions of the periodic table. Rost (1968) obtained neutron and proton potentials near ^{208}Pb from fits to single particle energies of particle and hole states and Elton (1968) showed that the charge density distribution calculated from the proton potential agreed well with the results of electron scattering experiments. Similar results were also obtained by Batty and Greenlees (1969) who studied in greater detail the Pb and Ca regions. However, serious problems were observed by Batty (1970) who compared the single particle potential for neutrons obtained from fits to level positions with the potential resulting from fits to neutron transfer reactions and to isobaric analogue resonances. Figure 2 demonstrates these difficulties by showing level positions of single neutron

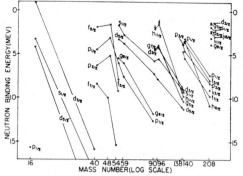

Fig. 1 Neutron single particle states throughout the periodic table. (Cohen et al. 1963)

states where the agreement between experiment and calculation is limited to particle states when calculations are made with potentials which fit cross sections. If potentials which best fit level positions are used to calculated cross sections for nucleon transfer reactions, then discrepancies between experiment and calculation as high as a factor of 3 are observed. These results were typical of some of the difficulties encountered when detailed comparisons were made between different experiments on the same nucleus, believed to be describable with the help of the single particle picture.

3. The "New" Era

The last 8-10 years may be regarded as the beginning of a new era in the study of single particle states in nuclei. With the accumulation in previous years of many experimental results and with their DWBA analyses it was not uncommon to have discrepancies in spectroscopic factors for the same reaction obtained from different sources which were far outside their respective quoted errors. That state of affairs naturally cast doubt on the validity of the single particle picture as a simple means of relating different experiments. It was only by properly distinguishing between the geometry of the problem and the nuclear spectroscopy that the single particle model could be retained as a meaningful and useful tool for correlating experimental results. This topic, together with other recent developments, are reviewed in the present section.

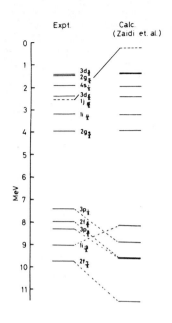

Fig. 2 Neutron single particle and single hole states in ^{208}Pb. Calculations are for a potential obtained from reaction data. (Batty 1970)

The differential cross sections for nucleon transfer reactions contain information on the transferred orbital angular momentum and on the degree of overlap between the nuclear states involved (the so-called spectroscopic factor). Whereas reliable information on the former was always obtained from the shape of the angular distribution, the latter is usually lumped together with the details of the radial wavefunctions of the transferred nucleon both in the target-residual nucleus system and in the projectile-ejectile one. This last point is usually related to the absolute normalisation of the reaction involved. Large sensitivity to the radial parameters of the single particle potential was noticed many years ago (e.g. Alty et al. 1966), but was usually ignored. The disentanglement of the three factors mentioned above was achieved only in recent years.

Figure 3 is an example of the strong dependence of cross sections calculated in the DWBA method on the root-mean-square (rms) radius of the wavefunction of the transferred neutron. In this example a change of 12% in the radius causes a change of almost a factor of 3 in the cross section. Obviously, any uncertainty in the details of that wavefunction is strongly reflected in the uncertainties in spectroscopic factors. To that, one should also add the uncertainty in the absolute normalisation. The comparative DWBA method offers a way of solving the problem, where by comparing the same nuclear transition induced by different reactions, one can deduce the absolute normalisation of one reaction relative to that of another one, in a way which is independent of the spectroscopic factors and of the wavefunction of the transferred nucleon. Figure 4 shows the results of Friedman et al.(1976) where the normalisation of the (t,d) reaction is determined relative to that of the well-known (d,p) one. Such experiments are usually performed close to the Coulomb barrier in order to reduce the dependence of the calculated cross sections on the optical potentials, and care must be taken to ensure that effects outside the DWBA, such as due to compound nucleus processes, are properly handled. The normalisations of other reactions with light particles were determined in a similar way (Friedman et al. 1977, Fink et al. 1982).

When the normalisation constant of a nucleon transfer reaction is known, the absolute scale of the cross section is determined by the spectroscopic factor and the rms radius of the wavefuntion of the transferred nucleon, and one can be determined if the other is known. Significant progress has been made in recent years on both these points. Clement and Perez (1973) have developed the Spin Dependent Sum Rule analysis (SDSR) which provides absolute spectroscopic factors from relative ones in a method which contains built-in checks on the consistency. However, they have not considered the dependence on the rms radius of the wavefunction of the transferred nucleon. That was done by Moalem (1977) who combined the comparative DWBA method with the SDSR one to provide a really comprehensive treatment of normalisations, radii and spectroscopic factors. Somewhat similar results were obtained by an elegant method of measuring neutron transfer reactions in a family of heavy ion reactions where the role of target and projectile nuclei is interchanged (Durell et al. 1976).

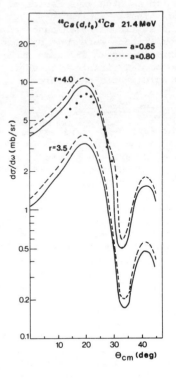

Fig. 3 Sensitivity of cross sections for neutron pickup reactions to the rms radius of the neutron orbital. (Friedman 1973)

The "magnetic" scattering of electrons by odd nucleons in nuclei is a dramatic recent development in the field of single particle states in nuclei (Sick et al. 1977, Platchkov et al. 1982). It is for the first time that an electromagnetic probe provides information on a single particle wavefunction of neutrons and good consistency is found between these results and results from nucleon transfer reactions, (Moalem and Friedman, 1978; Moalem and Vardi, 1979). It is promising that a consistent picture emerges from the comparative DWBA, the SDSR and the magnetic scattering.

As a last point one should mention the "new generation" of single particle calculations of Hodgson and collaborators (Malaguti et al. 1978) who use a single particle potential which is built to reproduce binding energies and size data (such as charge radius or spectroscopic factors). The potential turns out to be shell-dependent hence an orthogonalisation procedure is employed. Neutron densities calculated from this model have been successful in describing proton scattering in the 1 GeV region. (Ray and Hodgson 1979).

In conclusion, the single particle concept in nuclei is found to be capable of simultaneously describing several kinds of experiments and as such it is most useful in relating different types of experiment to a common basis.

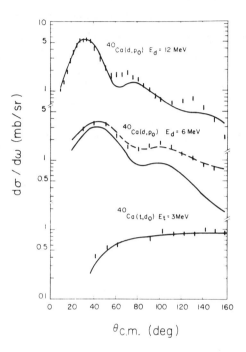

Fig. 4 Comparing neutron transfers near the Coulomb barrier. At 6 MeV compound nucleus effects are important. (Friedman et al. 1976)

References

Alty J L, Green L L, Jones J D and Sharpey-Schafer J F 1966 Nucl. Phys. 86 65
Batty C J 1970 Phys. Lett. 31B 496
Batty C J and Greenlees G W 1969 Nucl. Phys. A133 673
Burrows H B, Gibson W M and Rotblat J 1950 Phys. Rev. 80 1095
Butler 1950 Phys. Rev. 80 1095
Clement C F and Perez S M 1973 Nucl. Phys. A213 510
Cohen B L, Fulmer R H, McCarthy A L and Mukherjee P 1963 Rev. Mod. Phys. 35 332
Durell J L, Buttle P J A, Goldfarb L J B, Phillips W R, Jones G D, Hooton B W and Ivanovich M 1976 Nucl. Phys. A269 443
Elton L R B 1968 Phys. Lett. 26B 689
Enge H A and Buechner W W 1963 Rev. Sci. Instr. 34 155
Fink D, Friedman E, Paul M and Moalem A 1982 Nucl. Phys. A in press
Friedman E 1973 Phys. Rev. C 8 996
Friedman E, Moalem A, Suraqui D and Mordechai S 1976 Phys. Rev. C 14 2082
Friedman E, Moalem A, Suraqui D and Mordechai S 1977 Phys. Rev. C 15 1604
Malaguti F, Uguzzoni A, Verondini E and Hodgson P E 1978 Nucl. Phys. A297 287
Moalem A 1977 Nucl. Phys. A289 45
Moalem A and Friedman E 1978 Phys. Rev. Lett. 40 1064
Moalem A and Vardi Z 1979 Nucl. Phys. A332 195

Platchkov S K, Bellicard J B, Cavedon J M, Frois B, Goutte D, Huet M, Laconte P, Phan Xuan Ho, de Witt Huberts P K A, Lapikas L and Sick I 1982 Phys. Rev. C 25 2574

Ray L and Hodgson P E 1979 Phys. Rev. C 20 2403

Rost E 1968 Phys. Lett. 26B 184

Sherr R, Bayman B F, Rost E, Rickey M E and Hoot G C 1965 Phys. Rev. 139B B1272

Sick I, Bellicard J B, Cavedon J M, Frois B, Huet M, Leconte P, Nakada A, Phax X-H, Platchkov S, de Witt Huberts P and Lapikas L 1977 Phys. Rev. Lett. 38 1259

Takeuchi K and Moledauer P A 1969 Phys. Lett. 28B 384

Isotope shifts

Frank Träger

IBM Research Laboratory, San Jose, CA 95193, USA
and
Physikalisches Institut der Universität Heidelberg
Philosophenweg 12, D-6900 Heidelberg, Germany

1. Introduction
Isotope shifts in atomic spectra have been the main source of information about the distribution of charge in nuclear matter and its changes as more and more neutrons are incorporated. Investigations on isotope shifts date back to earlier than 1920 and were first observed in a spectral line of lead by Janicki (1909) and lateron e.g. by Merton (1919). This was many years before the neutron was discovered and the effect was not understood at that time. In 1922 Bohr suggested that isotope shifts could be due to different nuclear electric fields for different isotopes. However, theoretical calculations along these lines (Bartlett 1931, Racah 1932) could not explain the magnitude of the effect. Only after new, improved experiments had been carried out (Schüler and Keystone 1931) and Heisenberg had proposed (see Rosenthal and Breit 1932) that different numbers of a new particle called "neutron" in atomic nuclei give rise to different isotopes of a certain element, did isotope shifts begin to be fully understood. Since then, the impact of isotope shifts on nuclear physics became more and more important, e.g., Brix and Kopfermann (1949) concluded from unusually large isotope shifts in rare-earth spectra that strongly deformed nuclei must exist in this region, even if the nuclear spin is zero. In 1953, Fitch and Rainwater succeded in replacing one of the atomic electrons by another leptonic particle, namely a negative μ-meson, and measured the energies of these muonic atom X-rays. The investigation of muonic atoms has gained considerable importance, in particular for the calibration of optical isotope shifts. Measurements on hadronic, e.g. pionic or koanic atoms, have also been reported (see Backenstoss G 1975). However, interpretations of these X-ray data in terms of nuclear matter distributions are still difficult because of our limited understanding of the strong interaction.
The stock of nuclei accessible to investigations had been nearly exhausted in the 50's and 60's. However, the amount of data available on isotope shifts has rapidly increased in the last decade. This is for two main reasons: 1) The advent of dye lasers, which offer advantages like high intensity, narrow spectral bandwidth, tunability, and a small divergence of the light beam. They have replaced the classical light sources (see Heilig and Steudel 1978 for a review of the classical spectroscopic techniques). Lasers have made possible the application of new techniques (see e.g. Walther 1976, Shimoda 1976, Letokhov and Chebotayev 1977) such as saturation spectroscopy which allow for an unprecedented high resolution often limited only by the natural width of the transition under study. At the same time the detection sensitivity increased considerably, so that smaller and smaller amounts of atoms were required for isotope shift measurements. 2) The number isotopes accessible to

investigations has increased considerably. The ISOLDE facility at CERN, for example, provides many nuclei far from stability. Together with the modern laser techniques, this has initiated the study of very long chains of isotopes (see e.g. Kluge 1979, Otten 1981, Schüssler 1981, Klapisch 1981). In same cases, there is a change of the neutron number as large as 20-30 among the isotopes.

2. Electronic versus muonic isotope shifts

The shift in an atomic transition between isotopes with mass numbers A and A' is the sum of a mass shift term accounting for the change of the reduced mass of the electrons and the so-called field shift $\Delta\nu_f$:

$$\Delta\nu_{is} = \Delta\nu_f + K(A'-A)/AA'$$

The field shift is connected with changes of the nuclear charge distribution. In case of electronic transitions, it is to a good approximation proportional to the change of the mean square radius $\delta<r^2>$:

$$\Delta\nu_f = \text{const.} \ \Delta|\Psi(0)|^2 \delta<r^2>$$

where $\Delta|\Psi(0)|^2$ describes the change of the electron density at the nucleus in the transition. In muonic atoms the field shift also depends strongly on higher order moments of the charge distribution, because the large mass of the muon - 206 times the electron mass - leads to considerably smaller Bohr radii and the muon wavefunction varies considerably over the nuclear radius. The quantity determined by muonic X-rays is usually expressed as

$$<r^k e^{-\alpha r}> = 4\pi/Z \times \int \rho(r) r^k e^{-\alpha r} r^2 dr$$

These "Barrett moments" (Barrett 1970), however, can only be converted into model independent $\delta<r^2>$ values if the nuclear charge density $\rho(r)$ is explicitely known, i.e. from electron scattering experiments. If not, a Fermi distribution is often assumed. In addition, the change of the electron or muon density in the transition has to be known for the extraction of nuclear parameters. Whereas this is relatively simple for the hydrogenlike muon wavefunction, it turns out to be often difficult for optical lines. Here, empirical data can be helpful, i.e. derived from hyperfine structure splittings. In addition to $\Delta|\Psi(0)|^2$ the mass shift must be known in order to determine $\delta<r^2>$ values from experimental isotope shifts. Again, this is relatively easy for muonic transitions, where only the Bohr reduced mass correction has to be applied. For optical lines, however, a second effect comes into play which is kown as "specific mass shift" and which results from correlations in the motion of the electrons. Its calculation (Bauche 1974, Bauche and Champeau 1976) is difficult and only very recently new efforts for a more accurate determination have been made (Martensson and Salomonson 1982). Thus, the unknown or not precisely known specific mass shift often constitutes a major obstacle in extracting nuclear radii from optical isotope shifts, in particular for light nuclei where the mass shift is very large and the field shift only contributes a few percent to the total shift. The best way out of this unfavourable situation is to compare the optical with muonic or electronic X-ray isotope shifts (if available) in a normalization procedure known as a "King-plot" (King 1963), which allows for a determination of the total mass shift in the optical line. This requires data for both the optical and X-ray line. They are available in many cases for several stable isotopes. The calibration can then be extended to other, in particular very

scarce and/or short-lived isotopes far off the valley of stability, which are accessible to optical experiments only. In the "King-plot" procedure for calibrating optical shifts, data from muonic atoms play a more prominent role than measurements of electronic X-ray transitions, which are available for a rather limited number of isotopes only and often lead to a lower precision.
In summary, the combination of optical and muonic data leads to $\delta\langle r^2\rangle$ values of the highest accuracy. Both methods bring in their major advantages. For optical isotope shifts these are high experimental resolution and sensitivity. Even very scarce, in particular short lived isotopes, which cannot be studied by any other method, are accessible. In addition, optical techniques often provide further information, such as nuclear magnetic moments and electric nuclear quadrupole moments (Kopfermann 1958). Special advantages of muonic isotope shifts are that no problems with the specific mass shift are encountered. Thus nuclear radii can be determined without the help of other data. However, measurements are restricted to isotopes available in large quantities and imprecise nuclear polarization corrections can limit the final accuracy of the results.
More detailed treatments on isotope shifts have been given e.g. by Kopfermann (1958), Stacy (1966), Kuhn (1969), Barrett (1974) and Bauche and Champeau (1974). Isotope shift data together with the most important features have been compiled by Boehm and Lee (1974), Engfer et al (1974), and Heilig and Steudel (1974). For a recent experiment on electronic X-ray isotope shifts see Borchert et al (1981).

3. Examples of spectroscopic experiments

3.1 Muonic barium

The barium isotopes are an interesting case for the study of charge distributions in the vicinity of the neutron shell closure at N=82. Thus, laser spectroscopic measurements (Otten 1981, Rebel and Schatz 1982) of stable and radioactive nuclei were performed on a chain of isotopes with the neutron number changing by as much as 22 units between A=124 and A=146. However, they resulted in relatively imprecise $\delta\langle r^2\rangle$-values because the specific mass shift could only be estimated. Recently, refined investigations of X-ray isotope shifts in muonic barium for stable isotopes with mass numbers A=134-138 have been reported (Shera et al 1982). The new data provide a calibration for the optical shifts. The experiments on muonic Ba were carried out at the LAMPF facility in Los Alamos which provides a muon beam of high flux. The muons are stopped in isotopically separated targets followed by a cascade of decays from highlying levels to the ground state. The energies of the K and L X-rays were measured with a Ge(Li) detector. γ-ray samples provided known reference lines. After accounting for various corrections, such as nuclear polarization by the muon, values for the mean square radii of the nuclear charge distributions are obtained. Although a Fermi distribution for the nuclear charge was used in the analysis, the derived $\langle r^2\rangle$-values are largely model independent. This follows because both K and L X-rays, which are sensitive to different moments of the charge distribution, have been included in the analysis. The results obtained this way have then been combined with optical isotope shifts following the procedure outlined in the previous section. This leads to accurate $\delta\langle r^2\rangle$-values. Fig.1 shows their dependence on the mass number for all isotopes in the range from A=124 to A=146. One may attributes the changes of the mean square radii to a variation of $\langle r^2\rangle$ with the mass number (the nucleus being regarded as a homogeneously charged liquid drop) and to a second term accounting for changes of the nuclear deformation β at constant volume:

$$\delta\langle r^2\rangle = \delta\langle r^2\rangle_{sph} + 5/4\pi \langle r^2\rangle_{sph}\delta\langle\beta^2\rangle \tag{1}$$

In this way it is possible to account for the gross feature of Fig.1. The main shell effect at N=82 has to be attributed to variations of the nuclear deformation. It is interesting to note that other isotopic sequences in this region, such as Cs (Huber et al 1978) and Xe (Gerhardt et al 1981) show a very similar trend.

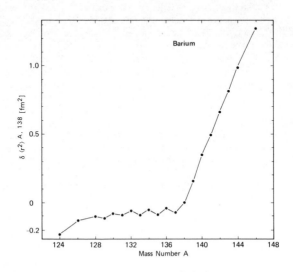

Fig.1: Nuclear charge radii versus mass number for Ba isotopes

3.2 Optical isotope shifts in mercury

In a series of experiments carried out on-line at the ISOLDE facility at CERN, mercury isotopes with mass numbers between A=181 and A=206 have been investigated (see e.g. Huber et al 1976, Kremmling et al 1979). Two experimental techniques have been applied: 1) Radiation detected optical pumping (RADOP), where the asymmetry of the β-decay or the anisotropy of the γ-radiation is used to monitor optical pumping in the atomic spectrum. Due to hyperfine coupling this pumping results in a polarization of the nuclear spins. Naturally, this technique is only applicable to nuclei with spin I≠0. Signals originating from different isotopes have been recorded by scanning a Zeeman component of the pumping line. The light source was placed in a variable magnetic field. 2) A tunable, frequency doubled, dye laser system was used to excite the Hg intercombination transition with λ=253.7 nm. A photomultiplier detected the emitted resonance fluorescence radiation. Quantities as small as 1×10^8 atoms in the resonance cell were sufficient for the measurements. Since the isotope shifts in Hg are rather large and the mass shift contributes only little, one does not have to rely on data from muonic atoms to extract $\delta\langle r^2\rangle$-values. The procedure described by Heilig and Steudel (1974) can be applied. Nevertheless, measurements on muonic Hg as well as electron scattering experiments are under way and will be combined with the optical data. Fig.2 displays the changes of the mean square nuclear charge radii for ^{181}Hg-^{206}Hg. It exhibits a rather smooth variation down to ^{186}Hg followed by tremendous changes of $\delta\langle r^2\rangle$. Here, the effect of one neutron on the radius is equivalent to the incorporation of 13 neutrons between ^{183}Hg and ^{196}Hg. This has been interpreted as a sudden change or "phase transition" from a weakly deformed oblate (for ^{184}Hg, ^{186}Hg and A>186) to a strongly deformed prolate nuclear shape for ^{181}Hg, ^{183}Hg, and ^{185}Hg.

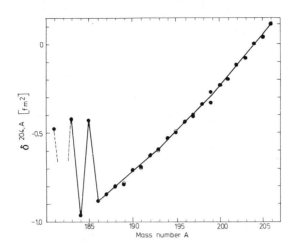

Fig.2: Nuclear charge radii versus mass number for Hg isotopes

3.3 Optical isotope shifts in calcium

The calcium isotopes have long been of major interest, since they exhibit several unique features: ^{40}Ca and ^{48}Ca have double magic nuclei, the only case where two stable isotopes of the same element with closed proton and neutron shells exist. In between the f(7/2) shell is completely filled up resulting in a change of the mass number of 20%, the neutron number varying by as much as 40%. Moreover, six stable nuclei exist in the natural composition and form the first long chain of isotopes in the periodic system. However, for an element as light as Ca the mass shift exceeds the field shift by a factor of 10 to 1000. Thus, highly accurate measurements and, in addition, a calibration by muonic data are essential for the extraction of $\delta\langle r^2\rangle$-values. Laser spectroscopic experiments (Träger 1979, (Bergmann et al 1980, Arnold et al 1981) have been carried out on the first Ca intercombination line. This transition has a natural width of only 410 Hz and permits measurements of the highest resolution if narrow band dye lasers are used. In addition to a frequency stabilized cw dye laser, an atomic beam of either natural calcium or small quantities (several μg) of radioactive isotopes was used. Due to the long life time ($\tau \approx 0.4\mu$sec) of the excited ^{3}P-state, a special trick could be applied. The reemitted resonance light can be monitored far away from the region of excitation some distance downstream along the atomic beam. This guarantees a high sensitivity because virtually all background light can be suppressed. The results of the experiments are consistent with muonic data (Wohlfahrt et al 1978) which have been used to derive $\delta\langle r^2\rangle$ values. Furthermore, good agreement is observed with measurements by classical optical spectroscopy (Brandt et al 1978) and laser experiments on the Ca resonance line (Andl et al 1982) which have recently determined the radius for ^{47}Ca. For an interpretation of the peculiar dependence of the $\delta\langle r^2\rangle$ values on the mass number one can rely again on Eq.(1). Then, the total variation of the charge radius has to be attributed to changes of the nuclear (dynamical) deformation, whereas the volume of the charge would remain constant (Träger 1981). This surprising result certainly constitutes a first order interpretation. More recent considerations indicate that higher order multipole vibrations might also influence the charge distribution (Andl et al 1981, Emrich et al 1981) of light nuclei.

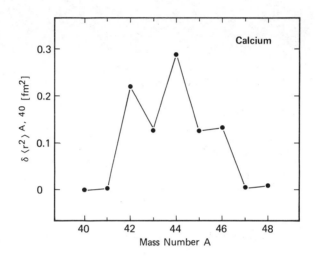

Fig.3: Nuclear charge radii versus mass number for Ca isotopes

4. Conclusions
This paper tried to put together the most important features of isotope shifts (without being a review) and to present a few examples of experimental work. However, there has also been progress in the theoretical description of nuclear radii (see e.g. Caurier et al 1980, Möller and Nix 1981, Epherre et al 1981, Campi 1982). The gross variations of the charge radii seem to be well reproduced with the droplet model (Myers and Schmidt 1981), if appropriate corrections for varying nuclear deformations are taken into account. Shell closure effects strongly correlate with changes in the permanent and/or vibrational deformation. Nevertheless, the origin of finer details of the variations, such as the "odd-even staggering", are not generally understood. Moreover, cases exist (Gerhardt et al 1979), where equation (1) does not provide a satisfactory description. Thus, more experimental and theoretical work is needed to understand better how neutrons in nuclei can influence the charge distribution.

The author gratefully acknowledges valuable discussions with G. zu Putlitz and J. Kowalski. A critical reading of the manuscript by P.S. Bagus is greatly appreciated. Most of the remarks on the history of isotope shifts are based on articles by P. Brix.
This work was primarily sponsored by the Deutsche Forschungsgemeinschaft.

5. References
Andl A, Bekk K, Goring S, Hanser A, Nowicki G, Rebel H, Schatz G, and Thompson

R C 1982 to be published
Arnold M, Bergmann E, Bopp P, Dorsch C, Kowalski J, Stehlin T, Träger F, zu Putlitz G 1981 Hyperfine Interact. 9 159 and to be published
Backenstoss G 1975 in Atomic Physics 4 ed zu Putlitz G, Weber E W and Winnacker A (New York, London: Plenum Press) p163
Barrett R C 1970 Phys. Lett. 33B 388
Barrett R C 1974 Rep. Prog. Phys. 37 1
Bartlett H J Jr 1931 Nature 128 408
Bauche J 1974 J. Physique 35 19
Bauche J and Champeau R-J 1976 in Advances in Atomic and Molecular Physics Vol. 12 ed Bates D R and Bederson B (New York, San Francisco, London: Academic Press), see also 1981 Comments Atom. Mol. Phys. 10 57
Bergmann E, Bopp P, Dorsch C, Kowalski J, Träger F and zu Putlitz G 1980 Z. Physik A292 401
Boehm F and Lee P E 1974 Atomic Data and Nuclear Data Tables 14 605
Bohr N 1922 Nature 109 746
Borchert G L, Schult O W B, Speth J, Hansen P G, Jonson B, Ravn H, and McGrory J B 1981 in Proc. 4th Int. Conf. on Nuclei far from Stability Vol.I (Geneva: Cern 81-09)
Brandt H-W, Heilig K, Knöckel H, Steudel A 1978 Z. Physik A288 241
Brix P and Kopfermann H 1949 Z. Physik 126 344
Campi X 1982 Nucl. Phys. A374 435c
Caurier E, Poves A and Zuker A 1980 Phys. Lett. 96B 11 and 15
Emrich H J, Fricke G, Hoehn M, Käser K, Mallot M, Miska H, Robert-Tissot B, Rychel D, Schaller L, Schellenberg L, Schneuwly H, Shera B, Sieberling H G, Steffen R, Wohlfahrt H D, Yamazaki Y 1981 in Proc. 4th Int. Conf. on on Nuclei far from Stability Vol.I (Geneva: Cern 81-09)
Engfer H, Schneuwly H, Viulleumier J L, Walter H K, Zehnder A 1974 Atomic Data and Nuclear Data Tables 14 509
Epherre M, Audi G, Campi X 1981 in Proc. 4th Int. Conf. on Nuclei far from Stability Vol. I (Geneva: Cern 81-09)
Fitch V L and Rainwater J 1953 Phys. Rev. 92 789
Gerhardt H, Matthias E, Rinneberger H, Schneider F, Timmermann A, Wenz R and West P J 1979 Z. Physik A292 7
Gerhardt H, Jeschonnek F, Makat W, Matthias E, Rinneberg H, Schneider F, Timmermann A, Wenz R and West P J 1981 Hyperfine Interact. 9 175
Heilig K and Steudel A 1974 Atomic Data and Nuclear Data Tables 14 613
Heilig K and Steudel A 1978 in Prog. in Atomic Spectroscopy Part A ed Hanle W and Kleinpoppen H (New York, London: Plenum Press) p263
Huber G, Bonn J, Kluge H-J and Otten E W 1976 Z. Physik A276 187 and 203
Huber G, Touchard F, Büttgenbach S, Thibault C, Klapisch R, Liberman S, Pinard J, Duong H T, Juncar P, Vialle J L, Jacquinot P and Pesnelle A 1978 Phys. Rev. Lett. 41 459
Janicki L 1909 Ann. Physik (4) 29 833
King H W 1963 J. Opt. Soc. Am. 53 638
Klapisch R 1981 in Atomic Physics 7 ed Kleppner D and Pipkin F M (New York, London: Plenum Press)
Kluge H-J 1979 in Progress in Atomic Spectroscopy Part B ed Hanle W and Kleinpoppen H (New York, London: Plenum Press) p727
Kopfermann H 1958 Nuclear Moments (New York: Academic Press)
Kremmling H, Dabkiewicz P, Fischer H, Kluge H-J, Kuhl T, and Schüssler H 1979 Phys. Rev. Lett 43 1376, see also 1977 Phys. Rev. Lett. 39 180
Kuhn H G 1969 Atomic Spectra 2nd edn (London: Longmans Green) p369
Letokhov V S, Chebotayev V P 1977 Nonlinear Laser Spectroscopy, Springer Series in Optical Sciences Vol. 4 (Berlin, Heidelberg, New York: Springer)
Martensson A-M and Salomonson S 1982 J. Phys. B: At. Mol.Phys. 15 2115

Merton T R 1919 Nature 104 406 and Proc. Roy. Soc. Series A 96 388
Möller P and Nix J R 1981 Nucl. Phys. A361 117
Myers W D and Schmidt K-H 1981 in Proc. 4th Int. Conf. on Nuclei far from
  Stability Vol.I (Geneva: Cern 81-09)
Otten E-W 1981 in Proc. 4th Int. Conf. on Nuclei far from Stability Vol.I
  (Geneva: Cern 81-09) and references therein
Racah G 1932 Nature 129 723
Rebel H and Schatz G 1982 Conf. on Lasers in Nuclear Physics Oak Ridge
  Tennessee, KfK 3344 and references therein
Rosenthal J E and Breit G 1932 Phys. Rev. 41 459
Schüler H and Keystone J E 1931 Naturwissenschaften 19 320 and Z. Phys. 70 1
Schüssler H A 1981 Physics Today Febr. p48
Shera E B, Wohlfahrt H D, Hoehn M V and Tanaka Y 1982 Phys. Lett. 112B 124
Shimoda K ed 1976 High Resolution Laser Spectroscopy, in Topics in Applied
  Physics Vol. 13 (Berlin, Heidelberg, New York: Springer)
Stacy D N 1966 Rep. Prog. Phys. XXIX 171
Träger F 1979 in Proc. Int. Discussion Meeting "What do we known about the
  radial shape of nuclei in the Ca-region?" ed Rebel H, Gils H J and
  Schatz G (Karlsruhe: KfK-Report 2830)
Träger F 1981 Z. Physik 299 33
Walther H ed 1976 Laser Spectroscopy of Atoms and Molecules, in Topics in
  Applied Physics Vol. 2 (Berlin, Heidelberg, New York: Springer)
Wohlfahrt H D, Shera E B, Hoehn M V, Yamazaki Y, Fricke G, Steffen R M, 1978
  Phys. Lett. 73B 131

Inst. Phys. Conf. Ser. No. 64: Section 3
Paper presented at Conf. on Neutron and its Applications, Cambridge, 1982

Wave properties of the neutron

C. G. Shull

Massachusetts Institute of Technology, Cambridge, Massachusetts 02139

Abstract. The wave properties of neutrons have been studied and exploited in many areas of physics almost from the time of Chadwick's discovery. Illustrations of these will be provided showing the extreme range of energy and de Broglie wavelength over which they have been observed. Attention will be directed to some of the characteristics associated with wave packet propagation.

1. Introduction

It was only a few years before Chadwick's epochal discovery that de Broglie gave revolutionary endowment of a wavelike character to all matter in motion in the form of his "pilot waves". It is difficult to overemphasize the importance of this concept of undulation to the great developments in quantum physics that were to follow in quick succession. Although originally developed to illuminate features in the electronic structure of atoms, the de Broglie marriage of "particle" momentum to the wavelength of "pilot waves" has been exploited in the physics of all constituents of matter. Nowhere has this been more dramatically effected, with the exception of electron radiation, than for the case of Chadwick's neutron radiation. It is our present theme to collect illustrative examples where the wave properties of neutrons have been useful in varied areas of physics and to discuss some of the classifying features of these neutron waves.

The realization that neutron radiation could be thermalized, i.e. brought into thermal equilibrium with matter by successive inelastic scattering processes that could occur without absorptive loss, led at an early date to the first demonstration of the de Broglie character of neutrons. Possessing normal thermal energy, the corresponding de Broglie wavelength of this thermal radiation came into the one angstrom region and this permitted wave interference demonstration by crystal diffraction techniques as first realized by von Halban and Preiswerk (1936) and by Mitchell and Powers (1936). Since that time, our thermal neutron sources have become more powerful by many orders of magnitude and it has become commonplace to exploit this wave interference in studies of condensed matter physics at our great national, neutron-physics centers such as Institut Laue-Langevin and elsewhere. The neutron technique has become a worthy competitor to the earlier established x-ray and electron diffraction techniques, indeed in many cases surpassing them, and the scientific literature abounds with such illustrations.

It is interesting to compare the actual observations in the 1936 demonstration experiments with what is typically available in present-day studies. Fig. 1 shows the von Halban-Preiswerk observations taken with a scattering sample of polycrystalline iron. It represents a "powder diffraction pattern" even though taken with full spectrum thermal neutrons approaching the sample over a 50° angular range. No Bragg reflection is recognizable but the existence of the diffraction process was indicated by the increase in forward scattered intensity when the mean neutron wavelength was reduced upon heating of the neutron source moderator. Anticipating what will be presented later, the reader may want to compare this wave interference demonstration with that available in a present-day powder diffraction pattern illustrated in later Fig. 9.

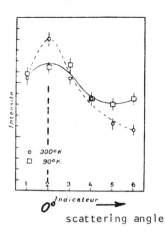

Fig. 1   Initial demonstration of neutron wave properties (von Halban and Preiswerk 1936).

scattering angle

Neutron wave interference effects are of course not confined to the thermal energy range but rather they have been explored and utilized over a very broad range. The diagram of Fig. 2 shows the relationship between the kinetic energy of neutrons and the associated de Broglie wavelength along with fiducial markers (crossed points) indicating the positions of experiments demonstrating neutron wave properties which have been selected for later discussion in this review. Relativistic contraction of the wavelength scale is to be noted in the high energy section of the diagram. One can hardly avoid being impressed by the extreme range encompassed by these wave observations, amounting to over eighteen orders of magnitude in energy and to over ten orders of magnitude in wavelength. Neutrons in the extreme energy regions are commonly referred to as ultra-cold-neutrons (UCN) with wavelengths approaching that of optical photons or ultra-hot-neutrons (UHN) where the wavelength is on a sub-nucleon scale as shown on the diagram.

It may be mentioned that the range of wave observability, broad as it is, must still be of limited extension. Certainly to the left of the diagram, a bound must be set by the decay period of a free neutron. Once the period of the neutron wave function oscillation becomes comparable to the decay period, one would hardly expect wave interference effects to be observable. This occurs at a kinetic energy of 10^{-18}ev, wavelength 2 cm,

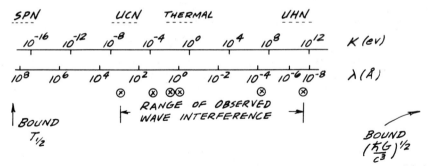

Fig. 2 Energy-wavelength classes of neutrons and range over which wave interference has been studied.

and such neutrons would be moving with a speed of 10^{-5} m/sec which is close to an average snail's pace. Accordingly they might be labelled snail's-pace-neutrons (SPN)! In the other extreme a high energy bound may be set by the Planck length $(\hbar G/c^3)^{1/2}$ of 10^{-25} Å, on which scale all of space become non-continuous or foam-like according to quantum geometrodynamical arguments. Neither of these bounds appears vulnerable to attack in the immediate future and we return to the real-cases of wave observation.

2. Single and Double Slit Interference

Perhaps the most elementary demonstration of the neutron wave properties comes from observations on the single and double slit interference patterns taken with neutron radiation. These have been studied recently by Zeilinger et al (1981) at Institut Laue-Langevin and are reproduced in Fig. 3 and Fig. 4. Monochromatic neutrons of wavelength about 20 Å were

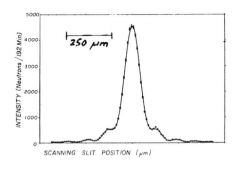

Fig. 3

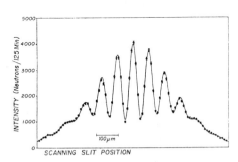

Fig. 4

Neutron interference patterns upon passage through a single slit (96 μm) and a double slit (22, 104, 22 μm). Neutron wavelengths were 19.3 and 18.45 Å respectively and the solid line patterns are calculated with standard wave treatment (Zeilinger et al 1981).

taken from a cold moderator source at the ILL reactor, passed through a 15 μm source slit and finally after flight over 5 m through the test single or double slit assembly. Detection of the interference pattern was accomplished through a scanning slit located 5 m beyond the test assembly. One sees in the patterns the characteristic maxima and minima that are expected in the interference action of overlapping wave fronts which originate at different positions within the slit assemblies. Particularly impressive is the close matching of the solid lines given by computer calculation using conventional optical wave theory and the experiment results represented by points. This has bearing on questions concerning the linearity of the Schrödinger wave equation as will be mentioned later.

3. Wave Interference with Ultra-Cold Neutrons

Of equivalent simplicity to the slit interference patterns as a demonstration of the wave character of neutron radiation are the thin-film multiple reflection observations of Steinhauser et al (1980) in which ultra-cold-neutrons were used. These experiments demonstrate the wave interference occurring between different multiply-reflected components from thin films and are analogous to the well known Fabry-Perot interference effects with optical photons. Typical results are reproduced in Fig. 5 showing a wave interference minimum in the reflected neutron intensity as a function of neutron wavelength from the double-film-sandwich shown in the inset diagram. In this assembly, which was prepared by evaporated deposition, thin films of Cu (240 Å) are separated by a layer of Al (860 Å) and interference occurs between rays reflected at the top and bottom films when the wavelength is appropriately selected. In the language of quantum mechanics, this resonance effect (and others similarly observed) represents the development of a quasibound state in the well between two potential barriers. The abscissa variable in Fig. 5 is the wavelength

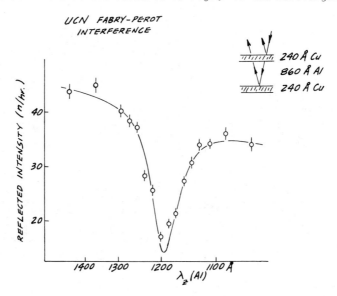

Fig. 5 Fabry-Perot interference in the reflection of ultra-cold-neutrons by Cu-Al sandwich films (Steinhauser et al 1980).

component normal to the films as established in the intervening layer and this could be adjusted by changing the gravity fall height of the incident neutrons. The peak is located at the expected 4/3 spacing value and very interestingly, the width of the peak suggests that the neutrons made about four round trips on the average within the sandwich. Equivalent transmission intensity effects have also been reported. It is to be noted that the neutron wavelength involved in these studies is approaching that of optical photons.

4. Neutron Wave Interference in the Scattering by Nucleons and Nuclei

Proceeding many orders of magnitude to higher neutron energy and shorter neutron wavelength, we can expect and we find interference effects in the scattering of neutrons by individual nucleons and nuclei. Two illustrations of this are provided in Fig. 6 and Fig. 7 showing neutron scattering by protons at extremely high neutron energy (360 GeV) in experiments of Dehaven et al (1979) and at high energy (24 Mev) by ^{116}Sn nuclei in experiments of Rapaport et al (1980). Neutrons of extremely high energy are available at the giant international accelerator centers, in the above case at Fermilab, and of course they are highly relativistic. The observations given in the figures show the variation of the differential scattering cross section with laboratory scattering angle and one sees characteristic wave interference maxima and minima. Although not particularly useful to workers in this field, this form of presentation serves our purpose of illustration. An over-simplified model of these scattering centers as being uniform spheres along with the simplest wave interference treatment would lead to a Bessel function relation between scattered intensity and scattering angle with maxima and minima. In

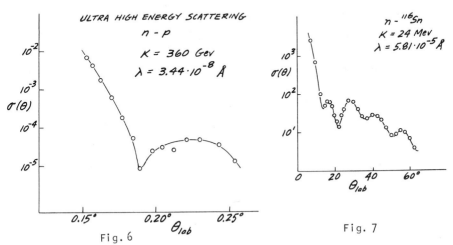

Fig. 6   Fig. 7

Differential scattering cross section as a function of scattering angle for neutron scattering by protons at 360 Gev (Dehaven et al 1979) and by ^{116}Sn nuclei at 24 Mev (Rapaport et al 1980).

spite of its naivety (particle exchange effects and internal structure are ignored), when this simple optical treatment is applied to interpret the angular position of the first minimum, one finds reasonable quantitative (within modest fractions) agreement with accepted values of the size of a proton or a Sn nucleus. The pronounced difference in the angular regions where interference occurs for the two cases arises of course because of the difference in de Broglie wavelength. There is little question that the same basic physics of wave interference is responsible for all of the observations of Figs. 3-7 spanning a dimensional scale of over ten orders of magnitude in wavelength.

5. Neutron Interferometry with Thermal Neutrons

Interferometer systems have been widely used over many years with many forms of electromagnetic radiation ranging above the ultraviolet in wavelength. A significant extension of this technique to the case of X-radiation with wavelengths on an angstrom scale was realized by Bonse and Hart (1965) who demonstrated that coherent splitting and recombination could be obtained through the Bragg diffraction process in perfect crystals. This was followed by successful operation of a thermal neutron interferometer, also operating with crystal diffraction processes in the wavelength range of an angstrom, by Rauch, Treimer and Bonse (1974). Such systems provide spatial separation of coherent components of the incident radiation over macroscopic laboratory dimensions before recombination to show wave interference intensity effects. Being available for laboratory manipulation, the coherent components can be modified in wave phase, in spin direction, in propagation direction, in momentum, and in amplitude.

These devices supply striking demonstration of the wavelike nature of neutron radiation as is shown in the fringe pattern of Fig. 8a taken from Kaiser et al (1979).

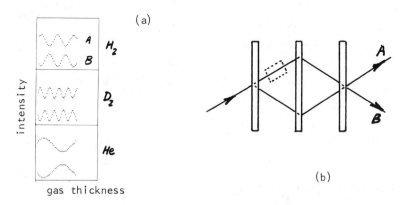

Fig. 8 Phase interference in a neutron interferometer upon changing the optical path length in one of the coherent beams with gas samples (Kaiser et al 1979).

The diagram of Fig. 8b illustrates the operational features of a typical crystal interferometer. Three identical perfect-crystal plates of silicon are used for the successive operations of coherent splitting, bending and recombination with the assembly extending over a length of perhaps 8 cm. Complementary interference information is carried in two beams (A and B) emanating from the last crystal and the intensity patterns show the effect of relative phase modification in the two coherent beams within the interferometer as sensed in either of the emerging recombined beams. The interference pattern of Fig. 8a was obtained by placing a gas cell in one of the coherent beams and changing the gas pressure. This changes the optical path length and modifies the relative phase of the separated coherent beams.

Studies with such interferometer systems are proceeding with a number of groups and new techniques and applications are developing. They have seen early application to fundamental problems of quantum mechanics, for instance in demonstrations of the wave function sign reversal for fermions upon 2π spin rotation as shown by Werner et al (1975) and by Rauch et al (1975) and of the effects of gravitational potential inclusion in the Schrödinger wave equation by Colella et al (1975). It may well be that some of the puzzling features about particle-wave duality that have been before us for many years may yield to illumination by these methods.

6. Crystal Diffraction with Thermal Neutrons

The illustrations of neutron wave properties that have been presented in the above sections have ranged over a very broad spectral range but this has perhaps been misleading in that most exploitation of neutron radiation has concentrated in the thermal energy region. Two reasons for this come to mind: namely that the neutron thermalization process inherent in nuclear reactor sources permits concentration of flux in the thermal energy region with consequent high source intensity and secondly that the neutron characteristics here--wavelength, momentum and energy--are admirably matched to atom spacings and atom excitations in ubiquitous matter. It is here that the usefulness of neutron radiation takes highest prominence and no outline would be complete without mention of this.

The very earliest demonstration of neutron wave properties as presented in Fig. 1 was possible because of the coincidental, but not unexpected, scale matching of the wavelength to atom separation in crystals. Since that period, we have of course seen great technological advances in our source characteristics and in our way of manipulating and exploiting the radiation. At the present time we are on the brink of seeing another great development through the use of pulsed, neutron spallation sources. These are being studied and constructed here in Great Britain and elsewhere and as witness to their potential for usefulness, Fig. 9 shows a time-of-flight powder diffraction pattern taken recently at the Intense Pulsed Neutron Source (IPNS) test facility in the U.S. by Jorgensen and Rotella (1982). This pattern represents scattering of full thermal spectrum neutrons at a fixed scattering angle by a randomly oriented crystal specimen. Each wavelength component finds its own Bragg diffracting set of atomic planes and all of the wealth of information shown in the figure is collected simultaneously. High speed computer processing of the data is obviously required. Not only are the relative intensities and positions of all of the Bragg reflections useful but as well the peak profiles are given computer scrutiny so that full crystallographic information becomes available. One can say that crystallography, so important in our understanding of condensed matter and its properties, will never be the same.

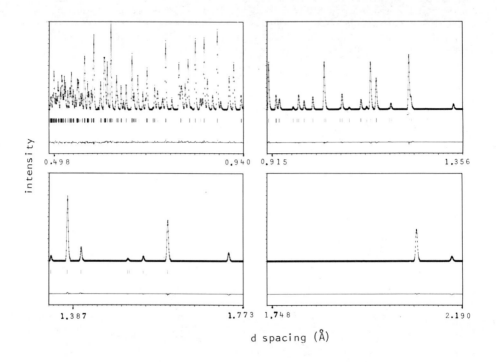

Fig. 9 Powder diffraction pattern of Al_2O_3 by flight time analysis at a fixed scattering angle of 160° using a pulsed neutron beam (Jorgensen and Rotella 1982).

A second illustration shown in Fig. 10 presents a Laue diffraction pattern taken with a single crystal of $Mn(CO)_3(C_7H_{11})$ by Schultz et al (1982) again with time-of-flight data collection from the pulsed IPNS neutron source. A state-of-the-art position-sensitive detector is used here to obtain Bragg reflection data at different angular positions in simultaneity. We are unable to do justice to the crystallographic information shown in the structural diagram of Fig. 11 that results from the data analysis other than to mention that the hydrogen atom crystallography is full established. The reader will be aware that this is an area which has received a great deal of attention by neutron wave analysis.

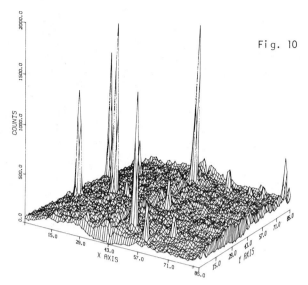

Fig. 10 Laue diffraction pattern taken for a crystal of $Mn(CO)_3(C_7H_{11})$ with a position-sensitive detector and pulsed neutron beam (Schultz et al 1982).

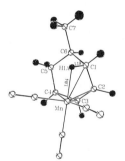

Fig. 11 Molecular structure of $Mn(CO)_3(C_7H_{11})$ from Laue pattern analysis. H atoms are shown as filled (Schultz et al 1982).

Although not serving for our purpose as particularly good illustrations of wave effects with neutrons because they are analyzed in terms of momentum and energy parameters, it must be mentioned that studies of inelastic scattering of thermal neutrons by crystals and atom assemblies, as pioneered by Brockhouse, have proven to be extremely effective in elucidating features of condensed matter states. Inelastic scattering can be treated as a modification of the wave vector, both in magnitude as well as direction, as discussed for example by Squires (1978). The matching of the kinetic energy and momentum of thermal neutrons to those of dynamic processes in crystals and molecular assemblies, positional or magnetic in nature, is of great significance and many researches have generated a wealth of information for solid state physics. It can be said that this technique of observation for studying dynamical processes in condensed matter has become the method-of-choice as will be evidenced in other presentations at this Conference.

7. Characteristics of de Broglie Neutron Waves

The illustrations of neutron wave character that have been presented above have emphasized the dominant feature of wave structure namely the periodicity with its quantifying parameter the wavelength. Without addressing the basic question of why this wave character exists, whose answer continues to be elusive after many decades, there are other features normally associated with propagating waves that are worthy of examination. These features would include classifying parameters such as wave amplitude, wave phase, wave packet character and stability, wave front characteristics and extent of wave coherency. Some of these elements have already received attention in studies of the type that have been presented and others are being given serious consideration. This outline would be neglectful without some mention of these topics.

Because of the brilliant successes of the Schrödinger wave mechanics in accounting for numberless physical observations, we are led to accept the travelling wave solutions of the Schrödinger wave equation as a description of particles in motion. In simplest form these travelling waves are infinite in extent and this presents problems with normalization and localization. These are surmounted however by allowing superposition of a band of waves to form a wave packet which now has the physical attractiveness of localization and normalizability. Presumably the longitudinal extent of such a wave packet is a measure of its coherence length and it becomes a challenge for experiment to quantify it. Indeed some of the experiment results or an extension of those that have been outlined offer some information on this--for example interferometer studies or the Fabry-Perot approach would seem useful in this respect. The data given in Fig. 5 would imply that the packet length must extend over at least several wavelengths but what is needed is a definite result rather than a lower limit so that it can be correlated with other parameters. It seems perfectly clear that neutrons prepared for one experiment must have different packet portraits than those for another and it remains as a challenge to experiment to draw such a distinction.

In parallel with the longitudinal coherency, questions arise about the transverse coherency character or the side-wise extent of the packet wave front. Again the observations shown in Figs. 3 and 4 would suggest that this extent must be greater than the transverse width of the slit assembly because of the agreement between observation and calculation. Of equal import to the scale of coherency is the question of the nature of the packet wave front, i.e. are neutrons to be described as plane waves, spherical waves, or otherwise. This again seems to be a futile or indeterminate question with the answer being that it depends upon prior-to-observation conditions. There have been experiments, extensions of those discussed, which suggest different answers but the critical and definitive experiments are yet to come.

A further implication that arises from this neutron wave packet model comes in the spreading of the packet in time and spatial postion. This has been cause for concern to many quantum physicists over the years and numerous attempts have been made to introduce stabilizing agents into the Schrödinger wave equation to overcome this spreading. Most recently, Bialynicki-Birula and Mycielski (1976) have pointed to the attractiveness of introducing a class of logarithmic, non-linear terms into the wave equation and some of the implications of this have been discussed by Shimony (1979). This has spurred experiment search for the existence of such terms by neutron optical and interferometric study with negative results even at the sensitivity level of 10^{-15} e.v. for the magnitude of such terms. This level is incredibly lower (twelve orders of magnitude) than the size of normal linear terms in the wave equation so that any perturbation of this type must indeed be on a small scale.

A very interesting question that arises with this travelling wave packet description of a neutron in motion is that associated with the response to an applied force. We know from early experiments that neutrons travelling in free space respond to the earth's gravitational force just as Newton's second law would suggest and indeed this feature has been extensively exploited in ultra-cold-neutron technology, in the gravity-fall-spectrometer of Maier-Leibnitz (1962) and in the interferometer experiment of Colella et al (1975) mentioned earlier. But for neutron wave packets travelling through a crystal medium under diffraction conditions, this is expected to be far from the truth and the anomalous response to an applied force leads to the novel concept that the neutron can be considered as possessing an *effective mass* that can be far different than its normal mass. This is worthy of discussion even though the definitive experimentation is yet to come.

In illustration of this, there is shown in Fig. 12 the case of neutrons approaching a crystal plate under exact Bragg diffraction conditions, namely that the approach angle to the diffracting atomic planes, which are normal to the plate surfaces, satisfies the Bragg law. For this case, dynamical diffraction treatment tells us that an incident packet is split coherently into two packets (α and β) which travel along the planes to the exit surface where they form two exiting beams travelling in the Bragg direction and back in the forward direction. It is known both from theory and direct experimental observation that the group velocities of these sub-packets within the crystal are slightly different, causing long-range Pendellösung beating, and that either one is significantly smaller than the group velocity of the approaching packet in spite of the fact that the index of refraction of the crystal medium is exceedingly close to unity. Moreover when a force is applied to these sub-packets, they are expected to respond with an anomalous acceleration as displayed in expressions of Fig. 12 derived by Horne and Atwood (1982). This suggests the novel concept that the neutrons in the crystal exhibit an *effective mass which can be far smaller than the normal mass and moreover can be of either sign* (the $\pm$ signs correspond to the α and β

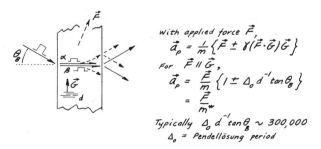

Fig. 12   Effect of applied force upon neutrons being diffracted by crystalline plate. An anomalous acceleration is produced which can be interpreted in terms of an effective neutron mass m*.

sub-packets). Anomalously large deflections are to be expected in the crystal upon force application (gravitational, Coriolis or magnetic gradient in origin) as suggested in the figure. Such effects are presently being searched for in experiments by Zeilinger and Werner.

Interesting questions that should be amenable to interferometer study arise regarding the distribution of wave function amplitude and phase upon coherent subdivision of the wave function. Likewise the unusual phenomenon of wave function reduction, at the heart of particle-wave duality, may yet yield some of its secrets to neutron study. The exploitation of neutron wave properties continues as a challenging area of study.

The preparation of this report was supported in part by the U.S. Department of Energy through Grant No. DE-AC02-76ER03342 and by the National Science Foundation through Grant No. DMR-8021057A01.

9. References

Bialynicki-Birula I and Mycielski J 1976 Ann. Phys. (N.Y.) 100 62
Bonse U and Hart M 1965 Appl. Phys. Lett. 6 155
Colella R, Overhauser A W and Werner S A 1975 Phys. Rev. Letters 34 1472
Dehaven C E, Ayre C A, Gustafson H R, Jones L W, Longo M J, Ramana Murthy P V, Roberts T J and Whalley M R 1979 Nucl. Phys. B148 1
von Halban H and Preiswerk P 1936 Compt. Rendus 203 73
Horne M A and Atwood D K 1982 unpublished
Jorgensen J D and Rotella F J 1982 J. Appl. Cryst. 15 27
Kaiser H, Rauch H, Badurek G, Bauspiess W and Bonse U 1979 Z. Physik A291 231
Maier-Leibnitz H 1962 Z. angew. Physik 14 738
Mitchell D P and Powers P N 1936 Phys. Rev. 50 486
Rapaport J, Mirzaa M, Hadizadeh H, Bainum D E and Finlay R W 1980 Nucl. Phys. A341 56
Rauch H, Treimer W and Bonse U 1974 Phys. Letters 47A 369
Rauch H, Zeilinger A, Badurek G, Wilfing A, Bauspiess W and Bonse U 1975 Phys. Letters 54A 425
Schultz A J, Teller R G, Beno M A, Williams J M, Brookhart M, Lamanna W and Humphrey M B 1982 unpublished
Shimony A 1979 Phys. Rev. A20 394
Squires G L 1978 "Introduction to the Theory of Neutron Scattering" Cambridge University Press, Cambridge
Steinhauser K A, Steyerl A, Scheckenhofer H and Malik S S 1980 Phys. Rev. Letters 44 1306
Werner S A, Colella R, Overhauser A W and Eagen C 1975 Phys. Rev. Lett. 35 1053
Zeilinger A, Gaehler R, Shull C G and Treimer W 1981 Sympos. on Neutron Scattering, Argonne National Laboratory, publ. by Amer. Inst. of Physics

The present state of neutron interferometry

H Rauch

Atominstitut der Österreichischen Universitäten, A-1020 Wien, Austria

Abstract. Widely separated coherent neutron beams are produced by perfect crystal neutron interferometry. Such devices represent a novel tool of research in the field of fundamental, nuclear and solid state physics. Mainly the features and the results of the interferometer setup at the high flux reactor at Grenoble, France, are described, where high degrees of coherence and high order interferences have been observed. The magnetic coupling permits the observation of the 4π-symmetry of spinor wave functions and recently the explicit verification of the spin superposition law. A measurement of the transverse coherence length using the monolithic interferometer crystal is reported. The features of the coherent inelastic energy shift produced by neutron magnetic-resonance systems are described.

1. Introduction

An important contribution to the progress in neutron optics has been achieved with the development of neutron inteferometry. Interferometry based on wave front and amplitude division has been realized in the past. The first attempt was made by Maier-Leibnitz and Springer (1962) using wave front division by single slit diffraction. Recently this technique has been developed further by Gähler et al. (1980) and has been used by Zeilinger et al. (1981a) and Klein et al. (1981) for some fundamental measurements of quantum mechanics. The characteristics of this technique are long beam paths but rather small beam separation, a limitation to low order interferences and an intensity limitation due to the narrow entrance slit.

Amplitude division interferometry was first demonstrated by Rauch, Treimer and Bonse (1974) using a perfect monolithic silicon crystal as shown in Fig. 1. Two widely separated coherent beams are produced, reflected and superposed by dynamical diffraction effects. The beam separation is up to 5 cm but the path length is limited to values below 10 cm due to the required perfection of the crystal. The roots of this technique are the slit neutron interferometer of Maier-Leibnitz and Springer (1962), the X-ray perfect crystal interferometer of Bonse and Hart (1965) and the observation of dynamical neutron diffraction by Shull (1968). Many textbook experiments of quantum mechanics dealing with matter waves have been performed with this technique. Typical examples are the verification of the 4π-symmetry of a spinor by Rauch et al. (1975) and Werner et al. (1975) and the observation of gravity effects on the phase of the neutron wave by Colella et al. (1975). Recent results on these subjects have been reported by Rauch et al. (1978a), by Staudenmann et al. (1980) and by Summhammer et al. (1982). The whole field of neutron interferometry up to 1978 is summarized in the

Proceedings of a Workshop at Grenoble (ed. Bonse and Rauch 1979a) and reviewed recently by Klein and Werner (1982).

In this article recent measurements performed by the Dortmund-Grenoble-Wien neutron interferometer cooperation at the high flux reactor at Grenoble are reported and the initial attempts of a Jülich-Wien cooperation to introduce optically active components by a neutron magnetic resonance system are discussed.

2. Interference Measurements

The well defined arrangement of the atoms within the monolithic perfect crystal guarantees the coherence of the neutron beams. The function of the perfect crystal interferometers is described by the dynamical diffraction theory for neutrons (e.g. Rauch and Petrascheck 1978b, Sears 1978), which has been applied to the triple Laue-case interferometer by Bauspiess et al. (1976) and Petrascheck (1976). For ideal geometry the wave function of the beam behind the interferometer in the forward direction is composed of equal parts coming from beam path I and beam path II

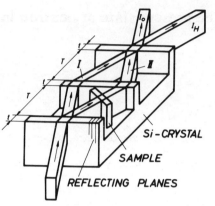

Fig. 1 Sketch of the perfect crystal neutron interferometer

$$\psi_0^I = \psi_0^{II}$$
$$I_0 = |\psi_0^I + \psi_0^{II}|^2 = 4|\psi_0^I|^2 \quad . \tag{1}$$

A sample of thickness D inserted into one of the coherent beams changes the wave function by a factor $\exp(i\chi)$, where the phase shift χ is determined by the index of refraction n, which is given by the mean optical potential $\overline{V}$ of the neutrons within this material

$$n = \frac{k}{k_0} = \sqrt{1 - \frac{\overline{V}}{E}} \simeq 1 - \frac{\overline{V}}{2E} \tag{2}$$

$$\chi = -(1-n)kD = -\lambda N b_c D \quad .$$

N is the particle density, b_c the coherent scattering length, E the kinetic energy and $\lambda = 2\pi/k$ the wave length of the neutron. Equations (1) and (2) yield a beam modulation as a function of the sample thickness or of the density

$$I_0 = 2|\psi_0^I|^2 (1 + \cos N b_c \lambda D) \tag{3}$$

The absorption and the incoherent scattering processes, the wave length distribution of the beam and the residual imperfections of the crystal cause an attenuation of the interference pattern, but the experimental results show that the idealized behavior can be reached rather closely (Fig. 2). Beam modulations up to 85% and high order interferences up to the 329th order have been observed (Rauch et al. 1976, Rauch 1979). It is worthwhile mentioning that the attenuation of the interference pattern at high order due to the wave length distribution of the beam is equally well described by a distribution function or by a wave packet representation of the wave functions (Rauch 1982a) In analogy to neutron spin echo systems, neutron phase echo systems become feasible (Badurek et al. 1980).

Equation (3) gives the basis for routine measurements of coherent neutron-nuclei scattering lengths (Rauch et al. 1976, Bauspiess et al. 1978, Kaiser et al. 1979, Hammerschmied et al. 1981, Bonse and Kischko 1982). As an example the experimental procedure and the results of the neutron-triton scattering length measurement is shown in Fig. 3 (Hammerschmied et al. 1981).

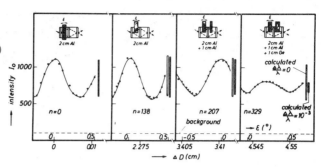
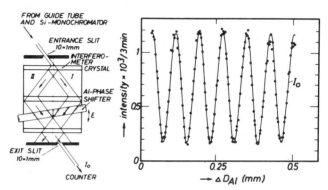

Fig. 2 Measurement of high order interferences (above) and of a high degree of beam modulation (below)

An appropriate balance technique has been used to account for the much larger phase shift of the tritium steel container, which was needed as a safeguard against radiation hazards of the tritium gas. The entire amount of tritium gas was 0.124 g ($\triangleq$ 1200 Ci) and the thickness of the container was 2 x 4 mm. The coherent neutron-triton scattering length could be extracted from the phase shift between the observed beam modulations measured by means of an auxiliary phase shifter to be b_c = 5.10 ± 0.10 fm. This quantity is, together with the earlier measured neutron-He-3 scattering length (b_c = 5.74 ± 0.07 fm, Kaiser et al. 1979), an important parameter for testing modern nuclear four-body theories. An extended compilation of all known coherent scattering lengths has been made recently by Koester et al. (1981).

3. Spinor Symmetry and Spin Superposition

The magnetic interaction (H = $-\vec{\mu}\vec{B}$) between the magnetic moment $\vec{\mu}$ of the neutron and the magnetic field $\vec{B}$ causes an additional phase factor in the form of a unitary rotation operator

$$U = \exp(-i\vec{\sigma}\vec{\alpha}/2) \qquad (4)$$

where $\vec{\sigma}$ are the Pauli spin matrices and $\vec{\alpha}$ is the rotation angle around the field. $\vec{\alpha}$ is numerically equal to the Larmor precession angle $\alpha = \gamma \int B d\tau$, where $\gamma = 2\mu/\hbar$ and τ is the time-of-flight of the neutrons within the field B. If the Zeeman splitting is small compared to the kinetic energy of the neutrons ($2\mu B \ll E$) the integral can be rewritten in the form $\alpha = (\gamma/v) \int B ds$, where v is the neutron velocity and ds is the path element. Using equations (1) and (4) one obtains the following beam modulation (Eder and Zeilinger 1979)

$$I_0 = 2|\psi_0^I|^2 (1 + \cos \frac{\alpha}{2}) \qquad (5)$$

showing the characteristic 4π-symmetry of a spinor wave function. This phenomenon has been verified experimentally by perfect crystal neutron interferometry (Rauch et al. 1975, Werner et al. 1975) and by Fresnel diffraction interferometry (Klein and Opat 1976). A precision measurement done later and using well defined magnetic fields within Mu-metal sheets yield a periodicity factor of (Rauch et al. 1980a)

$$\alpha_0 = 715.9 \pm 3.8 \text{ deg} \qquad (6)$$

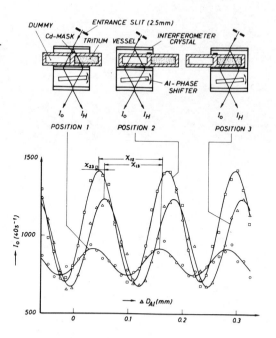

Fig. 3 Experimental procedure and results of the neutron-triton scattering length measurement

after correction for recent values of the physical constants. Bernstein and Zeilinger (1980) pointed out that for thermal neutrons the difference between static and dynamic rotations is far below the experimental error bars. Depending on the axis of quantization the effect is equally well understood by a spinor rotation around the magnetic field or by an index of refraction formalism. Additional references on the interpretation of this quantum mechanical phenomenon are Bernstein (1979), Mezei (1979), Byrne (1978), Silverman (1980), Zeilinger (1981b) and Santilli (1981).

Eder and Zeilinger (1976) also calculated the beat effects if nuclear (χ) and magnetic (α) phase shifts are applied simultaneously. The mutual intensity modulation and the polarization of the beam behind the interferometer if unpolarized incident neutrons are used have been demonstrated by Badurek et al. (1976).

Recently it was possible to extend the experiments to polarized incident neutron beams permitting an explicit verification of the so-called Wigner (1963) phenomenon. This predicts a final polarization after superposition of two oppositely polarized coherent beams which is perpendicular to both initial polarization states. The theoretical framework for the case of neutrons is given by Eder and Zeilinger (1976) and by Zeilinger (1979). The unitary operator describing a polarization reversal in one beam and an additional nuclear phase shift is

$$U = \exp(i\chi)\exp(-i\vec{\sigma}\vec{\alpha}/2) = -i\sigma_x \exp(i\chi) \qquad (7)$$

which predicts a final polarization

$$\vec{P} = \frac{\psi_0^{I+}(1+U^+)\vec{\sigma}(1+U)\psi_0^I}{I_0} = \hat{e}_x \sin\chi - \hat{e}_y \cos\chi \qquad (8)$$

lying in the (xy)-plane and rotatable within this plane by the purely scalar nuclear interaction. The experimental arrangement used for this experiment is shown in Fig. 4 (Summhammer et al. 1982a, Summhammer et al. 1982b). The spin reversal within the interferometer and the three-dimensional polarization analysis of the beam behind the interferometer is achieved by DC-flippers. Polarized incident neutrons are produced by magnetic prism deflection and the use of the nondispersive monochromator-interferometer arrangement (Badurek et al. 1979). The experimental results (Fig. 5) show that the final polarization lies in the (xy)-plane and that it can be rotated equivalently by the nuclear phase shift χ within the interferometer or by an additional Larmor rotation accelerator coil mounted behind the interferometer. This equivalence is shown additionally on the insert of this figure.

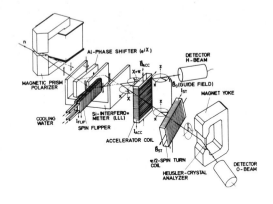

Fig. 4 Experimental arrangement for the spin superposition experiment

In a future experiment the influence of a HF-flipper will be investigated. In contrast to the DC-flipper which changes the kinetic energy E, the HF-flipper device changes at resonance the total energy E_t of the neutron and transforms the time dependent part of the wave function from

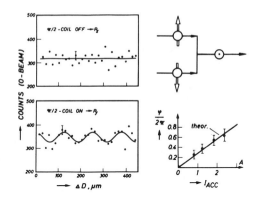

Fig. 5 Experimental result of the spin superposition experiment (left), its principle (right above) and the equivalence of scalar nuclear phase shift and additional Larmor precession (right below)

$\exp(-iE_t t/\hbar)$ to $\exp[-i(E_t \pm 2\mu B)t/\hbar]$ (Krüger 1980, Badurek et al. 1980). A time dependent measurement will be performed to observe the dependence of the phase of the neutron wave from the phase of the HF-field.

4. Coherent Multiple Laue-Reflections

The interferometer crystal (Fig. 1) is also suited for measuring multiple Laue-reflections whose rocking curves show a marked needle structure at the center (Bonse et al. 1977, Bonse et al. 1979b). This effect is described by the convolution of the well-known Pendellösung diffraction curve (e.g. Rauch and Petrascheck 1978)

$$I \propto |\psi|^2 = \frac{\sin^2 A \sqrt{1+y^2}}{1+y^2} \quad (9)$$

where for symmetrical Laue reflection A and y are given as

$$A = \frac{2\pi b_c N}{k \cos \Theta_B} t$$

$$y = \frac{(\Theta_B - \Theta) k^2 \sin 2\Theta_B}{4\pi b_c N} \quad (10)$$

t ... thickness of the crystal plate
Θ_B .. Bragg angle

The convolution yields for double Laue-diffraction a width of the central peak given as d_{hkl}/t where d_{hkl} is the lattice constant of the reflecting plane and which is on the order of 0.001 sec of arc (Bonse et al. 1979). The central peak to background ratio is further reduced for triple Laue diffraction (Rauch et al. 1982b). The high angular resolution is achieved by prism deflection by wedge shaped materials rotated around the beam axis. This deflection δ depends on the wedge angle β and the rotation angle α

$$\delta = (N b_c \lambda^2/\pi) \, tg(\beta/2) \sin \alpha \quad (11)$$

A recent experiment has observed the diffraction at a macroscopic slit introduced between the successive Laue-reflections (Fig. 6). The expected beam broadening is given by the Frauenhofer formula as $\beta_{1/2} \simeq 0.888 \, \lambda/a$, where a is the slit width (Shull 1969). The experimental results (Fig. 7) show this

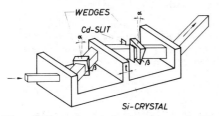

Fig. 6 Experimental arrangement for the observation of wide slit neutron diffraction

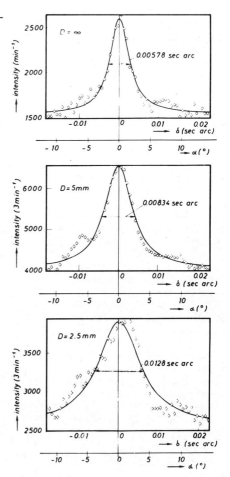

Fig. 7 Central peak of three successive Laue-reflections without (top) and with slits within the beam

broadening and indicate a
transverse coherence
length of 6.5 mm which
is 3.7×10^7 times the
neutron wave length.

5. Coherent Neutron Magnetic Resonance Systems

The Zeeman splitting of a
neutron within a magnetic
field can be doubled at
the exit if the polariza-
tion is inverted within
the magnetic field by a
time dependent interac-
tion as achievable by a
HF-flipper device
(Drabkin and Zhitnikov
1960, Badurek et al.

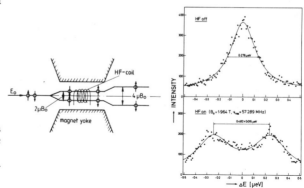

Fig. 8 Principle (right) and experimental verification of the inelasticity of neutron magnetic
resonance systems

1980). This phenomenon has been verified recently by Alefeld, Badurek and
Rauch (1981; see Fig. 8) and is now available as an optically active element
usable for the realization of a dynamical neutron polarization system or for
pumping of neutrons into a certain energy or angle interval. For most applications a multistage system is needed to make the coherent energy shift
larger than the energy width of the incident beam. For pulsed beams additional advantages arise due to their space-time dispersion along the flight
path which permits the application of proper energy, time or space focusing
and bunching procedures (Rauch 1982c). This beam tailoring can be done along
the guide tubes between the source and the experiment and needs no material
within the beam. These effects are described by the time-dependent Schrödinger equation and can be understood as interference phenomena along the beam
axis without spatial separation of the coherent beams.

The fruithful cooperation with G.Badurek, E.Balcar, G.Eder, S.Hammerschmied,
J.Summhammer (Wien), U.Bonse (Dortmund), W.Treimer (Berlin), D.Petrascheck
(Linz), A.Zeilinger (M.I.T.), U.Kischko (Grenoble) and B.Alefeld (Jülich)
is gratefully acknowledged.

References

Alefeld B, Badurek G and Rauch H 1981 Z.Physik B41 231
Badurek G, Rauch H, Zeilinger A, Bauspiess W and Bonse U 1976 Phys. Rev.
   D14 1177
Badurek G, Rauch H, Wilfing A, Bonse U and Graeff W 1979 J. Appl. Cryst.
   12 186
Badurek G, Rauch H and Zeilinger A 1980 Proc. Neutron Spin Echo ed F Mezei
   (Berlin, Heidelberg, New York: Springer) pp 136 - Lect. Notes in Phys.
   128 136
Bauspiess W, Bonse U and Graeff W 1976 J. Appl. Cryst. 9 68
Bauspiess W, Bonse U and Rauch H 1978 Nucl. Instr. Meth. 157 495
Bernstein J 1979 Neutron Interferometry ed U Bonse and H Rauch (Oxford:
   Clarendon) pp 231
Bernstein J and Zeilinger A 1980 Phys. Lett. 75A 169

Bonse U and Hart M 1965 Appl. Phys. Lett. 6 154 and 7 99
Bonse U, Graeff W, Teworte R and Rauch H 1977 phys. stat. sol. (a) 43 487
Bonse U and Rauch H (ed) 1979a Neutron Interferometry (Oxford: Clarendon)
Bonse U, Graeff W and Rauch H 1979b Phys. Lett. 69A 420
Bonse U and Kischko U 1982 Z. Physik A305 171
Byrne J 1978 Nature 275 188
Colella R, Overhauser A W and Werner S A 1975 Phys. Rev. Lett. 34 1472
Drabkin G M and Zhitnikov R A 1960 Sov. Phys. JETP 11 729
Eder G and Zeilinger A 1976 Nuovo Cim. 34B 76
Gähler R, Kalus J and Mampe W 1980 J. Phys. E13 546
Hammerschmied S, Rauch H, Clerc H and Kischko U 1981 Z. Physik A302 323
Kaiser H, Rauch H, Badurek G, Bauspiess W and Bonse U 1979 Z. Physik A291 231
Klein A G and Opat G I 1976 Phys. Rev. Lett. 37 238
Klein A G, Kearney P D, Opat G I, Cimmino A and Gähler R 1981 Phys. Rev. Lett. 46 959
Klein A G and Werner S A 1982 Rep. Progr. Phys. (in print)
Koester L, Rauch H, Herkens M and Schröder K 1981 Summary of Neutron Scattering Lengths Jül-report-1755 (Jülich: KFA)
Krüger E 1980 Nukleonika 25 889
Maier-Leibnitz H and Springer T 1962 Z. Physik 167 386
Mezei F 1979 Neutron Interferometry ed U Bonse and H Rauch (Oxford: Clarendon) pp 265
Petrascheck D 1976 Acta Phys. Austr. 45 217
Rauch H, Treimer W and Bonse U 1974 Phys. Lett. A47 369
Rauch H, Zeilinger A, Badurek G, Wilfing A, Bauspiess W and Bonse U 1975 Phys. Lett. 54A 425
Rauch H, Badurek G, Bauspiess W, Bonse U and Zeilinger A 1976 Proc. Int. Conf. Interaction Neutrons with Nuclei ed E Sheldon (Lowell: University) Vol. II pp 1027-CONF-760712-82
Rauch H, Wilfing A, Bauspiess W and Bonse U 1978a Z. Physik B29 281
Rauch H and Petrascheck D 1978b Neutron Diffraction ed H Dachs (Berlin, Heidelberg, New York: Springer) pp 303 - Top. Curr. Phys. 6 303
Rauch H 1979 Neutron Interferometry ed U Bonse and H Rauch (Oxford: Clarendon) pp 161
Rauch H 1982a Proc. Coll. Wave Particle Dualism ed G Loschak, D Fargue and S Diner (Dordrecht: Reidel) in print
Rauch H, Kischko U and Bonse U 1982b Z. Physik A in preparation
Rauch H 1982c Yamada Conf. Neutron Scattering of Condensed Matter Physica B in print
Santilli R M 1981 Hadronic J. 4 1166
Sears V F 1978 Can. J. Phys. 56 1261
Shull C G 1968 Phys. Rev. Lett. 21 1585
Shull C G 1969 Phys. Rev. 179 752
Silverman M P 1980 Eur. J. Phys. 1 116
Summhammer J, Badurek G, Rauch H and Kischko U 1982a Phys. Lett. A90 110
Summhammer J, Badurek G, Rauch H, Kischko U and Zeilinger A 1982b Phys. Rev. A in preparation
Staudenmann J L, Werner S A, Colella R and Overhauser A W 1980 Phys. Rev. A21 1419
Werner S A, Colella R, Overhauser A W and Eagen C F 1975 Phys. Rev. Lett. 35 1053
Wigner E P 1963 Am. J. Phys. 31 6
Zeilinger A 1979 Neutron Interferometry ed U Bonse and H Rauch (Oxford: Clarendon) pp 241
Zeilinger A, Gähler R, Shull C G and Treimer W 1981a Symp. Neutron Scattering Argonne Aug. 1981 in print
Zeilinger A 1981b Nature 294 544

Production and properties of ultracold neutrons

A Steyerl

Fakultät für Physik, Technische Universität München, D-8046 Garching

1. Introduction

Twenty three years ago, i.e. almost halfway back in the history of the neutron, it was pointed out by the Soviet physicist Zel'dovich (1959) that neutrons, in addition to their many useful - and in some "applications" less useful - properties, should exhibit another quite surprising property: You can fill them in a bottle and carry them home. One problem implied in this proposition is that it works only for extremely slow, the "ultracold", neutrons, and that it is not easy to produce these neutrons. This problem has been solved fourteen years ago in Garching (Steyerl 1969) and in Dubna (Lushchikov et al 1969). The first successful containment of ultracold neutrons (UCNs) was achieved in Moscow (Groshev et al 1971), and UCNs have, with increasing skill, been bottled up ever since. In the meantime, the small but growing community of those engaged in UCN research, seem to have solved another problem as well: They have, apparently, convinced a number of erstwhile skeptics that it is not only possible but also useful to produce, and do physics, with neutrons which can be bottled up. The opening of this new branch of research was tedious, as most of the methods and apparatus required were utterly unconventional, and they had to be developed starting from "zero". Moreover, the understanding of the physical phenomena involved in UCN storage was confused, for many years, by unclear experimental data. However, I think that the various contributions on ultracold neutron research presented at this "golden" anniversary of the neutron, demonstrate the significant progress made in all the diverse fields of UCN physics.

2. Production of ultracold neutrons

Although UCNs exist, as the low-energy fraction of the Maxwell-Boltzmann distribution, in the moderator of a reactor or other primary neutron source in small but appreciable numbers, it is well-nigh impossible to directly extract them from there. They are not able to penetrate the necessary structure material. Therefore it has become customary to re-create the UCNs from originally faster neutrons, either within the beam tube or immediately in front of the experiment. A number of different techniques have been applied:

    a Down-scattering in a "good" converter or "Cold Source" at low temperature, placed into the beam tube nose;
    b Deceleration by the action of gravity in a vertical or inclined neutron guide tube;
    c Spectral transformation by use of a mechanical "neutron turbine",

using as the receding mirrors either totally reflecting curved "turbine blades", or Bragg reflecting mosaic crystals;

d Down-scattering of 10 A neutrons in, probably, the "best" converter, namely helium-4 at a temperature below 1 K.

All these methods, and combinations thereof, have their special merits. The vertical guide tube, which had been pioneered in the original facility at Garching, is now being used in Leningrad in conjunction with an in-pile hydrogen Cold Source (Altarev et al 1980). It provides the strongest UCN beam available at present. The Garching-type total reflection "neutron turbine" (Steyerl 1975) transforms a continuous beam of faster neutrons, which is fairly easy to extract from the reactor, into an intense UCN beam of much wider cross section. We hope that the "Vertical UCN Source" which is presently in preparation at the Grenoble high-flux reactor - a vertical guide from the large D_2 Cold Source feeding a "neutron turbine" - will get close to the limit of possible stationary UCN beam intensity from a reactor. In the Doppler-shifting crystal device, which was pioneered in its principle by Shull and Gingrich (1963) and by Buras and Kjems (1973), among others, and which has now been realized at the Argonne Spallation Source (Brun et al 1980), a short burst of incident 10 A neutrons is stretched in time to produce a long intense UCN pulse. Thus, this technique is especially suited for pulsed neutron sources. Finally, the "Super-thermal Helium Source" invented by Golub and Pendlebury (1975) operates on the principle of accumulating a high density of UCNs in a liquid helium filled trap. Thus, it is capable of producing intense bursts of UCNs from a stationary primary neutron source.

3. Properties of ultracold neutrons

In a sense, UCNs are not different from other neutrons. They exhibit the same kinds of interaction with matter and fields. However, at very low neutron energies the usual nuclear, "electro-weak", and gravitational forces give rise to a number of unusual phenomena. The most prominent of these is total reflection of the neutron wave from the surface of many substances, at any angle of incidence, even normal to the surface. This phenomenon occurs for neutron energies below the locally averaged Fermi scattering potential for the wall, which is of the order of 10^{-7} eV for many substances. This peculiarity of UCNs is the basis for the possibility to store them in "neutron bottles". Storage has been the subject of many experimental endeavours, and when these failed to demonstrate the expected long storage lifetimes, it became the subject of various theoretical investigations, and when these failed to explain the "anomaly", even of bold speculations. Fortunately, I think, this problem has finally been settled by the recent experiments in Grenoble (Mampe et al 1981), which strongly suggest that the "anomalies" had simply been "dirt effects".

Since the original suggestion of Shapiro (1968) to use neutrons in a bottle for a highly sensitive search for an electric dipole moment of the neutron, this ambitious idea has been pursued vigorously in many laboratories (see, e.g., Pendlebury's contribution to this Conference).

Just as small as the Fermi nuclear scattering potential, but of similar importance for the behaviour of UCNs, are the magnetic interaction energy of 0.06 µeV per tesla and the gravitational potential of 0.1 µeV per metre of height. Ultracold neutrons can be confined in suitable magnetic field configurations, and a group of the University of Bonn has demonstrated that the neutrons live in a superconducting magnetic "neutron storage ring" for times consistent with the beta-decay time (Kügler et al 1978). This indicates the potential of UCNs for an improved measurement of the neutron's lifetime.

Since the present talk is classified under the heading "Neutron Optics" it may be appropriate to combine all the interactions mentioned, in the form of a spatially variable index of refraction for the neutron wave:

$$n(\underline{r}) = \left[1 - (U_{nucl} + U_{mag} + U_{grav} + ...)/E_0\right]^{1/2},$$

where U_{nucl} is the locally averaged nuclear scattering potential, $U_{mag} = \pm \mu_N B$ is the magnetic interaction between the neutron's magnetic moment μ_N and the B-field, and $U_{grav} = m_N gz$ is the Newtonian gravitational potential energy for the neutron mass m_N. E_0 denotes the kinetic energy of the neutron at an (arbitrary) point in space where the total potential is set equal to zero. Inclusion of the magnetic interaction in the form $\pm \mu_N B$ (allowing for the spin states parallel or antiparallel to B by the positive or negative sign) is, of course, justified only in adiabatic conditions where the neutron spin $\underline{\sigma}$ can "follow" orientational changes of the B-field. The dots indicate that other interactions, e.g. the parity violating ($\underline{\sigma} \cdot \underline{p}$)-type interaction giving rise to "neutron-optical activity" (Forte et al 1980), can be included as well.

As a consequence of the gravitational interaction even empty space, in the absence of magnetic fields, may be considered a refracting medium with a spatially variable index of refraction. It seems quite amusing to consider the effects of this interaction on the UCN wave field. Luschikov and Frank (1978) pointed out that discrete bound states with energies of the order of 10^{-12} eV should exist for neutrons on a flat mirror plate. Recently Berry (1982) described the "neutron rain bow", i.e., the diffraction pattern to be expected near the caustic surface (the "bounding paraboloid" of ballistics) for mono-energetic UCNs emerging from a narrow hole. An experimental demonstration of either of these effects would seem to be very difficult in view of the low beam intensities from existing UCN sources.

However, the effects of gravity on the neutron wave are very pronounced, and easy to observe, in the limit of geometrical neutron optics. This limit may be derived in different ways: Either by the application of Ehrenfest's theorem to the Schrödinger equation, or by using Fermat's principle (in the original form of stationary wave phase, not time!) in combination with the index of refraction given above. Either of these approaches is based on the concept of an extended coherent neutron wave field and yields the curvilinear rays consistent with the flight parabolas of classical particles. The beam curvature is so strong for UCNs that we may consider it another of their "special" properties. In the remainder of the present lecture I shall describe the methods we are using to exploit this feature.

The gravitational field is an extremely dispersive "medium" for ultracold neutrons. This provides the possibility to determine, or analyse, the energy of a UCN very precisely by measuring the maximum reach of its flight parabola, i.e., the distance between the launching point and an impact point which is chosen to lie on the caustic surface. The "UCN gravity spectrometer" NESSIE (for "NEutronen-Schwerkraft-SpektrometrIE") makes use of this principle (Steyerl 1978). With this instrument a FWHM of (16.9 ± 0.7) neV has been measured for an elastic scattering line. For a polymer in solution (polydimethylsiloxane in C_6D_6 at 70°C) a quasi-elastic line broadening by (7.0 ± 3.5) neV has been observed at a momentum transfer of 0.03 Å^{-1}/ℏ. This energy resolution is significantly higher than for other schemes of neutron spectroscopy, except the neutron spin-echo technique (see Mezei's contribution to this Conference).

The gravitational potential depends only on the vertical height z. Thus, the height of fall is a precise measure of the "vertical energy" (the kinetic energy corresponding to the vertical component of velocity) for a neutron travelling along its flight parabola. This simple principle which is the basis of the "neutron gravity refractometer" (Maier-Leibnitz 1962) has been employed also in the "UCN gravity diffractometer" (Scheckenhofer and Steyerl 1977). We have used the high resolution (1 neV for the "vertical energy") provided by this instrument, for demonstrating a few elementary quantum wave phenomena with UCNs, i.e., in a region of wavelength (∼100 nm) somewhere between the "microworld" of atoms and the "macroworld" of our more direct experience. Among these phenomena are diffraction by a ruled reflection grating (Scheckenhofer and Steyerl 1977) and the formation of resonances (or quasibound states) of the neutron in artificial potential wells created by a sequence of several thin films of different materials deposited on flat substrates (Steinhauser et al 1980). These resonances were observed to be split in more complicated sample structures where two identical resonators were coupled by a slightly transparent barrier (Steyerl et al 1981). All these "textbook experiments" provided demonstrations of standard, linear, quantum mechanics with precisions unmatched (in most cases) by other techniques.

Special mirrors can be used for three-dimensional focusing of UCN beams, and a novel "zone mirror" has been developed for nearly aberration-free imaging (Schütz et al 1980). Work on the development of a possible neutron microscope, using systems that admit of higher magnifications (Herrmann 1982), is in progress.

References

Altarev I S et al 1980 Phys. Lett. 80A 413
Berry M V 1982 J. Phys. A (Letters) in press
Brun T O et al 1980 Phys. Lett. 75A 223
Buras B and Kjems J 1973 Nucl. Instr. and Methods 106 461
Forte M et al 1980 Phys. Rev. Lett. 45 2088
Golub R and Pendlebury J M 1975 Phys. Lett. 53A 133
Groshev L V et al 1971 Phys. Lett. 34B 293
Herrmann P 1982 Thesis Techn. Univers. München unpublished
Kügler K-J, Paul W and Trinks U 1978 Phys. Lett. 72B 422
Luschikov V I and Frank A I 1978 Sov. Phys. Zh.E.T.F. Pis'ma 28 607
Lushchikov V I et al 1969 Sov. Phys. JETP Lett. 9 23
Maier-Leibnitz H 1962 Z. Angew. Phys. 14 738
Mampe W, Ageron P and Gähler R 1981 Z. Physik B 45 1
Scheckenhofer H and Steyerl A 1977 Phys. Rev. Lett. 39 1310
Schütz G, Steyerl A and Mampe W 1980 Phys. Rev. Lett. 44 1400
Shapiro F L 1968 Sov. Phys. Usp. 11 345
Shull C G and Gingrich N S 1963 J. Appl. Phys. 35 678
Steinhauser K-A et al 1980 Phys. Rev. Lett. 44 1306
Steyerl A 1969 Phys. Lett. 29B 33
Steyerl A 1975 Nucl. Instr. and Methods 125 461
Steyerl A 1978 Z. Physik B 30 231
Steyerl A et al 1981 Z. Physik B 41 283
Zel'dovich Ya B 1959 Sov. Phys. JETP 9 1389

Neutron spin echo

F. Mezei

Institut Laue-Langevin, 156X, 38042 Grenoble Cedex, and Central Research Institute for Physics, POB49, 1525 Budapest

Abstract. Paradoxically, the behaviour of the most non-classical angular momentum, the 1/2 spin, happens to be under certain circumstances very much like a classical one, showing no quantum effects. This fact provides the physical basis of neutron spin echo (NSE), whose principle was discovered just 10 years ago in Budapest. The fundamental aspects of the behaviour of neutron spin states in both situations leading to classical phenomena (Larmor case) and those showing real quantum effects (Stern-Gerlach case) will be discussed together with the basic concept of NSE high resolution inelastic spectroscopy. In addition a few representative experimental examples will be presented in order to illustrate the capabilities and the impact of the method.

1. Introduction

Neutron spin echo (NSE) is a conceptually unique approach to inelastic neutron scattering (Mezei 1972). Instead of the usual separate determination of the incoming and outgoing neutron velocities in the classical steps of monochromatization and analysis, in NSE the change of neutron energies is measured directly in a single step. This involves a very unusual situation in low energy particle scattering: the velocity of the same individual neutron has to be determined two times: once before and once after the scattering. This might appear as a very much classical mechanical idea, since in quantum mechanics the first observation is expected to perturb the particle state. In addition, what is even more an apparent contradiction, the quantum particle does not necessarily have a well defined velocity and in actual fact the expected, purely quantum mechanical scatter of each of these velocity measurements is orders of magnitude bigger than the measured changes. In the present talk I will focus attention on the question why the classical mechanical picture of pointlike neutrons with well defined velocities and classical spin vectors works perfectly well in NSE, and I will merely illustrate the capabilities of the method and the physical interest of high resolution neutron spectroscopy by a few recent experimental results. For more technical details and experimental examples the reader is referred to the proceedings of an ILL workshop (Mezei 1980a) and to more recent review papers. (Mezei 1982a, 1982b)

2. Classical approximation of neutron spin motion

The most general form of the neutron wavefunction reads:

$$\psi = \phi_+ |\uparrow\rangle + \phi_- |\downarrow\rangle \qquad (1)$$

where $\phi_\pm$ are spatial wave functions and $|\uparrow\rangle$ and $|\downarrow\rangle$ are the two basic spin states with respect to some quantization axis. The most important feature of this wave function is that ϕ_+ and ϕ_- can be very different, like due to the splitting of the beam in a Stern-Gerlach experiment. If however, the neutron did not experience "too high" magnetic fields, one can expect that there is no appreciable difference between the ϕ_+ and ϕ_- functions and the particle wave function can be written in the following approximate form:

$$\psi = \phi |\chi\rangle , \qquad (2)$$

where ϕ and $|\chi\rangle$ are now the spatial and spin wave functions. Erroneously, this special form, which is never strictly valid, is often considered as the most general case.

In a previous paper (Mezei 1980b) I have studied these types of states in a rigorous quantum mechanical manner, and I have shown that while eq.(1) (dubbed as Stern-Gerlach state) can describe specific quantum effects, eq.(2) (Larmor state) corresponds to a purely classical behaviour of the neutron beam polarization. This means, that in the latter case, independently of the true quantum mechanical states, the neutron beam polarization evolves as if the beam (or even a single neutron wave packet) would consist of an ensemble of classical point-like particles with well defined, classical three dimensional spin vectors. This analysis has also shown, that in polarized neutron beam work (including NSE) deviations of eq.(2) are nearly always completely negligible, i.e. a purely classical behaviour is to be expected.

In a NSE spectrometer, like IN11 at ILL (Dagleish et al 1980) shown in Fig.1, the incoming neutron velocity is determined by the angle of Larmor

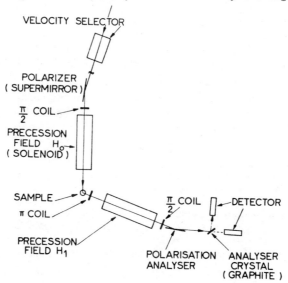

Fig. 1. The schematic lay-out of the IN11 NSE spectrometer at ILL. The length of the precession field solenoids is 2 m. The use of the graphite analyser crystal is optional (Dagleish et al 1980).

precession the neutron spin performs during the time of flight through the first precession field region. In view of the above conclusion, the precision of this determination is completely independent of, and it can be inferior to the momentum uncertainly of the quantum mechanical state of the neutron. (Note that in the conventional approach this uncertainty is the limit of the precision of velocity determination.) The information on its incoming velocity is remembered as Larmor precession angle by each "individual" neutron through the scattering process and it is then directly compared with the outgoing neutron velocity, which is in turn measured by the Larmor precession in the opposite sense through the second precession field region (cf. Fig.1)

Three aspects of this procedure are the most essential:
a) The individual neutron means the classical point-like particle, and not the quantum mechanical wave packet, whose momentum uncertainty is in general much bigger than the precision of the Larmor precession velocity determination. In view of the above mentioned analysis (Mezei 1980b) this is perfectly justified.
b) In quantum mechanical sense a measurement means the projection of the particle state onto a subspace of states by a macroscopic intervention. In this sense the conventional monochromatization and analysis are two quantum mechanical measurements. In NSE, and as far as the neutron velocity is concerned, there is only one such measurement, namely the analysis of the Larmor precessions at the detector end. The statistical uncertainties in quantum mechanics only apply to such measurements involving communication between the microscopic and macroscopic level. Therefore the incoming velocity determination in NSE, with the result only "known" to the neutron concerned (and to no "external agent"), does not perturb the neutron state.
c) The key to high resolution in NSE is the direct comparison of the incoming and outgoing velocities, or more pecisely, velocity components (viz. those perpendicular to the faces of the precession field regions, cf. Fig.1). On the one hand, this allows to observe minute changes in a simple and reliable manner. On the other hand, it makes the resolution independent of neutron beam monochromatization, i.e. intensity, which is the most important unique feature of NSE. In the conventional approaches the fundamental limitation is the dramatic loss of neutron intensity with increasing resolution.

3. Experimental examples

By now a considerable number (close to 50) of NSE experiments have been performed and well understood, and many of them produced significant results. The following few examples represent a small fraction of these, and they are intended to illustrate that, as expected, NSE provides substantially higher resolution in various types of inelastic experiments than the conventional methods, and that there is a considerable physical interest in such high resolution work.

The determination of the lifetimes, energy shifts, and other fine details of elementary excitation branches is the most sophisticated application of NSE. The example given in Fig.2 shows the temperature dependence of the roton linewidth in superfluid ^{4}He (Mezei 1980c). The resolution improvement with respect to previous neutron work is as much as 50 fold, and this allowed for the first time to probe really meaningfully the theoretical prediction (continuous line), since the light scattering data (dots) are indirect and give wrong results below 1.2 K.

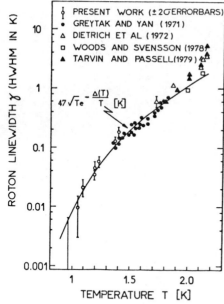

Fig. 2. Temperature dependence of the linewidth of the roton excitation in superfluid ^{4}He. The open circles represent the NSE results (Mezei 1980c), the dots indirect light scattering data, and the other symbols conventional neutron scattering results, respectively.

The IN11 instrument is the best adapted for quasielastic, small angle scattering work, and the majority of NSE experiments performed by now fall into this category. In quasielastic studies in most cases the line shape and the momentum dependence of the linewidth are investigated. NSE is a particularly powerful tool in the quantitative analysis of lineshapes, as illustrated by the two sample spectra in Fig.3. In NSE the time dependent Fourier transform of the inelastic scattering spectra is directly measured, which in the case of a Lorentzian line with halfwidth Γ corresponds to $S_q(t) = \exp(-\Gamma t)$ (q stands for the momentum variable). It is seen that the

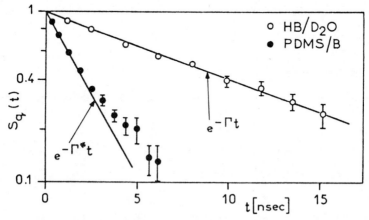

Fig. 3. Examples of Lorentzian (upper curve, $\Gamma = 60$ neV) and non-Lorentzian (lower curve, $\Gamma^* = 295$ neV) lineshapes. The NSE spectra shown were measured in 0.1 g/ml solution of hemoglobin (HB) in D_2O at T=18 °C and q = 0.152 Å^{-1} (Alpert and Mezei 1980), and in 0.05 g/ml solution of polydimethylsiloxane (PDMS) in deuterated benzene at T = 72 °C and q = 0.096 Å^{-1} (Richter et al 1978), respectively.

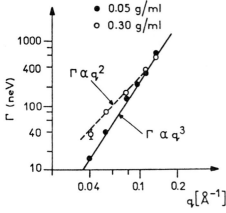

Fig. 4. Momentum dependence of the effective quasielastic linewidth in PDMS/benzene solutions at T=72 °C (Richter et al 1978)

lower curve (Richter et al 1978) deviates from this behaviour, which was predicted to be an important special feature of the dynamics of polymer solutions (de Gennes 1976). This effect is totally undetectable by the conventional techniques. Most strong deviations of the Lorentzian lineshape have been found to be a characteristic of spin glass relaxation dynamics, whose quantitative analysis (Mezei and Murani 1979) represents one of the unique results obtained by NSE.

The momentum dependence of the (effective) inelastic linewidth also provides important insight to the underlying dynamics, e.g. in polymer solutions it reflects the geometrical effects which become important at

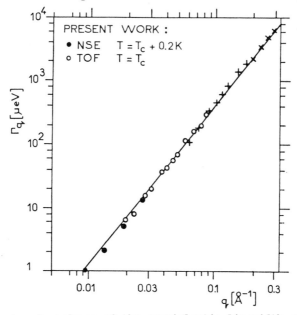

Fig. 5. Momentum dependence of the quasielastic linewidth of the critical fluctuations at the ferromagnetic Curie point T=1040 K in Fe. The time-of-flight (TOF) and NSE data are from recent work at ILL (Mezei 1982c). The crosses represent previous results of Collins et al (1969).

high concentrations, as illustrated in Fig.4. (Richter et al 1978). The neV scale energy resolution in this experiment is actually only accessible to NSE. Very significant results of similar type were also obtained on micelle solutions (Hayter and Penfold 1981).

The last example illustrates well the impact high resolution neutron scattering data can have due to the a priori model-independent character of neutron scattering information. At the ferromagnetic Curie point of Fe previous, conventional resolution neutron scattering experiments only confirmed the predicted $\Gamma \propto q^{5/2}$ law at $q \geq 0.05$ Å^{-1}. Anomalies in hyperfine field data (Chow et al 1980) were interpreted as evidence for a crossover to $\Gamma \propto q^2$ behaviour below this q value. The recent high resolution neutron results (Mezei 1982c) have shown (Fig.5), that this conclusion was wrong due to the indirect, model-dependent character of the hyperfine field data.

4. Conclusion

The just 10 years old neutron spin echo method is an interesting example of fruitful application of classical mechanical ideas in particle physics. The simplicity of the classical picture, and my intuitive faith in it, was certainly a key factor in the intellectual process which led to the discovery of the NSE principle. Much later rigorous quantum mechanical analysis showed that the classical picture is a perfectly justified approximation in this case, if this was still needed in view of the emerging volume of significant experimental results obtained by this uniquely high resolution inelastic neutron scattering technique.

References

Alpert Y and Mezei F 1980 unpublished
Chow L, Hohenemser C and Suter R M 1980 Phys. Rev. Lett. 45 205
Collins M F, Minkiewicz V J, Nathans R, Passell L and Shirane G 1969 Phys. Rev. 179 417
Dagleish P A, Hayter J B and Mezei F 1980 in: Neutron Spin Echo, ed F Mezei (Heidelberg: Springer Verlag) pp 66-71
de Gennes P G 1976 Macromolecules 9 587 and 594
Hayter J B and Penfold J 1981 J. Chem. Soc. Faraday Trans. I. 77 1851
Mezei F 1972 Z. Physik 255 146
Mezei F and Murani A P 1979 J. Magn. Magn. Mat. 14 211
Mezei F 1980a editor: Neutron Spin Echo (Heidelberg: Springer Verlag)
Mezei F 1980b in: Imaging Processes and Coherence in Physics, ed Schlenker M et al (Heidelberg: Springer Verlag) pp 282-294
Mezei F 1980c Phys. Rev. Lett. 44 1601
Mezei F 1982a Proc. Symp. Neutron Scattering, Argonne, Aug. 1981 (in press)
Mezei F 1982b Proc. Yamada Conf. VI, Hakone, Sept. 1982 (in press)
Mezei F 1982c Phys. Rev. Lett. (in press)
Richter D, Hayter J B, Mezei F and Ewen B 1978 Phys. Rev. Lett. 41 1484

Neutron diffraction

G E Bacon

University of Sheffield, Sheffield S3 7RH

Bearing in mind the contributions which are to follow later in this conference I am clearly not required to give a specialist review of what has been achieved over almost 40 years. Moreover, I note that I have been given the title of Neutron Diffraction, and not Neutron Scattering. This means, I think, that I am to talk about structure rather than about anything which might be called spectroscopy, though there is a very grey area between the two. In these terms I shall therefore try to give an account of how the study and use of neutron diffraction has developed, looking at some of the landmarks along the way. It will be a personal view.

In 1946 I was given the task of starting up work on neutron diffraction in this country - indeed the first work outside North America apart from Elsasser's original demonstration in Paris in 1936 that neutrons could be diffracted. As qualifications I had six months undergraduate experience of the first Metrovick continuously evacuated X-ray tube (just across the courtyard in the Cavendish from where Chadwick's neutron discovery was made), two month's subsequent experience as a research student working on FeCr alloys (mainly spent in trying to make β-filters for Mn radiation) and 6 years of wartime experience in the diffraction of radio waves (over a wavelength range of 15m to 3cm) and their scattering by aircraft.

Viewed at this time from the other side of the Atlantic the pioneer USA work had a strong flavour of nuclear physics and neutron optics. As a crystallographer (and one who as an undergraduate had known Bernal, W L Bragg, Bradley, Ewald, Fankuchen, Lipson, Dorothy Hodgkin, A F Wells and many others) I saw it rather differently. There was a promise of so much that X-ray diffraction had never been able to achieve, enshrined in the basic list of scattering lengths in Fermi and Marshall's paper of 1947 and which, in the hands of Shull and Wollan (1951), had extended to 60 elements and isotopes. This was the essential starting material, together with an awareness of the magnitudes of the energy quanta for X-rays, neutrons and photons. The latter was of more interest to Ray Lowde, who joined me and who very soon invented the name 'magnon'. Together, as we awaited the appearance of Harwell's reactor BEPO, we sought to understand the quantitative reflection of neutrons by large single crystals. This had emerged as a confusing topic in the early work in the USA and had led to their concentration of interest on powder diffraction. Our paper on secondary extinction in neutron crystallography (Bacon & Lowde (1948)), offered hope for the future but the extinction corrections necessary for intense reflections remain to this day one of the limitations of the accuracy of structural determinations. After a few practical measurements on the reactor GLEEP, the flux of which was adequate to study the performance of monochromators, but no more, we set up our first neutron diffractometer at BEPO in December 1949. To some extent we were second-class citizens at the reactor and we had been compelled to mount our instrument on a rail so that it could rapidly be removed from the reactor face if necessary. Our first results were presented at the British Association meeting at Birmingham in 1950. They concerned the electron structure of the carbon atom in graphite of interest for its use as a neutron moderator, and were a forerunner of what is now the X-N technique. Detailed comparison of the intensities of certain reflections for X-rays and neutrons (corresponding to scattering by electrons and nuclei respectively) established that the electrons were not distributed spherically and there was a concentration of electrons in the bonds. By present day standards the experimental technique was very primitive with hand-turning of the counter arm and use of stop-watch and scaling units, later replaced by continuous rotation, ratemeters and temperamental pen-recorders. Interference from neighbouring electrical apparatus, particularly overhead cranes, was an unpredictable hazard.

Meanwhile, in the USA a crucial milestone in the development of the subject had been the appearance of the classic paper on the magnetic structure of manganous oxide by Shull and Smart (1949), the practical demonstration of the magnetic scattering of neutrons as foretold by Bloch (1936) and Halpern & Johnson (1939). This was quickly followed by two other comprehensive papers from Oak Ridge on antiferromagnetic (Shull, Strauser & Wollan (1951)) and ferromagnetic (Shull, Wollan & Koehler (1951)) materials. This was surely the turning point at which magnetism became a realistic subject and magnetic architectures came to be identified and eventually shown to exist in almost as wide a variety of types as crystal structures. These papers gave great impetus to the study of black-and-white and colour space-groups, particularly by the Russian and French schools. Henceforth, neutron diffraction came well within the province of the solid state physicist, spurred on by intense studies of incommensurate structures, the metal chromium and the rare-earth metals. At this stage, in the later nineteen-fifties, two trends were evident. First, neutron diffraction had split into the two streams of chemical and magnetic applications which tended to be identified with single crystal and powder techniques respectively. Secondly, and largely because the advent of higher-flux reactors and data analysis equipment made the experimental work much faster and easier, the number of practitioners of the subject rapidly increased. For many years after the war, certainly in the United Kingdom, neutron diffraction had some aura of unrespectability as a subject carried out behind a high security fence so different from the freedom and open access of a University. With some notable exceptions, this attitude was slow to die but in due course the sheer value and versatility of the technique prevailed and it became a prime subject for support from the Science Research Council.

Returning to the chemical strand in the development of the subject, 1952 saw the first application of Fourier synthesis methods in neutron crystallography, to handle the single crystal data for KHF_2 (Peterson and Levy, 1952), giving a detailed picture of its very short hydrogen bond with the characteristic negatively-scattering contours of the hydrogen atom. The following year saw groups in both the UK and the USA working on the classic ferroelectric KH_2PO_4. By examining a single-domain sample in an electric field in the ferroelectric state it was possible to 'see' the movement of the hydrogen atoms between the ordered sites when

the direction of the field was reversed (Bacon &Pease (1955)).
(Admittedly, at the time there was a time-delay in the revelation set by
the need to complete the two Fourier plots with the aid of only Beevers-
Lipson strips.) The later nineteen-fifties, with the encouragement of
more reactors, higher-flux reactors and immense developments in the
technology of computers saw the extension of hydrogen atom location to
hydrates, organic materials and eventually to compounds of biological
importance. In particular, the advent of the third-generation research
reactor at the Brookhaven National Laboratory led to an extensive series
of studies of amino acids and their derivatives, mainly under the guidance
of W C Hamilton. It was noticeable that the accuracy of determination
of both atomic co-ordinates and thermal parameters had greatly increased.
Non-hydrogen atoms were located with an accuracy of 0.001 - 0.002 Å
and the hydrogen atoms to 0.002 - 0.003 Å, thus enabling the details
of the hydrogen bonds to be determined with precision.

So far we have been concerned almost entirely with direct structural
investigations, taking advantage of the favourable neutron scattering
by hydrogen (or, much better, by deuterium) and of the magnetic scattering
by atoms which have unpaired electronic spins. There are many refinements
to this work which have depended upon particular aspects of neutron
optics. For example, it is easy to obtain beams of neutrons with
wavelengths much longer than the usual 1-2 Å. The use of a cold source,
in which a subsidiary moderator is maintained at 20K, gives good
intensities up to 20Å and such neutrons, unlike X-rays of the same
wavelength, do not suffer high absorption. Accordingly, they can be
used to study defects in materials, bearing in mind that no Bragg
diffraction can take place if the wavelength is greater than twice the
maximum interplanar spacing in the scattering sample. Crossing the
border line between diffraction and scattering we may also note that these
long wavelength neutrons are of immense value in studies of small-angle
scattering, particularly for large biological molecules - and here the
power of the neutron technique is greatly increased by comparison
experiments with deuterated material, for the scattering properties of
deuterium are quite different from those of hydrogen.

To follow another branch of development we recall that neutron diffraction
and X-ray diffraction do not measure quite the same features of a solid.
Neutrons indicate the positions and motions of nuclei (apart from the case
of magnetic materials) whereas X-rays see electron clouds and define an
atomic position as the 'centre of gravity' of such a cloud. In the extreme
case of an O-H bond the length which emerges from X-ray data may be 0.2 Å
shorter than for neutrons. Combination of the two sets of experimental
data leads to a knowledge of the departure from spherical symmetry of the
electronic distributions around atoms. The power of this method is shown
most impressively in the classic picture of hexamethylene tetramine
(Duckworth, Willis & Pawley (1970)) which showed the lone-pair electrons
of the nitrogen atom.

We still have to regret that the most intense neutron beams are weaker
than X-ray beams, so that the amount of material needed in a diffraction
experiment with neutrons is greater. It is therefore fortunate that the
absorption coefficient of the neutron beam is smaller and the neutrons are
indeed able to penetrate the larger samples. This, in turn, has advantages
because it means that we can make texture measurements which are
representative of bulk materials, without any fear of confusion by surface
effects. Thus, in the nineteen-fifties, when the Wigner expansion of the

graphite moderator in reactors was of great importance we could make
convincing determinations of the preferred crystallite orientation in
extruded and compressed graphite blocks by using neutron samples several
centimetres in thickness. Likewise the preferred orientation in a uranium
rod used as a fuel element could be assessed in a single experiment in
which the neutrons traversed the full width of the rod. Far removed from
either of these applications, but depending on the same neutron optical
principles, I mention a current excursion into anatomy. The human skeleton,
like a block of graphite or a uranium fuel element, is neither a single-
crystal nor a randomly-oriented polycrystalline assembly. It is, in fact,
a carefully designed and engineered growth in which the basic components
of its bones, collagen and mineral apatite, are preferentially oriented in
direction so that they withstand the prevailing stresses. As the requirements change during the life-span of an individual the continuous regrowth
of the bones ensures that current requirements are met economically.
The c-axes of the apatite crystals preferentially choose to lie along the
directions of stress (Bacon, Bacon & Griffiths (1981)). The changes during
life are well-illustrated by the data for the human femur.

In the background to the whole of our discussion so far is the knowledge
that neutron diffraction is an "associate" of X-ray diffraction, that it is
not available in an ordinary industrial or university laboratory, it is
very expensive and accordingly it can only be used when X-rays fail.
This means that it is easy to lose sight of the distinctive techniques
which have been introduced. I will mention two of these. First, the
production and use of polarised beams. In the neighbourhood of a
diffracting sample which is in a magnetic field the incident neutrons can
be regarded as of two types, those whose magnetic spin lies in the
direction of the field and those with the contrary direction. By suitable
choice of a diffraction crystal the + and - groups can be separated and
used in subsequent experiments which are much more discriminating in
identifying and measuring magnetic scattering. Furthermore, if with a
polarised beam we can count separately those neutrons whose spin has, or
has not, changed as a result of the scattering process, then we gain a
direct means of distinguishing the different types of scattering which may
take place, eg nuclear or magnetic, coherent or incoherent. As our second
example of a technique we mention what is known as 'powder profile
refinement (Rietveld, 1969)'. In structural determination powdered
materials are at a great disadvantage compared to single-crystals because
of the loss of directional discrimination: reflected beams overlap and
structure factors cannot be separately identified. With neutrons,
because of the effective Gaussian shape of the diffracted beam, it
becomes practicable to direct attention to the ordinates of the intensity
profile of the diffraction pattern at, say, angular intervals of $0.05°$,
rather than attempting initially to identify specific reflections. If a
reasonably accurate structural model is available as a starting point,
then its parameters can be refined, though clearly only such features as
are postulated in the model can appear in the result - and there is a
limit to the total number of parameters.

In the whole of my account it has been implied that the neutron diffraction
experiments have been carried out with a beam of neutrons from a nuclear
reactor. For the great majority of the published work this has indeed been
the case, and, very largely, reactors which are continuously-running
rather than pulsed have been employed. However, work has been done using
neutron beams from linear accelerators since at least 1968 and future
improvements in neutron flux seem likely to depend on pulsed accelerator

sources. The present progress and prospects will be described in other sessions of this conference. How are these developments going to affect the future experimenter? There are an immense number of new operational techniques and data-analysis procedures to be developed. Will the queue in which the enthusiast waits with his sample disappear as the flux increases, or will it lengthen as possibilities proliferate and more-and-more sophisticated experiments are invented? Might everything become so expeditious that my femur studies will be able to extend to the living subject - achieving the triviality of a chest X-ray - so that we can follow directly the dependence of bone resorption of the life-style of an individual?

References

Bacon G E, Bacon P J and Griffiths R K (1981) New Scientist 92, 796-799
Bacon G E and Lowde R D (1948) Acta Crystallogr. 1, 303-314
Bacon G E and Pease R S (1955) Proc. Roy. Soc. London A320, 359-381
Bloch R (1936) Phys. Rev. 50, 259
Duckworth J A K, Willis B T M and Pawley G S (1970) Acta Crystallogr. A26, 263-271
Elsasser W M (1936) C.r. hebd. Séanc. Acad. Sci. Paris 202, 1029
Fermi E and Marshall L (1947) Phys. Rev. 71, 666-677
Halpern O and Johnson M H (1939) Phys. Rev. 55, 898-923
Peterson S W and Levy H A (1952) J. Chem. Phys. 20, 704-707
Rietveld H M (1969) J. Appl. Crystallogr. 2, 65-71
Shull C G and Smart J S (1949) Phys. Rev. 76, 1256
Shull C G, Strauser W A and Wollan E O (1951) Phys. Rev. 83, 333-345
Shull C G and Wollan E O (1951) Phys. Rev. 81, 527-535
Shull C G, Wollan E O and Koehler W C (1951) Phys. Rev. 84, 912-921

Slow neutron spectroscopy: an historical account over the years 1950-1977

B.N. Brockhouse

McMaster University, Physics Department, Hamilton, Canada L8S 4M1

Abstract The paper deals with development of technological and theoretical procedures for employment of inelastic scattering of slow neutrons in the study of molecular and magnetic dynamics of condensed matter and gases. The account covers mainly the years of the author's personal involvement in the subject from August 1950 to about 1977, first at Chalk River and then at McMaster University, Canada. But philosophical and historical preambles and an observer's postscript are also included.

1. Introduction

By 1950, nearly twenty years had passed since the first sightings of the neutron by Bothe and Becker in 1930, the studies of I. Curie and Joliot of 1931, the identification of the neutron as such by James Chadwick in 1932 and the appearance of the first North American papers on neutron physics about 1934 in Physical Review. By 1950 also, some five years had passed since the first neutron-induced nuclear chain reaction had been reported from Chicago and modern neutron physics had come to public attention; five years also since the awful threat of nuclear warfare had become public knowledge in Canada. There had been rapid progress in neutronics and many of the elementary items of neutronic technology with which we are familiar - Cd and B_4C shielding, Be filters, BF_3 proportional counters, etc. - were available,though not always commercially. There was complete confidence in the imminent emergence of the technology necessary for generation on a commercial scale of electricity from fission reactors, technology with its promised provision of ample energy for generations of humankind. This prospect was taken by many to balance (at least somewhat) the threat of Nuclear Armageddon and Götterdämmerung. But there are also the medical, biological and technological uses of neutron-induced radio-activity to be considered, each also with their pros and cons.

Slow neutron physics has played an important role in the development of the Physics of Condensed Matter and, latterly, of Dense Gases - as well a comparatively small role in the development of Nuclear Physics. It is not quite clear whether any of these roles are essential, that is, logically necessary to the development of the physics concerned. In this paper we are concerned with the first two roles and, in particular, with the development of Slow Neutron Spectroscopy, as opposed to (slow) Neutron Optics and Neutron Diffraction - the latter involving the static aspects of the atomic structure and the magnetic structure of the subject specimen and the former the dynamic aspects as well. So well are the properties of the neutron   (zero charge,near protonic weight and mass, nucleonic magnetic moment) adapted for use in study of the nuclear and magnetic dynamics of condensed matter that is is difficult to tell whether the adaption could be improved. It might well be said that, if the neutron did not exist, it would need to be invented!

Most of the development of Slow Neutron Spectroscopy occurred as a "by-product" of the operation of fission reactors whose existence and operation were for other purposes - particularly for experiments connected with power reactor design and for isotope production. This appears to be the case for Argonne, Oak Ridge, Chalk River, Harwell and Saclay. The major exceptions would be the Graphite and HFBR reactors at Brookhaven (since c1950 and c1965) and (since c1972) the ILL reactor at Grenoble. The existence of the two latter, very expensive, facilities is, in large part, a tribute to the perceived importance of slow neutron spectroscopy and of neutron diffraction and optics - and of neutron physics generally.

Normally, physics and physical sciences develop along four "fronts" - scientific, technological, theoretical and experimental, in more or less cyclic fashion. Scientific categories, concepts and structures, technological *materiel* and devices, theoretical techniques, experimental apparatus and methods - these develop in sequence historically, in response to challenges and opportunities from outside the subject concerned as well as response to inner drives within the subject itself. Thus physics (and the physical sciences: chemistry, electronics, metallurgy and so on) have developed, for the most part, in reasonably coherent fashion. Physical sciences are often seen as applications of physics and this view has a certain validity. But equally, a physics will not develop until the apposite physical sciences have reached a certain level of development. An example for us is the close coupling to be observed in the development of neutron physics with that of digital electronics, particularly with that of electronic digital computation. Another example involves certain Solid State Physics and chemical methods for preparation of high purity chemicals.

Theoretical Physics comprises techniques*, which we normally think of as mathematics, and models, which we think of metaphorically as physical, and if - physics which, if physics, are physics. The concepts in a physics are themselves physics; physics exist. Physics is about physics which are the foundations of physical existence. Like the proton and the electron, the neutron is an if-physic or an if-atomic or a message.

Results from Theoretical Physics and from Experimental Physics are scientific to the extent that the results can be expressed as IF-THEN statements of either determinate or statistical form. Results from a Natural Science, which purport to be physical, are IF-THEN statistics; operations and errors and uncertainties are inherent in physical science. Like any physics, Slow Neutron Spectroscopy is, at once, experimental and theoretical and technological. Like any physics, Slow Neutron Spectroscopy crosses fields; applied to crystals it is, at once, Neutron Physics, Crystal Physics and Chemical Physics. That may be the reason that broad-based meetings on Neutron Physics, such as this one, seem to be held only at intervals of several years.

The categories, concepts and structures of Experimental Physics come from Natural Philosophy; those of Theoretical Physics may be those accepted by Experimental Physics at the time - or may be new inventions by physicists, invented perhaps in analogy with those from other areas of physics or from physical science. Thus analogy has taken a major role in the development of Slow Neutron Physics, particularly analogies with Photon Physics in the shape of its crafts of X-ray Diffraction and Optical

---

* The if-physic, Quantum Mechanics, is a technique of Theoretical Quantum Physics and a structure of Natural Philosophy.

Spectroscopy. Thus, perhaps also, the neutron was invented in Natural Philosophy, discovered by James Chadwick, characterized within Quantum Mechanics by Theoretical Physics and if-sciented by Neutron Physics for use as a constituent of Nuclear Physics and for employment as a probe in the study of the physics of condensed matter.

Acceptance by Experimental Physics involves incorporation in the apparatus and in presentation of results. Thus Slow Neutron Spectroscopy currently accepts the notions of wavelength and momentum for the neutron and the DeBroglie relation between them. Thus also, the notions of reciprocal space and the crystal reciprocal lattice and of their validity for description of the scattering of slow neutrons has been accepted. Thus also, the quantum concepts: phonon, magnon etc. and their decays in time and space, as well as the more recent generic notion of quasi-particle (of which the phonon and the magnon are approximate special cases), seem to have been accepted.

Contrariwise, Slow Neutron Spectroscopy is not much interested in the lifetime (τ_n) of the neutron, nor yet in the constancy of the velocity(c) of light - provided that these are "large enough". And so on, for other Sciences of Physics also. A concept in an if-physics, which is to say, an if-physic such as "a neutron" or τ_n or c, exists in context within Physics as Language and Physics as Game. An if-physics exists also in a context, but not yet in a universal context. An if-physics is a more-or-less-worked-out-notion of a religion towards some aspect of the physically present and is subjective heresy, at risk of being wrong. An if-physics is about if-physics - which are at risk of being non-existent, in future potentiality as well as in actuality. We can use a name, Iff-Physics, to express this metaphysical problem. Physics, that is to say the totality of all real and true if-physics, would be, I think, a metaphysic - which is to say, would be an objective heresy, correct here and now but possibly not correct elsewhere and to be employed with caution there if employed at all. As it is, an Iff-Physics is a metaphorical Island in Metaphysical Archipelago Now. There we will leave it, for this introduction at least.

Slow Neutron Spectrometry (manifestly a technological physics) questions the physical specimen in the language of that Physics as Game, getting answers in that "language" in the shape of experimental intensity distributions. Slow Neutron Spectroscopy, an if-physics, questions those answers in whatever Physics as Language seems appropriate. Part of the craft of experimental physics lies in arranging that the two languages be concordant - that apparatus has the proper configuration, that the specimen has a suitable space and size, that beam cross sections and magnetic fields are uniform, and so on. If this is not done, it will probably not be possible to interpret the experimental results, though one may yet be able to compare the experimental data with theoretical model calculations. In each new field this must be learned anew, though analogy can help - and a new field has the benefit of more analogies than had an old one in the days of its development.

Slow Neutron Spectroscopy has made decisive contributions to Phonon and Magnon Physics and to Quasi-Particle Physics generally. It has made major contributions in many fields: to the Physics of Classical Fluids, of Perfect Crystals, of Defect Crystals, to the Physics of Phase Transitions, to the Physics of Hydrogen, of Helium, of normal metals and of electrical superconductors. It has contributed to our knowledge of the molecular and magnetic dynamics of many hundreds of chemical substances, in many cases decisively. As the end result of much investment, time and effort, it is these contributions which are of importance technologically and theoretically.

The size of the present scientific literature relevant to Slow Neutron
Diffraction and Spectroscopy is illustrated by the fact that the bibliography compiled by Larose and Vanderwal (1974) of publications up to that
year contains more than 8000 entries. A similar bibliography to 1982
would probably contain twice that number. Thus no one person any longer
knows what is in that literature, though some could probably hazard reasonable guesses.

So, how is one to approach the history of the subject?

2. Pre-History and Iff-History to 1950 of Slow Neutron Physics and Historical Categories to 1977.

How ought one to approach the History of Slow Neutron Spectroscopy? One
requires a philosophy of history as well as a philosophy of physics and a
philosophy of atomics. As a strictly amateur and beginning historian, my
philosophy of history is not well-developed - what there is is of an
existential strain, as is my philosophy of physics. One requires also a
set of historical categories and one requires <u>time</u>. The avalanche of new
papers and new books in the scientific literature would overwhelm the wistful historian even as he wrote. So there has to be a cut-off date; in the
set of categories presented herein, that date is 1977 and the occasion is
the Conference on Neutron Inelastic Scattering of the International Atomic
Energy Agency (I.A.E.A.) held in Vienna in October of that year. By reason
of time and space however, comments about the Categories of this history
are ridiculously compressed and references to persons and literature very
much limited.

2.1 Pre-History of Neutron Physics and Neutronics to 1932:

An account of the pre-history of the concept "neutron" has been set down
by Kröger (1980) to the year 1932 of Chadwick's discovery.

2.2 Historical Development, 1933-39:

Recollections of the early years of the if-scientification of the neutron
have been given in a symposium of 1962 by Purcell, Feather, Segré, Chadwick
and others (1964). A reference is given by Kröger (1980).

2.3 Historical Development, 1939-45:

Peruse the literature or ask your "Ancient Mariner", if you know one.

2.4 Historical Development, 1945-50:

Peruse the literature or ask your Student of the History of Physics. Some
of the literature of this period dates from the era 2.3.

2.5 Neutron Diffraction Present, 1946-50:

An account of the first year or so of neutron diffraction work at Oak
Ridge National Laboratory by Wollan, Shull and others, has been set down
by Shull (1976) in the Proceedings of the Gatlinburg Conference.

2.6 Thermal Diffuse Scattering of Slow Neutrons

Here we meet inelastic neutron scattering, in two guises: Between the
Debye-Scherrer lines of a powder pattern obtained by neutron diffraction,
there is observed diffuse intensity which is temperature dependent. The
total cross section $\sigma_T(\lambda)$ for materials with small nuclear absorption has
at long wavelengths (λ) a temperature-dependent λ-proportional component as

shown by Cassels (1950). Both were related approximately to the vibrational spectrum of the powder via a quasi-Debye Temperature. For both of these related phenomena the inelasticity of the scattering was not demonstrated until several years later.

3. *In Media Res*: Setting the Stage for Slow Neutron Spectrometry: 1951-54

3.1 The State of Electronic, Neutronic and Computational Technology in 1951.

3.2 The State in 1951 of Condensed Matter and of Neutron Scattering Theory.

3.3 Neutron Diffraction, a Discipline by 1951: Tables of bound scattering amplitudes in existence. The reduced variable $<|\vec{Q}|> \simeq \frac{4\pi}{\lambda} \sin(\frac{\phi}{2})$ in use and understood to require correction for inelastic effects.

3.4 Slow Neutron Spectroscopy, an Idea whose Time seemed to be coming: Spring 1951. Genesis of the Idea at Chalk River in a study-group meeting in December (1950) in which D.G. Hurst, G.H. Goldschmidt, N.K. Pope and the author participated. (The Idea is known to have germinated elsewhere at about the same time.) Decision that the experiments were feasible with the NRX reactor, in operation then for several years.

3.5 Estimation of scattered energy distributions from measurements of absorption curves: 1951-2. The work of Egelstaff (1951) and of Brockhouse and Hurst (1952) was discussed.

3.6 Setting in motion of plans for Neutron Spectrometry: first crude efforts.

3.7 Monochromator oh monochromator! 1946 → ∞ ? Sodium chloride to treated metal crystals to pyrolitic graphite...?

4. Slow Neutron Spectrometry an Accomplishment: 1954-58

4.1 Demonstration Experiments in late 1954: at NRX a variety of experiments e.g. Brockhouse (1955) with a crude 3-axis instrument; at Saclay an experiment by Jacrot employing a crude double chopper.

4.2 The Van Hove Transformations: Van Hove (1954). $S(\vec{Q},\omega) \leftrightarrow G(\vec{r},t)$

4.3 The Triple-Axis Crystal Spectrometer: Crude and *ad hoc* (1954); mechanically sound (1956); Variable incoming energy (1958).

4.4 The Filter-chopper, Time-of-flight Spectrometer: Brookhaven (1955).

4.5 If-scientification of the phonon and the magnon through **v**erification of the existence of their dispersion relations.

4.6 Theory of Cohen and Feynman (1957); if-scientification of the phonon cum roton of liquid HeII from the observation of a continuous dispersion relation in work at Stockholm, Los Alamos and Chalk River.

4.7 If-scientification of the notion: types of self-diffusion in liquids "at atomic distances" in H_2O by observation and study of quasi-elastic scattering.

4.8 Digital computation fairly readily available; Fourier transforms no longer done mechanically!

4.9 Conception of "cold sources" and "hot sources" and "tailored spectra".

5. Slow Neutron Spectrometry a Discipline: 1958-64

5.1 The Constant $\vec{Q}$ Method; constant ΔE and other special methods; making the language of experiment concordant with that of theory.

5.2 Other Instruments: The Multiple Chopper Spectrometer; The Rotating Crystal Spectrometer; The Beryllium Detector Spectrometer; Liquid Hydrogen sources. See especially IAEA proceedings, Neutron Inelastic Scattering(1961).

5.3 Phonon and Magnon Dispersion Relations; Kohn Anomalies in $\nu(\vec{q})$ related to the Fermi Surface; Phonon line-broadening;"soft modes"in $SrTiO_3$.

5.4 Dispersion Relations for Liquid Helium 4 as function of T and P.

5.5 Experimental Van Hove Correlation Functions (in liquid Pb, Ar, H_2O etc.) Search for phonons in classical liquids.

5.6 IAEA Conferences: Neutron Inelastic Scattering, Vienna (1960), Chalk River (1962).

6. Maturation of Slow Neutron Spectroscopy: 1964-74

Magnons and crystal fields; magnon and phonon branch crossing; line-broadening and spectra in defect and anharmonic crystals; isotopic substitution in study of Van Hove correlation functions; polarized slow neutron spectrometry. Texts and monographs now available.

6.1 IAEA Conferences: Bombay (1964); Copenhagen (1968); Vienna (Instrumentation, 1969); Grenoble (1972).

7. Concluding Scientific Postscript: 1974-77

The Institut Laue-Langevin in Grenoble, its remarkable instrumentation and its impact on the subject. The *tour de force* experiments on liquid Helium 3 at Grenoble and Argonne.

7.1 Conferences: The 30th Anniversary of Neutron Diffraction, ORNL Gatlinburg (1976); Neutron Inelastic Scattering IAEA Vienna (1977).

7.2 Bibliographies: Larose and Vanderwal Scattering of Thermal Neutrons IFI/PLENUM (1974); Sakamoto et al, JAERI-M6857 (1976).

8. Concluding Postscript

Because of time considerations, the paper presented to the Conference did not include the whole of the Introduction nor did it include discussion of most of the considerable number of topics enumerated. The discussions were illustrated with slides from the author's publications, in particular from Brockhouse and Hurst (1952) and Brockhouse (1955) and (1958). They included reminiscences of the genesis in 1951 of interest in inelastic scattering at Chalk River and of the genesis (1958) of the Idea of the Constant $\vec{Q}$ and constant ΔE techniques - the latter as a confluence of a remark by R. Stedman (re a mechanical apparatus which achieved constant ΔE) and a remembrance of hearing of computer control of experiments in another field at another time.

9. References: Most of these may be obtained from the bibliographies and Conference proceedings cited or from texts such as Bacon (1962, 1975).
Bacon   G E   1962, 1975 Neutron Diffraction, 2nd, 3rd Ed.(London: Oxford.)
Brockhouse   B N   and Hurst   D G   1952   Phys.Rev.88   547.
Brockhouse   B N   1955   Can. J. Phys. 33   889.
Brockhouse   B N   1958   Suppl. Nuovo Cimento Ser X 9   45.
Cassels   J M   1950   Progress in Nuclear Physics 1   185.
Egelstaff   P A   1951   Nature 168   290.
Kroger   B   1980   Physics 22   175.

Inst. Phys. Conf. Ser. No. 64: Section 4
Paper presented at Conf. on Neutron and its Applications, Cambridge, 1982

Neutron scattering and magnetism

A. R. Mackintosh

H.C. Ørsted Institute, University of Copenhagen, DK-2100 Denmark

Abstract. Those properties of the neutron which make it a unique tool for the study of magnetism are described. The scattering of neutrons by magnetic solids is briefly reviewed, with emphasis on the information on the magnetic structure and dynamics which is inherent in the scattering cross-section. The contribution of neutron scattering to our understanding of magnetic ordering, excitations and phase transitions is illustrated by experimental results on a variety of magnetic crystals.

1. Introduction

The discovery of the neutron fifty years ago presaged a revolutionary development in our understanding of magnetism. It was not however until about twenty years later that this revolution really began, when intense beams of thermal neutrons from nuclear reactors became available for research. The effort which has been made since the early 1950's in studying magnetic materials by neutron scattering has led to the determination of the magnetic structures of countless crystals. In many cases the dynamics of the atomic moments has also been elucidated, leading to a detailed description of magnetic excitations, interactions and phase transitions.

The neutron has a number of special properties, on which its utility as a tool for the study of magnetic materials depends. Because it is a neutral particle, it can penetrate deeply into most crystals, interacting through its magnetic moment with the electronic moments strongly enough to be measurably scattered, but without perturbing the magnetic system too severely. Thermal neutrons have wavelengths comparable with interatomic spacings and energies comparable with those of magnetic excitations in solids, so that they are ideally suited for studying both the spatial arrangement and the dynamics of the magnetic moments. As we shall see, the neutron scattering cross-section contains precisely that information which is needed to characterize a magnetic material and to make a stringent comparison with theoretical calculations of its properties. The neutron is indeed the key which allows us to open a magnetic crystal and peer inside.

The purpose of this article is to provide a succinct review of the theory and practice of neutron scattering, a brief account of the results of some of the more important experiments which have been performed, and an indication of the significance of this information for our understanding of magnetism. We begin with the question of what is measured in a neutron scattering experiment and how the scattering cross-section is related to the magnetic properties of the scattering crystal. The next section is

0305-2346/83/0064-0199 $02.25 © 1983 The Institute of Physics

devoted to elastic scattering, or neutron diffraction, and the determination of magnetic structures and magnetization distributions. The connection between inelastic scattering and the dynamics of the magnetic system is outlined, and applied to the determination of the magnetic excitations of crystals. The relation between neutron scattering experiments on the dynamical behaviour near magnetic phase transitions, and in low-dimensional systems, and the predictions of statistical mechanics is then discussed. Finally an attempt is made to put in perspective the contribution of neutron scattering to the study of magnetism.

2. Principles of Neutron Scattering

The essence of a neutron scattering experiment is the determination of the probability that a neutron which is incident on the sample with momentum $\hbar \vec{k}_o$ is scattered into the state with momentum $\hbar \vec{k}$. The intensity of scattered neutrons is thus measured as a function of the energy transfer

$$E = \hbar\omega = \frac{\hbar^2}{2M}(k_o^2 - k^2)$$

where M is the neutron mass, and the momentum transfer

$$\hbar\vec{K} = \hbar(\vec{k}_o - \vec{k})$$

$\vec{K}$ is known as the scattering vector. This measurement may be carried out in various ways, but by far the most effective experimental technique for magnetic studies is generally the triple-axis crystal spectrometry developed by Brockhouse (1961).

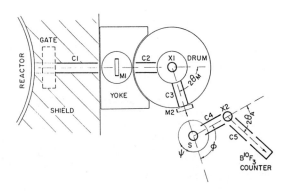

Fig.1 Principle of the triple-axis spectrometer, after Brockhouse (1961).

The principle of this method is indicated in Fig.1. An incident beam of neutrons with a well defined $\vec{k}_o$ is selected from the continuous reactor spectrum by the monochromator crystal X_1, and scattered from the sample S. The intensity of the scattered beam is measured as a function of $\vec{k}$ by the analyser crystal X_2 and the neutron detector. Although most of the results which we shall discuss were obtained with crystal spectrometers, it may be advantageous for some experiments to use time-of-flight or neutron spin-echo techniques (Mezei 1972), especially if very high energy resolution is required.

The scattering of neutrons from a magnetic sample was first considered in detail in the classic paper of Halpern and Johnson (1939). A rather complete exegesis of the theory of neutron scattering has been given by Marshall and Lovesey (1971). Starting with the Born approximation, a lengthy but straightforward calculation gives for the differential scattering cross-section per unit solid angle and energy transfer, for unpolarized neutrons

$$\frac{d^2\sigma}{d\Omega dE} = (\gamma r_0)^2 (k/k_0) [\tfrac{1}{2} gf(\vec{\kappa})]^2 e^{-2W(\vec{\kappa})} \sum_{\alpha\beta} (\delta_{\alpha\beta} - e_\alpha e_\beta)$$

$$\times \sum_{\lambda\lambda'} p_\lambda \sum_{\ell\ell'} e^{i\vec{\kappa}\cdot(\vec{R}_\ell - \vec{R}_{\ell'})} \langle\lambda|J_\ell^\alpha|\lambda'\rangle \langle\lambda'|J_{\ell'}^\beta|\lambda\rangle \delta(\hbar\omega + E_\lambda - E_{\lambda'})$$

(1)

γ is the neutron magnetic moment in nuclear magnetons, r_0 the classical electron radius and g the Landé factor for the ions with total angular momentum J, situated on sites ℓ with lattice vectors $\vec{R}_\ell$. $f(\vec{\kappa})$ is the magnetic form-factor and the scattering by the orbital component of the moment is taken into account by the so-called dipole approximation. $W(\vec{\kappa})$ is the Debye-Waller factor, which accounts for the reduction in scattered intensity due to the thermal vibration of the lattice. $\vec{e}$ is a unit vector in the direction of $\vec{\kappa}$, while α and β are Cartesian coordinates. The initial state of the scatterer is $|\lambda\rangle$, with energy E_λ and thermal population factor p_λ, and its final state is $|\lambda'\rangle$.

A useful and physically transparent form for the cross-section was derived by Van Hove (1954). Using the integral-representation of the energy δ-function in (1), he showed that it can be transformed to

$$\frac{d^2\sigma}{d\Omega dE} = (\gamma r_0)^2 (k/\hbar k_0) [\tfrac{1}{2} gf(\vec{\kappa})]^2 e^{-2W(\vec{\kappa})} \sum_{\alpha\beta} (\delta_{\alpha\beta} - e_\alpha e_\beta) S^{\alpha\beta}(\vec{\kappa},\omega)$$

(2)

where the scattering function is given by

$$S^{\alpha\beta}(\vec{\kappa},\omega) = \frac{1}{2\pi} \sum_{\ell\ell'} \int_{-\infty}^{\infty} e^{i\vec{\kappa}\cdot(\vec{R}_\ell - \vec{R}_{\ell'})} e^{-i\omega t} \langle J_\ell^\alpha(0) J_{\ell'}^\beta(t)\rangle dt$$

(3)

$\langle J_\ell^\alpha(0) J_{\ell'}^\beta(t)\rangle$ is the thermal average for the Heisenberg operators at the specified times. For an unbounded perfect lattice we can sum over ℓ' and find

$$S^{\alpha\beta}(\vec{\kappa},\omega) = \frac{N}{2\pi} \sum_\ell \int_{-\infty}^{\infty} e^{i(\vec{\kappa}\cdot\vec{R}_\ell - \omega t)} G^{\alpha\beta}(\vec{R}_\ell, t) dt$$

(4)

where N is the number of atoms. The Van Hove time-dependent pair-correlation function $G^{\alpha\beta}(\vec{R}_\ell, t)$ is just $\langle J_0^\alpha(0) J_\ell^\beta(t)\rangle$, and gives essentially the probability that, if the moment at the origin has some specified (vector) value at time zero, then the moment at the position $\vec{R}_\ell$ has some other specified value at time t. A neutron scattering experiment measures the Fourier transform in space and time of $G^{\alpha\beta}(\vec{R}_\ell, t)$, which is clearly just what is needed to describe the dynamics of the magnetic system.

There is yet another very useful way of expressing the cross-section, which may be derived from linear response theory. If a magnetic field which varied periodically in space and time were applied in the direction β, an oscillatory moment would result, given by

$$J^\alpha(\vec{q},\omega) = \chi^{\alpha\beta}(\vec{q},\omega)H^\beta(\vec{q},\omega)$$

or in tensor notation

$$\vec{J}(\vec{q},\omega) = \bar{\bar{\chi}}(\vec{q},\omega)\vec{H}(\vec{q},\omega) \tag{5}$$

where $\bar{\bar{\chi}}(\vec{q},\omega)$ is the (complex) generalized susceptibility tensor (the magnetic susceptibility divided by $g\mu_B$). The fluctuation-dissipation theorem gives a relation between the thermal fluctuations of the system and the dissipative component of the dynamic susceptibility; specifically

$$S^{\alpha\beta}(\vec{q},\omega) = (N\hbar/\pi)(1 - e^{-\hbar\omega/k_B T}) \mathrm{Im}\chi^{\alpha\beta}(\vec{q},\omega) \tag{6}$$

The cross-section is then given by

$$\frac{d^2\sigma}{d\Omega dE} = (\gamma r_0^2)(Nk/\pi k_0)[\tfrac{1}{2}gf(\vec{k})]^2 e^{-2W(\vec{k})}(1 - e^{-\hbar\omega/k_B T})$$
$$\times \sum_{\alpha\beta}(\delta_{\alpha\beta} - e_\alpha e_\beta)\mathrm{Im}\chi^{\alpha\beta}(\vec{k},\omega) \tag{7}$$

If the imaginary part of the susceptibility is measured for all ω, the real part may be derived from the Kramers-Kronig relations.

The neutron may therefore be considered as a magnetic probe which effectively establishes a frequency- and wavevector-dependent magnetic field in the scattering sample, and detects its response to this field. No other experimental technique can aspire to producing such detailed microscopic information about magnetic systems.

3. Magnetic Structures

From equations (2) and (3) it is clear that elastic neutron scattering is determined by the static magnetic structure. For a simple ferromagnetic structure in a Bravais lattice, for example, with the moments oriented in the z-direction, the neutron diffraction cross-section is

$$\frac{d\sigma}{d\Omega} = \frac{8\pi^3 N}{V}(\gamma r_0)^2[\tfrac{1}{2}gf(\vec{k})]^2 e^{-2W(\vec{k})}(1 - e_z^2)<J^z>^2 \sum_{\vec{\tau}}\delta(\vec{k}-\vec{\tau}) \tag{8}$$

where V is the volume of the unit cell and $\vec{\tau}$ a reciprocal lattice vector. In this case Bragg reflections are observed when $\vec{k} = \vec{\tau}$, and are superposed on the reflections due to the interaction of the neutrons with the nuclei. The temperature dependence of the magnetic reflections traces out the magnetization curve.

In antiferromagnetic structures, the cross-section is more informative. For example, for the helical structure

$$J_\ell^x = J \cos \vec{Q}\cdot\vec{R}_\ell \quad : \quad J_\ell^y = J \sin \vec{Q}\cdot\vec{R}_\ell \tag{9}$$

with $\vec{Q}$ in the z-direction, it is

$$\frac{d\sigma}{d\Omega} = \frac{2\pi^3 N}{V}(\gamma r_o)^2 [\tfrac{1}{2}gf(\vec{\kappa})]^2 e^{-2W(\vec{\kappa})}(1 + e_z^2)J^2$$
$$\times \sum_{\vec{\tau}}[\delta(\vec{\kappa} + \vec{Q} - \vec{\tau}) + \delta(\vec{\kappa} - \vec{Q} - \vec{\tau})] \tag{10}$$

In this case, extra Bragg reflections are observed, from which the repeat distance of the periodic structure may be deduced. If this repeat distance is commensurable with the lattice, a simple antiferromagnetic structure results as in Fig.2, which

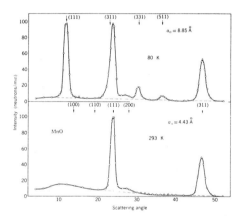
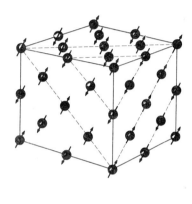

Fig.2 Neutron diffraction in MnO below and above the Néel temperature, after Shull, Strauser and Wollan (1951). The extra reflections at 80 K correspond to the antiferromagnetic structure shown on the right.

shows the results of Shull, Strauser and Wollan (1951) on MnO. These historic measurements were the first to reveal the unique power of neutron diffraction in elucidating complex magnetic structures, and provided the genesis for innumerable magnetic-structure determinations. Neutron diffraction is still very much alive and flourishing; as examples of recent interesting areas of study we might mention nuclear magnetism (Goldman 1979) and magnetic superconductors (Moncton, Shirane and Thomlinson 1979).

Incommensurable magnetic structures are quite common, particularly in magnetic metals. Koehler (1972) and his collaborators have investigated the magnetic structures of the rare earth metals in great detail and some of what he calls their "panoply of exotic spin configurations" are shown in Fig.3. These structures can all be described by a combination of the helix (9) and the longitudinal wave-structure

$$J_\ell^z = J' \cos \vec{Q}' \cdot \vec{R}_\ell \tag{11}$$

allowing the propagation vectors to be zero, and including higher harmonics.

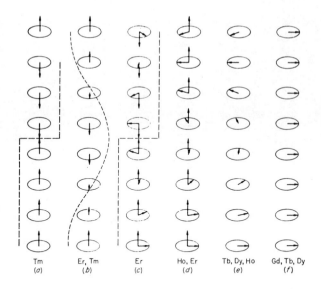

Fig.3 The magnetic structures of the heavy rare-earth metals, after Koehler (1972). The moments in a particular plane normal to the hexagonal c-axis are all parallel, and the relative orientations between planes are indicated.

Magnetic structures can usually be determined from a rather small number of Bragg reflections but, if the intensities of many such reflections are measured, the form-factor $f(\vec{\kappa})$ and hence the distribution of the magnetization in the unit cell can also be deduced. If there is no orbital moment

$$f(\vec{\kappa}) = \int e^{i\vec{\kappa}\cdot\vec{r}} s(\vec{r}) d^3r \qquad (12)$$

where $s(\vec{r})$ is the normalized spin density in the unit cell, over which the integral is taken. If there is an orbital contribution, the formulation becomes considerably more complicated (see Marshall and Lovesey 1971) but the magnetization density can still be determined from the absolute intensities of the Bragg peaks.

Such a measurement of the magnetization density may be performed with high precision. In Fig.4 are shown, for example, the rings of weak negative-spin density deduced by Shull and Mook (1966) from their polarized-neutron measurements on Fe. It is frequently advantageous, and sometimes essential, to use polarized neutrons, particularly when a precise separation between nuclear and magnetic scattering is required. In a full polarization-analysis experiment, the scattering cross-section is measured as a function of the spin-states of both the incident and scattered neutrons (Moon, Riste and Koehler 1968). Such measurements contain a wealth of information which unpolarized beams cannot provide, but the relatively low efficiency of available neutron polarizers has so far restricted the practical utility of the method, especially for inelastic-scattering experiments.

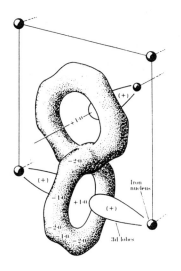

In a disordered magnetic alloy, elastic diffuse scattering occurs between the Bragg peaks, due to the different moments at different sites. In a dilute alloy the diffuse-scattering cross-section gives essentially the Fourier transform of the disturbance in the spin density due to a single impurity. The diffuse scattering in transition-metal alloys has been extensively studied by Low (1969) and his collaborators, who were able to determine the magnetization distribution around the impurities in ferromagnetic hosts. They were able to show, for example, that the Fe impurities in a dilute ferromagnetic Pd alloy polarize several hundred neighbouring host atoms, giving rise to a "giant moment".

Fig.4 A sketch of the spin-density in the unit cell of Fe, after Shull and Mook (1966). The magnetization is measured in kilogauss and the rings of negative spin-density originate primarily from the sp-electrons.

4. Magnetic Excitations

The inelastic-scattering cross-section contains an abundance of information on the excitations of magnetic crystals, which may be extracted by suitably designed experiments. A moment $\vec{J}_\ell$ localized on an ion at a lattice site $\vec{R}_\ell$ is subjected to single-ion forces originating from the crystalline electric field, and is coupled to neighbouring moments by two-ion exchange and other interactions. If we first consider an isolated ion, the crystal field will in general lift the m_J-degeneracy, giving rise to $(2j+1)$ eigenfunctions $|i\rangle$. The dynamic response of such an ion is then readily calculated by perturbation theory to be

$$\chi_o^{\alpha\beta}(\omega) = \lim_{\varepsilon \to 0^+} \left[\sum_{\substack{i,j \\ E_i \neq E_j}} \frac{\langle i|J^\alpha|j\rangle\langle j|J^\beta|i\rangle}{E_i - E_j - \hbar\omega - i\varepsilon} (n_j - n_i) \right.$$

$$\left. + \frac{1}{k_B T} \frac{i\varepsilon}{\hbar\omega + i\varepsilon} \left\{ \sum_{\substack{i,j \\ E_i = E_j}} \langle i|J^\alpha|j\rangle\langle j|J^\beta|i\rangle n_i - \langle J^\alpha\rangle\langle J^\beta\rangle \right\} \right] \qquad (13)$$

where $n_i = e^{-E_i/k_B T}/\sum_j e^{-E_j/k_B T}$ is the thermal population factor. The cross-section may now be obtained from (6) (or(1)) and per ion is

$$\frac{d^2\sigma}{d\Omega dE} = (\gamma r_o)^2 (k/k_o)[\tfrac{1}{2}gf(\vec{K})]^2 e^{-2W(\vec{K})} \sum_{\alpha\beta}(\delta_{\alpha\beta}-e_\alpha e_\beta)$$

(14)

$$\times \sum_{ij} n_i <i|J^\alpha|j><j|J^\beta|i>\delta(\hbar\omega+E_i-E_j)$$

Inelastic neutron scattering thus directly measures the energy levels of the ion, and hence gives information on the crystalline electric fields. The results of Kjems, Touborg and de Jong (1979) on isolated ions of Er in Mg are shown in Fig.5 which, incidentally, illustrates the sensitivity of experiments carried out with modern techniques.

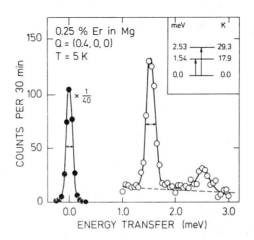

Fig.5 Inelastic-scattering spectrum for isolated Er ions in Mg, after Kjems, Touborg and de Jong (1979).

In a concentrated system, two moments at $\vec{R}_\ell$ and $\vec{R}_{\ell'}$ will be coupled by an exchange interaction, which we may approximate by the Heisenberg form $- J(\vec{R}_\ell-\vec{R}_{\ell'})\vec{J}_\ell\cdot\vec{J}_{\ell'}$. The application of a magnetic field periodic in space and time then gives rise to an oscillating moment at site ℓ

$$\vec{J}_\ell(\omega) = \bar{\bar{\chi}}_o(\omega)\left[\vec{H}_\ell(\omega) + \sum_{\ell'} J(\vec{R}_\ell-\vec{R}_{\ell'})\vec{J}_{\ell'}(\omega)\right]$$

Fourier transforming in space

$$\vec{J}(\vec{q},\omega) = \bar{\bar{\chi}}_o(\omega)\left[\vec{H}(\vec{q},\omega) + J(\vec{q})\vec{J}(\vec{q},\omega)\right]$$

where $J(\vec{q}) = \sum_\ell J(\vec{R}_\ell)e^{i\vec{q}\cdot\vec{R}_\ell}$, and using the definition (5) of $\bar{\bar{\chi}}(q,\omega)$

$$\bar{\bar{\chi}}(\vec{q},\omega) = \bar{\bar{\chi}}_o(\omega)\left[1 + J(\vec{q})\bar{\bar{\chi}}(\vec{q},\omega)\right]$$

which may be written

$$\bar{\bar{\chi}}(\vec{q},\omega) = \frac{\bar{\bar{\chi}}_o(\omega)}{1-J(\vec{q})\bar{\bar{\chi}}_o(\omega)} \qquad (15)$$

In this random-phase approximation, the frequency- and wavevector-dependences are separated, and determined respectively by the single-ion and two-ion interactions.

A straightforward example of the application of this theory is provided by the light rare-earth metal Pr, which has been extensively studied by Houmann et al. (1979). In the paramagnetic phase, the ground state on the hexagonal sites is the singlet $|J_z = 0\rangle$ and the first excited state is the doublet $|\pm 1\rangle$ at energy Δ. The single-ion susceptibility at low temperatures is from (13)

$$\chi_o^{xx}(\omega) = \chi_o^{yy}(\omega) = \lim_{\varepsilon \to 0^+} \frac{2|\langle 1|J^x|0\rangle|^2}{\Delta - \hbar\omega - i\varepsilon} \qquad (16)$$

so that, from (7) and (15), the scattering cross-section has δ-function resonances at the excitation energies

$$E_{x,y}^2(\vec{q}) = \Delta^2 - 2\Delta|\langle 1|J^x|0\rangle|^2 J^{x,y}(\vec{q}) \qquad (17)$$

These excitations, sometimes called magnetic excitons, may be considered as linear combinations of single-ion crystal-field excitations, with a specific relative phase, which are coupled by the exchange and hence propagate through the lattice. In (17) we have allowed for the anisotropy of the exchange interaction, which is written in the form $-\sum_\alpha J^\alpha(\vec{R}_\ell - \vec{R}_{\ell'}) J_\ell^\alpha J_{\ell'}^\alpha$, and lifts the double degeneracy of the excitations, as illustrated in Fig.6.

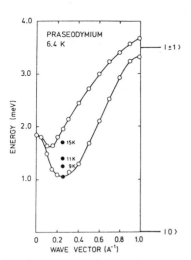

Fig.6 Magnetic excitations in Pr, after Houmann et al.(1979). The exchange produces a dispersion of the transitions between the lowest two crystal-field levels.

Also shown in this figure is the temperature-dependence of the mode of lowest energy. McEwen, Stirling and Vettier (1978) have shown that the energy of this incipient soft mode can be driven to zero by a suitable uniaxial stress, resulting in the longitudinal wave antiferromagnetic structure (11). The hyperfine interaction between the electrons and nuclei is also sufficient to induce a transition to an antiferromagnetic state at about 60 mK (Bjerrum Møller et al. 1982). Even at low temperatures, the random-phase approximation is not quite adequate, since experimentally the excitations are observed to have a finite lifetime and are not therefore δ-functions in ω. This problem and the lack of complete self-consistency in the calculations have stimulated many attempts to develop the theory further.

The random-phase approximation may also be used in magnetically ordered phases, provided that the average exchange field acting on the single ion is included in the molecular-field approximation. If the crystal field is negligible, the dispersion relation for the spin-wave (or magnon) excitations in a ferromagnetic crystal is found to be

$$E(\vec{q}) = J\left[J(0) - J(\vec{q})\right] \tag{18}$$

where the first term in the brackets represents the average exchange, or molecular field. The magnon dispersion relations in the anisotropic ferromagnet Tb are shown in Fig.7. In contrast to (18), the excitation energy at zero wavevector is finite because of the work required to turn the moments against the anisotropy fields in the crystal. On the other hand, the energy of a magnon in the helical structure (9) does go to zero at long wavelengths, since it corresponds only to a change in the phase of the helix, which costs no energy. Such phasons are a general feature of incommensurable structures; they were first studied in the helical magnetic phase of Tb (Bjerrum Møller, Houmann and Mackintosh 1967) in which the periodicity of the helix is incommensurable with that of the lattice. A plenitude of information about the magnetic forces in crystals, and the interactions between magnetic and other excitations, is contained in magnon dispersion relations, especially when they are measured as a function of temperature and magnetic field (Mackintosh and Bjerrum Møller 1972).

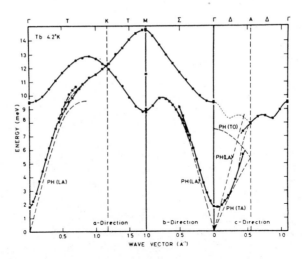

Fig.7 Magnon dispersion relations in Tb, after Mackintosh and Bjerrum Møller (1972). The splitting of some of the branches is due to interaction with the phonons.

If the magnetic electrons are itinerant, as in the ferromagnetic transition metals Fe, Ni and Co, the details of the theory are quite different, but the neutron-scattering cross-section can still be calculated via $\overline{\overline{\chi}}(\vec{q},\omega)$. In a magnetically polarized non-interacting gas of conduction electrons, a magnetic excitation of wavevector $\vec{q}$ is formed by promoting an electron in a Bloch state $E(\vec{k},\uparrow)$ to the state $E(\vec{k}+\vec{q},\downarrow)$. The spectrum of such Stoner excitations has a minimum energy, which is a function of $\vec{q}$, and at $\vec{q} = 0$ is equal to the minimum splitting, for the same $\vec{k}$, between up- and down-spin bands. Above this minimum, the Stoner excitations for a given $\vec{q}$ form a continuum up to some maximum energy. The susceptibility transverse to the polarization direction has the form

$$\chi_o(\vec{q},\omega) = \lim_{\varepsilon \to 0^+} \frac{1}{N} \sum_{\vec{k}} \frac{f(\vec{k},\uparrow) - f(\vec{k}+\vec{q},\downarrow)}{E(\vec{k}+\vec{q},\downarrow) - E(\vec{k},\uparrow) - \hbar\omega - i\varepsilon} \qquad (19)$$

where $f(\vec{k})$ is the Fermi-Dirac function.

The interaction between the conduction electrons may be taken into account with a random-phase approximation, yielding the susceptibility

$$\chi(\vec{q},\omega) = \frac{\chi_o(\vec{q},\omega)}{1 - I(\vec{q})\chi_o(\vec{q},\omega)} \qquad (20)$$

where $I(\vec{q})$ is the Fourier transform of the effective electron interaction. The cross-section (7) now has contributions not only from the Stoner excitations, but also from collective modes, corresponding to bound-state linear combinations of Stoner excitations. These spin-wave modes have an energy proportional to q^2 at low q-values and rise rapidly in energy. When the spin-wave dispersion relation enters the Stoner continuum, the magnon modes are expected to decay into Stoner excitations, and consequently to have reduced lifetime and intensity.

Since the pioneering measurements of Sinclair and Brockhouse (1960) on Co, many groups have studied the excitations in the ferromagnetic transition metals. In particular Lowde and Windsor (1970) have made extensive measurements on Ni, and Mook and his collaborators have studied the spin waves in Fe and Ni to high energies. As shown in Fig.8, the dispersion relations in Fe do indeed rise rapidly with q and the intensity decreases abruptly above about 100 meV. These results have been discussed in terms of numerical calculations based on energy-band theory by Cooke (1979).

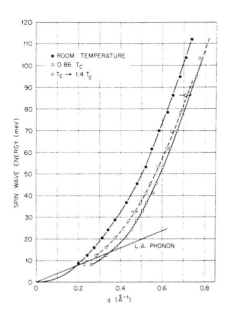
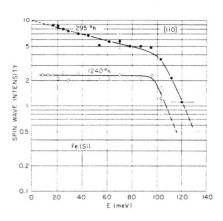

Fig.8 The energies and intensities of spin waves in Fe , after Lynn (1975).

It should be mentioned that it is not yet clear whether a localized or an itinerant picture provides a more appropriate description of the magnetic excitations in the rare-earth mixed-valence compounds. Studies of these compounds by inelastic neutron scattering have only recently been started, and it may be anticipated that excitation measurements of the type reported by Loewenhaupt and Bjerrum Møller (1981) will contribute substantially to our understanding of the mixed-valent state.

As indicated in Fig.8, rather little change is observed in the neutron scattering from Fe when it is heated through its Curie temperature. Many examples are known in which propagating modes are observed in regions of short-range order, but perhaps the most striking are in quasi-1-dimensional magnetic crystals. In such systems, infinite chains of magnetic atoms are separated so effectively from neighbouring chains that the exchange between them is negligible compared with that along the chains. If this separation were perfect, such systems would not order; in fact they usually order at very low temperatures. A very good example is the one-dimensional antiferromagnet $(CD_3)_4NMn\, C\ell_3$ (TMMC), whose dynamics have been studied in detail by Hutchings et al. (1972). Although the three-dimensional ordering temperature is around 0.8 K, very strong one-dimensional correlations persist up to tens of K and clear spin-wave excitations, with negligible dispersion normal to the chain direction, are observed throughout the whole Brillouin zone. Spin waves have also been extensively studied in other one-dimensional systems (see Steiner, Villain and Windsor 1976) and in the one-dimensional ferromagnet $CsNiF_3$, evidence has in addition been observed by Kjems and Steiner (1978) for highly non-linear magnetic excitations. These soliton modes, which can be considered as dynamic domain boundaries between different ordered regions, may be thermally excited and scatter neutrons like a gas of magnetic particles. The study of such non-linear modes, which have also been observed in other one-dimensional magnetic crystals, is a new and interesting departure from the traditional experiments on linear magnetic excitations.

5. Phase Transitions

The strong enhancement of the neutron-scattering cross-section near a magnetic phase transition, illustrated in Fig.9, was first observed in Fe by

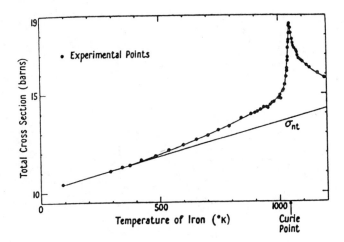

Fig.9 The critical scattering near the Curie point of Fe, after Squires (1954).

Palevsky and Hughes (1953) and Squires (1954), and ascribed to the critical fluctuations of the moments which occur near the transition temperature. As T_c is approached, either from above or below, these fluctuations become strong, long-ranged and slowly-varying in time. At a second-order phase transition, the component $\chi^{\alpha\alpha}(\vec{Q})$ of the susceptibility, which characterizes the nature and periodicity of the ordering, diverges and a spontaneous magnetization with the wavevector $\vec{Q}$ develops below T_c. The correlation-range ξ, defined at temperature T by the expansion

$$\chi_T^{-1}(\vec{q}) = \chi_T^{-1}(\vec{Q})(1 + \xi^2(q-Q)^2 + \ldots) \tag{21}$$

also diverges. In the critical region, each of these quantities has a power-law variation like $|T-T_c|^\zeta$, where ζ is the critical exponent which characterizes its behaviour near T_c. These exponents can readily be measured by neutron scattering and detailed studies have been made on numerous magnetic materials. The behaviour in the critical region may also be calculated with modern techniques in statistical mechanics, series expansions and scaling and renormalization-group theories. The comparison between these theories and the experimental results on the spatial correlations has been thoroughly reviewed by Als-Nielsen (1975). On the whole, the agreement between theory and experiment is very good. The dynamics of the critical fluctuations have been less comprehensively studied, but inelastic scattering measurements have been made on a number of materials near T_c and, for example, Tucciarone et al. (1971) and Dietrich, Als-Nielsen and Passell (1976) found that their results for respectively the Heisenberg antiferromagnet $RbMnF_3$ and ferromagnet EuO could be well accounted for by dynamical scaling theory. To return to Fe for the last time, Collins et al. (1969) found their extensive measurements to be consistent with dynamic scaling, but Mezei (1982) has observed evidence that the theory breaks down at small wavevectors.

It is characteristic of critical phenomena that the details of a system undergoing a phase transition are of little significance for the behaviour near T_c, which instead depends only on its general properties. The dimensionality of a magnetic system and the symmetry of the exchange or other two-ion interactions is thus sufficient to determine its critical properties. The two-ion coupling may be uniaxial (Ising model), planer (x-y model) or isotropic (Heisenberg model) and may in addition be restricted to near neighbours or of long range. During the last decade there has been remarkable progress in identifying and growing crystals in which the effective dimensionality of both the magnetic lattice and the magnetic interactions can be chosen to be 1, 2 or 3. All of these crystals order in 3-dimensions at sufficiently low temperatures so that care must be taken in interpreting the measurements, especially on the quasi-2-dimensional crystals in which the magnetic ions are arranged in widely separated sheets, but many striking results have been obtained and theories tested to an extent which had not previously been possible. The critical phenomena in these systems have recently been reviewed by Lindgård (1978), and we will only give a brief discussion of a few particularly interesting examples of systems which may be accurately described by exactly solvable models.

Birgeneau et al. (1971) studied the spatial correlations in the previously mentioned TMMC, which is a very good example of a 1-dimensional Heisenberg antiferromagnet. They found that both the form of the correlations and their temperature dependence from 1.2 K to 40 K were in very good agreement with the predictions of Fisher for the classical 1-dimensional nearest-neighbour Heisenberg model. These measurements were thus a direct experimental verification of an exactly solvable model in the theory of magnetic phase transitions.

The spatial correlations have also been studied in the 1-dimensional Ising ferromagnet $CsCoCl_3$ by Yoshizawa and Hirakawa (1979)and were again found to be in good agreement with the classical theory at higher temperatures. Onsager's famous solution of the 2-dimensional Ising model remains the only exact calculation available for 2-dimensional systems. The layer compound Rb_2CoF_4 is a good example of such a system and Samuelsen (1973) discovered that the magnetization below the ordering temperature indeed follows Onsager's prediction very closely, as illustrated in Fig.10. The similar compound K_2CoF_4 has been comprehensively studied by Ikeda et al. (1975) who deduced several critical exponents and found excellent agreement with Onsager's predictions. No 3-dimensional model has been solved exactly over a wide range of temperature. However, the Ising ferromagnet with dipolar interactions manifests a special form of critical fluctuations, which allow the behaviour near T_c to be described by mean-field theory, with corrections which depend on $\ln[|T-T_c|/T_c]$. The system is said to have an upper marginal dimensionality of 3, and the corrections can be calculated with arbitrary accuracy by renormalization-group theory. Neutron scattering experiments on $LiTbF_4$ (Als-Nielsen 1976) confirm the results of the renormalization-group calculations with high precision.

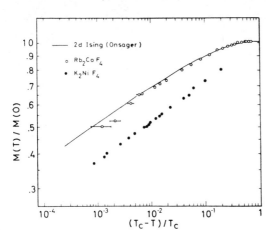

Fig.10 Temperature-dependence of the sublattice magnetization in 2-dimensional systems, compared with Onsager's theory, after Samuelsen (1973).

If the magnetic ions in a crystal are randomly replaced by nonmagnetic ions, the properties of the system may be drastically modified. Of particular interest are 2-dimensional systems with short-range interactions. There is a critical concentration of magnetic ions, known as the percolation concentration c_p, at which a continuous chain of magnetic neighbours can just be traced through a sheet of ions. Below this percolation threshhold, the system breaks up into isolated clusters of magnetic ions and long-range order cannot be sustained. The point (c_p, T = 0) on a phase diagram is a multicritical point, terminating a line of second-order transitions. The properties of 2-dimensional Heisenberg and Ising antiferromagnets near the percolation concentration have been studied by neutron scattering by Cowley et al. (1980), with special attention to the critical exponents. In related experiments, Yoshizawa et al. (1982) have examined the effects of an applied magnetic field on randomly diluted 2- and 3-dimensional Ising antiferromagnets. The external field generates a random field at the magnetic sites in the crystal and destroys the magnetic ordering in both cases, in accordance with theories which predict that the random field lowers the effective spatial dimensionality by two. The lower marginal dimensionality, below which the system cannot order, thus becomes 3 for an Ising system.

Below the critical concentration for magnetic ordering, a 3-dimensional system with long-range interactions, as are typically found in magnetic metals, may form a structure at low temperatures in which the moments are

"frozen" into locally favourable configurations. Such spin-glasses are
extremely intractable, both experimentally and theoretically. However,
Mezei and Murani (1979) have used the neutron spin-echo technique with
polarized neutrons to investigate spin-glass dynamics over a very wide
range of time-scales, from 10^{-12}s to 10^{-9}s. They find that the dynamical
behaviour at long wavelengths may be characterized by relaxation times
which span this range. Around the freezing temperature these relaxation
times increase rather rapidly, and a characteristic frequency-dependent
cusp is observed in the susceptibility.

Conclusion

The preceding brief review has perhaps provided some justification for the
initial assertion that the neutron has been the instrument of a revolution
in magnetism. No reasonable understanding of a magnetic crystal can be
attained without a precise knowledge of its magnetic structure and, although
this structure can sometimes be inferred or guessed in favourable cases,
the only systematic way of deducing it is by neutron diffraction. The magnetic structure is determined by the interactions, but is not in itself adequate to reveal more than their general form. However, the comprehensive
study of the dynamics which inelastic neutron scattering allows gives information on the interactions with a detail which fundamental theories of
magnetism in solids cannot at present match. The structure and interactions
together determine the temperature-dependence of the magnetic properties,
which may be calculated by the methods of statistical mechanics. The study
by neutron scattering of magnetic systems near phase transitions has opened
up new possibilities for the confrontation between theory and experiment
in statistical physics.

The nature of this revolution does not however correspond precisely to those
with which historians of science are normally concerned. Most of the fundamental ideas about magnetism which were developed in the 1930's have
survived relatively unscathed, but they have been augmented by such an
increase in the detailed knowledge of magnetic crystals that our understanding of magnetism can justifiably be said to have undergone a revolutionary
change, thanks to the neutron. This type of development, in which a quantitative increase of knowledge leads to a qualitative improvement in understanding, but without overthrowing the basic principles established earlier,
is, of course, very common in modern physics.

For those who have been working with neutrons and magnetism during the last
decades, it may be natural to suppose that the heroic days are over. However,
a constant stream of new results appear on systems of ever increasing complexity, and it would be difficult to maintain that they are inherently any less
interesting than those which have been produced earlier. It also seems likely
that we are on the brink of a decisive technical breakthrough in the manufacture of highly efficient polarizers in the form of magnetic layer-crystals.
These would allow a range of new experiments which are at present quite
impracticable. All in all it is reasonable to suppose that the neutron will
still be making an important contribution to magnetism at its centenary.

Acknowledgements

While preparing this review, I have benefitted greatly from discussions with
my colleagues Jens Als-Nielsen, Per Bak, Hans Bjerrum Møller, Jens Jensen,
Jørgen Kjems, Per-Anker Lindgård and Bente Lebech. However, any explicit or
implicit judgements which it may contain are my own responsibility.

References

Als-Nielsen J 1975 Phase Transitions and Critical Phenomena Vol.5 ed C Domb and M S Green (London-New York: Academic) pp87-164
Als-Nielsen J 1976 Phys.Rev.Lett. **37** 1161
Birgeneau R J, Dingle R, Hutchings M, Shirane G and Holt S L 1971 Phys.Rev. Lett. **26** 718
Bjerrum Møller H, Houmann J G and Mackintosh A R 1967 Phys.Rev.Lett. **19** 312
Bjerrum Møller H, Jensen J Z, Wulff M, Mackintosh A R, McMasters O D and Gschneidner K A Jr. 1982 Phys.Rev.Lett. **49** 482
Brockhouse B N 1961 Inelastic Scattering of Neutrons (Vienna: IAEA) pp113-51
Collins M F, Minkiewicz V J, Nathans R, Passell L and Shirane G 1969 Phys. Rev. **179** 417
Cooke J F 1979 J.Magn.Magn.Mat. **14** 112
Cowley R A, Birgeneau R J, Shirane G, Guggenheim H J and Ikeda H 1980 Phys. Rev. **B21** 4038
Dietrich O W, Als-Nielsen J and Passell L 1976 Phys.Rev. **14** 4923
Goldman M 1979 J.Magn.Magn.Mat. **14** 105
Halpern O and Johnson M H 1939 Phys.Rev. **55** 898
Houmann J G, Rainford B, Jensen J and Mackintosh A R 1979 Phys.Rev. **B20** 1105
Hutchings M T, Shirane G, Birgeneau R J and Holt S L 1972 Phys.Rev. **B5** 1999
Ikeda H, Hatta I, Ikushima A and Hirakawa H 1975 J.Phys.Soc.Japan **39** 827
Kjems J K and Steiner M 1978 Phys.Rev.Lett. **41** 1137
Kjems J K, Touborg P and de Jong M 1979 J.Magn.Magn.Mat. **14** 277
Koehler W C 1972 Magnetic Properties of Rare Earth Metals ed R J Elliott (London-New York: Plenum) pp81-128
Lindgård P-A 1978 Neutron Diffraction ed H Dachs (Berlin-Heidelberg: Springer) pp197-242
Loewenhaupt M and Bjerrum Møller H 1981 Physica **108B** 1349
Low G 1969 Adv.Phys. **18** 371
Lowde R D and Windsor C 1970 Adv.Phys. **19** 813
Lynn J W 1975 Phys.Rev. B **11** 2624
Mackintosh A R and Bjerrum Møller H 1972 Magnetic Properties of Rare Earth Metals ed R J Elliott (London-New York: Plenum) pp187-244
Marshall W and Lovesey S W 1971 Theory of Thermal Neutron Scattering (Oxford: Clarendon)
McEwen K A, Stirling W G and Vettier C 1978 Phys.Rev.Lett. **41** 343
Mezei F 1972 Z.Physik **255** 146
Mezei F 1982 in Proceedings of ICM82 J.Magn.Magn.Mat.
Mezei F and Murani A P 1979 J.Magn.Magn.Mat. **14** 211
Moncton D E, Shirane G and Thomlinson W 1979 J.Magn.Magn.Mat. **14** 172
Moon R M, Riste T and Koehler W C 1968 Phys.Rev. **181** 920
Palevsky H and Hughes D J 1953 Phys.Rev. **92** 202
Samuelsen E J 1973 Phys.Rev.Lett. **31** 936
Shull C G and Mook H 1966 Phys.Rev.Lett. **16** 184
Shull C G, Strauser W A and Wollan E O 1951 Phys.Rev. **83** 333
Sinclair R N and Brockhouse B N 1960 Phys.Rev. **120** 1638
Squires G L 1954 Proc.Phys.Soc. **A67** 248
Steiner M, Villain J and Windsor C G 1976 Adv.Phys. **25** 87
Tucciorone A, Lau H Y, Corliss L M, Delepalme A and Hastings J M 1971 Phys. Rev. **B4** 3206
Van Hove L 1954 Phys.Rev. **93** 268
Yoshizawa H and Hirakawa K 1979 J.Phys.Soc.Japan **46** 448
Yoshizawa H, Cowley R A, Shirane G, Birgeneau R J, Guggenheim H J and Ikeda H 1982 Phys.Rev.Lett. **48** 438

Neutron scattering review — magnetic studies

W. C. Koehler

Oak Ridge National Laboratory, Oak Ridge, Tennessee 37830

My charge, for this Review Session on Neutron Scattering is to present to you a survey of the contributions that neutron scattering has made to studies of magnetism. Some aspects of this broad subject have been covered in the preceding two papers (Early History, B. N. Brockhouse and Neutron Diffraction, G. E. Bacon) and specialized areas of this topic will have been treated in detail in the session, on Wednesday afternoon, that is devoted exclusively to magnetism. In addition, on Thursday morning, Professor Mackintosh will address a considerably more general audience on the same topic.

It may therefore be a little difficult to avoid some repetition, but in the hope of so doing I have decided to follow a somewhat historical approach, and I apologize, in advance, for any omissions I make. Obviously I cannot hope to make a complete survey.

The earliest applications of neutron scattering to magnetism were concerned in part with the determination of the arrangement and orientation of magnetic moments in ordered magnetic configurations. Generally these investigations were associated with the understanding of the bulk properties of crystalline solids; the temperature variations of susceptibility and resistivity, specific heat anomalies, etc. A strong motivation in these, and indeed in all subsequent magnetic scattering studies, was the confrontation with and test of, theory which often ran far ahead of experiment. The Halpern Trilogy (1939, 1941a, 1941b) on neutron scattering was published before the relatively intense sources of neutrons provided by the Oak Ridge and Argonne Reactors became available, for example, and theoretical predictions of the antiferromagnetic state (Néel 1936) and of the structure of ferrites and the ferrimagnetic (Néel 1948) state predated their experimental verification at Oak Ridge in the work on MnO (Shull et al 1951a) and Fe_3O_4 (Shull et al 1951b).

In the course of these early experiments, as well as in magnetic reflection studies at Argonne, (Hughes and Burgy 1951) the nature of the interaction between the neutron and the electronic moments of atoms was clarified. The question was raised in 1936 and 1937 in papers by Bloch (1936, 1937) and Schwinger (1937) as to whether the neutron moment should be treated as arising from an elementary dipole or from an elementary Amperian current. In each case the magnetic cross section was strongly angularly dependent varying as $\cos^2\alpha$ in the first case and $\sin^2\alpha$ in the second where α is the angle between the scattering and magnetization vectors. As you all know it is the latter interpretation that is the accepted one.

Since the first experiments carried out in 1948-1951, a vast literature of magnetic structures has evolved. In a few cases it might have been possible to postulate microscopic structures from considerations of bulk properties, as for instance in the Néel model for ferrites, but in general it was only with neutron diffraction that such information become available.

A good example of the interplay between theory and experiment is the discovery of the helical or more generally the non-collinear configurations. Almost at the same time, in 1959, Villain, working at Saclay with Herpin and Mériel on $MnAu_2$, (1959) and Yoshimori (1959) interpreting data on MnO_2 obtained at Oak Ridge by Ericson (1952) arrived at a theoretical model involving long range interactions that successfully provided an explanation of the experimental data. Subsequently, the collection of structures exhibited by the rare earth metals; helical, sinusoidally modulated moments, conical, antiphase domain, etc., was discovered (Koehler 1965) and, for a time, there was an outburst of activity among theorists; Elliott, Nagamiya, Cooper, de Gennes, Freeman, Loucks, and Lindgard to name only a few.

At present it seems certain that a magnetic structure determination will be a part of any program in which the macroscopic magnetic properties are investigated. A knowledge of the magnetic structure is always necessary before any dynamical studies are undertaken; it gives the first clues toward understanding the nature of the exchange and anisotropy energies of the moments of the atoms in the structure.

I cannot leave this section of my review without mentioning that the solution of even very complex magnetic structures has been aided greatly by the group theoretical methods developed by E. F. Bertaut (1963) and his students and colleagues.

An important contribution of magnetic neutron scattering to knowledge is the information it has provided on magnetic moment densities, and hence to electronic wave functions, in condensed matter. This comes about because the elastic magnetic neutron scattering is given by the Fourier transform of the time-averaged magnetization density. In general this is the sum of the spin density and the moment density due to unquenched orbital currents. In many cases of interest, the 3-d metals, for instance, the orbital contribution is small but it is seldom negligible. In the rare earth metals and compounds it can be exceedingly important.

Measurements of moment densities have undergone a natural evolution since the first experiments at Oak Ridge. The first such measurements were made on polycrystalline specimens, with unpolarized neutrons, of substances either in the paramagnetic or in the ordered state. For paramagnets it was necessary to estimate other contributions to the incoherent diffuse scattering; spin, isotope, multiple, etc., in order to isolate the paramagnetic diffuse scattering. For ordered systems, the nuclear and magnetic contributions to the intensity of a Bragg peak are additive, except for very simple antiferromagnets where the magnetic intensities are free of nuclear contamination. These first measurements were concerned with Mn^{++} (Shull et al 1951a).

The strong orbital moment scattering from rare earth ions was first demonstrated in a study of the paramagnetic scattering from rare earth oxides (Koehler and Wollan 1953) and the theory was advanced in papers by Trammell (1953) and Kleiner (1953). Knowledge of the wave functions was relatively primitive, even for the free rare earth ions and comparison was made with calculations based on screened hydrogenic wave functions with the screening constant taken as an adjustable parameter. Measurements with unpolarized neutrons are still satisfactory today when p>b, the case for the inner reflections of some materials, but for the outer reflections where p<<b the magnetic scattering is completely dominated by the nuclear intensities. There are cases where it is necessary to make significant measurements in the range $p/b \lesssim .01$. It is here that polarized neutrons are essential.

The standard polarized beam method, developed first by Shull and his students (Shull and Yamada 1962, Moon 1964, Mook 1966) depends on the fact that one can measure the interference between magnetic and nuclear amplitudes. One measures the intensity of a Bragg peak as a function of the direction (+ or −) of the beam polarization. The ratio of intensities $R = I^+/I^-$ can be determined to high statistical accuracy and its sensitivity to small values of p (really the component p_z parallel to the polarization) is greatly enhanced. Each observation is self-calibrating, assuming the nuclear amplitude is known. Peak intensities rather than integrated intensities are usually measured to make efficient use of counting time.

Precise measurements of the magnetic form factors have been reported for a large number of ferromagnetic and ferrimagnetic materials; Fe, Co, Ni, ^{160}Gd, Tb, Er, etc. as well as for a number of intermetallic compounds. The results have provided a strong impetus to theorists to re-examine their wave function calculations and have thus produced great improvements in our knowledge of the electronic structures of transition metals. In particular, the results on Gd have led to a reevaluation of the importance of fully relativistic calculations for the rare earth wave functions. The same experiments exhibited clearly the distribution of conduction electron density in the rare earth metals (Moon et al 1972, Freeman and Desclaux 1972).

These techniques have been applied to paramagnetic metals. In these measurements a magnetic moment of several milli-Bohr magnetons is induced in the specimen by means of a strong magnetic field. The distribution of induced magnetization is inferred from the observed polarization ratios. The expanded distribution predicted long ago for electrons near the bottom of the 3-d band (Wood 1960) was observed in this way for the first time in Sc (Koehler and Moon 1976). Generally, the theoretical position is improving but there are great deficiencies in our knowledge of the orbital moments in the paramagnetic transition metals (Gupta and Freeman 1976).

A new application of the standard polarized beam method involves the measurement of polarization ratios of the coherent phonon peaks for a ferromagnet, the magneto-vibrational scattering. Magnetic scattering amplitudes can thereby be measured, in principle, at any point in reciprocal space, not just at the Bragg peak positions. Measurements on Ni and Fe (Steinsvoll et at 1981) have been carried out that provide additional testing of the currently accepted electronic structures of these metals. A very important application of magnetic neutron scattering is to the measurement of moment distributions in disordered systems. Of these the greatest effort has been in the study of binary ferromagnetic alloys. Some of the first experiments of this type were carried out at Oak Ridge by Shull and Wilkinson (1955). Very extensive measurements were later made by Collins and Low (1965, 1967). This type of experimentation is being done at present at ORNL by Cable and his co-workers and by scientists at Saclay and at the ILL.

In the formation of an alloy, the electron shells of the atoms exert a mutual influence that is revealed by a change in the magnetic properties of the atoms. Neutron scattering measurements on disordered systems can provide quantitative information about the influence of the surroundings on a given moment, and about the interactions between the components.

Several experimental methods are in use to study disordered ferromagnetic systems. They differ in the method used to extract the magnetic disorder scattering, free of errors, from the competing nuclear disorder scattering multiple scattering, inelastic effects, and short range order effects. One may use polarized neutrons to measure separately the sum of nuclear and magnetic cross sections, and a spin-dependent nuclear-magnetic interference term. If the magnetization is perpendicular to the scattering plane, the magnetic cross section is proportional to a moment-moment correlation, the nuclear cross section to a short range order function, and the interference term to a site-occupation magnetic moment correlation. If an unpolarized beam is used, the interference term vanishes and the magnetic cross section can be reduced to zero by applying a sufficiently large field parallel to the scattering vector. By working at long wavelengths, beyond the Bragg cut-off, multiple Bragg scattering can be eliminated, and one may use energy analysis to ensure that only elastic scattering is detected. The theory for interpreting disorder scattering data is due primarily to W. Marshall (1968).

The transitions that magnetic materials undergo when they become ordered ferromagnets or antiferromagnets have attracted great interest from both experimentalists and theorists. The fundamental idea of a second-order transition is that it is describable by an order parameter which serves as a measure of the amount and kind of ordering that arises in the critical region. In the case of a ferromagnet, as an example, the order parameter is the magnetization, M. The order parameter must be non-zero below $T_{Critical}$ and must approach zero continuously as $T \rightarrow T_C$ from below. The approach to zero is generally given by a power law and the corresponding exponent is a so-called "critical exponent." The singular behavior of other equilibrium properties such as susceptibility and specific heat is also described by power laws with critical exponents. Scaling theory leads to relations among the various critical exponents. The concept of universality relates to the classification of second-order transitions. These, according to the universality concept, may be divided into classes according to the dimensionality of the system and the symmetry of the order parameter; and, within a given class, the critical properties are supposed to be identical. Neutron scattering experiments on magnetic systems have provided the most comprehensive evidence for the validity of these theories. This area of research represents a rare instance in which calculations can be made for a realistic many-body problem.

It will not be possible for me to review or even to mention all the experimental and theoretical work that has gone on in the field. Most of the early experimental studies were done with iron. The critical scattering was first observed in the total cross section by Palevsky and Hughes (1954), and by Squires (1954), in the near forward direction by Shull, Wilkinson, and Gersch (1956) and near a Bragg peak by Lowde (1958). McReynolds and Riste (1954) observed the phenomenon in magnetite. Probably the most extensive modern measurements were carried out at Brookhaven beginning in about 1968 with the work of Nathans and of Corliss and Hastings and their collaborators and more recently by Shirane and his associates. The foundations of the theory were given by Van Hove 1953a,b and extended by Elliott and Marshall (1958) and Marshall and Lowde (1968).

Inelastic neutron scattering has played an extremely important role in furthering our understanding of the dynamics and of phase transitions in condensed matter. Professor Brockhouse, who with his associates, is the inventor of the triple-axis crystal spectrometer, a time-of-flight spectrometer, and the constant q and constant E techniques, has already traced the early history of inelastic neutron scattering.

As concerns magnetic systems, one of the important applications has been the study of magnon dispersion relations in the heavy rare metals. The measurements that have been made on the ferromagnets, in particular Tb, Dy, and Gd, have shown that the indirect exchange coupling of the RKKY type between the spins is the dominant spin-spin coupling. In the helical structures, Ho, Dy, Tb-Ho alloy, the measurements clearly demonstrate how the nature of the exchange influences the stability of a particular magnetic structure. The magnitudes and positions of the maxima in $J(\vec{q})$, the Fourier transformed exchange energies, for example, correlate quite well with the wave vectors of the periodic structures (Mackintosh and Bjerrum Møller 1972).

With the measurements on Er in the low-temperature conical state, (Nicklow et al 1971) and measurements on Tb (Jensen et al 1975a,b,c) in an applied magnetic field, the usual methods of analysis and interpetation of the data led to an unusually large anisotropic exchange interaction. This in turn led the theorists to re-examine the basic Hamiltonian, taking into account the fact that the exchange and crystal field terms are of the same order of magnitude (Cooke and Lindgard 1977, Lindgard 1978). The present status is that no giant anisotropy terms are required.

Generally there are two approaches to understanding the magnetic properties of matter; the local moment model and the band theoretical model. The rare earths are well represented by the first of these. As concerns the 3-d ferromagnets it seems that the neutron inelastic scattering data for iron and nickel at low temperatures are best described by calculations of the generalized susceptibility $\chi(\vec{q},\omega)$ where electron correlations in the band structure are properly taken into account. At temperatures above T_c, well defined spin-wave excitations have been observed for relatively large values of q and some features of both localized and itinerant models must be invoked (Lynn and Mook 1981).

Many other applications of neutron scattering to magnetism could be cited. I will mention in conclusion the use of neutron polarization analysis (Moon et al 1969), to separate magnetic from nuclear events, and the application of neutron spin echo methods (F. Mezei 1972) to very high resolution studies of critical dynamics.

This research was sponsored by the Division of Materials Sciences, U.S. Department of Energy under Contract No. W-7405-eng-26 with the Union Carbide Corporation.

REFERENCES

1. Bertaut E F 1963 Magnetism III ed Rado and Suhl (New York: Academic Press) p150
2. Bloch F 1936 Phys. Rev. $\underline{50}$ 259; 1937 Phys. Rev. $\underline{51}$ 994
3. Collins M F and Low G G 1965 Proc. Phys. Soc. $\underline{86}$ 535
4. Cooke, J F and Lindgard P A 1977 Phys. Rev. $\underline{16}$, 408
5. Elliott R J and Marshall W 1958 Rev. Mod. Phys. $\underline{30}$ 75
6. Erickson R A 1952 Thesis Agricultural and Technological College of Texas
7. Freeman A J and Desclaux J P 1972 Int. J. Magnetism $\underline{3}$ 311
8. Gupta R J and Freeman A J 1976 Phys. Rev. Lett. $\underline{36}$ 613
9. Halpern O and Johnson M 1939 Phys. Rev. $\underline{55}$ 898
10. Halpern O, Hammermesh M and Johnson M 1941 Phys. Rev. $\underline{59}$, 981
11. Halpern O and Holstein T 1941 Phys. Rev. $\underline{59}$ 960
12. Herpin A, Mériel P and Villain J C 1959 C. R. Acad. Sci. Paris $\underline{249}$ 1334
13. Hughes D J and Burgy M T 1951 Phys. Rev. $\underline{81}$ 498
14. Jensen J, Houmann J C and Møller H Bjerrum 1975a Phys. Rev. B $\underline{12}$ 303, 1975b $\underline{12}$ 320, 1975c $\underline{12}$ 332
15. Kleiner W H 1953 Phys. Rev. $\underline{90}$ 168
16. Koehler W C and Wollan E O 1953 Phys. Rev. $\underline{92}$ 1380
17. Koehler W C 1965 J. Appl. Phys. $\underline{36}$ 1078
18. Koehler W C and Moon R M 1976 Phys. Rev. Lett. $\underline{36}$ 616
19. Lindgard P A 1978 Phys. Rev. B $\underline{17}$ 2348
20. Low G G 1967 Proc. Phys. Soc. $\underline{92}$ 938
21. Lowde R D 1958 Rev. Mod. Phys. $\underline{30}$ 69
22. Lynn J W and Mook H A 1981 Phys. Rev. $\underline{23}$ 198
23. Mackintosh A R and Møller H Bjerrum 1972 Magnetic Properties of Rare Earth Metals ed. R J Elliott (New York: Plenum Press) p188
24. Marshall W 1968 J. Phys. C $\underline{1}$ 88
25. Marshall W and Lowde R D 1968 Reports on Prog. in Phys. $\underline{31}$ 705
26. Mezei F 1972 Z. Physik $\underline{255}$ 146
27. Moon R M 1964 Phys. Rev. A $\underline{136}$ 195
28. Moon R M, Riste T and Koehler W C 1969 Phys. Rev. $\underline{181}$ 920
29. Moon R M, Koehler W C, Cable J W and Child H R 1972 Phys. Rev. B $\underline{5}$, 997
30. Mook H. A. 1966 Phys. Rev. $\underline{148}$ 495
31. Néel L 1936 Comptes Rendus $\underline{203}$ 304
32. Néel L 1948 Ann. Phys. $\underline{3}$ 137
33. Nicklow R M, Wakabayashi N, Wilkinson M K and Reed R E 1971 Phys. Rev. Lett. $\underline{27}$ 334
34. Palevsky H and Hughes D J 1954 Phys. Rev. $\underline{93}$ 268
35. Schwinger J 1937 Phys. Rev. $\underline{51}$ 544
36. Shull C G, Strauser W and Wollan E O 1951a Phys. Rev. $\underline{83}$ 333
37. Shull C G, Wollan E O and Koehler W C 1951b Phys. Rev. $\underline{84}$ 912
38. Shull C G and Wilkinson M K 1955 Phys. Rev. $\underline{97}$ 304
39. Shull C G, Wilkinson M K and Gersch H A 1951 Phys. Rev. $\underline{103}$ 525
40. Shull C G and Yamada K 1962 J. Phys. Soc. Japan $\underline{17}$, Suppl. B-III, 1
41. Squires G L 1954 Proc. Phys. Soc. A $\underline{67}$ 248
42. Steinsvoll O, Moon R M, Koehler W C and Windsor C 1981 Phys. Rev. B $\underline{24}$ 4031
43. Trammell G T 1953 Phys. Rev. $\underline{92}$ 1387
44. Van Hove L 1954a Phys. Rev. $\underline{93}$ 202, b 1374
45. Wood J H 1960 Phys. Rev. $\underline{117}$ 714
46. Yoshimori A 1959 J. Phys. Soc. Japan $\underline{14}$, 807

Linear and non-linear magnetic excitations in perfect and disordered crystals

W.J.L. Buyers

Atomic Energy of Canada Limited, Chalk River Nuclear Laboratories, Chalk River, Ontario, K0J 1J0, Canada

Abstract. A discussion is given of the use of neutrons to study the longitudinal response from spin waves, from the central mode at phase transitions, and from two-magnon processes, as well as the transverse response in random systems, and the longitudinal and transverse response of propagating solitons in one-dimensional systems.

1. Introduction

The neutron is a sinusoidal probe of collective excitations in condensed media. When the excitations in the medium also propagate in a sinusoidal fashion both in time and space, sharp peaks are observed in the neutron inelastic scattering. Examples of such linear excitations are spin-waves and magnetic excitons in perfect crystals at low temperatures.

When the excitations are not sinusoidal either in time or space, broad neutron scattering is observed. Examples in the time domain are the relaxation of spin configurations at elevated temperatures or of local moments coupled to conduction electrons in Kondo or mixed valence systems.

In the spatial domain, examples are the propagation of spin excitations through random systems or the spin excitations and propagating solitons in Ising-like antiferromagnetic chains.

Three types of excitations will be discussed, longitudinal linear spin excitations in localized spin systems, transverse excitations in random systems, and non-linear transverse spin-waves and longitudinal solitons in a one-dimensional system.

2. Longitudinal Excitations of Localized Spins

In contrast to the transverse excitations of ordered spin systems, whose analogue is the spin precession of classical spins, linear longitudinal excitations, for which there is no classical picture, occur when the ionic spin, $\vec{S}$, is coupled to the orbital angular momentum, $\vec{L}$. In d-electron systems the spin-orbit coupling, λ, produces low-lying manifolds of states described by $|\tilde{\ell}, \vec{S}, j, j^z\rangle$, where $\tilde{\ell} = \vec{L}/\alpha$ is the effective orbital angular momentum of the lowest orbital state produced by the crystal field. Neutron scattering can probe the transitions $j \to j + 1$ between different spin-orbit manifolds with spacing $\alpha\lambda(j+1)$. Those transitions with the same j^z are known as Davydov excitations, in analogy with the theory of molecular excitons. The Davydov excitations in an antiferromagnet, unlike the doubly degenerate transverse excitations, are singly degenerate and periodic over the nuclear rather than magnetic Brillouin zone. Their splitting was shown by Holden et al (1977a) and by Cowley et al (1973) to provide a unique measurement of the zz part of the exchange interaction, not available from ordinary spin-wave measurements.

In f-electron systems the crystal field interaction provides a low-lying manifold of states, $|L,S,J,\Gamma,J^z_\Gamma\rangle$, corresponding to the Hund's-rule value of the total angular momentum $\vec{J} = \vec{L} + \vec{S}$. Transitions between states belonging to different crystal field levels, Γ, but having the same J^z then give rise to the longitudinal inelastic scattering. In cubic crystals with a singlet ground state such as TbSb, Pr_3Tl or $PrAl_2$, the longitudinal excitation to the $J^z = 0$ component of the excited triplet has a strength and dispersion relation similar to the transverse spin-waves.

In contrast to theoretical expectations that the divergence of χ^{zz} at the phase transition should involve an instability in the longitudinal mode, the modes observed in neutron scattering were found by Birgeneau et al (1972) to be relatively temperature independent. This surprising result prompted the development by a number of authors of a combined molecular field, random phase approximation. It predicted a gap in the longitudinal spectrum at all temperatures, and a central peak whose intensity diverged at the phase transition (Fig.1). The theory for the singlet-triplet model, which does not show a gap, was first developed by Smith (1972). It was later generalized by Buyers (1975) for any many-level system. The divergent central mode intensity was subsequently observed by Als-Nielsen et al (1977a) in Pr_3Tl; it had already been observed in TbSb and $PrAl_2$ (Holden et al 1977b). Because the exchange fluctuations are small in Pr_3Tl the theory can be used for critical effects: it showed correctly that the correlation range in a singlet ground state system is much longer than in pure spin systems (Als-Nielsen et al 1977b).

Longitudinal scattering also arises from two-magnon processes through the Bose expansion $S^z \approx S - a^+a$. Two-magnon creation occurs in an antiferromagnet. The observations of Cowley et al (1969) are shown in Fig. 2. In an antiferromagnet it provides a unique determination of the spectrum of zero-point spin fluctuations. In a ferromagnet at $T = 0$ two-magnon scattering cannot take place, but at elevated temperatures the difference process, corresponding to annihilation and creation of a magnon, gives rise to a thermally activated central peak. Two-magnon scattering has complicated the interpretation of measurements of the soliton response in the 1d XY ferromagnet $CsNiF_3$ (Kjems and Steiner 1978 and Reiter 1981).

3. Transverse Spin Excitations in Random Systems

Although spin excitations are strongly scattered by magnetic impurities or vacancies, they remain essentially linear excitations of the spin system. The broad spin-wave response observed in systems with two dispersive impurity bands (Cowley and Buyers 1972) spawned a number of theories for the apparent non-linearity in $(\vec{q},\omega)$ space, as reviewed by Cowley (1982). Theories such as the coherent potential approximation, in which the non-linearity was described by a complex self-energy ran into serious difficulties when self-consistency was required (Leath 1982). The calculations were also complicated and required much numerical work. Nowadays, for the same or less amount of computational work, the equations of motion for the spin excitations of a large finite system can be integrated forward in time (Thorpe and Alben 1976) or directly solved in (i,ω) space, where i labels the lattice sites, as reviewed by Buyers (1982). The

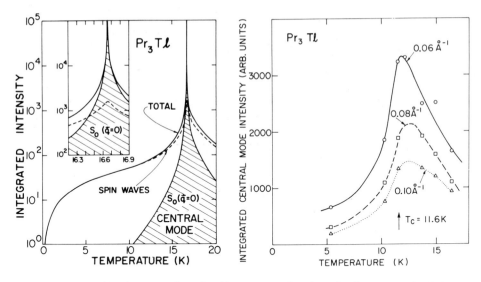

Fig. 1 Longitudinal central mode divergence in the singlet-ground-state system Pr_3Tl as predicted (left) and observed (right).

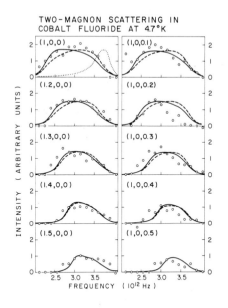

Fig. 2 Two-magnon scattering at 4.7K in CoF_2. The solid (broken) lines are the spin-wave theory with (without) interactions. The density of states (dotted) disagrees with the results.

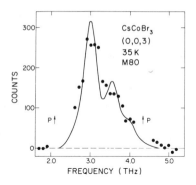

Fig. 3 Spin-wave continuum scattering at zone centre (Nagler et al 1983). The theoretical curve includes the effect of intra- and interchain exchange mixing.

linearity of the spin response is evident in (i,ω) space, for the dynamical susceptibility, $g_i(\omega)$, at site i is linearly related by the exchange coupling, V_{ik}, to that at other sites, leading to a crystal response

$$G_{i\vec{Q}}(\omega) = g_{i\vec{Q}}(\omega) + g_{i\vec{Q}}(\omega) \sum_k V_{ik} G_{k\vec{Q}}(\omega). \tag{1}$$

The second subscript $\vec{Q}$ indicates a Fourier transform over only the second spatial label. The equations may be solved iteratively, by diagonalization or by the Lanczos method to give an exact set of excitation frequencies. When the spatial Fourier transform measured by neutron scattering is calculated, it is broad in frequency because of the lack of translational invariance of the random system.

4. Transverse and Longitudinal Scattering in One-Dimensional Systems

Perhaps the most complex but most interesting system is the S = 1/2 Ising-Heisenberg antiferromagnet whose Hamiltonian is

$$H = 2J \sum_i [S_i^z S_{i+1}^z + \epsilon(S_i^x S_{i+1}^x + S_i^y S_{i+1}^y)]. \tag{2}$$

The spins point along the chain as the ratio, ε, of transverse to longitudinal coupling is only about 0.1. Examples are $CsCoCl_3$ and $CsCoBr_3$.

The history of the subject goes back to the year before the neutron was discovered with the celebrated solution to the isotropic chain by Bethe (1931). The ground state consists largely of the Néel state. The first excited states at ≈ 2J consist of the degenerate soliton states formed by creating a domain beginning at site i that extends over ν consecutive spins; the transverse interaction splits these into a band of states in which the solitons hop two steps at a time down the chain. Transitions from the ground state to odd-length domains are transverse, and those to even-length domains are longitudinal. The longitudinal spin excitations occur in the same frequency regime as those of transverse symmetry, because it takes the same energy, 2J, to form a domain of any length for small ε. The longitudinal scattering is the analogue of two-magnon creation scattering shifted down from 4JzS by the binding energy 2J, where z = 2, S = 1/2 for the 1d system. The spin excitations therefore form a triple continuum as shown by Fowler (1978).

In the spin-wave response solitons of wave-vector k, and length ν, are created; they are equally coupled to solitons of lengths ν ± 2. The spin-wave neutron scattering was shown by Buyers et al (1980), and by Satija et al (1980) to be in the form of a continuum widest at the zone centre (Fig.3) and narrowest at the zone boundary.

The soliton response (Fig. 4) consists of transitions within the thermally activated 2J band between states $|k,\nu\rangle$ and states $|k+Q+\pi,\nu'\rangle$. It can be regarded as a two-excitation difference process, provided the excitation is taken as a domain rather than a single-spin excitation. However, the analogy with longitudinal two-magnon scattering is not particularly useful as there is no linear spin-wave theory on which to base it, and transverse as well as longitudinal scattering occurs (Nagler et al 1982). Villain (1975) showed that a description in terms of almost independent solitons of the type described above provided the appropriate basis for a theory. Although the longitudinal response from the soliton motion is a continuum, it gives rise to a well-defined peak in $S^{zz}(Q,\omega)$ at

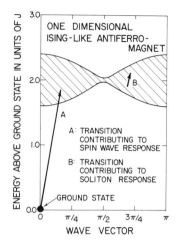

Fig. 4 Spin waves and solitons.

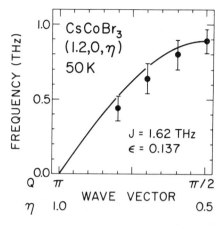

Fig. 5 Dispersion relation for solitons in CsCoBr$_3$. The line is the free soliton dispersion of Villain (1975).

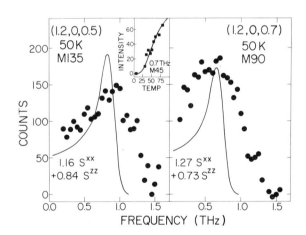

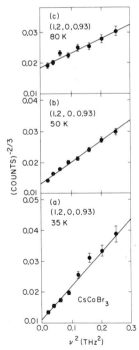

Fig. 6 (above) Peaks in the neutron scattering compared with the theory for independent solitons. The peaks are thermally activated (inset) and collision broadened.

Fig. 7 (right) The spectrum of long-wavelength soliton scattering is consistent with a Lorentzian to the 3/2 power (Nagler et al 1983). The slope decreases as the soliton density increases with temperature.

$$\omega_Q = 4\varepsilon J \sin(Q), \quad (3)$$

where Q is the inter-spin phase difference. The inelastic soliton peak was observed for the first time in $CsCoBr_3$ by Nagler et al (1982) as shown in Figs. 5 and 6. The quasielastic peaks seen in $CsNiF_3$ by Kjems and Steiner (1978) and in TMMC by Regnault et al (1982) correspond to the diffusion of solitons. In $CsCoBr_3$ the solitons are unique in that, despite collisions with other solitons, they survive at short wavelengths as a propagating excitation. At the zone centre, $Q = \pi$, the solitons behave as a gas, analogous to the solitons of the ϕ^4 and sine-Gordon systems. Their spectrum has the unusual form of a Lorentzian to the 3/2 power (Maki 1981). This spectrum (Fig. 7) has been observed by Nagler et al (1983). Neutron scattering has thus proved, as it has with linear excitations, to be a most powerful probe of the non-linear dynamics of the soliton.

5. Epilogue

Disturbances that involve non-linear evolution of a sinusoidal may be nearly linear within a soliton basis and vice-versa. Non-linearity probed with the neutron brings new insights. Although the paper of Bethe is one year older than the neutron, 50 years later the neutron and the physics of one dimension are enjoying a rich and productive partnership.

References

Als-Nielsen J, Kjems JK, Buyers WJL and Birgeneau RJ 1977a J Phys C10 2673
Als-Nielsen J, Kjems JK, Buyers WJL and Birgeneau RJ 1977b Physica 86-88B+C 1162
Bethe HA 1931 Z Physik 71 205
Birgeneau RJ, Als-Nielsen J, and Bucher E 1972 Phys Rev B62 2724
Buyers WJL 1975 AIP Conf Proc 24 27
Buyers WJL, Yamanaka J, Nagler SE and Armstrong RL 1980 Solid State Comm 33 857
Buyers WJL, Bertrand D, Locke KE and Stager CV 1982 Excitations in Disordered Systems ed Thorpe MF (New York: Plenum) 411
Cowley RA, Buyers WJL, Martel P and Stevenson RWH 1969 Phys Rev Lett 23 86
Cowley RA and Buyers WJL 1972 Rev Mod Phys 44 406
Cowley RA, Buyers WJL, Martel P and Stevenson RWH 1973 J Phys C6 2997
Cowley RA 1982 Excitations in Disordered Systems ed Thorpe MF (New York: Plenum) 373
Fowler M 1978 J Phys C11 L977
Holden TM, Buyers WJL and Svensson EC 1977a Physica 86-88B+C 1041
Holden TM, Buyers WJL, Svensson EC and Purwins H-G 1977b Crystal Field Effects in Metals and Alloys ed A Furrer (New York: Plenum) 189
Kjems JK and Steiner M 1978 Phys Rev Lett 41 1137
Leath PL 1982 Excitations in Disordered Systems ed Thorpe MF (New York: Plenum) 109
Maki K 1981 Phys Rev B24 335
Nagler SE, Buyers WJL, Armstrong RL and Briat B 1982 Phys Rev Lett 49 590
Nagler SE, Buyers WJL, Armstrong RL and Briat B 1983 (to be published)
Regnault LP, Boucher JP, Rossat-Mignod J, Renard JP, Bouillot J and Stirling WG 1982 J Phys C15 1261
Reiter G 1981 Phys Rev Lett 46 1981
Satija SK, Shirane G, Yoshizawa H and Hirakawa K 1980 Phys Rev Lett 44 1548
Smith SRP 1972 J Phys C5 L157
Thorpe MF and Alben R 1976 J Phys C9 2555
Villain J 1975 Physica B79 1

Neutron scattering studies of itinerant electron magnets

Y. Ishikawa

Physics Departments, Tohoku University, Sendai 980, Japan

1. Introduction

This paper makes a critical review of the neutron scattering studies of the itinerant electron magnets in order to see how neutron scattering has contributed to recent understanding of itinerant electron magnetism and also to give an insight into the future aspect of the neutron scattering studies of this subjct.

It was the famous theory by Izuyama, Kim, and Kubo (1963) which gave the fundation of study of the itinerant electron magnet by neutron scattering. By using the Kubo's linear response theory, they showed that the neutron scattering cross section $d^2\sigma/d\omega d\Omega$ is proportional to the imaginary part of the generalized susceptibility $\chi(\vec{Q},\omega)$;

$$\frac{d^2\sigma}{d\omega d\Omega} = A(k_i k_f) \Sigma^{\alpha\beta}(\delta_{\alpha\beta} - \hat{e}_q^{\alpha} \hat{e}_q^{\beta}) \frac{\hbar}{\pi} \frac{Im\chi^{\alpha\beta}(Q,\omega)}{1-\exp(-\hbar\omega/kT)}. \qquad (1)$$

Since $\chi(Q,\omega)$ can be calculated by the band theory based on the random field approximation (R.P.A) as

$$\chi(\vec{Q},\omega) = \frac{\chi_0(\vec{Q},\omega)}{1-I\chi_0(\vec{Q},\omega)} , \qquad (2)$$

with $\chi_0(Q,\omega)$, the susceptibility of the non-interacting system and I the intra-atomic interaction, neutron scattering can provide direct information on the adequacy of the band theory based on R.P.A.

Encouraged by this theory, Lowde and Windsor (1970) made a pioneering neutron scattering work on metallic Ni and found that the neutron scattering intensities over a wide range of ω and $\vec{Q}$ spaces and at various temperatures across the Curie temperature T_c were compatible with the R.P.A. band theory provided that a proper $\vec{q}$ dependence was assumed for I. Although their results were less convincing because of lack of scattering intensity and resolution, the work is highly appreciated, because it paved a good way for later investigations.

A simple calculation of eq(2) for the ferromagnetic electron gas (Thompson 1965) has shown that the magnetic excitations in the itinerant ferromagnet has a unique characteristic that, in addition to the sharp excitations corresponding to the spin wave excitations, there exists wide region of excitations called the Stoner continuum and the spin waves disappear when the dispersion crosses the Stoner boundary. Therefore the later effort of neutron scattering studies was mainly directed to the observation of the Stoner continuum.

2. Magnetic Ground State and R.P.A. Band Theory.

Neutron scattering studies of the itinerant electron system were promoted with the advent of high-flux reactors (> 1968). Mook and others (1969, 1973a) found that the intensity of the spin wave scattering from metallic Ni and Fe decreased substantially when the spin wave energies approach to 100 meV and they attributed it to the presence of the Stoner boundary around this energy. The magnetic excitations they observed at low temperatures were then shown to be in excellent quantitative agreement with theoretical calculations by Cooke et al (1976) based on the generalized R.P.A. where realistic energy bands were used and the band and wave vector dependences of the electron-electron interaction matrix element were taken into account. The calculations also predicted that the [100] acoustical spin wave encounters an optical branch at 128 meV as shown in Fig.1, which has also been confirmed by neutron scattering (Mook et al 1979) as plotted by open circles in the same figure. The results provided a definite experimental evidence that the P.R.A is a good approximation for describing the magnetic excitations at the lowest temperature.

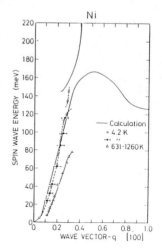

Fig.1 Calculated magnon dispersions along [100] in Ni based on R.P.A. band theory(Cooke 1976) and comparison with observations, (Mook et al 1979) (Lynn et al 1981)

We should note here that the adequacy of the R.P.A. is not restricted to the ideal itinerant electron system. The band theoretical calculations of the magnetic excitations have also been performed for a number of Heusler alloys (Cu_2MnAl, Ni_2MnSn and Pb_2MnSn) (Ishida et al 1980, Kubo et al 1981) based on the generalized R.P.A. developed by Cooke et al 1973) and the results agree satisfactorily with those ovserved by neutron scattering (Noda and Ishikawa, 1976a, Tajima et al 1977). An example is shown in Fig.2 where the calculated spin wave dispersion for Pd_2MnSn (closed circles and triangles are compared with the obsevation (open circles).

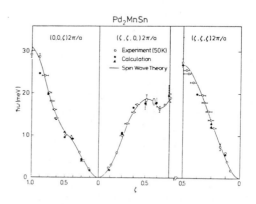

Fig.2. Calculated magnon dispersions in Pn_2MnSn based on R.P.A. band theory(o) and comparison with observations(o) (Kubo et al 1980)

The result is quite important because these Heusler alloys have a definite large localized moment and their thermodynamical properties can well be described by the two magnon theory based on the Heisenberg model (Noda and Ishikawa 1976b). Therefore

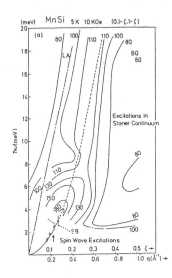 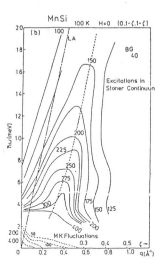

Fig 3. Magnetic excitations in MnSi, (a) at 5K and (b) at 100 K. Low energy excitations are measured with high resolutins. (Ishikawa et al 1977)

we may conclude that the R.P.A. calculation based on the realistic band is reasonably successful even for describing the magnetic ground state of the localized spin system.

Another important contribution which neutron scattering made to this problem is the succesful observation of the magnetic excitations in the Stoner continuum in MnSi (Ishikawa et al 1977). Although MnSi has a helical spin structure below T_N=29 K, it becomes ferromagnetic in an applied field of 6.2 KOe and exhibits typical characteristics of the weak itinerant ferromagnet; the paramagnetic moment of 1.4 μ_B, and is much higher than the saturation magnetization at 0 K, 0.4 μ_B, and the sample shows a strong magneto-volume effect. Neutron inelastic scattering from MnSi has revealed that well defined spin waves exist only below 2 meV and substantial increase of the linewidth of the excitations was found above 2.5 meV as displayed in Fig.3(a), suggesting the spin wave dispersion merges into the Stoner continuum. The variation in intensity and linewidth of spin waves along the dispersion across the Stoner boundary is consistent with calculations based on the R.P.A. for the ferromagnetic electron gas, thus providing other experimental evidence for the adequacy of the R.P.A. for describing the magnetism at 0 K.

It is, however, remarked that there is a distinct deviation of the observation from the R.P.A. calculation. The observation has shown as seen in Fig.3(a), the magnetic excitations in the Stoner continuum develop along the extrapolated spin wave dispersions, while in the calculations the ridge of the excitations in the continuum has a slope different from that of the spin wave dispersion. A similar descrepancy has also been found in the plasmon dispersion in Li by X-ray inelastic scattering (Platzman and Eisenberger 1974) which show the dispersion merges into the particle-hole continuum without any change of slope, in disagreement with the R.P.A. calculation. Since this effect has been interpreted as a many body effect where the plasmons couple with the multi-pair excitations (Awa et al 1981), a similar interpretation may also be made for MnSi.

3. Finite Temperature Magnetism

Although the finite temperature magnetism is still a controversial subject for the itinerant electron system, there has been a remarkable progress in understanding it in this decade (Moriya, 2979, 1981, 1982) and the contribution of neutron scattering to this subject is also invaluable. The observation of sloppy spin waves in Ni and Fe above T_c (Mook et al 1973b, Lynn 1975) as shown in Fig.1 is a clear evidence that the R.P.A. does not work at finite temperatures.

According to recent theoretical and experimental investigations (Moriya, 1981 Ishikawa, 1979), the magnetic properties of itinerant electron systems at finite temperatures can be understood in terms of the temperature variation of the local magnetic moment $<M_L^2>$ on the atoms. In contrast with the Heisenberg system where $<M_L^2>$ is constant, independent of temperature, $<M_L^2>$ in the itinerant electron magnet exhibits a variety of behavior, depending on the strength in the electron correlation. In the extreme case of the strong correlation limit, which is the case for the Heusler alloys, $<M_L^2>$ keeps a constant value even above the Curie temperatures. On the other hand, in another extreme case of the weak electron correlation as the weak itinerant ferromagnet $ZrZn_2$ or $MnSi$, the Moriya's theory (1981) has predicted that $<M_L^2>$ decreases with increasing temperature, but it takes a finite value of $<M_L(T_c)^2>=3/5M(0)^2$ at T_c and increases again with increasing temperature. The Curie-Weiss relation of the magnetic susceptibility is a result of the linear increase of $<M_L^2>$ with temperature, quite a contrast with the conventional mechanism of the susceptibility.

Neutron scattering has provided good experimental support for this expectation. We have shown that the temperature dependence of magnetization $m_z(T)$ of a Heusler alloy Pd_2MnSn can well be reproduced up to 0.8 T_c by the two magnon theory for the Heisenberg system using the exchange parameters determined at OK (cf. solid lines in Fig.2) (Noda and Ishikawa 1976b). The energy spectra of the paramagnetic scattering are also compatible with the Heisenberg system with these exchange parameters. More direct experimental proof has been obtained by polarized neutron scattering (Brown et al 1982) who showed that the paramagnetic moment remains almost unchanged up to 4 T_c. Therefore we may conclude that in the case of the strong correlation limit as for the Heusler alloys, the magnetism at 0 K is explained by the R.P.A. band theory, while its thermodynamics can be treated by the Heisenberg model.

The most direct way to determined $<M_L^2>$ is to measure the neutron magnetic scattering over the whole Brillouin zone, because $<M_L^2>$ is related with the scattering cross section as

$$N<M_L^2> = \sum_q <M_q M_{-q}>$$
$$= 3kT\sum_q \chi(q)$$
$$=2C\sum_q \int_{-\infty}^0 (d^2\sigma/d\Omega d\omega)d\omega \ . \ (3)$$

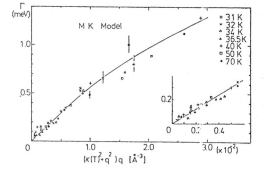

Note that the zero point motion term should be subtracted from the cross section. The polarized neutron scattering with the polarization analysis of the scattered neutrons is the best way to separate the para-

Fig.4. Linewidth Γ of M K fluctuations plotted against $q/\chi(q)$

magnetic scattering from the background. Brown and others (1982) determined by this method $<M_q^2>$ of various materials (Fe, CeFe$_2$, MnSi etc) above T_c though their results include the zero point motion, which has a significant effect for MnSi at low temperatures. The Moriya theory (1981) also has shown that $\chi(q)$ is further related with $<M_L^2>$ for small wave vector q as

$$\frac{1}{\chi(q)} = \frac{1}{\chi_0(q)} - 2I + \frac{5}{3}\frac{\gamma}{N}<M_L^2(T)> \qquad (4)$$

, where $\chi_0(q)$, I and γ are the non-interacting susceptibility, intra-atomic exchange energy and a constant representing the mode-mode coupling or longnitudinal stiffness of the moment. If $\Sigma_q \chi(q)$ is independent of temperature, $<M_L^2>$ increases linearly with temperature,(cf.eq(3)) resulting in a linear increases of $1/\chi(q)$ with temperature, $1/\chi_0(q)-2I$ being weakly temperature dependent. Therefore $\Sigma_q \chi(q)$=constant is an important condition for the new Curie-Weiss law,which should be checked by neutron scattering.

MnSi is a good candidate to examine this mechanism, because the magnetovolume effect above T_c has shown that $<M_L^2>$ takes a minimum value of 2.8/5 $M(0)^2$ at T_c and then it increase as the theory suggested. (Matsunaga et al 1982) Therefore a detailed study of the spin fluctuations has been carried out by neutron scattering to see whether the Moriya's theory really holds for this material . In Fig.3(b) is shown the magnetic excitations measured at 100 K (3.3 T_c) (Ishikawa et al 1977). The excitations in the Stoner continuum remain almost unchanged even up to room temperature (10 T_c), while the spin wave excitations observed at 5 K are collapsed into critical scattering above T_c, and we could see the quasi-elastic scattering with small energy transfers as plotted by broken lines in Fig.3(b). Since the quasi-elastic scattering is not accounted for by the R.P.A., we call the scattering as the "M K fluctuations". The theory predicts (Ishikawa 1979b) that the M K fluctuations has the scattering function $S(q,\omega)$ of the form $S(q,\omega) \propto 1/\kappa^2+q^2 \cdot \Gamma\omega/\Gamma^2+\omega^2$ just as the critical scattering of the Heisenberg system, but with $\Gamma=Cq/\chi(q)$, in contrast with the Heisenberg system where $\Gamma=cq^2/\chi(q)$. In Fig.4 is plotted the linewidths of the M K fluctuations against $(\kappa^2+q^2)q$ (Ishikawa et al 1982) which shows that $\Gamma=Cq/\chi(q)$ relation really hold for low q values, suggesting that the observed scattering is different from the fluctuations of the localized spin systems. Furthermore the generalized syceptibility $\chi(q)$ obtained by integrating the M K fluctuations shows a strong q dependence, $\chi(q)=C/(T-T_c)+900q^2$ over a temperature range between 33 K and 70 K and for q less than 0.4 A. Such a strong q dependence is a typical trend of the M K fluctuations. Since there is no other excitation at q=0, this fluctuation at q=0 should be responsible for the Curie-Weiss law. The high energy components of the spin fluctuations was also estimated by integrating the excitations in the Stoner continuum. The high energy component is absent below T_c!, but it increases almost linearly with increasing temperature and the contribution to the spin correlation $<M_q M_{\bar{q}}>$ from this component becomes almost the same order of magnitude as the M K fluctuations at T=10 T_c. Since the high q components are distributed more widely in q space, it is possible that $\Sigma_q \chi(q)$ is constant of temperature, resulting in the new Curie-Weiss mechanism predicted by Moriya(1981). The most important conclusion we got from our analysis is that the increase of $<M_L^2>$ above T_c is a result of thermal excitations in the Stoner continuum.

In case of the intermediate correlations,as Fe and Ni, there still exists a strong controvercy on the model to explain the mechanism of the sloppy spin waves. It is now generally admitted that the short range order persists above T_c for Fe and Ni as have been observed by polarized neutron scattering (Brown et al 1982) and forbidden magnon scattering (Lowde et al 1982), but no consensus has been reached about the fundamental question whether amplitude fluctuations or angular fluctuations or both are responsible for a transition from ferromagnetism to paramagnetism. For a critical understanding of the present state of this problem, the reader should refer to a recent review article by Moriya(1983).

In conclusion, a brief survey of the neutron scattering studies of the itinerant electron magnet has revealed the invaluable contribution of neutron scattering for understanding the magnetism at 0 K as well as at finite temperatures. In order to get, however, denifite answers to the important problems in the quasi-localized spin systems as the sloppy spin waves or the invar problem (Ishikawa 1979a, 1979b), the measurememts of magnetic correlations by polarized neutron scattering as has been done by Brown et al (1982) but <u>with energy analysis in a wide ω range to eliminate the zero point motion</u> is highly required. Since it requires the high energy polarzed neutrons, the advent of the intense pulsed neutron sources (>1986) would be a start of new development for the itinerant electron magnetism.

References
Awa K, Yasyhara H, and Asahi T 1981 Solid State Commu, 38 1285
Brown P.J, Cappellman H, Deportes J. Givord D, Ziebeck K.R.A. 1982 J. Appl Phys 53 1973
Cooke J.F. and Davis H.L. 1973 AIP conf. Proc 24 329
Cooke J.F. 1976 Proc. Conf. Neutron Scattering (Gatlinburg) 723
Ishida S, Yasuhara H, and Asano S 1980, Jphys. Soc Jpn 48 814
Ishikawa Y, Shirane G, Tarvin J and Kohgi M 1977 Phys Rev B 16 4956
Ishikawa Y, Onodera S, Tajima K 1979a J. Mag.Mag.Mater, 10 183
Ishikawa Y 1979b J. Mag.Mag.Mater 10 123
Ishikawa Y, Noda Y, Fincher C and Shirane G 1982a Phys. Rev. B25 254
Izuyama T, Kim O, Kubo R 1963 J. Phys Soc Jpn 18 1025
Kubo Y, Ishida S and Ishida J. 1981 J. Phys. Soc. Jpn 50 47
Lowde R.D and Windsor C.G . 1970 Adv. in Phys 19, 813
Lowde R.D, Moon R.M., Pagonis B, Perry C.H. Sokotoft J.B., Vaughan-Watkins R.S., Wiltshire and Crangle J 1982 to be published.
Lynn J.W. 1975 Phys. Rev. B11, 2624
Lynn J.W. and Mook H.A. 1981 Phys. Rev B23 198
Matunaga M, Ishikawa Y and Nakajima T 1982, J. Phys Soc Jpn 51 1153
Mook H.A., Nicklow R.M., Thompson, E.D. and Wilkinson M.K 1969 J. Appl Phys 40, 1450
Mook H.A. and Nicklow R.M. 1973a Phys. Rev B7 336
Mook H.A., Lybb, J.W, and Nicklow R.M. 1973b Phys. Reb.30 556
Mook H.A. and Tochetti 1979 Phys Rev Letters 43, 2029
Moriya T, 1979 J. Mag.Mag.Mat 14 1
Moriya T, 1981 Magnetism in Narrow Band Systems ed by T. Moriya (Springer-verlag; Berlin) P2-27
Moriya T, 1983, Proc. Intern Conf. Mag. 1982(Kyoto)
Noda Y and Ishikawa Y 1976a, 1976b J. Phys. Soc Jpn 40 690, 699
Tajima K, Ishikawa Y, Webster P J, Stringtellow M W, Tochetti O and Ziebeck K R A 1977 J. Phys. Soc Jpn 43 483
Thompson E.D. 1965 Adv. in Phys. 14 213

Magnetism in the actinides: the role of neutron scattering

G. H. Lander

Argonne National Laboratory
Argonne, Illinois 60439, U.S.A.
and
Institut Laue Langevin
156X, 38042 Grenoble, France

Abstract

Neutron scattering has played a crucial and unique role of elucidating the magnetism in actinide compounds. Examples are given of elastic scattering to determine magnetic structures, measure spatial correlations in the critical regime, and magnetic form factors, and of inelastic scattering to measure the (often elusive) spin excitations. Some future directions will be discussed.

1. Introduction

Magnetism in actinide systems presents a particularly intriguing challenge for solid-state physics. There is a simple physical reason for this complexity. The unpaired $5f$ electrons, which give rise to the magnetic behavior, have a relatively large spatial extent but, since they are f electrons, also have a large inherent orbital moment. What this means in practice is that we find a plethora of magnetic behavior in actinide systems. This ranges from classical "localized" systems such as UPd_3, to itinerant magnets, e.g. UN, $NpOs_2$, $NpSn_3$, to spin fluctuation systems such as UAl_2 and $PuAl_2$, to unusual multi-k magnetic structures in UAs, and to metamagnetic almost "devil's staircase" behavior such as found in UAs-USe solid solutions and Np pnictides.

2. Magnetic Structures

The elements U, Np, Pu and Am do not exhibit spontaneous magnetic order. In the first three the f-electrons participate in the bonding so that the bandwidths are large and the interatomic exchange correlations are insufficient to overcome the large intrinsic bandwidth. Americium has a $5f^6$ configuration and forms a J=0 ground state. Cm is the first magnetic actinide element, but its antiferromagnetic structure is not yet determined. Berkelium is antiferromagnetic whereas no magnetic order has yet been found in element 98, californium.

---

* Present address; operated by U. S. Department of Energy.

The number of compounds examined is of course immense. Tabulations are given in the Actinides Volumes edited by Freeman and Darby (1974), and by Brodsky (1978). A number of compounds are ferromagnets. The unusual feature of many of these is the anisotropy present. Many of the antiferromagnetic structures are complex. Some of those exhibited by the neptunium pnictides (Aldred et al 1974) are shown in Fig. 1. The structures are basically built up of (001) planes of ferromagnetically coupled spins, but the sequence of these planes along the direction of the spins is variable.

Examples such as shown in Fig. 1, and those from a wide variety of U systems, led to suggestions that the coupling within the (001) planes was somehow stronger than that between the planes. Such a coupling would be unusual since the crystal symmetry is cubic.

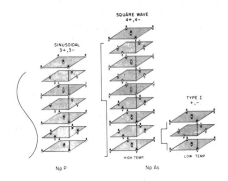

Fig. 1 Magnetic structures of NpX compounds. NpP has sinusoidal modulation of the magnetic moment along the propagation axis with a repeat of 3 unit cells, 6(001) planes. At high temperature, NpAs has a 4+, 4-, structure but has a first-order transition at 142K to the type-I, + -, structure, which is also the structure of NpSb.

3. Critical Scattering

These ideas about anisotropic exchange that arose from the magnetic structures received a more direct confirmation from measurements of the spatial correlations in the critical regime just above the ordering temperature. An example of the diffuse scattering from UAs is shown in Fig. 2 taken from Sinha et al., 1981. One interesting feature of this scattering, is that it is much more diffuse in the $[00\eta]$ direction than in the $[\xi\xi 0]$ direction, see Fig. 2. When translated into real space this implies that the spatial correlations are much weaker along the [001] direction, than perpendicular to it. Since the structure consists of ferromagnetic sheets stacked along the [001] axis in the scheme +-+- this means that within each sheet the spins are strongly correlated, but that there is only weak correlation between them. Similar measurements relating to the spatial correlations are reported for USb (Lander et al, 1978) and for UN (Holden et al, 1982).

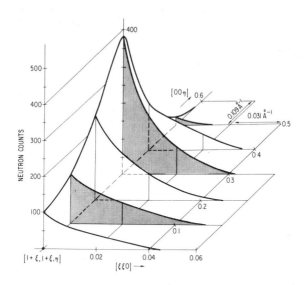

Fig. 2 Schematic representation of the diffuse scattering at T_N+1.2 K around the 110 point. The reciprocal-lattice projection is 1$\bar{1}$0. Note the difference in scale in the [ξξ0] and [00η] directions. The instrumental resolution functions are Δξ=0.012 and Δη=0.005 in appropriate reciprocal-lattice units.

4. Magnetic form factors

The interaction of the neutron with the unpaired 5f electrons gives rise to the possibility of measuring the spatial extent of these electrons. Many experiments of this nature have been done, including some on Np (Aldred et al 1976) and Pu (Lander and Lam 1976) systems. However, for our purposes here the most interesting experiment is that on USb (Lander et al, 1976). The aim of the experiment was to determine the ground-state wavefunctions of the 5f electrons. Two possible shape functions for the quadrupole moment are shown in Fig. 3. The measurement of the form factor in USb showed unequivocally that the magnetization density was oblate with respect to the spin quantization direction. This was an unexpected result since the distribution in the analogous 4f material NdSb is prolate.

We can now see a common thread in these experiments. The strong correlations in the (001) planes, i.e. perpendicular to the spin direction, were first suggested by the magnetic structure work, demonstrated directly in the critical scattering, and their microscopic single-atom consequence shown by the form-factor studies. If we think of the oblate, or pancake-like, distribution of spin density in the (001) plane then it is clearly much easier for it to interact strongly with charge densities <u>within</u> the (001) plane than with those out of the plane. This qualitative picture has now been put on a quantitative scale by Sinha and Fedro (1979) and by Cooper (1982).

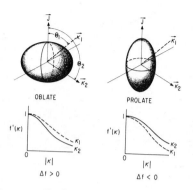

Fig. 3 Schematic distribution of the magnetization densities and corresponding form factors for oblate and prolate densities. The function Δf is the difference between two form factors at the same $Q \equiv K$ but different directions $\hat{Q}$. The distribution shown here are quadrupolar, i.e. we have neglected deviations related to ϕ, the azimuthal angle.

5. Neutron Inelastic Scattering

Neutron inelastic scattering has not yet played a major role in the study of the actinides. There are good reasons for this. The large amount of material needed (at least 0.5 cm^3) has confined studies to uranium systems and the complexity of the dynamical response function in the actinides has made experiments very difficult. Initially, it was thought that transitions between crystal-field levels could be seen easily with neutrons in the actinides in the same way as in lanthanide systems. This has not been the case, in fact the only system found so far in which crystal-field levels appear well defined is UPd$_3$ (Buyers et al, 1981).

The most extensively studied systems are the uranium rocksalt systems. To illustrate the complexities we show the magnetic and phonon dispersion of USb in Fig. 4 from Lander and Stirling (1980). This is one of the few materials that do exhibit well-defined spin waves. Their dispersion is steep and their polarization unusual, although the latter may be related to the triple-$\vec{k}$ nature of the magnetic ordering in USb (Rossat-Mignod et al 1980, Jensen and Bak 1981). What is not understood yet is why the excitations essentially vanish by ~100K, which is less than 0.5T$_N$, into a continuum of magnetic scattering. Such wide distributions of magnetic scattering in these systems was first intimated by Wedgwood in 1974, but it has taken a long time to learn to measure them reliably.

An example is UAs. Here no spin-waves are found (Stirling et al, 1980), but a measurement of the magnetic inelastic scattering was recently completed by Loewenhaupt et al, 1982. The scattering can be characterized in the first approximation by a Lorentzian centered at E=0 but with a certain half width that usually is temperature dependent. These response functions have also been found in materials that exhibit valence fluctuations (Loewenhaupt et al, 1979 and Shapiro, 1981) but the situation is unusual in UAs because the system orders magnetically. The neutrons can, of course, measure the susceptibility as a function of Q, the wavevector, and in Fig. 5 we show the results for the total susceptibility at a wavevector of 0.7Å^{-1}, which is about half-way to the first magnetic zone center at (110) of 1.6Å^{-1}. Notice that at this point X(Q) is essentially constant. This shows that at Q=0 we are measuring correlation effects - even in an antiferromagnet. They decrease, of course, when the material orders, whereas X(Q) increases at the positions of antiferromagnetic wavevectors. Such experiments give a direct picture of what happens microscopically and are therefore useful in constructing theoretical models. Similar experiments have also been done on UN and UTe, Buyers et al (1981).

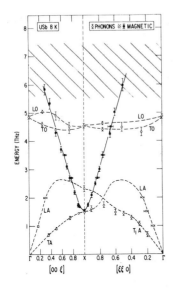

Fig. 4 The dispersion curves for USb; energy plotted against wave-vector transfer $\vec{Q}$ (in units of $2\pi/a$). The dashed lines represent the phonons. The magnetic modes are represented by solid squares (the collective excitation) and the hatched area (excitonic level).

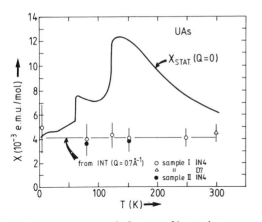

Fig. 5 Temperature dependence of bulk susceptibility of UAs Troc and Lam, 1974) compared to the values as calculated from the integrated (over energy) neutron intensity of magnetic scattering at $Q = 0.7 \text{ Å}^{-1}$.

6. Conclusions and future directions

I have tried in this very brief paper to describe some of the ways in which neutrons have played a key role in characterizing actinide magnetism. Two series of future experiments are particularly worthy of mention. The first is related to the recent development of transuranium single-crystal growth capabilities at the Euratom Center, Karlsruhe, under the guidance of O. Vogt and J. C. Spirlet. Detailed neutron experiments will soon be undertaken on materials such as PuSb, PuAs, and the neptunium pnictides. These experiments will do more than extend our knowledge, they will test whether the theoretical concepts we have developed for uranium are valid as the actinide contraction, and subsequent localization of the wavefunctions, occurs. The second development is related to pulsed neutron sources and using the epithermal neutrons to measure the dynamical response function (see Sec. 5) to higher energies. These experiments will be very difficult but they represent the cutting edge of both neutrons and actinides - a symbiotic relationship that from a historical perspective goes back to the first neutron and the production of the first transuranium elements.

7. Acknowledgements

All aspects of actinide research are truly collaborative and for the work discussed here it is my pleasure to thank P. Burlet, W. J. L. Buyers, B. R. Cooper, T. M. Holden, M. Loewenhaupt, M. H. Mueller, J. Rossat-Mignod, S. M. Shapiro, S. K. Sinha, W. G. Stirling, and O. Vogt for much stimulation over the years.

References

Aldred A T, Dunlap B D, Harvey A R, Lam D J, Lander G H and Mueller M H (1974) Phys. Rev. B9 3766
Aldred A T, Dunlap B D, Lander G H (1976) Phys. Rev. B14 1276
Brodsky M B (1978) Rep. Progr. Phys. 41 103
Buyers W J L, Murray A F, Jackman J A, Holden T M, duPlessis P de V, and Vogt O (1981), J. Appl. Phys. 52 2222
Cooper B R, Proceedings of Rare-Earth and Actinide Conference, Durham, UK, March 1982, to be published in J. Mag. Mag. Matls. See also Siemann R and Cooper B R (1980) Phys. Rev. Letters 44 1015
Freeman A J and Darby J B, "The Actinides: Electronic Structure and Related Properties", Academic Press, New York (1974)
Halg B, Furrer A, Halg W and Vogt O (1982) J. Appl. Phys. 53 1927
Hill H H (1970) in "Plutonium 1970 and Other Actinides" ed. Miner W N, AIME, New York, 2
Holden T M, Buyers W J L, Svensson E C, and Lander G H (1982) Phys. Rev. (to be published)
Jensen J and Bak P (1981) Phys. Rev. B23 6180
Lander G H, Sinha S K, Sparlin D M and Vogt O (1978) Phys. Rev. Letters 40 523
Lander G H and Lam D J (1976) Phys Rev. B14 4064
Lander G H and Stirling W G (1980) Phys. Rev. B21 436
Lander G H, Mueller M H, Sparlin D M and Vogt O (1976) Phys. Rev. B14 5035
Loewenhaupt M, Lander G H, Murani A and Murasik A (1982) J. Phys. C (to be published)
Loewenhaupt M and Holland-Moritz E (1979) J. Appl. Phys. 50 7456
Rossat-Mignod J, Burlet P, Quezel S and Vogt O (1980) Physica 102B 237
Shapiro S M (1981) J. Appl. Phys. 52 2129
Sinha S K and Fedro A J (1979) J. de Physique 40 C4-214
Sinha S K, Lander G H, Shapiro S M and Vogt O (1981) Phys. Rev. B23 4556
Stirling W G, Lander G H and Vogt O (1980) Physica 102B 259
Troc R and Lam D J (1974) Phys. Status Solidi 65b 317
Wedgwood F A (1974) J. Phys. C7 3203

Elementary excitations in one-dimensional magnets

J. Villain

Département de Recherche Fondamentale, CENG, 85X, 38041 Grenoble, France

Abstract. Theories and inelastic neutron scattering experiments in quasi-one dimensional magnetic materials are reviewed and compared.

1. Introduction.

One dimensional (1-D) and two-dimensional (2-D) systems have been extensively studied in the last 15 years, both experimentally and theoretically. In our opinion, the main reason of this interest is probably the existence of collective excitations without long range order (LRO). Absence of LRO implies the impossibility of using standard linearisation procedures, and therefore low-dimensional systems are a test of the ability of theoretical physicists to treat non-linear problems. Actually, a large body of theoretical work has been devoted to non-linear quantum and statistical physics of low-dimensional systems in recent years. For an experimental check, two things are necessary, a tool, and samples. The most complete tool for the investigation of collective excitations is neutron scattering, although magnetic resonance is also very powerful. The preparation of samples is a difficult problem. It is clearly impossible to prepare a strictly 1-D system, with a size of order 1 Å in two directions. What can be prepared is a magnetic, 3-D system in which the exchange interactions are much weaker in two directions than in the third one. The best known example is $(CD_3)_4NMnCl_3$, alias TMMC or tetramethylammonium-manganese chloride.

The realisation of nearly 2-D magnetic systems is much more difficult, and in our opinion there is no really satisfactory solution. Ideally 2-D systems are hardly accessible to neutron spectroscopy, and quasi-2D systems are always complicated by crossover effects. For this reason, the present study will be exclusively devoted to magnetic chains. Detailed lists of quasi-1-D or quasi-2-D magnets can be found in the reviews by De Jongh and Miedema (1974), Hone and Richards (1974) Steiner et al (1976) and De Jongh (1981).

2. Elementary excitations of an easy axis magnetic chain.

2.1 The model.

As a simple introduction to theoretical problems related to 1-D systems, we shall consider a chain of spins which like to be parallel to a definite axis z. The Hamiltonian may be

$$H = \sum_{j=1}^{N-1} \left[-JS_j^z S_{j+1}^z - \lambda J(S_j^x S_{j+1}^x + S_j^y S_{j+1}^y) + AS_j^{z^2} \right] \tag{1}$$

with either $|\lambda| \ll 1$ or (if the spin modulus s is at least 1) $-A \gg |J|$
Formula (1) may be written in terms of $S_j^{\pm} = S_j^x \pm S_j^y$:

$$H = \sum_{j=1}^{N-1}\left[-JS_j^z S_{j+1}^z - \frac{\lambda}{2}J\ (S_j^+ S_{j+1}^- + S_j^- S_{j+1}^+) + AS_j^{z^2}\right] \quad (2)$$

2.2 Ground state and neutron scattering in the ground state.

The ground state of (1) exhibits LRO, so that elementary excitations may be obtained by standard linearisation procedures. Our study will be limited to the ferromagnetic case (J > 0), which is simple because the state

$$|F\rangle = |+++++\ldots+++++\rangle \quad (3)$$

defined by $S_j^z|F\rangle = s|F\rangle$ (for any j) is an eigenstate. It is expected to be the ground state if $|A|$ is large enough or λ small enough. Inelastic neutron scattering at T=0 is a measurement of

$$\tilde{S}^{\perp}(q,\omega) = \int_{-\infty}^{\infty} dt\ \langle F|S_q^+ \exp(itH/\hbar)S_q^- \exp(-itH/\hbar)|F\rangle \exp(-i\omega t) \quad (4)$$

where
$$S_q^{\pm} = N^{-1/2}\sum_{j=1}^{N-1} S_j^{\pm}\ \exp(\pm iqj) \quad (5)$$

A straightforward calculation yields

$$S(q,\omega) = \langle F|S_q^+ S_q^-|F\rangle \delta(\omega-\omega(q)) \quad (6)$$

with
$$\hbar\omega(q) = 2s(A+J - J\cos q) \quad (7)$$

2.3 Transverse scattering at T≠0 (ferromagnetic case).

At temperature T≠0, there is no LRO, but short range order persists at low temperature. It is described by the equal-time correlation function

$$\langle S_j^z S_{j+r}^z\rangle = s^2\ \exp(-\kappa r) \quad (8)$$

where κ is given for large $|A|$ (or small λ if s=1/2) by

$$\exp(-\kappa) = \tanh(Js^2/K_B T) \quad (9)$$

where K_B is the Boltzmann constant.

In analogy with (6), the transverse neutron scattering function

$$\tilde{S}(q,\omega) = \int_{-\infty}^{\infty} dt\langle S_q^x S_{-q}^x(t) + S_q^y S_{-q}^y(t)\rangle\ \exp(-i\omega t) \quad (10)$$

is expected to be sharply peaked at $\omega(q)$ provided $q \gtrsim \kappa$. The expected modifications are : i) a slight shift with respect to (7). ii) a broadering. iii) a weak energy-gain peak at $\omega = \omega(q)$. This is a general property : magnons with a wave vector <u>larger</u> than the reciprocal correlation length κ <u>can be observed</u> by inelastic neutron scattering <u>even in the absence of LRO</u>. This intuitive prediction is confirmed by more sophisticated theories (e.g. Tomita and Mashiyama 1979, Reiter and Sjölander 1980) and experiment (e.g. Birgeneau et al 1971)

2.4 Propagating domain walls (Villain 1975, Shiba and Adachi 1981).

The one-magnon states $S_q^-|F\rangle$ are not the lowest excited states. The lowest excited states are in the ferromagnetic (F) case

$$|++++\ldots\ldots++++----\ldots\ldots------> \qquad (11)$$

and in the antiferromagnetic (AF) case

$$|+-+-+-\ldots\ldots+-+-+--+-+-+\ldots\ldots-+-+-+> \qquad (12)$$

In these states a domain wall is present between sites j and (j+1). Let the corresponding excited state be denoted by $|j+1/2\rangle$. In the F case, the position (j+1/2) of the domain wall does not move because the total spin $M = \sum S_i^z = (2j-N)s$ commutes with the Hamiltonian (2). The AF case is more interesting because domain walls propagate like quasi-particles, with constant velocity. The (approximate) eigenstates are not $j + 1/2\rangle$, but

$$|q\rangle = N^{-1/2} \sum_j |j+1/2\rangle [\exp iq(j+1/2)] \qquad (13)$$

The energy is $\hbar\Omega(q) = J/2 + \lambda J \cos 2q$ (14)

or, for small q $\hbar\Omega(q) = \hbar\Omega_o + \hbar^2 q^2/2M$ (15)

where $M = \hbar^2/4J\lambda$ may be regarded as the mass of the quasi-particle. In principle, propagating domain walls can be observed in the <u>longitudinal</u> scattering function

$$S_{//}(q,\omega) = \int_{-\infty}^{\infty} dt \langle S_q^z S_q^z(t) \rangle \exp(-i\omega t) \qquad (16)$$

However this function never reduces to a delta function, in contrast with (6). The reason is that neutrons cannot create or destroy walls as they create or destroy magnons. They are only scattered by domain walls. Detailed calculation of (16) predicts square root singularities at $\omega = \pm[\Omega(q) - \Omega(o)]$ at very low temperature in an infinite, ideally isolated chain. Experiment (Yoshizawa et al 1981, Nagler et al 1982) shows that the damping is severe, perhaps because of pinning by impurities. Propagating domain walls of the easy-axis, AF chain provide a simple example of rather general features in 1-D systems : elementary excitations can be defined but, in contrast with 3-D magnons, they are not visible in a first glance at the neutron scattering spectrum. Their characteristics can only be derived from a careful analysis of experimental data from various sources.

3. The easy-plane magnetic chain in zero field.

In 1973, Steiner and Dorner performed inelastic scattering experiments on $CsNiF_3$. This material is a system of ferromagnetic chains, weakly coupled together by an antiferromagnetic interaction and the spins like to be parallel to a definite plane (xy, say). The generally accepted Hamiltonian is (1), with $\lambda = 1$ and $A > 0$. As in the nearly isotropic case studied by Birgeneau et al (1971, Steiner was expecting a broad magnon observable for $q \gtrsim x$, as explained in § 2.3. The observed magnon was much narrower than expected. The explanation is simple, at least for large spins (Villain 1974). The Hamiltonian (1) may be written as

$$H = \sum_{i=1}^{N-1} \left[-JS_i^z S_{i+1}^z - J\sqrt{(s^2 - S_i^{z^2})(s^2 - S_{i+1}^{z^2})} \cos(\phi_i - \phi_{i+1}) + AS_i^{z^2} \right] \qquad (17)$$

where ϕ_i is the angle between the x direction and the projection of S_i^z into the xy plane. At low T, S_i^z and $(\phi_i - \phi_{i+1})$ are small and (17) may be replaced by a quadratic expression. Furthermore, S_i^z and ϕ_i can be shown to be conjugate variables. A Fourier transformation reduces (17) to a sum of harmonic oscillators, and it follows that the "out-of-plane" scattering function defined by (16) consists of two narrow peaks at $\omega = \pm\omega(q)$ with

$$\hbar\omega(q) = 2sJ\sqrt{(1-\cos q)(1 - \frac{J}{|J|}\cos q + A/|J|)} \qquad (18)$$

What about the "in-plane" scattering function defined by (10) ? In terms of the ϕ_i's it may be approximated by

$$S^\perp(q,\omega) = \frac{s^2}{N}\sum_{r,j}\int_{-\infty}^{\infty} dt \langle \exp[i\phi_j(t)-i\phi_{j+r}(0)]\rangle \exp(iqr-i\omega t) \qquad (19a)$$

$$= \frac{s^2}{N}\sum_{r,j}\int_{-\infty}^{\infty} dt \exp\left\{-\frac{1}{2}\langle [\phi_j(t)-\phi_{j+r}(0)]^2 \rangle\right\} \exp(iqr-r\omega t) \qquad (19b)$$

In the 3-D systems at low temperature, $S^\perp(q,\omega)$ is the correlation function of a quadratic function of the harmonic oscillator operators, and for this reason it reduces to a delta function (if damping is neglected). Expression (19) is more complicated, and its calculation yields a broad peak centred around $\pm\omega(q)$ (if $q \gtrsim \kappa$) or around 0 (if $q \lesssim \kappa$), as expected (Villain 1974, Ciepłak and Sjölander 1981).

4. The easy-plane chain in a field : solitons (Steiner 1982)

Consider the Hamiltonian (1) with $\lambda = 1$, and add a term $-B\sum_i S_i^x$. A/J is assumed to be large enough to oblige the spins to lie within the xy plane. The field B favours the spins being oriented along x (if J>0) or along y (if J<0).

However, above the 3-D ordering temperature, the field is generally not strong enough to prevent the spins to undergo at some places (Fig.1) a rotation of 2π (if J>0) or π (if J<0) within the plane. These excitations are domain walls,

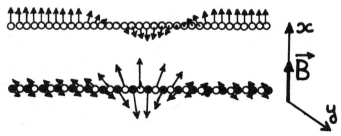

Fig. 1. Solitons of an easy-plane chain with a field in the plane. a) J>0 (ferromagnetic case). b) J<0 (antiferromagnetic case).

similar to (11) and (12). If they satisfy certain additional conditions (Bullough and Caudrey 1980, p.6, Lamb 1980, Eilenberger 1981) they are called "solitons". Essentially, solitons are quantised excitations of a classical, continuous system which are not scattered by other solitons or quasi-particles : they can be retarded or accelerated during the collision, but the velocity (and shape) at the end of the collision is the same as before. Since solitons are defined as excitations of a continuous system, the narrow walls (11) and (12) may not be called solitons. However, broad walls (Fig.1) are authentic solitons (Enz 1964). It was first suggested by Mikeska (1978) that solitons of a ferromagnetic chain scatter neutrons and therefore can be observed by inelastic neutron spectroscopy. The experiment was carried out by Kjems and Steiner (1978), but the cross section is weak and it is difficult to separate the soliton contribution from multimagnon effects. Also, the application of classical, continuous approximations to $CsNiF_3$ is questionable and the interpretation of the experiment is still controversial (Loveluck et al 1980, Reiter 1981, Pynn et al 1982). A rather different experiment was done by Regnault et al (1981) to prove the existence of solitons in the AF chain TMMC. This system may be considered easy-plane at low enough temperature. The purpose of the experiment was not to observe the scattering of neutrons by solitons, but to investigate the lifetime τ_q of AF fluctuations. This lifetime is due to the motion of domain walls (Fig.1b). The q-dependence of τ_q is not the same if walls have a diffusive motion or if they propagate freely. The experiment indicates that they propagate, and therefore they deserve the title of solitons. The difficulty of the experiment is by no means the intensity, which is quite large, but the energy analysis of a very narrow peak. It was only possible by means of the refined instrument IN 12 at the ILL, and (unfortunately) for a limited range in q.

5. Quantum effects.

Most of the theories mentioned in the previous Sections are classical theories valid for large spins. Real systems are quantum, and the theory is difficult. For instance, in the case A > 0, the formula (18) should be modified. According to Lindgård and Kowalska (1976), Balucani et al (1980, Cibert (1981) it is sufficient to replace s by sR and A by AK, where R and K are numbers. For $CsNiF_3$ (s=1, J>A>0) Cibert finds $R \approx 1$ and $K \approx 1/2$, in reasonable agreement with experiment. The case A 0 was investigated by Rastelli and Tassi (1982). The case $\lambda=1$, s=1/2, J$\approx$0 can be treated exactly (Des Cloizeaux and Pearson 1962). One finds R=$\pi/2$ in agreement with neutron scattering experiments (Hutchings et al 1979). More generally, a large number of exact results may be derived for s=1/2. The magnon spectrum has been obtained by Johnson et al (1973) and the spin pair correlation function has been investigated in various special cases by Beck et al (1981) and Perk and Capel (1977). The difficulty of the theory is illustrated by a recent controversy (Faddeev and Takhtajan 1981) around Des Cloizeaux and Pearson's work. For s>1 and A 0, a recent suggestion by Haldane (1982) looks like a challenge to neutron scattering people. Haldane finds that for even s, the spin wave spectrum has a gap, and the equal time correlation function decays exponentially as in (8), while it decays as $1/|r|$ for odd s.

Similarly, the soliton mass, defined by relation (15), is sensitive to quantum effects (Maki 1981, Zotos and Fowler 1982, Mikeska 1982). The agreement with neutron scattering data (Regnault 1982) and AFMR data (Tuchendler et al 1981) is only qualitative. In fact the adequacy of the usual theories in which A is assumed to be very large is questionable in real systems (Loveluck et al 1980) and more realistic (classical) models give a better agreement with experiment (Harada et al 1981).

6. Conclusion

In some sense, theory is more difficult in D=1 than in D=3 because equations cannot so often be linearised. On the other hand, non-linear equations are more easily solved in D=1. More easily does not mean very easily, and theories are **elaborate**. Experimental check also requires elaborate instruments. The present (certainly incomplete) review is an attempt to give an idea of both theoretical and experimental point of view. The agreement is satisfactory, but there are still open questions in spite of a remarkable collaboration between experimentalists and theorists. On the other hand, one-dimensional Physics is a rather artificial and somewhat academic topic, in contrast with two-dimensional Physics which finds a natural application in surfaces. One-dimensional Physics is a good exercise from which much has been learnt. In our opinion, it is essentially an exercise.

References

BalucaniU, Pini M G, Rettori A and Tognetti V 1980 J. Phys. C13 3895
Beck H, Puga M W and Müller G 1981 J. Appl. Phys. 52 1998 and 1968
Birgeneau R G, Dingle R, Hutchings M T, Shirane G and Holt S L 1971 Phys. Rev. Lett. 26 718
Bullough R K and Caudrey P J 1980 ed "Solitons" (**Berlin**: Springer)
Cibert J 1981 Thesis (University of Grenoble)
Cieplak M and Sjölander A 1981 J. Phys. C14 4861
De Jongh L J 1981 in "Recent developments in condensed matter physics" ed J T Devreese (Niw York: Plenum) Vol.1 p 343
De Jongh L J and Miedema A R 1974 Adv. Phys. 23 1
Des Cloizeaux J and Pearson J J 1962 Phys. Rev. 128 2131
Eilenberger G 1981 "Solitons" (Berlin: Springer)
Enz U 1964 Helv. Phys. Acta 37 245 Phys. Rev. 131 1392-1963
Faddeev L D and Takhtajan L A 1981 Phys. Lett. 85A 375
Haldane F D M 1982 Preprint and Bull. Am. Phys. Soc. 27 181
Harada I, Sasaki K and Shiba H 1981 Solid State com. 40 29
Hone D W and Richards P M 1974 Ann. Rev. Mat. Sc 4 337
Hutchings M T, Ikeda H and Milne J M 1979 J. Phys. C12 1739
Johnson J D, Krinski S and Mc Coy B M 1973 Phys. Rev. 18 2526
Kjems J D and Steiner M 1978 Phys. Rev. Lett. 41 1137
Lamb G L Jr 1980 "Elements of Soliton theory" (New York: Wiley)
Lingard P A and Kowalska A 1976 J. Phys. C9 2081
Loveluck J M, Schneider T, Stoll E and Jauslin H R 1980 Phys. Rev. Lett. 45 1505
Maki K 1981 Phys. Rev. B24 3991
Mikeska H J 1978 J. Phys. C11 L29
Mikeska H J 1982 Preprint. See also Z. Phys. B43 209 1981
Nagler S E, Buyers W J L, Armstrong R L and Briat B.1982 Preprint
Perk J H H and Capel H W 1977 Physica 89A 265
Pynn R, SteinerM, Knop W, Kakurai K and Kjems J K 1982 in "Non-linear phenomena at phase transitions and instabilities" ed T Riste (New York: Plenum) p 97
Rastelli E and Tassi A 1982 J. Phys. C15 509
Regnault L P 1981 Thesis (University of Grenoble unpublished)
Regnault L P, Boucher J P, Rossat-Mignod J, Renard J P, Bouillot J and Stirling W G 1982 J. Phys. C15 1261
Reiter G 1981 Phys. Rev. Lett. 46 202
Reiter G and Sjölander A 1980 J. Phys. C13 3027
Tomita K and Mashiyama H 1979 J. Phys. C12 3059
TuchendlerJ, Magariño J and Renard J P 1981 Phys. Rev. B24 5363
Shiba H and Adachi K 1981 J. Phys. Soc. Japan 50 3278
Steiner M 1982 Proc. of the 22nd Scottish Universities Summer school (to be published)
Steiner M and Dorner B 1973 Solid State Com. 12 537
Steiner M, Villain J and Windsor C G 1976 Adv. Phys. 25 87
Villain J 1974 J. Physique 35 27
Villain J 1975 Physica 79B 1
Yoshizawa H,Hirakawa K, Satija S K and Shirane 1981 Phys. Rev. B23 2298
Zotos X and Fowler M 1982 Phys. Rev. B25 to be published.

> # Neutron scattering and phase transitions

R.A. Cowley

Department of Physics, University of Edinburgh, Mayfield Road,
Edinburgh EH9 3JZ.

Abstract. Neutron scattering experiments play a particularly important role in the study of phase transitions because they permit the study of the spatial and time dependence of the critical fluctuations, and because they can be applied to magnetic and structural phase transitions. In the former case magnetic systems can be chosen which very closely approximate to the simple models studied in statistical mechanics, and many important tests have been performed. In the latter case the situations are frequently more complicated but neutron scattering provides essential information.

1. Introduction

One of the first applications of neutron scattering techniques was the determination of the antiferromagnetic structures of several materials. In any detailed study of a phase transition, it is essential to know the structure of both the low and high temperature phases. Neutron scattering has played a unique role in the determination of magnetic structures and of many crystal structures, but I do not wish to discuss this aspect of the contribution which neutrons have made to the study of phase transitions. I wish to concentrate on the way in which neutrons have contributed to the study of the effects of the fluctuations, particularly close to continuous phase transitions.

I believe the first neutron scattering experiment to see the effect of the fluctuations was that of Latham and Cassels (1952) who found that the total scattering cross-section of iron increased near the Curie temperature, and they correctly identified this as due to scattering by the magnetic fluctuations. It is perhaps worthy of note, in view of the development of accelerator sources of neutrons, that they used a cyclotron to produce the neutrons. This work was followed up in more precise measurements using reactor sources by Squires (1954). A major step forward was the theoretical work of Van Hove (1954) which worked out the theory of the scattering from the magnetic fluctuations close to a phase transition.

2. Critical Phenomena and Neutron Scattering

Van Hove (1954) showed that the neutron scattering cross-section was proportional to the wavevector and frequency transform of the space and time dependent pair correlation function. Neutron scattering provides therefore a direct measure of the pair correlation function of the fluctuations close to a phase transition. Of particular interest is the intensity at fixed wavevector transfer and integrated over all frequency transfers. This gives the static pair correlation function and in many cases close to a structural phase transition this can be written as

$$J(\underline{Q}) = A\delta(\underline{Q} - \underline{\tau}) + \frac{c}{\kappa^2 + \underline{q}^2}, \qquad (1)$$

where the first term gives the long range order associated with the wavevector, $\underline{\tau}$, and the second term represents the fluctuations and $\underline{q} = \underline{Q} - \underline{\tau}$.

Close to a continuous phase transition the parameters in equation (1) are temperature dependent and can be written as

$$A = A_o(T_c - T)^{2\beta}, \quad \kappa = \kappa_o |T - T_c|^\nu \quad \text{and} \quad \frac{c}{\kappa^2} = \chi = \chi_o |T - T_c|^{-\gamma},$$

where β, ν and γ are known as critical exponents.

The theory of phase transitions is too complex to review here in detail (Ma, 1976). There have been tremendous strides in the understanding of phase transitions in the last ten years and in no small part has this been because of the contribution made by experimental neutron scattering work. I do however firstly comment on the results of the theory and then subjectively choose a few experiments to illustrate the behaviour. In the fourth section I discuss dynamical behaviour and the central peak and in the final section some recent experiments on systems in random applied fields.

It is now believed that all systems with particular types of interactions and of particular dimensionalities have the same 'universal' values of the exponents. The simplest theory of phase transitions is the well known mean field theory and neglects the fluctuations. This theory gives $\beta = \frac{1}{2}$, $\nu = \frac{1}{2}$ and $\gamma = 1$. These results are correct for systems for which the dimensionality is so large that the fluctuations play only a minor role. This limiting dimensionality is known as the upper critical dimension. For lower dimensionalities the fluctuations play a more important role and the values for the exponents are modified. Typically β is reduced while ν and γ are increased, until at a still smaller dimensionality, called the lower critical dimension, the critical fluctuations prevent the establishment of long order except possibly at $T = 0$. This behaviour is illustrated by the Ising model with short range forces for which the upper critical dimensionality is 4 and the lower critical dimensionality is 1. More subtle and complex behaviour can occur at the upper and lower critical dimensions such as logarithmic corrections to the behaviour at the upper critical dimension and possibly complex phases at the lower critical dimension, such as occur in the two-dimensional XY model.

By measurement of $J(\underline{Q})$ neutron scattering can determine the exponents and hence whether a system is above, below or at its marginal dimensionality and further provide information about the values of these exponents, and other detailed properties of the phase transition.

3. Some Illustrative Experiments on Static Behaviour

In my subjective view the modern era of neutron scattering experiments began with the work by Als-Nielsen and Dietrich (1967) on β-brass. This order-disorder transition is an example of a three-dimensional Ising model and they clearly showed that all the exponents β, γ and ν were different from their mean field values, and agreed reasonably well with the values which had been obtained theoretically from series expansions.

The study of the static phenomena received a considerable development from the discovery that in some three-dimensional crystals, the magnetic ions might be magnetically coupled in chains (one-dimensional systems), planes (two-dimensional systems) or throughout the system (three-dimensional systems). This development meant that it became possible to study the properties of one- and two-dimensional systems while using three-dimensional samples. Classic experiments were performed, particularly at Brookhaven, in the early 1970's which showed the effect of dimensionality on the properties of the system and gave good agreement with theoretical results. The experiments which still appeal as significant to me were those of Hutchings et al (1972) on TMMC and of Ikeda et al (1974) on K_2CoF_4. In the former material the spins are largely coupled only in chains and down to 1.1 K the material shows paramagnetic behaviour, although the molecular field theory would predict a Nèel temperature of 76 K. In detail the results for the static properties are in excellent agreement with earlier calculations of Fisher (1964) for a Heisenberg linear chain. In the latter case the exchange interactions are of Ising character and the magnetic ions are coupled in sheets. The results of the experiments are in strikingly good agreement with the exact solution of the two-dimensional Ising model by Onsager (1944).

In these magnetic systems the magnetic interactions are mostly between nearest neighbours. Particularly in the field of structural phase transitions the interatomic forces may be more complicated and alter the nature of the critical fluctuations and hence of the critical properties. Neutron scattering showed this particularly elegantly in the case of DKDP. The ferroelectric fluctuations were measured in the a - c plane and it was found (Paul et al, 1970, Skalyo et al, 1970) that the fluctuations were suppressed for wavevectors along and near to the c axis. This suppression of the fluctuations is due to the macroscopic electric field which increases the restoring forces for these fluctuations. This reduction of the fluctuations means that they play a less important role and the upper critical dimension of a uniaxial ferroelectric is, because of these forces, reduced from 4 to 3. The physical reason for this reduction was clearly shown in the neutron scattering results. Measurements of the fluctuations in the a - b plane (Skalyo et al, 1970) showed (Cowley, 1976) that the coupling between the ferroelectric fluctuations and the transverse acoustic mode propagating along the a-axis further reduces the fluctuations for wavevectors not along these axes. This has the effect of reducing the influence of the critical fluctuations on the phase transition of DKDP even further and the upper critical dimension is less than 3. Consequently mean field theory can be used to give a good description of this phase transition.

These examples illustrate the power of the neutron scattering in the study of phase transitions. Direct measurements of the fluctuations can be used to demonstrate the appropriate model for the critical properties, and then detailed measurements of the exponents, for example, can test the predictions of these models.

By now very many systems have been studied. The measurement of exponents still requires considerable experimental care. At first glance it would seem easiest to measure the order-parameter exponent β, but this is frequently made difficult by extinction problems. In practice it is usually easier to measure the correlation length exponent, ν. In order to do this it is essential to perform the frequency integral of the intensity at fixed wavevector. This can be done with a two-axis spectrometer provided that the energy of the incident neutrons is much larger than the energy of the fluctuations. Although detailed tests of this approximation were made around 1970, it has been all too often taken for granted more recently. Finally the exponents are calculated theoretically only as T approaches T_C asymptotically. Clearly in practice there is an experimental limitation on how close to T_C it is possible to make measurements. A measurement of the exponent is however reliable only if the reduced temperature $t = |T - T_c|/T_c$ is measured over two decades preferably for $t < 0.1$. Even in this case the result will be dismissed by theorists as not being in the asymptotic region if it disagrees with their theory!

4. Dynamical Experiments

The theory of the dynamics of the critical fluctuations is less well developed than that of the statics. The predictions are different for different dimensionalities and forces as with the static properties, but also differ depending on whether or not the fluctuations must satisfy any conservation conditions. Several experiments on the dynamical behaviour of the ideal magnetic systems have been performed and the appropriate dynamic exponents have been determined.

The dynamical behaviour has been studied far more extensively in the case of structural phase transitions. In most magnetic systems and also order-disorder systems like β-brass the phase transition occurs from a high temperature disordered phase to a low temperature ordered phase. Although similar structural phase transitions occur when molecular groups become ordered; $NaNO_2$ is an example, many structural phase transitions are transitions from one ordered phase to another ordered phase. It has proved useful to consider these in terms of the soft mode theory (Cochran, 1960) which treats the transition as an instability against a particular normal mode of vibration. The frequency of this mode then decreases to zero as the phase transition is approached from either above or below. Neutron scattering has played an important role in testing this model and in many systems a soft mode has been studied in detail although probably the most careful work has been performed on $SrTiO_3$ (for a review see Cowley, 1980).

Although the soft mode theory provides a very useful picture it was found (Shapiro et al, 1972) that close to T_C the dynamical response consisted of two components; a phonon component and a narrow quasi-elastic component or central peak. Similar results have been obtained in magnetic systems which transform from one ordered state to another ordered state but these central peaks are not observed in transitions of the order to disorder type. Despite extensive investigations over the last 10 years we do not yet have a good understanding of the central peak. We know that it can be produced and enhanced by the presence of impurities or defects, but it seems difficult to reconcile its properties with the expected results for pure systems. (A detailed review of the present situation is given by Bruce & Cowley, 1980.)

Although there has been much effort using the refined techniques of the spin echo spectrometer and the back-scattering spectrometer to determine the dynamical response of the central peak, there has been surprisingly little effort towards determining its spatial or wavevector properties. Indeed there have been few studies of the static properties close to order-order phase transitions, and it would seem to be an appropriate topic for more study. The results on simple magnetic systems give us considerable confidence in our understanding of the static properties for many different types of system, and the principle of universality asserts that the static

properties of the appropriate structural phase transitions should be the
same. Detailed tests of this and in particular of the wave-vector dependence of the central peak would be very useful.

5. Recent Developments

There have been many developments in the study of phase transitions over
the past few years. In particular the study of incommensurate phase transitions has been very rewarding but these phases will be reviewed at this
meeting by Axe.

My own particular interest has been largely on the study of disordered
systems and in particular on simple disordered magnetic systems, in which
the interactions are between nearest neighbours and in which it is possible
to change the dimensionality and the nature of the magnetic interactions.

If the disorder changes only the exchange interactions the consensus of
both theoretical and experimental work is that the properties of the phase
transition are altered only slightly if at all. The predicted changes in
the exponents are either zero or small and it is difficult for experiment
to test these predictions quantitatively.

The effect of the disorder is much more marked if magnetic fields which are
randomly in different directions are applied. At first sight this seems a
difficult experiment but Fishman and Aharony (1979) pointed out that a uniform magnetic field applied to a random antiferromagnet such as $Co_xZn_{1-x}F_2$
produces a randomly directed staggered field. Briefly if a particular unit
cell of the diluted system contains two Co ions the field produces little
effect. If, however, it contains one Co ion the effect is large and
dependent on whether the Co ion is on the 'up' or 'down' sub-lattice.

The first theoretical study of the problem was by Imry and Ma (1975) who
showed that an Ising system in a random field was unstable against the
formation of domains for dimensionalities $d \leqslant 2$. The lower critical dimension was raised by the random field from $d=1$ to $d=2$ and this result has
been supported recently by Grinstein and Ma (1982). Then followed a number
of theoretical studies which suggested that the properties of a system in a
random field were the same as those of a normal system but in two higher
dimensions (Pytte et al, 1981). The lower critical dimension of the Ising
system in a random field is then 3 while the upper critical dimension is 6.
In view of these dramatic changes in the properties and the theoretical
controversy we (Yoshiyawa et al, 1982) have conducted an extensive series
of experiments on $Co_xZn_{1-x}F_2$, $Fe_xZn_{1-x}F_2$, $Mn_xZn_{1-x}F_2$, $Rb_2Co_xMg_{1-x}F_4$. In
each system the effect of the applied field is to drastically modify the
critical phenomena. In the presence of a field and at high temperature the
scattering is characteristic of a normal paramagnetic phase. On cooling
the inverse correlation length decreases but the intensity profiles can no
longer be described by eqn. 1. We find that we can obtain a good description of the results with the form:

$$J(\underline{Q}) = B \frac{\kappa}{(\kappa^2 + \underline{q}^2)^2} + \frac{c}{(\kappa^2 + \underline{q}^2)} \qquad (2)$$

When the data for the Co and Fe salts is analysed using this form we find
that on cooling at constant field κ decreases steadily until a temperature
just below T_N, κ becomes almost independent of temperature. At low
temperatures we find that B is almost independent of field and that κ
varies as H^ν with $\nu = 1.6 \pm 0.2$ in $Rb_2Co_xMg_{1-x}F_4$, 2.1 ± 0.3 in $Fe_xZn_{1-x}F_2$

and 3.3 ± 0.5 in $Co_xZn_{1-x}F_2$.

These results seem to suggest that in all these systems the field does destroy the long range order, and from the power law behaviour of κ, it is reasonable to infer that the lower critical dimension is larger than 3. The only reservation about this is that at low temperatures in these Ising systems the spins are frozen, and so we may not be studying the true equilibrium low temperature state. Nevertheless we believe the systems are in equilibrium at temperatures below which κ has become almost constant.

The results for $Mn_xZn_{1-x}F_2$ can also be fitted by eqn. 2 and are similar for $T \geqslant T_N$. The value of κ then slowly decreases and for temperatures below $0.6T_N$ there is true long range order on the scale for which we can determine it. An unresolved question is whether the low energy spin waves in the Mn salt enable us to reach the true ground state of the Ising system or whether they so modify the behaviour that we no longer sample the ideal Ising behaviour.

Clearly neutron scattering is and will continue to play an essential role in the study of phase transitions. Because of its ability to study the microscopic fluctuations and because the clearest tests of statistical mechanics can often be best made with magnetic systems, neutrons will remain an irreplaceable tool in testing the statistical mechanics of phase transitions.

This paper was written while the author was at the Brookhaven National Laboratory. Financial support was provided by the Division of Materials Science, U.S. Department of Energy under contract DE-AC02-76CH00016.

References

Als-Nielsen J and Dietrich O W 1967 Phys. Rev. 153 706 711 717
Bruce A D and Cowley R A 1980 Adv. in Phys. 29 219
Cochran W 1960 Adv. in Phys. 9 387
Cowley R A 1976 Phys. Rev. Lett. 36 744
Cowley R A 1980 Adv. in Phys. 29 1
Fisher M E 1964 Am. J. Phys. 32 343
Fishman S and Aharony A 1979 J. Phys. C 12 L729
Grinstein G and Ma S K 1982 (to be published)
Hutchings M T, Shirane G, Birgeneau R J and Holt S L 1972 Phys. Rev. B5 1999
Ikeda H and Hirakawa K 1974 Solid State Commun. 7 529
Imry Y and Ma S 1975 Phys. Rev. Lett., 37 1367
Latham R and Cassels J M 1952 Proc. Phys. Soc. A65 241
Ma S K 1976 Modern Theory of Critical Phenomena, Benjamin Inc.
Onsager L. 1944 Phys. Rev. 65 117
Paul G L, Cochran W, Buyers W J L and Cowley R A 1970 Phys. Rev. B2 4603
Pytte E, Imry Y and Mukamel D 1981 Phys. Rev. Lett. 46 1173
Shapiro S M, Axe J D, Shirane G and Riste T 1972 Phys. Rev. B 6 4332
Skalyo J, Frazer B C and Shirane G 1970 Phys. Rev. B 1 278
Van Hove L 1954 Phys. Rev. 95 249
Yoshizawa H, Cowley R A, Shirane G, Birgeneau R J, Guggenheim H J and
  Ikeda H 1982 Phys. Rev. Lett. 48 438 and further work in collaboration with M Hagen, S Satija, W J L Buyers, V Jaccarino which will be published.

Phonons and their interactions

R. M. Nicklow

Oak Ridge National Laboratory, Oak Ridge, Tennessee 37380

The phonon energy spectra $\nu(\vec{q})$ of crystalline materials contains key information about the interatomic interactions. However, it is generally not possible to fully understand the phonon spectra without also understanding the influence on phonon energies and lifetimes caused by interactions with defects, electrons and other excitations. The study of several of these types of interactions have grown over the years so as to now constitute subfields of solid state physics and the contributions of neutron scattering research to each has been, if not of paramount importance, at least very significant. In the present review we can merely touch on a few high lights, and even those selected represent areas of personal interest to the author and in no way represents a thorough review of these exciting area of research.

Perhaps the largest research effort is expended on electron-phonon interactions. These interactions are, of course, fundamental to the properties of metallic solids. They are "seen" in the phonon $\nu(\vec{q})$ of metals in a wide variety of effects. We shall mention three: the relatively small fine structure produced by Kohn singularities, large "anomalies" and phonon lifetimes measured in some superconductors and in materials with fluctuating valence.

All of these phenomena can be related to the behavior of the electronic susceptibility $\chi(q)$ and the strength of the electron-ion interaction. Soon after the prediction by Kohn (1959) that singular behavior of $d\chi(q)/dq$ for $q = 2k_F$, where k_F is the Fermi wave number, should produce images of the Fermi surface in the $\nu(q)$ for a metal, such "Kohn anomalies" were observed by Brockhouse et al. (1961) in neutron scattering experiments on lead. Systematic searches for such singularities can provide direct experimental information on the Fermi surface geometry which complements that obtained from de Haas-van Alphen experiments (Stedman et al. 1967, Weymouth and Stedman 1970). However, such work can be extremely tedious since the magnitude of the electron-ion interaction can be quite small for nearly free electrons, leading to very small singularities in spite of considerable structure in $\chi(q)$. Consequently, only a few such studies have been carried out in detail on free electron metals. For some metals no Kohn anomalies are observed at all.

However, the measured $\nu(q)$ for some transition metals and transition metal carbides and nitrides, especially those which have high superconducting transition temperatures, can exhibit quite unusual behavior (e.g., Nakagawa and Woods 1963, Smith and Glaser 1970). For these metals the existence of partially localized d electron states near the Fermi energy makes the calculation of χ very difficult. Consequently, the interpretation of unusual structure in $\nu(q)$ in terms of the behavior of χ and the electron-ion interaction is still a challenging problem. The existence of strong electron-phonon interactions in these metals is certainly consistent with their superconductivity, but the detailed relationship is not yet clear (Allen 1980).

Electron-phonon interactions can also lead to significant phonon-lifetime effects. Phonon lifetimes for Nb have been studied in some detail by neutron scattering. In the superconducting state phonons with energies below 2Δ, the electronic energy gap, do not interact with electrons since they do not have enough energy to break Cooper pairs. The changes observed in the lifetime between phonons with energies just below and just above 2Δ provides a direct measure of only the electron-phonon coupling for these phonons since experimental resolution, anharmonic and defect broadening cancel (Shapiro et al. 1975). One also obtains in such neutron scattering experiments the magnitude, temperature dependence and crystallographic anisotropy of Δ.

For phonons with energies well above 2Δ, the measured phonon linewidths must be corrected for instrumental resolution effects by calculation. Fortunately, this can be done with good accuracy and in addition, for Nb, the linewidths are quite large; $\Delta\nu/\nu$ is nearly 0.1 for some phonons. The experimental results are in very good agreement with microscopic theoretical calculations which were based on the rigid muffin-tin approximation and a realistic band structure (Butler et al. 1977).

Some materials possess 4f electron states very near the Fermi energy E_F. In these materials temperature changes and/or modest pressures can produce valence changes resulting from $f \rightleftarrows d$ electronic transitions which are accompanied by large changes in volume and in the electronic density of states at E_F. This strong electron lattice coupling produces very anomalous phonon softening and large phonon linewidths (Mook et al. 1978).

Impurities and defects in crystals destroy translational symmetry and the point symmetry for atoms nearby. Consequently, the vibrational displacements of atoms cannot "rigorously" be written in terms of noninteracting normal modes having well-defined wave vectors and frequencies. Neutron scattering measurements of the changes in mode frequencies and lifetimes due to defects provides information on the microscopic defect-lattice interactions and also on the validity of theoretical approaches used to describe not only lattice dynamical properties but electronic behavior as well.

The first neutron scattering studies of phonon-defect interactions were carried out on Cu(Au) by Svensson et al. (1965) and on Cr(W) by Møller and Mackintosh (1965). A definite resonant perturbation of the phonons by the heavy impurities was observed in both materials, but the observed behavior differed quantitatively from the predictions of mass-defect (only) theory. The high frequency localized vibrations of light impurities has also been observed in many materials (Nicklow et al. 1968), but again the agreement with theory was only qualitative. Attempts to improve agreement between theory and various experiments has led to considerable theoretical work on force-constant disorder and on the treatment of systems having large defect concentrations exemplified by alloys (see reviews by Taylor 1975 and Elliott and Leath 1975). For some alloys, comprised of elements having greatly differing masses, e.g. Ni_xPt_{1-x} for $x \simeq 0.5$ (Tsunoda et al. 1979), very remarkable gaps and "extra" branches appear in the phonon dispersion relation. Some of this behavior is in qualitative agreement with theoretical predictions.

Gaps can also occur in the phonon dispersion relation for crystals containing very low concentrations ($x \simeq 0.01$) of complex impurities; i.e., those which possess internal energy states, at phonon energies which coincide with the energy separations of the internal states. A strong coupling between the phonons and the internal excitations occurs which can give rise to a mixed-mode spectrum for phonons having the same symmetry as the internal excitation. These internal energy states can be the vibrational or librational levels of molecular impurities (Nicklow et al. 1980) or the vibronic levels of impurities which produce Jahn-Teller distortions of the host lattice (Challis et al. 1979).

Mixed-mode spectra involving phonons are also observed in some magnetic systems, particularly those containing rare earth or transition metal ions having significant anisotropy in their charge distributions, thereby giving rise to strong crystal field interactions. The phonon modulation of crystal field potentials often leads to a phonon-magnon coupling which can result in "anticrossing" behavior for the dispersion relations of the phonons and magnons. This was first observed in a neutron scattering study of UO_2 (Dolling and Cowley 1966). One of the most detailed studies, theoretically and experimentally, was carried out on FeF_2 for which much was known about the crystal field parameters from independent studies (Rainford et al. 1972 and S. W. Lovesey 1972).

In the case of magnons and phonons the dispersion relation of both types of excitations can be individually measured completely since both have appreciable neutron cross sections. In some systems, for example, those showing cooperative Jahn-Teller effects, there exist electronic excitations (excitons) which may have little or no neutron cross section, but which may nevertheless be studied by neutron scattering because of their strong coupling to phonons. Measurements of the coupled exciton-phonon spectrum then gives information about the exciton disperion relation at wave vectors where it crosses the phonon dispersion relation, and about the wave-vector dependence of the ion-ion interaction parameters which cannot be obtained in any other measurement (see review by Kjems 1977).

This research was sponsored by the Division of Materials Sciences, U.S. Department of Energy under Contract No. W-7405-eng-26 with the Union Carbide Corporation.

REFERENCES

1. Allen P B 1980 Dynamical Properties of Solids Vol. 3 ed. G K Horton and A A Maradudin (Amsterdam: North Holland) pp95-196
2. Brockhouse B N, Rao K R and Woods A D B 1961 Phys. Rev. Lett. 7 93
3. Butler W H, Smith H G and Wakabayashi N 1977 Phys. Rev. Lett. 39, 1004
4. Challis L J, Fletcher J R, Jefferies D J, Sheard F W, Toombs G A, de Goer A M and Hutchings M T 1979 Phys. Rev. B 19, 296
5. Dolling G and Cowley R A 1966 Phys. Rev. Lett. 16, 683
6. Elliott R J and Leath P L 1975 Dynamical Properties of Solids Vol 2
7. Kjems J K 1977 Electron-Phonon Interactions and Phase Transitions ed Tormod Riste (New York: Plenum) pp.302-322
8. Kohn W 1959 Phys. Rev. Lett. 2 393

9. Lovesey S W 1972 J. Phys. C: Solid State Phys. $\underline{5}$ 2769
10. Møller H B and Mackintosh A R 1965 Phys. Rev. Lett. $\underline{15}$ 623
11. Mook H A, Nicklow R M, Penney T, Holtzberg F and Shafer W M 1978 Phys. Rev. B $\underline{18}$, 2925
12. Nakagawa Y and Woods ADB 1963 Proc. International Conference on Lattice Dynamics ed R F Wallis (New York: Pergamon) p. 39
13. Nicklow R M, Crummett W P, Mostoller M and Wood R F 1980 Phys. Rev. B $\underline{22}$ 3039
14. Nicklow R M, Vijayaraghavan P R, Smith H G and Wilkinson M K 1968 Phys. Rev. Lett. $\underline{20}$ 1245
15. Rainford B D, Houmann J G and Guggenheim H J 1972 5th IAEA Symposium on Neutron Inelastic Scattering, Grenoble, France (Vienna: IAEA)
16. Shapiro S M, Shirane G and Axe J D 1975 Phys. Rev. B $\underline{12}$ 4899
17. Smith H G and Gläser W 1970 Phys. Rev. Lett. $\underline{25}$ 1611
18. Stedman R, Almquist L, Nilsson G and Raunio G 1967 Phys. Rev. $\underline{163}$ 567
19. Svensson E C, Brockhouse B N and Rowe J M 1965 Solid State Commun. $\underline{3}$ 245
20. Taylor D W 1975 Dynamical Properties of Solids Vol 2 ed G K Horton and A A Maradudin (Amsterdam: North-Holland) pp285-384
21. Tsunoda Y, Kunitomi N, Wakabayashi N, Nicklow R M and Smith H G 1979 Phys. Rev. B $\underline{19}$ 2876
22. Weymouth J W and Stedman R 1970 Phys. Rev. B $\underline{2}$ 4743

Incommensurate structures

J.D. Axe

Brookhaven National Laboratory, Upton, New York 11973[*]

Abstract. A review is given of neutron scattering studies of displacive incommensurate structures, and of the instabilities of crystalline solids that lead to their formation.

1. Introduction

Incommensurate structures are peculiar quasi-crystalline substances that lack periodic translation symmetry not in a haphazard amorphous way but because two (or perhaps more) elements of translational symmetry are present which are mutually incompatible. Suppose $A(\vec{r})$ and $B(\vec{r})$ represent the spatial distribution of two characteristic properties of a material and that

$$A(\vec{r}) = \sum_{\{\vec{G}\}} A_G e^{i\vec{G}\cdot\vec{r}} \quad ; \quad B(\vec{r}) = \sum_{\{\vec{G}'\}} B_{G'} e^{i\vec{G}'\cdot\vec{r}} \tag{1}$$

The structure is incommensurate if the sets of reciprocal lattice vectors $\{\vec{G}\}$ and $\{\vec{G}'\}$ have only the trivial elements $\vec{G} = \vec{G}' = 0$ in common. Various cases are possible depending on what A and B represent, as shown in Table 1.

Neutron scattering has been instrumental in the study of incommensurate structures of all of the above types, but this is particularly true for the magnetic case, since neutrons are uniquely sensitive to magnetic spatial distributions on an atomic scale. The essential role of competing forces (in this case due to magnetic exchange interaction) of various ranges was emphasized in the models of the phenomena. Major reviews by Mackintosh (1982) and by Koehler (1982) in this volume make further comment here unnecessary.

The history of incommensurate intergrowth/overgrowth structural studies with neutrons is, by comparison, brief. The study of adsorbed surface monolayers, which depended crucially on the advent of exfoliated graphite substrates is reviewed by Nielsen (1982) in the volume. The best studied example of an incommensurate intergrowth compound to date is $Hg_{3-\delta}AsF_6$. This material has many remarkable properties, not the least of which is that of a one dimensional liquid, and the results of neutron studies have been recently reviewed (Axe, 1980). The present review will concentrate

---

[*]Supported by The Division of Materials Sciences, U.S. Department of Energy under Contract No. DE-AC02-76CH00016.

Table 1

Incommensurate Structure Type	A(r)	B(r)	Example
1. magnetic	magnetic density, M	nuclear density, ρ	Cr, r.e. metals
2. compositional	average $\langle \rho_1 + \rho_2 \rangle$	differential density, $\langle \rho_1 - \rho_2 \rangle$	CuAuII, feldspars
3. intergrowth/ overgrowth	lattice ρ_1	lattice ρ_2	adsorbed monolayers
4. displacive	displacement field, $\vec{u}(\vec{r})$	average	quasi-1d & 2d metals

on the last of the four types, displacive modulation, which involves periodic displacements,

$$\vec{u}(\vec{r}) = \vec{A} \cos(\vec{q} \cdot \vec{r} - \phi) \qquad (2)$$

of the atoms away from an average regular lattice site. Not uncommonly the modulation amplitude, $\vec{A}$, disappears above a certain temperature, T_o, the material thereby transforming from an incommensurate structure into a commensurate (normal) phase with the average structure.

2. Charge Density Waves

In metals an incommensurate transformation may result from a charge density wave (CDW) instability in the conduction electrons near the Fermi surface. Neutrons couple not directly to the CDW but rather to the distortions that result as the atomic cores adjust by coulomb screening of the CDW. CDW instabilities are favored by large portions of the Fermi surface separated by the special wavevector, $\vec{q}_o$, (Fermi surface nesting) which is more probable for quasi-one- or two-dimensional metals, as the Fermi energy then becomes independent of some component(s) of the electron momenta. For reviews of neutron studies of quasi-one-dimensional metals see Comès and Shirane (1979). Additional interest in the study of these materials derives from the fact that there are strongly competing tendencies among the conduction electrons to form charge density waves, spin density waves and superconductivity. The successful culmination of the latter tendency has been recently confirmed in certain quasi-one-dimensional organic metals.

Of the various quasi-two-dimensional CDW systems, the 2H polytype of TaSe$_2$ has been the most thoroughly investigated. Early electron diffraction studies (Wilson et al. 1975) showed satellite Bragg reflections appearing at $T < T_o = 123K$, and presumed to be commensurate with the average lattice. A subsequent higher resolution neutron study (Moncton et al. 1977) showed the satellites to be incommensurate near T_o with a temperature dependent wavevector, $\vec{q}_\delta = (1-\delta)\vec{a}^*/3$. Moncton et al. went on the study the temperature dependent amplitude of the CDW and to discuss the mechanism for $\delta(T)$ and the "lock-in" transformation at T_c=90K below which $\delta(T)$=0. The latter discussion was extended by McMillan (1976). The key point to recognize is that purely sinusoidal modulation (eqn. 2 with constant ϕ) cannot take advantage of the periodic potential of the average lattice. However with the proper spatial variation of $\phi(r)$ it is possible to produce large commensurate regions separated by narrow domain walls in which $\phi(r)$ changes rapidly. Model calculations suggest that under certain conditions an ordered array of these domain walls, which McMillan called discommensurations (D.C.'s) would be the state of lowest energy. The position of the dominant Bragg sattelite would be determined by the density of D.C.'s.

Figure 1 shows $\delta(T)$ as obtained in an even higher resolution study using x-rays by Fleming et al. (1980). In addition to confirming the lock-in transformation found with neutrons, the new study established the existance of two distinctly different incommensurate phases. The fully incommensurate phase obtained on cooling from high temperatures incorporates three incommensurate wavevectors directed along alternate 6-fold axes of the hexagonal reciprocal lattice with magnitude q_δ. On warming from the commensurate phase an additional phase is found in which one of the three wavevectors remains commensurate. In terms of a real space description, the latter phase involves a parallel stripe D.C. pattern, whereas the former requires a hexagonal honeycomb array of D.C.'s. Direct confirmation of the D.C. model has come from recent dark field electron microscopy studies (Fung et al. 1981) and (Chen et al. 1981).

3. Dynamics

Neutron studies have also been instrumental in studying the dynamics of displacive incommensurate phase transformations. The form of the periodic displacement in eq. (2) suggest that we view the transformation as a condensation or "freezing-in" of a phonon with wavevector q_o, which migh occur because the phonon frequency $\omega(q_o)$ vanishes. Such "soft mode" transformations are reviewed by Cowley (1982) in this volume. In the case of a CDW transformation, the screening effects discussed previously can be reflected in $\omega(q_o)$ through giant Kohn anomalies (Kohn, 1959) which have been observed spectacularly in the quasi-one-dimensional metal KCP ($K_2Pt(CN)_6Br_x$). (Renker et al. 1974) and (Carneiro et al. 1976). Substantial, but incomplete phonon softening is observed in the quasi-two-dimensional CDW materials as well (Moncton et al. 1977). However, the best example of a soft mode incommensurate transformation occurs not in a CDW metal but rather in the insulator K_2SeO_4 (Iizumi et al. 1977) as shown in Fig. 2. This dispersion relation is essentially the Fourier transform of the interplanar forces, from which we can deduce that the instability in K_2SeO_4 results from anomalously large and temperature dependent forces between planes of atoms which are third-nearest neighbors. On a fundamental microscopic level, coulomb (dipolar) forces presumably play a decisive role, although model calculations are far from simple. (See, for example, Hague et al. 1978).

But what about the dynamics of the incommensurate phases themselves? Because it should be possible to add an overall phase shift to eq. 2 while leaving the energy unchanged it follows that there should appear a new type of gapless long wavelength excitation (a phason) which represents slow spatial variations in ϕ (Overhauser 1968). In the simple soft mode picture presented above the added presence of static displacements below

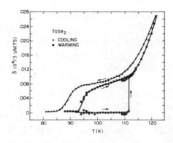

Fig. 1. Temperature dependence of incommensurability, δ, in 2H-TaSe$_2$. On cooling only the fully incommensurate phase appears. On warming an additional stripe phase is present between $T_{cs}=93K$ and $T_{si}=112K$. (After Fleming et al. (1980)).

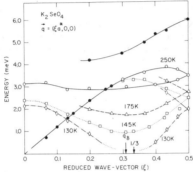

Fig. 2. Dispersion of the soft phonon branch in K$_2$SeO$_4$, leading to an incommensurate phase below $T_o=130K$. After Iizumi et al. (1977).

T_o require new harmonic modes to be constructed from linear combinations of phonons with wavevectors $(\vec{q}+\vec{q}_o)$ and $(\vec{q}-\vec{q}_o)$ respectively. The result is an upper branch for which

$$\vec{u}_+(\vec{r},t) \propto \cos(\vec{q}\cdot\vec{r}-\phi)e^{i(\vec{q}\cdot\vec{r}-\omega_+ t)} \qquad (3)$$

and a lower branch for which

$$\vec{u}_-(\vec{r},t) \propto \sin(\vec{q}\cdot\vec{r}-\phi)e^{i(\vec{q}\cdot\vec{r}-\omega_- t)} \qquad (4)$$

Comparision of eqs. (2-4) shows that the upper branch is equivalent to a modulation of the amplitude of the static displacements, while the lower branch represents a modulation of their phase. Although phasons exhibit linear acoustic-like dispersion, $\omega(q)=vq$, the velocity v has no relation to the velocity of sound and the phasons are not to be confused with acoustic phonons, as Fig. 3 makes clear. It has proven difficult to unambiguously identify these phason modes with neutron scattering, but recently there have been reports of their observation in two incommensurate insulators, biphenyl (Cailleau et al. 1980) and ThBr$_4$ (Bernard et al. 1981).

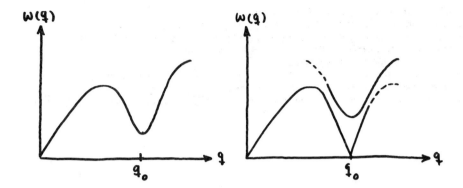

Fig. 3. Schematic illustration of the dispersion relation for a material undergoing an incommensurate displacive phase transformation. (a) soft branch with minimum at q_0 above T_0; (b) splitting of modes into a gapless 'phase' branch and an upper 'amplitude' branch below T_0.

The foregoing discussion concerned phason modes in displacive incommensurate structures. In intergrowth/overgrowth structures the phason does represent (essentially) an acoustic mode of the sublattice. These modes have been studied extensively in $Hg_{3-\delta}AsF_6$ (Heilmann et al. 1979) and detailed discussions of their dispersion and damping have appeared recently (Axe et al. 1982).

References

Axe J D (1980). Ordering in Strongly Fluctuation Condensed Matter Systems, T. Riste, ed. (New York, Plenum) pp. 399-414.
Axe J D and Bak P (1982). To appear in Phys. Rev. B.
Bernard L, Delamoye P, Currat R and Zeyen C (1981). Inst. Lane Langevin Progress Report, p. 72.
Cailleau H, Moussa F, Zeyen C M E and Bouillet J (1980). Sol. St. Comm. 33, 407-409.
Carniero K, Shirane G, Werner SA and Kaiser S (1976). Phys. Rev. B13, 4258-4265.
Chen C H, Gibson J M and Fleming R M (1981). Phys. Rev. Lett. 47, 723-738.
Comès R and Shirane G (1979). Highly Conducting One-Dimensional Solids, J.T. Devreese et al. ed. (New York, Plenum) pp. 17-67.
Cowley R A (1982). This volume. Kohn W (1959). Phys. Rev. Lett. 2, 393-396.
Fleming R M, Moncton D E, McWhan D B and DiSalvo F J (1980). Phys. Rev. Lett. 45, 576-579.
Fung K K, McKernan S, Steeds J W and Wilson J A (1981). J. Phys. C14, 723-740.
Hague M S, Hardy J R, Kim Q and Ullmann F G (1978). Solid State Comm. 27, 813-817.
Heilmann I U, Axe J D, Hastings J M, Shirane G, Heeger A J and MacDiarmid A G (1980). Phys. Rev. B20, 751-760.
Iizumi M, Axe J D, Shirane G and Shimaoka K (1977). Phys. Rev. B15, 4392-4400.
Koehler W (1982). This volume.
Kohn W (1959). Phys. Rev. Lett. 2, 393-395.

Mackintosh A R (1982). This volume.
MacMillan W L (1976). Phys. Rev. B14, 1496-1508.
Moncton D E, Axe J D, DiSalvo F J (1977). Phys. Rev. B16, 801-812.
Overhaus A W (1968). Phys. Rev. $\underline{167}$, 691-700.
Nielsen M (1982). This volume.
Renker B, Pintschovius L, Gläser, W, Reitschel H, Comès R, Liebert L and Drexel W (1974). Phys. Rev. Lett. $\underline{32}$, 836-842.
Wilson J A, DiSalvo F J and Mahajan S (1975). Adv. In Phys. $\underline{24}$, 117-140.

Liquid helium

E C Svensson

Atomic Energy of Canada Limited, Chalk River, Ontario, Canada K0J 1J0
and Brookhaven National Laboratory,* Upton, New York 11973, USA

Abstract. Recent neutron-scattering studies on liquid ^{4}He which have led to a simplified interpretation of the dynamics and to new values for the condensate fraction are summarized.

1. Introduction

Liquid ^{4}He has been studied far more extensively by neutron-scattering techniques than any other material. To date, experimental results have been reported from at least 13 different centres by approximately 70 different authors. The results of the first transmission studies were reported more than three decades ago (Goldstein et al 1951). These were soon followed by total-scattering measurements such as those of Henshaw and Hurst (1953). Then came the very important inelastic-scattering measurements of Palevsky et al (1957,1958) and Yarnell et al (1958), true landmark experiments which confirmed that superfluid ^{4}He indeed had a dispersion relation of the "phonon-roton" form envisaged by Landau (1947).

Such a prodigious wealth of information has been accumulated in the 31 years of neutron measurements on liquid helium, that only a small fraction can be covered in the space allowed for this article. Since the earlier work has already been extensively reviewed (see, e.g., Woods and Cowley 1973, Cowley 1976, 1978, Price 1978 and Stirling 1978), I will concentrate on the work of the past 4-5 years. A considerable amount that is new and exciting has been learned in this recent period as a result of very extensive and detailed studies of the temperature dependences of $S(Q)$ and $S(Q,\omega)$. I will first discuss the studies of $S(Q,\omega)$ in the phonon-roton region which have led to a simplified interpretation of the dynamics and then turn to the studies which have led to new values for the condensate fraction.

2. A simplified interpretation of $S(Q,\omega)$ for liquid ^{4}He

Neutron studies of the temperature dependence of $S(Q,\omega)$ prior to 1978 appeared to indicate that nothing dramatic happened on passing through T_λ = 2.17 K, and further suggested that "roton-like" excitations persisted above T_λ. This was puzzling since rotons were originally thought to be associated with the superfluid. The dynamics of liquid ^{4}He appeared to be much more complex than had been expected.

---

*Visiting Physicist, September 1, 1981 - August 31, 1982.

Studies by Svensson et al (1978) and Woods and Svensson (1978) for $0.8 \leq Q \leq 1.926$ Å^{-1} finally showed that something "dramatic" did happen at T_λ - the "one-phonon" peak in $S(Q,\omega)$ disappeared. As can be seen from Fig. 1, this peak is a clearly identifiable feature for all temperatures below T_λ, but above T_λ it is gone leaving only a single broad component in $S(Q,\omega)$. The one-phonon peak is thus indeed a signature on the superfluid. Further, Woods and Svensson found that $S(Q,\omega)$ could be very well described as the sum of a superfluid component with a weight $n_s = \rho_s/\rho$ and a normal-fluid component with a weight $n_n = 1 - n_s$, namely,

$$S(Q,\omega) = n_s S_s(Q,\omega) + n_n S_n(Q,\omega). \qquad (1)$$

Here $S_s(Q,\omega)$ consists of a one-phonon peak plus a broad multiphonon component, while $S_n(Q,\omega)$ consists of only a single broad peak characteristic of non-superfluid ^{4}He. This interpretation resolved several previously puzzling problems. Most important of all, it showed that there were no "one-phonon" excitations above T_λ. It also gave roton linewidths which agreed very well with the predictions of the theory of Landau and Khalatnikov (1949) whereas earlier neutron values were considerably larger especially near T_λ. Further, it gave roton energies which agreed very well with the values inferred from thermodynamic measurements whereas earlier neutron values were lower by about 25% near T_λ (see Brooks and Donnelly 1977). The earlier roton linewidths and energies were taken from the whole distribution and hence were distorted by the effect of the broader $S_n(Q,\omega)$ centered at a lower energy than the roton peak in $S_s(Q,\omega)$.

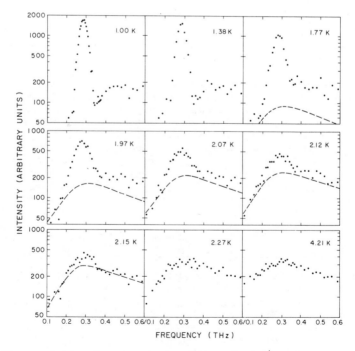

Fig. 1. The resolution broadened $S(Q,\omega)$ for liquid ^{4}He for $Q = 1.13$ Å^{-1} at nine temperatures (from Woods and Svensson 1978). The dashed curves show the "normal-fluid" components, i.e., the last term in equation (1).

Equation (1) was originally proposed simply as an empirical relationship. It has since received considerable direct support from theoretical studies by Griffin (1979) and Griffin and Talbot (1981), and somewhat more indirect support from other studies. Mezei (1981) determined roton linewidths and energy changes at low temperatures where one would expect the Landau-Khalatnikov theory to be valid. His results confirmed this, and extrapolation to higher temperatures gave good agreement with the results of Woods and Svensson but not with the results of Dietrich et al (1972) and Tarvin and Passell (1979) who used the whole distributions to determine the roton parameters. Tarvin and Passell in fact showed that energies and linewidths obtained by fitting to the whole distributions depended strongly on the analytical form assumed for the lineshape. This is exactly what one would expect from equation (1) since one is trying to fit a "single-peak" lineshape to a distribution which is actually the sum of two components characterized by different energies and linewidths. The recent theory work of Bedell et al (1982) has placed a bound on the temperature variation of the roton energy which is satisfied by the results of Woods and Svensson but not by the results of Dietrich et al and Tarvin and Passell. Further support for equation (1) has come from the work of Svensson et al (1981) who pointed out that this relationship implied analogous relationships for $S(Q)$ and $g(r)$ which were very well satisfied by the experimental results of Svensson et al (1980). (See also Svensson and Murray 1981.)

There is clearly a great deal of evidence in support of equation (1). In the final analysis, it will probably turn out to be somewhat too simple, but it has, I believe, put us back on the right track to understanding the dynamics of liquid ^{4}He. By changing the picture from depressingly complex to appealingly simple, it has stimulated a renewed interest on the part of theorists which hopefully will soon lead us to a better understanding.

3. The condensate fraction n_0

Ever since Fritz London (1938) proposed that the λ transition in liquid ^{4}He was connected with the phenomenon of Bose-Einstein condensation, scientists have been searching for a way to experimentally determine the supposed macroscopic occupation of the zero-momentum state. Following the proposal by Hohenberg and Platzman (1966) that n_0 might be determined by neutron measurements at large Q where $S(Q,\omega)$ should directly reflect the momentum distribution $n(\vec{p})$, there were several studies which, however, led to values of n_0 ranging from 0.02 to 0.17 in apparent disagreement with each other. (For references see Sears and Svensson 1979.) With the work of Martel et al (1976) it became clear that these disagreements were largely attributable to inadequacies in analysis. Martel et al also proposed a method for obtaining more reliable values of $n(\vec{p})$ from the neutron measurements, and subsequent application (Woods and Sears 1977) to the results of Cowley and Woods (1971) gave $n_0(1.1 \text{ K}) = 0.069\pm0.008$. This value has since been revised (Sears et al 1982) to 0.109 ± 0.027.

In their very recent study, Sears et al (1982) have, on the basis of high resolution measurements of $S(Q,\omega)$ at large Q, determined the $n(\vec{p})$ values shown in Fig. 2. There is essentially no change in $n(\vec{p})$ between 4.27 and 2.27 K, but on going to 2.12 K there is a large increase at low p which is a direct consequence of a finite condensate fraction. The surprisingly large increase at 2.12 K (just 0.05 K below T_λ) reflects an "enhancement" caused by a p^{-2} singularity in the $n(\vec{p})$ for the non-condensate atoms which is only present if n_0 is finite and then strongest near T_λ. By applying an improved analysis procedure to the results of Fig. 2 and the results of

Woods and Sears (1977), Sears et al (1982) obtained the n_0 values for 1.00, 1.1 and 2.12 K shown by solid symbols in Fig. 3. These are, I am quite confident, the most reliable values of n_0 obtained to date.

Hyland et al (1970) proposed that n_0 could also be determined via the relationship

$$g(r) - 1 = (1 - n_0)^2 [g_n(r) - 1] \tag{2}$$

where $g_n(r)$ refers to a temperature just above T_λ. The first measurements sufficiently accurate and complete for a critical test and application of equation (2) were those of Svensson et al (1980). Some features of the pair correlation functions obtained by these authors are shown in Fig. 4. Note that the amplitudes of the oscillations, which are a direct measure of the correlations between the atoms, increase with decreasing temperature above T_λ, but then at T_λ there is a dramatic reversal and the amplitudes decrease continuously with further cooling. This anomalous behavior is, according to Hyland et al, a direct consequence of atoms entering the condensate state. The n_0 values (open symbols in Fig. 3) obtained by application of equation (2) to experimental g(r) values are clearly in excellent agreement with the values (solid symbols) obtained from $S(Q,\omega)$ via $n(\vec{p})$ by Sears et al. Since there is, however, considerable controversy (see Sears et al 1982) concerning the method of Hyland et al, this agreement could be fortuitous. It is, nonetheless, support for equation (2) at least as an empirical relationship. There is also agreement, to within the combined uncertainties, between the best theoretical estimates of $n_0(0)$, shown by x's in Fig. 3, and the extrapolation to T = 0 of the best fit to all the experimental values shown by the solid curve (see Sears et al 1982 for details).

As a result of these studies, especially the work of Sears et al (1982), we can now conclude, 44 years after London's original proposal, that there

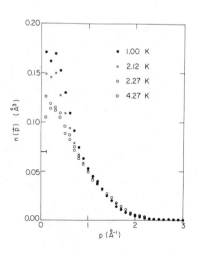

Fig.2. $n(\vec{p})$ for liquid ^{4}He at four temperatures (from Sears et al 1982).

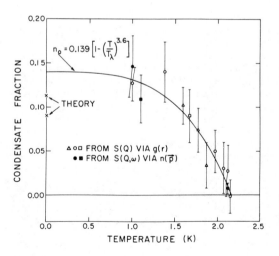

Fig.3. Results for the condensate fraction in superfluid ^{4}He given by Sears et al (1982).

really is a finite occupation of the zero-momentum condensate state in superfluid ^{4}He. The best results indicate that approximately 13% of the ^{4}He atoms are in this state at 1 K.

4. Concluding remarks

Equation (1) together with the analogous relationships for $S(Q)$ and $g(r)$ proposed by Svensson et al (1981) and, to a large extent, also equation (2) constitute a set of intuitively appealing but still essentially empirical relationships which have been found to give a very good description of the highly accurate neutron-scattering results obtained in the last 4-5 years. We are clearly in need of theoretical work to confirm, modify, or even refute these relationships.

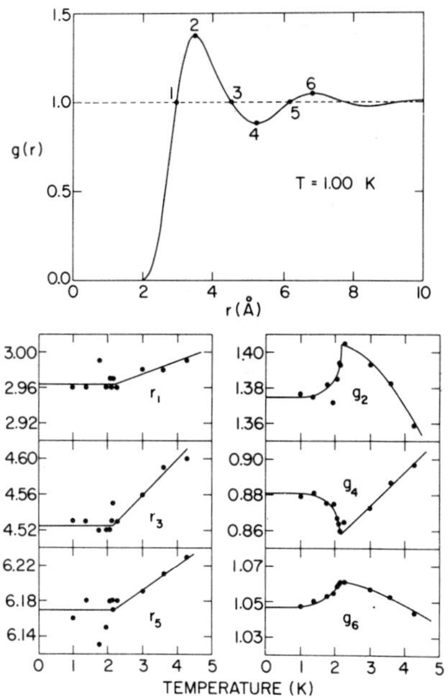

Although we have now obtained a reasonably accurate set of values for $n_0(T)$, the condensate saga has certainly not come to an end. No one has yet observed a separate sharp condensate component directly in $S(Q,\omega)$, and I'm sure we will see a succession of attempts to do so over the next few years using the new spallation sources. It is not going to be easy to obtain sufficiently high energy resolution at the Q values that are required (at least 100 Å^{-1}), but hopefully it will prove to be possible.

Fig. 4. $g(r)$ for liquid ^{4}He at 1.00 K and the temperature dependence of various features (from Sears and Svensson 1980).

Before ending, I must mention the work on liquid ^{3}He which, by comparison with the work on ^{4}He, is still in its early youth (barely 8 years old). This isotope was long avoided by neutron scatterers because of its dreadfully high absorption cross section. By using a statistical chopper and a very clever design for the sample container, Sköld et al (1976) were able, using the modest-flux CP-5 reactor at Argonne National Laboratory, to make the first observations of both the zero-sound modes and the lower-frequency spin-fluctuation peaks in liquid ^{3}He. This achievement certainly ranks as one of the great feats in the history of neutron scattering. Much work on liquid ^{3}He has also been done at the Institut Laue-Langevin (see, e.g., Stirling 1978 and Hilton et al 1980).

In spite of the inordinate amount of work that has already been done on ^{4}He and the inordinate difficulty of doing measurements on ^{3}He, both isotopes are currently commanding a great deal of attention from neutron scatterers as well as from the scientific community at large. I'm sure they will continue to do so for many years to come. The fascination with liquid ^{4}He which began 74 years ago still shows no sign of waning.

Acknowledgments

I have benefited greatly from discussions with R A Cowley, A Griffin, P Martel, L Passell, R Scherm, V F Sears and A D B Woods. Work at Brookhaven National Laboratory was supported by the Division of Materials Sciences, US Department of Energy, under contract DE-AC02-76CH00016.

References

Bedell K, Pines D and Fomin I 1982 J. Low Temp. Phys. 48 417
Brooks J S and Donnelly R J 1977 J. Phys. Chem. Ref. Data 6 51
Cowley R A 1976 Proc. of the Conf. on Neutron Scattering ed R M Moon (Springfield, Virginia, Nat. Tech. Inf. Service) Vol. II, pp. 935-53
Cowley R A 1978 Quantum Liquids ed J Ruvalds and T Regge (Amsterdam: North Holland) pp 27-61
Cowley R A and Woods A D B 1971 Can. J. Phys. 49 177
Dietrich O W, Graf E H, Huang C H and Passell L 1972 Phys. Rev. A 5 1377
Goldstein L, Sommers H S Jr., King L D P and Hoffman C J 1951 Proc. of the Int. Conf. on Low Temp. Phys. ed R Bowers, pp. 88-9
Griffin A 1979 Phys. Rev. B 19 5946
Griffin A and Talbot E 1981 Phys. Rev. B 24 5075
Henshaw D G and Hurst D G 1953 Phys. Rev. 91 1222
Hilton P A, Cowley R A, Scherm R and Stirling W G 1980 J. Phys. C 13 L295
Hohenberg P C and Platzman P M 1966 Phys. Rev. 152 198
Hyland G J, Rowlands G and Cummings F W 1970 Phys. Lett. 31A 465
Landau L D 1947 J. Phys. USSR 11 91
Landau L D and Khalatnikov I M 1949 Zh. Eksp. Teor. Fiz. 19 637
London F 1938 Nature 141 643
Martel P, Svensson E C, Woods A D B, Sears V F and Cowley R A 1976 J. Low Temp. Phys. 23 285
Mezei F 1980 Phys. Rev. Lett. 44 1601
Palevsky H, Otnes K, Larsson K E, Pauli R and Stedman R 1957 Phys. Rev. 108 1346
Palevsky H, Otnes K and Larsson K E 1958 Phys. Rev. 112 11
Price D L 1978 The Physics of Liquid and Solid Helium Part II ed K H Bennemann and J B Ketterson (New York: Wiley) pp. 675-726
Sears V F and Svensson E C 1979 Phys. Rev. Lett. 43 2009
Sears V F and Svensson E C 1980 Int. J. Quant. Chem: Quant. Chem. Symp. 14 715
Sears V F, Svensson E C, Martel P and Woods A D B 1982 Phys. Rev. Lett. 49 279
Sköld K S, Pelizzari C A, Kleb R and Ostrowski G E 1976 Phys. Rev. Lett. 37 842
Stirling W G 1978 J. de Physique 39 C6-1334
Svensson E C, Scherm R and Woods A D B 1978 J. de Physique 39, C6-211
Svensson E C, Sears V F, Woods A D B and Martel P 1980 Phys. Rev. B 21 3638
Svensson E C, Sears V F and Griffin A 1981 Phys. Rev. B 23 4493
Svensson E C and Murray A F 1981 Physica 108B 1317
Tarvin J A and Passell L 1979 Phys. Rev. B 19 1458
Woods A D B and Cowley R A 1973 Rept. Prog. Phys. 36 1135
Woods A D B and Sears V F 1977 Phys. Rev. Lett. 39 415
Woods A D B and Svensson E C 1978 Phys. Rev. Lett. 41 974
Yarnell J L, Arnold G P, Bendt P J and Kerr E C 1958 Phys. Rev. Lett. 1 9

Neutron studies of liquids and dense gases

P.A. Egelstaff

Department of Physics, University of Guelph, Guelph, Ontario N1G 2W1

1. Background

At the time the neutron was discovered our knowledge of the liquid state was well developed, the reason for the existence of a third state of matter was understood, and the broad outline of the static and dynamic properties of fluids had been established. During the thirties these general ideas were given a more satisfactory theoretical basis (e.g. Yvon 1966, or Hill 1956 for static properties and Frenkel 1946 for dynamic properties) and the principal experimental results from X-ray and light scattering were improved and unambiguously interpreted. Nevertheless the field was a frustrating one because each advance, however stimulating, produced on balance only a small step towards the expected goal. This goal was to understand and predict the static and dynamic properties (on both microscopic and macroscopic scales) starting from intermolecular potentials. These predictions should be reasonably accurate, say ~ 1%. The theoretical and computation difficulties were so large that even the decision to discover the better of two alternatives to solving a problem could involve a formidable program. Consequently by the fourties and fifties, there was little contact between theoretical and experimental investigations in many cases.

Fermi et al's (1934) famous experiment on the production of thermal neutrons, was the first neutron inelastic scattering experiment and the first use of neutrons of wavelength ~ 2 Å. It was realised quickly that these features were useful in the study of condensed matter, including fluids. The discovery of a new technique was welcomed eagerly because of the possibility that some of the many difficulties might be eased, but in this case its implementation was delayed by many years because of low neutron source fluxes. Because neutrons could detect both the positions and motions of nuclei on suitable scales, it was hoped that neutron data would provide more basic information and better contact with theoretical results than was available from other techniques.

Modern reviews of static properties by Barker and Henderson (1976), of dynamic properties by Copley and Lovesey (1975) give the recent background. Reviews of the neutron scattering studies were given at the Gatlinburg Conference by Narten (1976) for static and by Egelstaff (1976) for dynamic properties. In the limited space available only a few examples can be given for each category, and some topics will be omitted (e.g. liquid He which has been reviewed by Cowley 1976).

2. Static Structure Factors

After the development of the nuclear reactor it was realised that neutron diffraction techniques, similar to those employed for X-rays, would yield useful data on the structure of fluids. Alcock and Hurst (1949) studied gases and Chamberlain (1950) studied liquids. Data was taken slowly, sometimes recorded by handwriting, and the precision was not high because corrections for side effects were not well known and the statistical errors were significant. But at this stage greater precision than that provided by X-ray methods (typically several per cent) was required. Although capable in principle of achieving this goal, neutron techniques needed many years of development to reach it. Partly for theoretical and partly for experimental reasons liquid metals were often studied by experimentalists. North et al (1968) made useful advances and it began to be possible to recognise real differences between X-ray and neutron data (Egelstaff 1966). With further improvements in data collection and reduction, good data on a liquified noble gas near the triple point (Yarnell et al 1973) were obtained. After another round of technical improvements structure factors for a noble gas at a number of states at low and medium densities were measured (Teitsma and Egelstaff 1980). Realistic and useful comparisons between theory and experiment have become possible at this stage. It has also become possible to exploit the differences between X-ray data (i.e. electronic correlation functions) and neutron data (i.e. nuclear correlation functions) for molecular fluids (Egelstaff 1982). Over this period the computer simulation of the pair correlation function has had a stimulative function which will be discussed later. Finally neutrons have unique advantages for systems containing more than one kind of atom (Enderby et al 1966).

3. Dynamic Structure Factors.

A definition may be given in terms of the neutron cross section (for an atomic system of unique scattering length b):

$$b^{-2} d^2\sigma/d\Omega d\omega = (k/k_0)S(q,\omega) \quad (1)$$

where k_0, k are wave numbers before and after scattering and

$$\underline{q} = \underline{k}_0 - \underline{k}; \quad \omega = \left(k_0^2 - k^2\right)\hbar/2m \quad (2)$$

These conservation equations limit the range of (q,ω) that may be studied. Thus if the angle of scatter (θ) and the energy transfer ($\hbar\omega$) are chosen, the minimum value of q that it is possible to obtain is $\sqrt{\pm 2m\omega}\sin\theta/\hbar$. This occurs for an incident energy of $E_0 = \hbar\omega(1 \pm \mathrm{Cosec}\theta)/2$, and the $\pm$ refer to $\pm$ values of ω. Following van Hove (1954) $S(q,\omega)$ is usually divided into a self term (for the motion of one nucleus) and a distinct term (for the relative motion of two different nuclei), and these may be separated for systems in which the scattering amplitude may be varied. Thus van Hove's space-time correlation functions may be determined in principle.

Because the frequency moments of $S(q,\omega)$ are known (Placzek 1952), and for the self term the classical zero and second moments do not depend on interatomic interactions while the higher moments depend strongly on these interactions only for low q, the general shape of $S(q,\omega)$ is similar for all atomic systems. Models in which the lowest two moments are correct have been used for many years (e.g. Vineyard 1958), and improvements in computing ability allowed some data on higher moments (4th and 6th) to be added later. To extract useful information from $S(q,\omega)$ for a fluid, it is necessary to have excellent data and a sophisticated theory. The latter has involved computer molecular dynamics calculations in many cases.

The technical development of suitable instruments occupied many years (see Egelstaff 1976), and early experiments were directed towards establishing the new ideas. Brockhouse (1958 and 1959) measured part of $S(q,\omega)$ for water and lead, and laid an experimental and interpretive foundation for incoherent and coherent specimens on which many later workers were to build. Meanwhile the self term of van Hove was studied extensively for reactor moderators (e.g. Egelstaff and Poole 1969); an endeavour which involved the development of technical, interpretive and theoretical ideas. Prominent among these was the use of a spectral density for fluids, defined as the Fourier transform of the velocity correlation function, and the understanding of the significance of the detailed balance factor.

Throughout this period improvements were made in data collection, analysis and correction for unwanted side effects. During the 60's and 70's many experiments on the dynamic structure factor were reported, and with time they grew more sophisticated. For argon at the triple point Sköld et al (1972) measured both the self and distinct parts of $S(q,\omega)$ to study its general behaviour, while for dense argon gas Postol and Pelizzari (1978) measured the low q region to study the propagation of long wavelength modes. Other states have been studied by Verkerk (1977). For liquid metals Copley and Rowe 1974 measured $S(q,\omega)$ of rubidium at the triple point over a wide range of q and ω and observed propagating modes up to $q \simeq 1 \text{ Å}^{-1}$, and Söderström et al (1980) removed an earlier controversy by observing similar modes in liquid lead. By exploiting the high data collection rate (with good statistics) from a new instrument at the I.L.L. Egelstaff et al (1982) were able to study the variation of $S(q,\omega)$ along an isotherm of dense krypton gas. The contribution of different components of the interatomic forces has now become evident through the interpretation of these different experiments.

4. Thermodynamic Derivatives and Variation with (ρ,T)

The interpretation of a single $S(q)$ or $S(q,\omega)$ function is usually dependent on models and their relationship to the underlying interatomic forces is not always clear. In addition the contribution of 3-body or higher order forces to these functions is not known properly. To clarify these questions additional information is required and it has been proposed that the thermodynamic derivatives or the detailed (ρ,T) dependence of $S(q)$ and $S(q,\omega)$ be measured (Schofield 1966, Egelstaff et al 1971, Raveche et al 1972, Gubbins et al 1978). The accumulation of such data is difficult and their interpretation has required extensive computer simulations. An example was cited above (Kr gas).

5. Discussion

This brief survey has covered the development of (neutron) experimental and interpretive techniques to the point where they can become a "spearhead" in the study of fluids. Any new method must pass through a development stage before it can become an exciting frontier research tool. For slow neutrons that period was extended partly by the technical difficulties in sources and instruments, partly through the problems of handling the wealth and detail of the potential information and partly by the need to improve on the sophisticated data available through complementary methods.

Modern developments in this field rest on a triad of activities, illustrated by the diagram:

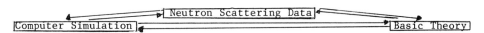

The interaction between these three activities over the past 10 years has improved our understanding of the relation between intermolecular forces and nuclear correlation functions, and it is likely to continue to do so in the future. From neutron scattering experimentalists, precise data as a function of ρ and T will continue to be needed, and the most appropriate ranges of the four variables (q, ω, ρ, T) need to be chosen with guidance from basic theory and simulations. Even with four variables to cover a wise initial choice coupled with the power of modern instruments (e.g. I.N.4 at the I.L.L.) allows good and interesting work to be done in experiments of ~ 2 weeks duration. Thus the future of neutron studies of fluids appears to be more rewarding than the past.

References

Alcock N Z and Hurst D G 1949 Phys. Rev. 75 1609
Barker J A and Henderson D 1976 Rev. Mod. Phys. 48 587
Brockhouse B N 1958 Nuovo Cimento Suppl. 9 45
Brockhouse B N and Pope N K 1959 Phys. Rev. Lett. 3 259
Chamberlain O 1950 Phys. Rev. 77 305
Copley J R D and Rowe J M 1974 Phys. Rev. A 9 1656
Copley J R D and Lovesey S W 1975 Rep. Prog. Phys. 38 461
Cowley R 1976 Proc. Conf. on Neutron Scattering ed R̄ Moon Conf.-760601-P2 (O.R.N.L.-U.S.E.R.D.A.) p 935
Egelstaff P A 1966 Properties of Liquid Metals ed P D Adams et al (Taylor and Francis) p 147
Egelstaff P A and Poole M J 1969 Experimental Neutron Thermalisation (Pergamon Press)
Egelstaff P A, Page D I and Heard C R T 1971 J. Phys. C 4 1435
Egelstaff P A 1976 Proc. Conf. on Neutron Scattering ed R̄ Moon Conf.-760601-P2 (O.R.N.L.-U.S.E.R.D.A) p 909
Egelstaff P A, Gläser W, Litchinsky D, Schneider E and Suck J-B 1982 – to be published
Egelstaff P A 1982 J. Phys. and Chem. Liq. 11 353
Enderby J E, North D M and Egelstaff P A 1966 Phil. Mag. 14 961
Fermi E, Amaldi E, Pontecorvo B, Rasetti F and Segrè E 1934 Ric. Sci. 5 282
Frenkel J 1946 Kinetic Theory of Liquids (Oxford University Press)
Gubbins K E, Gray C G and Egelstaff P A 1978 Molec. Phys. 35 315
Hill T L 1956 Statistical Mechanics (London: McGraw-Hill)
Narten A H 1976 Proc. Conf. on Neutron Scattering ed R Moon Conf.-760601-P2 (O.R.N.L.-U.S.E.R.D.A.) p 889
North D M, Enderby J E and Egelstaff P A 1968 J. Phys. C 1 784
Postol T A and Pelizzari C A 1978 Phys. Rev. A 18 2321
Placzek G 1952 Phys. Rev. 86 377
Raveche H J, Mountain R D and Street W B 1972 J. Chem. Phys. 51 4999
Schofield P 1966 Proc. Phys. Soc. 88 149
Sköld K, Rowe J M, Ostrowski G and Randolph P D 1972 Phys. Rev. A 6 1107
Söderström O, Copley J R D, Suck J-B and Dorner B 1980 J. Phys. F 10 151
Teitsma A and Egelstaff P A 1980 Phys Rev A 21 367
van Hove L 1954 Phys. Rev. 95 249
Verkerk P 1977 Neutron Inelastic Scattering II (IAEA, Vienna) p 53
Yarnell J L, Katz M J, Wenzel R G and Koenig S H 1973 Phys. Rev. A7 2130
Yvon J 1966 Les Corrélations et l'Entropie en Mechanique Statisique Classique (Paris: Dunod)

The structure and dynamics of ionic solutions

Professor J.E. Enderby

H.H. Wills Physics Laboratory, University of Bristol, Royal Fort, Tyndall Avenue, Bristol BS8 1TL.

1. Introduction

This review is concerned with the impact that neutron diffraction and quasi-elastic neutron spectroscopy has had on the study of aqueous solutions of strong electrolytes. Consider a liquid formed by dissolving an electrolyte MX_n in H_2O so that the solution contains $M^{n(+)}$ cations, X^- anions and water molecules (fig. 1).
In order to describe this system structurally, ten pair correlations are required $(g_{MO}(r), g_{MH}(r) \ldots g_{XX}(r))$ and to generalise the description to include the dynamical aspects of the liquid ten distinct Van Hove correlation functions $(G_{MO}(r,t)\ G_{MH}(r,t) \ldots G_{XX}(r,t))$ and four self-correlation functions $(G_H(r,t), G_O(r,t)\ G_M(r,t), G_X(r,t))$ have to be invoked. It is this plethora of correlation functions which makes a fundamental description of aqueous solutions so difficult. Single diffraction experiments tend to yield a gross average over several correlation

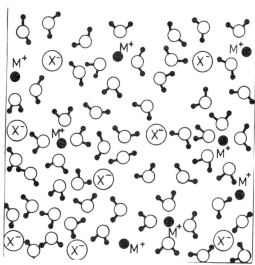

Fig. 1. A microscopic picture of an aqueous solution containing cations (M^+), anions (X^-) and water molecules.

functions. However, by exploiting two key properties of neutron scattering, we have shown that it becomes possible to isolate individual correlation functions or combinations of them which are of chemical or physical interest (Enderby and Neilson, 1981). The two properties are (a) the isotope effect whereby the scattering amplitude b depends on the number of neutrons in the nucleus for a given Z and (b) the large incoherent scattering cross-section associated with the protons in (light) water.

2. Hydration of Cations

ion and solute	molality	ion-oxygen distance/Å	ion-deuterium distance/Å	θ/deg	coordination number	ref.
Ni^{2+}						
$NiCl_2$	0.086	2.07±0.03	2.80±0.03	0±20	6.8±0.8	1
	0.46	2.10±0.02	2.80±0.02	17±10	6.8±0.8	1
	0.85	2.09±0.02	2.76±0.02	27±10	6.6±0.5	1
	1.46	2.07±0.02	2.67±0.02	42±8	5.8±0.3	1
	3.05	2.07±0.02	2.67±0.02	42±8	5.8±0.2	1
	4.41	2.07±0.02	2.67±0.02	42±8	5.8±0.2	1
$Ni(ClO_4)_2$	3.80	2.07±0.02	2.67±0.02	42±8	5.8±0.2	2
Ca^{2+}						
$CaCl_2$	1.0	2.46±0.03	3.07±0.03	38±9	10.0±0.6	3
	2.8	2.39±0.02	3.02±0.03	34±9	7.2±0.2	3
	4.5	2.41±0.03	3.04±0.03	34±9	6.4±0.3	3
Cu^{2+}						
$CuCl_2$	4.32	2.05±0.02	2.56±0.10	57±5	2.3±0.3	4
Li^+						
	3.57	1.95±0.02	2.55±0.02	40±10	5.5±0.3	5
$LiCl$	9.95	1.95±0.02	2.50±0.02	52±15	3.3±0.2	5

References for Table 1 : Hydration of cations

1. Neilson and Enderby (1978)
2. Newsome et al (1981)
3. Hewish et al (1982b)
4. Neilson (1982)
5. Newsome et al (1980)

The neutron first-order difference method (Soper et al 1977) yields a weighted distribution function $G_M(r)$ which is dominated by $g_{MO}(r)$ and $g_{MD}(r)$ i.e. the correlation functions related to cationic hydration. The isotopic substitution is of the form $MX_n \rightarrow M^*X_n$ and the solvent is normally D_2O. In a similar way, the substitution $MX_n \rightarrow M^*X_n$ yields $G_X(r)$ which is related to anionic hydration. The results of these investigations have been fully reported in the literature (Enderby and Neilson 1981; Hewish et al 1982b) and have recently been put into a general solution context by Hunt and Friedman (1982). We shall therefore highlight three important findings of this research.

(a) $Ni^{2+}(aq)$

In solution the hydration complex is $Ni(H_2O)_6^{2+}$, a fact known for many years (Taube, 1962). The lifetime of this inner sphere complex is very long ($\sim$ 30 µs); on the other hand NMR studies suggest that the symmetry of the complex is less than cubic (Friedman et al 1979). It is therefore very satisfying that $G_{Ni}(r)$ (fig. 2 and table 1) derived by the first order difference method "decisively support the generally accepted stoichiometry for the complex" (Hunt and Friedman 1982). A new feature of the data, however, is that they show that the tilt angle θ is reduced

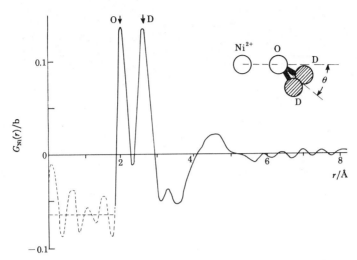

Fig. 2. $G_{Ni}(r)$ for a 4.41 molal solution of $NiCl_2$ in D_2O.

at low ionic strengths. This remains something of a mystery as model calculations based on relatively simple potentials favour $\theta \sim 55°$ in the limit of infinite dilution.

(b) $\underline{Ca^{2+}(aq)}$

Although Ca^{2+} is, like Ni^{2+} doubly charged this ion displays, in nature, a remarkable ability to adjust its local coordination. In solution this characteristic is also evident in that the coordination number increases from ~ 6 to ~ 10 as the concentration is reduced from 4.5 molal to 1.0 molal. Whether or not this increase is a characteristic of all weakly hydrating ions remains to be seen. Experiments on K^+ and Li^+ (at molalities less than 3) are being planned to test this.

(c) $\underline{Cu^{2+}(aq)}$

The ionic radii of Cu^{2+} and Ni^{2+} are similar (0.72 and 0.69Å respectively). Calculations based on simple pair potentials in which the ionic size is essentially the only adjustable parameter would, of course, predict that $G_{Ni}(r)$ and $G_{Cu}(r)$ should be similar. Neilson (1982) has shown that this is not the case (figure 3, table 1) so far as $CuCl_2$ solution is concerned. Interestingly, $G_{Ni}(r)$ and $G_{Cu}(r)$ are remarkably similar for $Cu(ClO_4)_2$. We see here a need to invoke a specific chemical effect associated, presumably, with the Jahn Teller character of Cu^{2+}.

3. The Hydration of Anions

The method has been applied to Cl^- and NO_3^- and in a wide range of solutions (tables 2 and 3). Typical results for $G_{Cl}(r)$ are shown in figure 4 where the $Cl^- - D_2O$ conformation consistent with these data is shown. When we compare the nature of the Cl^- hydration among the various solutions, several conclusions emerge. First, the general form of G_{Cl} is similar in all cases. This apparent lack of sensitivity of anionic hydration to the nature of the counter-ion emphasizes the importance of

Table 2. Cl^--water coordination

Solute	Molality	$r_{ClD(1)}$ (Å)	r_{ClO} (Å)	$r_{ClD(2)}$ (range) (Å)	ψ^* (range) (deg)	Coordination number
LiCl	3.57	2.25±0.02	3.34±0.04	-	0	5.9±0.2
LiCl	9.95	2.22±0.02	3.29±0.04	3.50-3.68	0	5.3±0.2
NaCl	5.32	2.26±0.03	3.20±0.05	-	0-10	5.5±0.4
RbCl	4.36	2.26±0.03	3.20±0.05	-	0-10	5.8±0.3
$CaCl_2$	4.49	2.25±0.02	3.25±0.04	3.55-3.65	0-6	5.8±0.2
$NiCl_2$	4.35	2.29±0.02	3.20±0.04	3.40-3.50	6-11	5.7±0.2
$NiCl_2$	3.00	2.23±0.03	3.25±0.05	-	0-6	5.5±0.4
$BaCl_2$	1.10	2.24±0.04	3.26±0.05	-	0-6	6.2±0.4

*Assuming that r_{OD} is 1Å. Reference for table 2 : Cummings et al (1980).

Table 3. NO_3^--water coordination

		Nearest neighbour			Next-nearest neighbour		
Solute	Molality	$r_{ND_{11}}$ (Å)	r_{NO_1} (Å)	Coordination number	$r_{ND_{21}}$ (Å)	r_{NO_2} (Å)	Coordination number
$NaNO_3$	7.80	2.05(2)	2.65(10)	1.3(2)	2.65(5)	3.40(10)	2.4(3)

Reference for Table 3 : Neilson and Enderby (1982).

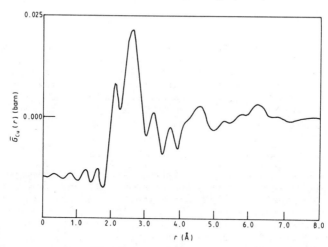

Fig. 3. Real space Cu^{2+} distribution function, $\bar{G}_{Cu}(r)$ for 4.32 molal $CuCl_2 \cdot D_2O$ solution.

local effects. The coefficients of the partial radial distribution functions are more favourable in some cases than others, and the enhanced resolution thereby obtained allow us to define closely the Cl^--D_2O geometry. The data clearly favour the linear configuration tentatively proposed by Soper et al (1977), but with a tilt angle generally less than 7°. The Cl-O distance is in good agreement with the value predicted by quantum mechanical cluster calculations (Schuster et al 1975). The nature of water coordination around NO_3^- has been discussed by Neilson and Enderby (1982).

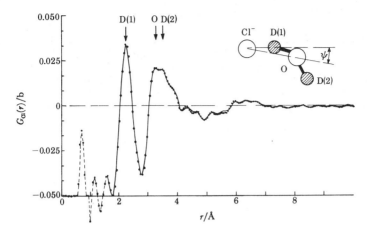

Fig. 4. $G_{Cl}(r)$ for a 9.95 molal solution of LiCl in D_2O. (J.R. Newsome, personal communication).

4. Ion-Ion Correlations

An extensive review of our work in this area is in preparation and it would be unnecessary duplication to give a detailed discussion here. The experiments depend on obtaining a second-order difference and are at the edge of current technology. Nevertheless, several new insights have been obtained. For example, the nature of $g_{ClCl}(r)$ has been found to depend strongly on the hydrating strength of the cation.

5. Proton Dynamics

A recent application of QENS to the solution problem which exploited the high resolution available on the IN10 spectrometer at Grenoble has been reported elsewhere (Hewish et al 1982a). The incoherent cross-section of the hydrogen nucleus is so large that $S(k,\omega)$ is dominated by the term $S_H^S(k,\omega)$ the other terms being sufficiently small that they may be neglected in the data analysis. Knowledge of $S(k,\omega)$ thus leads directly to the proton dynamics, which, provided that observations times τ_{ob} are long compared with those characteristic of vibrational and rotational motions, may be identified with those of the water molecule itself. By implication, the lower time-limit for diffusive behaviour mentioned above, $10^{-11}s$, satisfies this constraint. In addition, the experimental conditions must obey

$$Dk^2 \tau_c \ll 1$$

where τ_c is the rotational correlation time for a water molecule and D is the translational diffusion coefficient. Typically in aqueous solutions $D \sim 10^{-9} m^2 s^{-1}$ and $\tau_c \sim 10^{-11}s$ so that k must be $\ll 1\text{Å}^{-1}$, a severe constraint which is satisfied in the work reported here. Previous measurements involved values for Q of 0.5Å^{-1} to 6.0Å^{-1}, in which case proper account must be taken of the detailed sub-diffusive dynamics of the proton.

For a particle obeying the diffusion equation, the correlation function

G(r,t) has a Gaussian form, and S(k,ω) is a Lorentzian of the form

$$S_H^S(k,\omega) = \frac{1}{\hbar} \frac{Dk^2}{(Dk^2)^2 + \omega^2}$$

This is the form to be expected when the longest ionic binding time, τ_b, present is short relative to τ_{ob}, so that during this observation time any proton is able to sample the entire range of environments present in the solution. We refer to this situation as the 'fast exchange limit'. When the opposite is the case, and $\tau_b > \tau_{ob}$ for one ionic species, usually the cation, S(k,ω) take the form

$$S_H^S(k,\omega) = \frac{1}{\hbar} \left[\frac{c_1 D_1 k^2}{(D_1 k^2)^2 \omega^2} + \frac{c_2 D_2 k^2}{(D_2 k^2)^2 + \omega^2} \right]$$

This behaviour has been observed in the cases of Ni^{2+} and Mg^{2+} in solutions of their respective chlorides (Hewish et al, 1982a). The key aspect of their analysis of the data is that one of the diffusion coefficients (D_1) is equal to that of the cation (D_{ion}), a quantity accessible from independent tracer measurements, and the relative weighting of the two Lorentzians is known from structural measurements, so that the fitting of a curve to the data involves an absolute minimum of adjustable parameters. The values derived for D_2 cannot be equated with D_o, the diffusion coefficient of pure water and this demonstrates clearly that a simple two state model of the sort used by Sakuma et al (1979) and Martel and Powell (1981) is not adequate.

Acknowledgements

I wish to thank my colleagues Spencer Howells, George Neilson, Nicholas Hewish, Sue Langron and Philip Salmon for many helpful conversations and for carrying out the experiments I have described.

References

Cummings S, Enderby J E, Neilson G W, Newsome J R, Howe R A, Howells W S and Soper A K 1980 Nature 287 714
Enderby J E and Neilson G W 1981 Rept. Prog. Phys. 44 593
Friedman H L, Holz M and Hertz H G 1979 J. Chem. Phys. 70 3369
Hewish N A, Enderby J E and Howells J S 1982 Phys. Rev. Letts. 48 756
Hewish N A, Neilson G W and Enderby J E 1982b Nature 297 138
Hunt J P and Friedman H L 1982 Prog. Inorg. Chem. 30 (to appear)
Martel P and Powell B M 1981 Sol. State Comm. 39 107
Neilson G W 1982 J. Phys. C. 15 L233
Neilson G W and Enderby J E 1978 J. Phys. C. 11 L625
Neilson G W and Enderby J E 1982 J. Phys. C. 15 2347
Newsome J R, Neilson G W and Enderby J E 1980 J. Phys. C. 13 L923
Newsome J R, Sandström M, Neilson G W and Enderby J E 1981 Chem. Phys. Letts. 82 399
Sakuma T, Hoshino S and Fujii Y 1979 J. Phys. Soc. Jap. 46 617
Schuster P, Jakubutz W and Marius W 1975 Topics Curr. Chem. 30 215
Soper A K, Neilson G W, Enderby J E and Howe R A 1977 J. Phys. C. 10 1793
Taube H 1962 Progress in Stereochemistry 3 95

Disorder and correlations in molecular and liquid crystals

A J Leadbetter

Chemistry Department, The University, Exeter EX4 4QD

Abstract. Examples are given of the study of static and dynamic aspects of disorder and related phase transitions in two types of system using a variety of neutron scattering techniques. The first is an order-disorder transition in $(CH_3)_3CCN$ and the second is the study of the modulated hexagonal B phase of the liquid crystal $C_5H_{11}OC_6H_4CHNC_6H_4C_7H_{15}$ and its relation to the monoclinic G phase.

1. Introduction

Neutron scattering methods provide powerful probes for investigating static and dynamic aspects of molecular disorder and phase transitions in molecular systems. Two examples have been chosen to illustrate the kind of studies currently being undertaken. Both are of systems in which rapid rotation occurs about one molecular axis and they have been studied by both coherent scattering to determine structure and collective motions, using fully deuteriated samples, as well as incoherent scattering to determine the single particle motions on hydrogenous samples.

The first example is of t-butyl cyanide $[(CH_3)_3CCN]$ for which a very detailed understanding has been reached of the order-disorder transition which appears to be triggered by the molecular rotations. The second is of a much more complex liquid crystal system in which orientational disorder in the so-called B and G phases has also been fairly well characterised but where the transition between the phases is associated rather with longitudinal displacements of the molecular long axes, of similar type to that occurring in the modulations of the layers in the B phase.

2. Tertiary Butyl Cyanide $(CH_3)_3CCN$

This compound has a first order phase transition ($\Delta S \sim 9 JK^{-1} mol^{-1}$ at 233K and melts ($\Delta S_f \sim 32 JK^{-1} mol^{-1}$) at 292 K. The structure of the low temperature phase has been determined by powder profile refinement using a fully deuteriated speciment (Frost et al 1982a). The very low temperature completely ordered structure is shown in Fig. 1. In this phase the molecules, which have C_{3v} symmetry, reorient by jump motions between 3 equivalent potential wells, the barrier height decreasing from a value of $\sim 10 kJ mol^{-1}$ at 5K to $\sim 4 kJ mol^{-1}$ at the transition. This was determined by incoherent neutron quasi elastic and inelastic scattering experiments on hydrogenous samples (Frost et al 1980). The EISF demonstrating the 3-fold whole molecule rotation about the C-C-N axis is shown in Fig. 2. The internal rotation of the methyl groups was determined in separate experiments and is up to two orders of magnitude slower than the whole molecule rotation because of the relatively high ($\sim 18 kJ mol^{-1}$) intramolecular potential barrier.

Above the phase transition the molecules are still rotating about their three fold axes with a barrier height $\lesssim 4$ kJ mol^{-1} and the transition is clearly not associated with the <u>onset</u> of this uniaxial rotation. The detailed INQES measurements showed, however, that additional motions are occurring which are presumably connected with the disorder required to produce the average tetragonal (P4/n) structure in which the 3-fold molecular axes are nearly parallel to the crystal 4-fold axes. The average structure of the high temperature phase was solved by single crystal neutron diffraction experiments (Gane et al 1982) and the local structural correlations were studied directly using the diffuse scattering spectrometer (D7) and the cold neutron triple axis spectrometer (INI2), all on fully deuteriated samples.

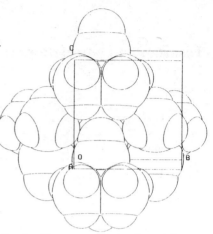

Fig.1. The monoclinic (P2$_1$/m) low temperature structure viewed down the a axis.
At 5 K, a = 6.12Å, b = 6.88Å, c = 6.72Å, β = 95.6°, z = 2.

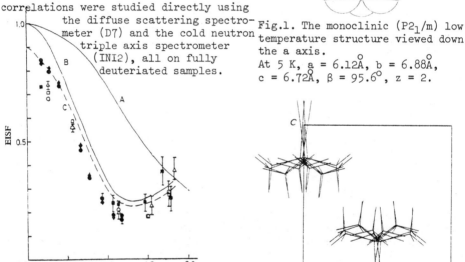

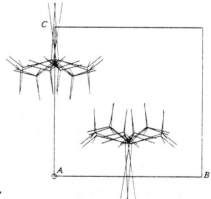

Fig.2. EISF for low temperature phase. Points denote experimental data at different T and resolution. Lines are model calculations. A - CH$_3$ rotations, B - uniaxial molecular rotations, C - B plus multiple scattering.

Fig.3. The tetragonal (P4/n) high temperature structure viewed down the a-axis (a = 6.85Å, c = 6.77Å, z = 2)

The average structure of the low temperature phase is shown in Fig. 3. The molecules have four equivalent positions of equal ($\frac{1}{4}$) occupancy related by 90° rotations about the four fold axis. The EISF determined from incoherent quasielastic data such as those illustrated in Fig. 4 is in excellent agreement with that calculated from the structural results, including both the uniaxial and four fold motions, as shown in Fig. 5. The four-fold disorder is not, of course, completely random and there exist pronounced correlations between the configuration of neighbouring molecules. These local structural correlations resemble very closely the low temperature monoclinic structure and extend over about two unit cells, as shown by the diffuse scattering peaks located near reciprocal lattice points which are present in the ordered monoclinic structure but symmetry disallowed in the high temperature phase (Fig. 6). The local structures relax very rapidly among the four different possible configurations, the lifetime being $\sim 10^{-12}$s at 270K (Fig. 7) which is essentially identical to that of the uniaxial molecular rotations, showing that the motions are strongly coupled, although there is some evidence that the cooperative relaxations become slower than (and therefore decoupled from) the uniaxial rotations as the temperature is decreased towards the transition. Analysis of powder diffraction data on the low temperature phase, in the light of knowledge of the structure of the disordered phase, showed the order parameter for the transition to be the relative occupancy of the four different local configurations. The temperature dependence of the order

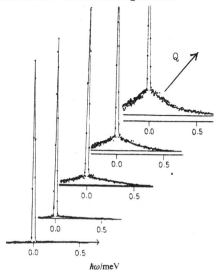

Fig.4. Incoherent quasielastic scattering $S(Q,\omega)$ from the high T phase.

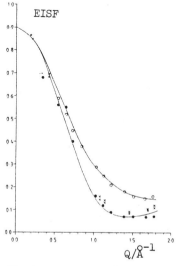

Fig.5. EISF for high T phase. Sample at 45° to beam (●,X) and at 135° to beam (O). Lines calculated from structural data with one adjustable parameter describing the preferred sample orientation.

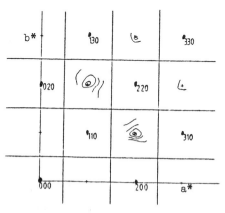

Fig.6. $\vec{a}^*\vec{b}^*$ Reciprocal lattice plane in high T phase showing diffuse scattering peaks as contours around h+k=2n+1

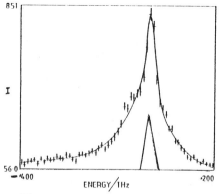

Fig. 7. Energy profile of diffuse scattering peak at (1, 2, 0)

parameter is in reasonable agreement with mean field theory (Frost et al 1982b) with $T_c \sim 245K$ (cf. $T_{tr} \sim 233K$).

The basic driving force for the transition is believed to be the increasing frequency of molecular C_3 rotations which causes local instabilities when neighbours are rotating simultaneously. This in turn leads to increasing occupancy of the different local structures related to the parent structure by 90° rotations about $\vec{c}*$.

3. N-(4-n-pentyloxybenzylidene)-4-n-heptylaniline. (50.7)

The compound 50.7 ($C_5H_{11}O$ Ph CH=N Ph C_7H_{15}) is typical of the no.m series in showing a rich polymorphism, its phase behaviour is as follows:

Crystal 15°C S_G 36°C S_B 52°C S_C 53°C S_A 64°C N 78°C Liquid

($S_X \equiv$ smectic X, N $\equiv$ nematic)

It is now clear that the so called smectic G and B phases in these compounds are crystalline in the sense that the molecules have long range positional 3-d order. The G phase is C-centred monoclinic while the B phase is hexagonal with both ABA and ABCA layer stackings being found at different temperatures (Leadbetter et al. 1980, Gane et al. 1981). The molecules have orientational disorder both about short and long axes, among a set of configurations consistent with the average lattice symmetry.

A variety of experiments has been carried out on fibre-aligned specimens of these phases in which the crystal $\vec{c}$ axis (approximately coincident with the long molecular axes) is aligned, but the specimen is disordered about this axis. Fig. 8 shows the scattering intensity along a path perpendicular to $\vec{c}$ measured on a fully deuteriated sample using the IN6 time of flight spectrometer with an energy resolution of ~ 50 μeV(HWHM). Strong diffuse scattering is observed under the Bragg peaks, which has been attributed to very soft shear modes of the smectic layers (Pershan et al. 1981), and a diffuse peak centred at $Q \sim 1.9$ Å^{-1} (and more pronounced for the G phase) which is associated with local orientational ordering of a herringbone nature about the long molecular axes (Levelut et al. 1981). The strong quasi elastic component of the diffuse scattering shows the correlations to be dynamic with a time scale of about 3×10^{-12}s which appears to get longer with increasing temperature into the smectic A phase.

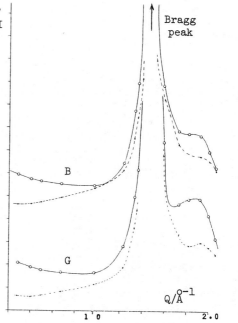

Fig. 8 Coherent scattering intensity for $Q \perp c$ in B and G phases, —O—O— elastic; —•—•— quasielastic.

Incoherent quasi-elastic scattering experiments on hydrogenous

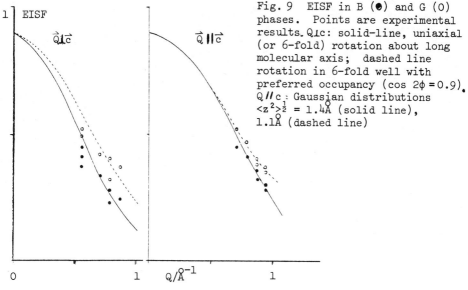

Fig. 9 EISF in B (●) and G (○) phases. Points are experimental results. $Q \perp c$: solid-line, uniaxial (or 6-fold) rotation about long molecular axis; dashed line rotation in 6-fold well with preferred occupancy ($\cos 2\phi = 0.9$). $Q \parallel c$: Gaussian distributions $\langle z^2 \rangle^{\frac{1}{2}} = 1.4\text{Å}$ (solid line), 1.1Å (dashed line)

50.7 show that single particle motions are also observed on a similar time scale, which is consistent with a rather short correlation length for the fluctuations ($\sim 10\text{Å}$). Higher resolution studies ($\Gamma \sim 10\ \mu\text{eV}$ HWHM) for which the experimental EISF data are shown in Fig. 9, reveal localised diffusive motions of two types in both phases, each having $\tau \sim 5 \times 10^{-11}$s:

i) rotational diffusion about the long axes with no preferred orientation in the B phase (the Q-range was inadequate to determine the extent of localisation into 6 equivalent orientations) but with evidence of some preferred orientation in the G phase consistent with the monoclinic symmetry: this can be described by the order parameter $\langle \cos 2\phi \rangle = 0.9$

ii) a localised diffusive motion of the molecules parallel to $\vec{c}$ adequately describable by a Gaussian distribution with $\langle z^2 \rangle^{1/2}/\text{Å} \simeq 1.4$ for B and 1.1 for G.

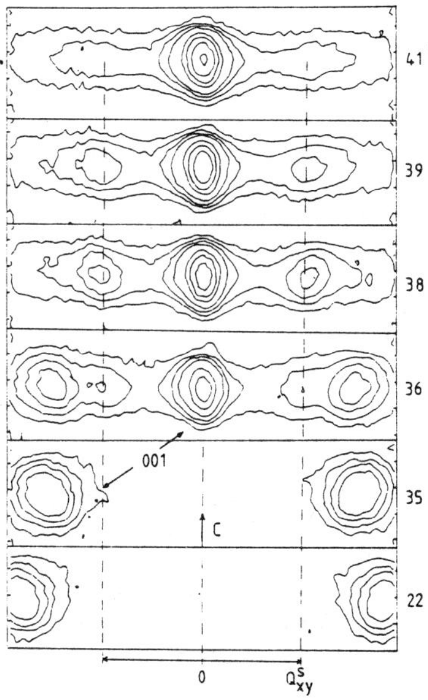

The major novel feature of the B phase is the appearance at about 6°C above the transition to the G phase of satellite reflections at $\Delta Q_z = 0$, $\Delta Q_{xy} \sim 0.08\text{Å}^{-1}$ which grow in intensity with decreasing temperature to reach a maximum at the 1st order B → G transition (Fig. 10). These show that the

Fig. 10. Neutron diffraction results for 50.7 showing the development of satellite reflections associated with the first layer reflection (001) in the B phase, and the B → G transition.

"smectic" layers have a simple sinusoidal modulation of maximum amplitude 4 Å as illustrated in Fig. 11 which also illustrates the amplitude lock-in transition to the G phase. This seems to occur when the amplitude and wavelength of the modulation (together with the rapid longitudinal displacements described above) make the rippled B-layers unstable relative to the tilted G phase ($\beta \simeq 115°$), the transition involving simply longitudinal displacement of the molecules by not more than half a molecular length. Because experiments have only been performed on fibre-aligned samples no information has been obtained on the direction of the modulation wave vector in the layer planes.

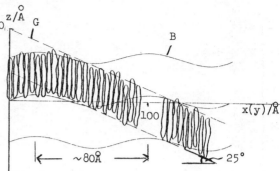

Fig.11. Modulated B structure and its relation to the tilted G structure.

Using both a high resolution triple axis spectrometer (IN12) and the spin-echo spectrometer (IN11) we have sought, but failed to find, excitations or relaxations directly associated with the modulations on a time scale of up to 10^{-8}s.

Acknowledgements

All of the experiments described here were carried out at the Institut Laue Langevin, Grenoble. In addition to the coworkers named in the references J B Hayter and W G Stirling participated in some of this work.

References

Frost J C, Leadbetter A J and Richardson R M 1980 Farad. Discuss. Chem. Soc. 69 32
Frost J C, Leadbetter A J, Richardson R M, Ward R C, Goodby J W, Gray J W and Pawley G S 1982a J. Chem. Soc. Farad. Trans. 2 78 179
Frost J C, Leadbetter A J, Ward R C and Richardson R M 1982b J. Chem. Soc. Farad. Trans. 2 78 1009
Gane P A C, Leadbetter A J, Ward R C, Richardson R M and Pannetier J. 1982 J. Chem. Soc. Farad. Trans. 2 78 995
Gane P A C, Leadbetter A J, Wrighton P G 1981 Mol. Cryst. Liq. Cryst. 66 247
Leadbetter A J, Mazid M A and Richardson R M 1980 Liquid Crystals (Heydon & Son, London, Ed. S Chandrasekhar) pp 65-79
Levelut A M, Moussa F, Doucet J, Benattar J J, Lambert M and Dorner B 1981 J. Phys. (Paris) 42 1651
Pershan P S, Aeppli G, Litster J D and Birgeneau R J 1981 Mol. Cryst. Liq. Cryst. 67 205

Neutron scattering in chemistry (Scattering from layer lattices and their intercalation compounds — an illustration)

J. W. White

Physical Chemistry Laboratory, South Parks Road, Oxford, OX1 3QZ

I was asked to speak about Neutron Scattering in Chemistry but this cannot be done in thirty minutes any more since the variety of experiments in diffraction and spectroscopy is so vast. (see Egelstaff-White rules, Egelstaff 1976). Moreover many of the areas of interest to chemists are covered in later sessions of this meeting (Polymers, Plastic & Liquid crystals, surfaces, magnetism, for example) and so I should like to highlight one of the principal aims of physicochemical experiments - namely the determination of intermolecular (and molecule surface) potentials through the study of the molecular excitations in ordered phases and the diffusive motions in thermally disordered substances. The cases of molecules adsorbed on a free surface or inside layer lattice intercalation compounds are typical of the problems and the methodology of attack and are here taken as an example.

For molecular crystals, pioneering theoretical work (Pawley 1967) and experiments (Reynolds, Kjems and White 1974; Pawley, Dorner et al 1979) has led to the suitable potentials for calculating, for example, the elastic properties of polymer crystals (Reynolds, Twisleton and White 1982). The same progress can now be expected for the adsorbed state by a combination of neutron measurements with computing as powerful as that presently being deployed in other areas of chemistry (see for example Catlow and Cormack 1982; Richards and Sackwild 1982).
Three cases are illustraive of what behaviour can be expected (and what a good potential should model) for adsorbed molecules as a function of the relative strengths of the molecule-molecule and molecule-surface parts of the potential and multilayer formation is allowed to begin. Three cases will be described here:

(a) The second stage alkali metal-graphite intercalation compounds (Pietronero and Tosatti 1981) such as $C_{24}Cs$ which adsorb hydrogen, methane and other gases (Watanabe 1974). Here the molecules in the compounds of stoichiometry $C_{24}Cs(X)_1$ are kept well apart by the alkali sublattice and we can study the evolution of the molecular dynamics with rather weak interadsorbent forces compared to the substrate-molecule interactions which cause adsorption. (Beaufils et al 1981)(Trouw and White 1981).

(b) Methane physisorbed on the basal plane of graphite, by contrast, shows the effect of relatively strong methane-methane interactions. Here rotational tunnelling spectroscopy is sensitive to the parameters of the potential.

(c) Water physisorbed on clay minerals such as vermiculite or montmorillonite where the layer thickness can be changed from one to fifty layers. This work will not be presented here in detail as there is a recent review (White 1982).

A. The statistical submonolayer - isolated molecules in a surface potential

In the second stage alkali metal-graphite intercalation compounds there is a repeating sandwich of alkali metal ions between graphite bilayers and though the detailed structures are not know for sure the best evidence for the $C_{24}Cs$ compound (Solin 1981) is that there is at least approximately threefold "coordination" of the caesium ions in the layer. Two possible structures have respectively the $2\sqrt{3} \times 2\sqrt{3}$ and the $\sqrt{7} \times \sqrt{7}$ arrangement with respect to the graphite layer lattice.

These compounds absorb gases, acting as molecular sieves. Here we are concerned with $C_{24}Cs(CH_4)$ which forms readily at 77K with a small expansion of the C-axis lattice parameter and an enthalpy of formation of some 20 kJ mol^{-1} - approaching the region of heats for chemisorption. As a first step to understanding this phenomenon we have calculated the potential energy of an isolated methane molecule in $C_{24}Cs(CH_4)_x$ where x<1 for the two favoured crystal structures. (Trouw and White 1982). Maps of this potential are shown in Figure 1 for the case where the orientation of the molecule was allowed to relax at each point in the crystal structure at which the potential was calculated. The electrostatic part of the

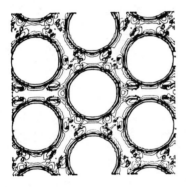

$\sqrt{7} \times \sqrt{7}$ $2\sqrt{3} \times 2\sqrt{3}$

potential was modelled using the self consistent octapole moment of methane (James & Keenan 1959), a unit charge on the Cs ions and a negative charge distribution in the layers calculated by assuming charging of only the nearest layer and L.C.A.O. theory. Finally the van der Waals part was modelled for the Cs using a known methane-krypton atom-atom potential and that for the carbon layers used the best atom-atom parameters found in our methane-graphite experiments (v. infra).

The potentials are by no means perfect yet but predict clear differences between the behaviour of methane sorbed in the two cases; the sorption sites can be seen and the enthalpy of formation is approximately correct.

With this model in mind it is of interest to see just how the methane does behave in $C_{24}Cs(CH_4)$. The adsorption isotherm is known (Watanable et al 1973) and shows a single well developed step after a threshold adsorption pressure (BET Type V). Figure 2 shows that even at low temperatures (20K) there is both a narrow elastic component and quasielastic scattering. In contrast to methane sorbed on graphite no rotational tunnelling transitions have yet been seen - presumably much lower temperatures and higher resolutions will be needed.

As the temperature is raised to 160K the width of the quasielastic scattering increases uniformly. Above 120K the signal intensity begins

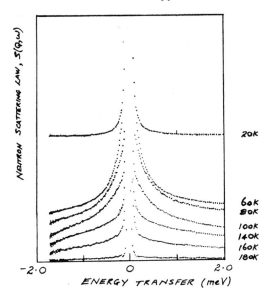

Figure 2

Quasielastic scattering from $C_{24}Cs(CH_4)$ between 20K and 180K.

to drop as CH_4 is lost from the lattice. At all temperatures the quasielastic line is well fitted by a single Lorentzian line whose width is essentially independent of momentum transfer. The line width as a function of absolute temperature follows a perfect Arrhenius law with a slope of 1.8 kJ mole^{-1}.

Thus methane in $C_{24}Cs$ sublimes out of the solid state and we must presume that the methane-methane interactions are so weak as to not allow a liquid phase to form.

B. The Effects of Lateral Interactions between adsorbed molecules

When methane is adsorbed on the clean basal plane surface of graphite the diffraction pattern immediately reveals the importance of lateral attractive force between the molecules. At less than one monolayer coverage the molecules are not dispersed in a "statistical sub monolayer" but cling together in "rafts" giving the characteristic long tailed 2D layer lattice diffraction pattern (Vora et al 1979; Bomchil et al 1980) first seen for nitrogen on Graphite by Kjems et al (1976). Further evidence for the intermolecular forces was the observation of compression of the registered $\sqrt{3} \times \sqrt{3}$ structure at coverages above c. 0.9 monolayers and melting of the layer at about 55K. How then to get a more quantitative estimate of the relative magnitudes of the lateral and the molecule-surface forces in this system?

C. Rotational tunnelling of adsorbed molecules

It is possible to calculate the effects of a hindering potential on the rotational states of methane (Smalley et al 1981) and in particular the effect of the physisorption potential for the methane graphite system. Figure 3 compares the energy level diagrams for the rotational tunnelling states so obtained with that expected for a potential of tetrahedral symmetry (as in solid methane for example). The matrix elements h and h_4 are parameters of the (trigonal) crystal field at the molecule and I denotes the nuclear spin angular momentum. All of the transitions marked on Figure 3 have been observed by high resolution incoherent inelastic scattering at energy transfers between 10 µeV and 150 µeV, with the

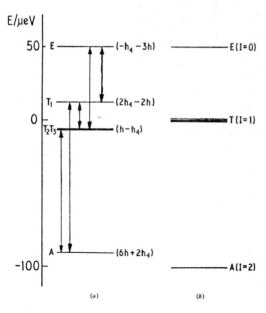

Figure 3

Rotational tunnelling levels for CH_4, (a) on the surface of graphite, and (b) in a potential of tetrahedral symmetry. The observed transitions are marked, together with the nuclear spins, and the values of the energy in terms of the parameters h and h_4.

instruments at the Institute Laue Langevin (Grenoble) and their intensities and energies compared with values calculated from a model potential which takes account of the librational and vibrational excitations also seen. From these data one can conclude that the in plane potential has C_{3v} symmetry and that the methane layer is rotationally ordered below 20K. Best fit values for a 6-exp atom-atom potential have been deduced (Bomchil et al 1980). The methane-methane interactions account for approximately one third of the adsorption enthalpy at half coverage and present a rotational hindering potential about ten times that presented by methane surface repulsive interactions (Table 1)

Comparison of Atom-Atom Potential Calculations and Experimental Measurements for the $\sqrt{3} \times \sqrt{3}$ Phase of Methane Adsorbed on Graphite

Source of empirical parameters	Binding energy of isolated molecule (kJ mol^{-1})	Methane-methane interaction energy (kJ mol^{-1}) [cm^{-1}]	Isosteric heat of adsorption at 130K (kJ mol^{-1}) [cm^{-1}]	Distance from surface (C-C) (nm)	Vibration frequency perpendicular to surface (cm^{-1})	Barrier to rotation about perpendicular axis (cm^{-1})			Barrier to rotation about C-H bond (cm^{-1})
						Methane-Surface only	Methane-Methane	Total	
Kitaigorodskii[a]	17.6(17.3)	6.6[505]	20.3[1550]	0.325(0.327)	102	57(2)	582	639(580)	614(367)
Taddei(A)[b]	17.5(17.4)	5.2[400]	18.9[1450]	0.333(0.333)	98	40(2)	359	399(357)	447(288)
Taddei(B)[b]	15.7(15.8)	3.2[270]	15.4[1180]	0.335(0.335)	93	26(1)	143	169(142)	291(196)
Williams 4[c]	16.6(16.8)	3.8[290]	16.6[1270]	0.332(0.331)	86	20(1)	261	281(260)	207(173)
Williams[d]	14.9(15.0)	3.5[270]	14.5[1110]	0.333(0.332)	93	19(1)	207	226(206)	208(173)
Experiment	-	-	11.2[e]	0.330±0.005	100	-	-	100-150	100-150

[a]Kitaigorodskii (1973), [b]Taddei et al. (1977), [c]Williams (1967), [d]Williams (1970), [e]Kiselev & Poshkus (1976). The barrier to rotation about C-H bond refers to one of the three equivalent C-H bonds pointing towards the surface. Values in brackets are calculated for methane over the centre of a carbon hexagon, the others are for methane directly over a carbon atom. The isosteric heat is obtained from the binding energy by adding 7/2 RT (three rotational, two translational, and one vibrational degree of freedom). The calculated vibration frequencies are those appropriate to a Morse potential.

As concerns the dynamical coupling between adsorbed molecules and the surface the temperature dependence of the tunnelling spectra of both methane on graphite as well as hydrogen on $C_{24}Cs$ (Beaufils et al 1981) are very weak - the tunnelling states persisting up to about 120K. These cases are the reverse of CH_4 in $C_{24}Cs(CH_4)$ and as yet these phenomena are unexplained though the effects may not be unconnected to the low density of states of both graphite and the intercalate basal plane phonons in the frequency range of the (narrow) tunnelling peaks in the first two cases.

D. At what thickness does a multilayer become a liquid?

This question has been most thoroughly addressed in respect of the dynamics of thin layers of water at the silicate surfaces of clay minerals such as montmorillonite. These surfaces are composed of planes of oxygen atoms - parts of the boundary layer SiO_4 tetrahedra and had been supposed to induce "ice like" structure possibly out to many hundrends of angstroms for water and other liquids adsorbed upon them.

That this is not so - at least as far as the water diffusion coefficient is concerned is shown by Figure 4.

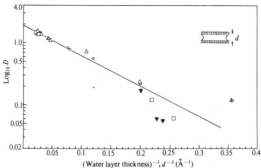

Fig. 4. Logarithm of self diffusion constont for water (*D*) in (△) sodium montmorillonite, (○) lithium montmorillonite, (□) lithium vermiculite and (▼) sodium vermiculite as a function of inverse water layer thickness. The full line shows a theoretical curve derived from the Kelvin equation using a model for the clay water film shown by the inset.

Here the logarithm of the water diffusion coefficient has been plotted as a function of the inverse water layer thicknesses (Olejnik and White 1972). The diffusion coefficient was determined from the low momentum transfer slopes of the quasielastic neutron scattering line widths plotted as a function of squared momentum transfer.

These early measurements have now been extended to thinner layers (Cebula, Thomas and White 1981) and the anisotropy of the water diffusion studied. The measurements have been extended to bivalent counter ion systems (Hall and Ross 1978) but the "story" is the same. The water in the first one or two layers is strongly coordinated to the counter ions but at larger thicknesses it develops its macroscopic diffusion coefficient in a range characterised by a distance of about 15 Å. A most surprising feature - not yet explained is the low anisotropy of the water diffusion tensor in these systems. This may be connected to the great flexibility of the 10 Å thick aluminosilicate sheets and relates to recent theories of diffusion in two phase systems (Weissman 1980, Pusey 1981).

E. A future direction

There is much to do to apply the above approaches systematically to a variety of surfaces and adsorbates in order to build up a library of useful potentials for physisorption. But it is already clear that neutron methods lend themselves to the study of the solid-liquid interface and its adsorption properties. (Cebula et al. 1978; Harris et al 1982).

Small angle neutron scattering lends itself to studies of the structure, dimensions and possible phase transitions in the adsorbed monolayer through the beneficial effects of contrast matching. By combining these studies with neutron spin echo measurements (Hayter et al 1982) it should be possible to study the dynamics of the adsorbed monolayer films.

Conclusions

This brief review is meant to have illustrated one aspect of the use of neutron scattering in chemistry - giving only, in summary, the physics and methodology of the experiments. More detail will be found in the references and recent reviews (e.g. Thomas 1982).

References

Beaufils J P, Crowley T, Rayment T, Thomas R K and White J W, 1981 Mol. Physics, 44 1257.
Bomchil G, Huller A, Rayment T, Roser S J, Smalley M V, Thomas R K and White J W, 1980 Phil. Trans. Roy.Soc. (Lond) B290, 537
Catlow C R A and Cormack A N, 1982 Chem.in Britain, September 1982, p627.
Cebula D J, Thomas R K and White J W, 1981 Clays and Clay Minerals 29, 241
Cebula D, Thomas R K, Harris N M, Tabony J and White J W, 1978, Faraday Disc. Chem. Soc., 65, 76.
Egelstaff P A, 1976 in Neutron Scattering for the Analysis of Biological Structures, BNL 50463 1976, p. I-26.
Englert G and Saupe A 1963 Phys. Rev. Lett. 11, 462.
Hall D and Ross D K 1972, Procedings of the Vienna Conference on Neutron Scattering I.A.E.A.
Harris N M, Ottewill R H and White J W, 1982 in "Adsorption from Solution" R.H. Ottewill,(Ed), Academic Press 1982).
Hayter J B and Penfold J, 1981, Mol. Physics, 109.
James H M and Keenan T A 1959, J.Chem.Phys. 31, 12.
Kitiagorodskii A E 1973, Molecular Crystals, New York Academic Press.
Kjems J K, Passell L, Taub H, Dash J G and Novaco A D, 1976, Phys. Rev. B13, 1446.
Olejnik S and White J W, 1972, Nature (Phys. Sci) 236, 15.
Pawley G S, Mackenzie G A, Dorner B, Kalus J, Natkaniec I, Schmelzer U, 1979, Mol. Phys. 39, 251.
Pietronero L and Tosatti E, 1981, "The Physics of Intercalation Compounds" Springer 1981
Pusey P N and Tough R J A, 1981 "Dynamic Light Scattering" Ed. R. Pecora, Plenum.
Reynolds P A, Kjems J K and White J W, 1974, J. Chem. Phys., 60, 824.
Reynolds P A, Twisleton J F and White J W, 1982, Polymer 23, 578.
Richards G and Sackwild V, 1982 Chemistry in Britain, September 1982 p. 635.
Solin S A, 1982, Advances in Chemical Physics, 49, p. 455, Wiley and Sons.
Taddei G, Righini R, Manzetti P, 1977, Acta Crystalog. A.33, 626.
Thomas R K 1982, Prog. Solid State Chem., 14, 1.
Trouw F and White J W, 1981. Institut Laue Langevin Annual Report 1982, p. 359.
Trouw F and White J W, 1983, Molecular Physics. To be published.
Vora P, Sinha S K and Crawford R K, 1979, Phys. Rev. Lett. 43, 704.
Watanabe K, Kondow T, Soma M, Onishi T and Tamaru K, 1973. Proc. Roy. Soc. A.333, 51.
Weissman, M B, 1980, J. Chem. Phys., 72, 231.
White J W, 1982. Proceedings of NATO Advanced Study Institute on Mobility at Interfaces, Heyden, 1982.
Williams D E, 1967, J. Chem Phys., 47, 4680.
Williams D E, 1970, Am. Crystallog. Ass., 6. 21.

Diffraction measurements on physisorbed films

M. Nielsen, J. Bohr, K. Kjaer, Risø National Laboratory, DK-4000
Roskilde, Denmark, and
J.P. McTague' NSLS Brookhaven National Laboratory, Upton, L.I.,
N.Y. 11973, U.S.A.

Abstract. Neutron and X-ray diffraction have been applied to the study of two dimensioned lattice structures of films of rare gases or simple molecular gases which are physisorbed on graphite substrates. The main goal is to measure the properties of the films at their two dimensional phase transitions and much progress has been done recently towards this. Typical examples will be discussed.

1. Introduction

Graphite forms, due to the very strong bonds between the carbon atoms in the hexagonal (002) planes, almost ideally flat surfaces of large extensions. Therefore, as substrate, graphite is unique; films of monoatomic physisorbed layers of the simple gases form at low temperatures two dimensional (2-d) crystals which are undisturbed by defects over large areas.

The first neutron scattering measurement on these systems was performed by Kjems et al. (1976) on N_2 adsorbed on Grafoil, a graphite product from Union Carbide. They observed both a commensurate (see below), an incommensurate and a fluid phase. This was followed by similar neutron studies on Ar, Kr, D_2, He, Ne, O_2 and several molecular gases (Taub et al. (1977), Thorel et al. (1976), Nielsen et al. (1975), (1980) and Lauter et al. (1980). Inelastic measurements of the phonon spectra have been done on Ar and D_2 films (Taub, Nielsen) and magnetic ordering was formed in O_2 layers (Nielsen et al. (1979)). Recently ordering of molecular orientations have been observed in several systems, the simplest is N_2 films (Eckert et al. (1979)). Of particular interest have been the neutron studies on superfluid helium films, but we shall not discuss them in this article; see Carneiro et al. (1976), Lauter et al. (1980).

Experimentally the most significant recent development has been the application of synchrotron x-ray diffraction and the use of the high quality substrate UCAR-ZYX. When the first synchrotron diffraction study in this field was done by Birgeneau et al. (1980) they found using the ZYX substrate undistorted film crystal grains larger than 2000 Å; almost an order of magnitude larger than seen before. In fact, with synchrotron x-ray diffraction, momentum resolutions corresponding to coherent scattering across more than 10000 Å is readily achieved, so, at present we are severely limited by substrate quality even when using the ZYX substrate.

Another severe limitation is due to the powder nature of the substrates which makes impossible the study of angular epithaxy. Such studies are done with the LEED technique, but so far these measurements have only been done with momentum resolution corresponding to a coherence length of 1-200 Å.

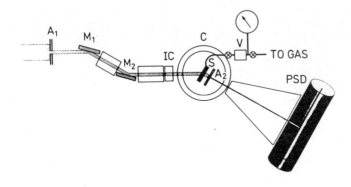

Fig. 1. Schematic top view of x-ray spectrometer installed in HASY laboratory, Hamburg.
A1, A2: slits
M1, M2: Ge(111) crystals
C: cryostat
S: sample cell
IC: beam monitor
PSD: position sensitive detector

2. Technique

In the neutron diffraction measurements on films conventional triple axis spectrometers have in general been used. Similarly, crystal spectrometers are used in the x-ray studies and Fig. 1 shows schematically a spectrometer installed at the storage ring Doris in HASY Laboratory in Hamburg. Two perfect Ge (111) crystals are used as monochromaters. Due to substrate scattering it is essential that higher order contamination of the incoming beam is suppressed and this is, for the second order, achieved by using the (111) reflection and, for the third order, by missetting one of the Ge crystals by about the Darwin-width of the (111) reflection. The analyser consists of a position sensitive detector (PSD). With this setup the instrumental resolution is predominantly determined by the width of the sample slit as seen from PSD. In general a 1 mm slit is mounted as close to the sample as possible, and the detector arm is typical 600 mm. This gives a resolution corresponding to a coherence length of 1200 Å.

3. Experimental Results

As examples of diffraction results we will in the following discuss the phase diagrams of the rare gas films of Ar, Kr and Xe, then show data for the melting of "free floating solids" and for a commensurate incommensurate (C-I) transition.

3.1 Phase Diagrams of Ar, Kr and Xe

Fig. 2 shows schematically the honeycombe graphite surface and two examples of an adsorbed structure. In the upper part the particles are in registry with the substrate lattice: one is located above the center of every third hexagon giving the $\sqrt{3} \times \sqrt{3}$, $30°$ commensurate structure. The lower part shows a denser monolayer with no registry with the graphite lattice; it is a triangular incommensurate structure. A commensurate 2-d

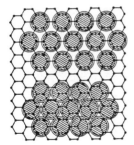

Fig. 2.
The hexagons illustrate the surface structure of the graphite substrates. The circles show adsorbed atoms and in the upper structure they are located above every third hexagon center and form a √3 x √3 commensurate phase. The lower structure illustrates an incommensurate "free floating" phase.

solid is translationally locked to the substrate surface, whereas the incommensurate solid is not and it is characterized as "free floating".

In Fig. 3 we show phase diagrams for Ar, Kr and Xe films. Ordinates are the 2-d layer densities in units of density of the commensurate √3 structure. The temperatures are normalized by the respective 3-d triple point temperature. In three dimensions the corresponding phase diagrams are almost identical to each other. For the 2-d films they are quite different. Kr-films form the √3 x √3, 30° commensurate structure and the substrate field stabilizes this solid such that it melts at higher temperature than Ar and Xe films. Ar atoms are too small by about 8% to fit ideally the √3 structure and Xe atoms are too large by about the same amount. They both are incommensurate at submonolayer coverages and have neighbour distances close to those of their 3-d solids. Yet they also have different melting behaviour. Xe-films have like 3-d systems a first order melting transition including a triple point line but at high layer densities the melting becomes continous as indicated by the broken line in the figure. Ar-films have continous melting even at submonolayer coverages and the triple point

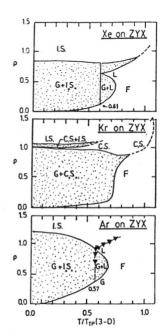

Fig. 3.
The phase diagrams for Xe, Kr and Ar monolayers on graphite. The symbols are:
ρ: averaged layer density in the units of the density of the √3 x √3, 30° commensurate structure.
G: **2-D gas**
F: 2-D fluid
L: 2-D liquid
I.S.:   Incommensurate solid
C.S.:   2-D commensurate solid (√3 x √3, 30°) and T/T_{Tp}(3-D): the temperature in units of the triple point temperatures of the respective 3-D solids.

is replaced by a critical endpoint. The nature of the continous melting has been the subject of great theoretical (Halperin et al. (1978)) and experimental interest (Heiney et al. (1982), McTague et al. (1982)), and we show examples of experimental data below.

3.2 Melting of a "Free Floating Solid"

True long range order does not exist in free floating 2-d solids and the ideal diffraction response from such films is not δ-functions at the Bragg positions but power law functions around the reciprocal lattice points. However, the "quasi-long-range-order" persists at low temperatures over such long distances that the power law nature is very difficult to observe. For the melting of this 2-d solid there exists (in contrast to what is the case for 3-d solids) theoretical predictions including the structure factors according to which these should change in a continous way from power laws to Lorentzian-like functions (see Halperin et al. (1978)). This behaviour was observed by Heiney et al. (1982)in a synchrotron x-ray diffraction measurement on dense Xe-films on ZYX substrate. Here we show in Fig. 4 results for Ar films obtained by Moncton et al. (1982) using the same technique. The diffraction groups are measured around the (10) reflection of the triangular incommensurate structure and the coverage was such that the solid coexists with a dilute gas during the melting to a dense fluid. The groups in Fig. 4 demonstrate that the structure factor develops in a continous fashion from a narrow group with a resolution limited width of the law Q side and into a broad group where the shape is determined by the intrinsic (Lorentzian) scattering function. The analysis of the data is not yet completed.

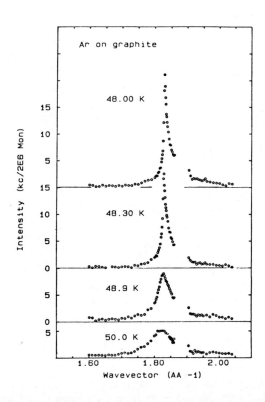

Fig. 4.
Experimental diffraction groups measured around the (10) reflection of the triangular incommensurate structure of Ar monolayers on graphite. The coverage is in the solid + gas coexistence region (see Fig. 3) and the melting temperature is 47.9 K. The data are measured by D. Moncton et al. at SSRL, Stanford.

3.3 Commensurate-Incommensurate Transitions

As for melting of 2-d films theoretical predictions exist for the commensurate - incommensurat (C-I) transition inducing the expected diffraction response. Here the transition is driven by the formation of domains and different behaviour is expected dependent on the domain wall configuration (Bak (1982)). It should be noted that this applies only in the region close to commensuration and it may be preempted by a strong first order transition. In Fig. 5 is shown an example of diffraxtion data for a continous C-I transition observed in CF_4 monolayers on the ZYX substrate also by synchrotron x-ray diffraction (Kjaer et al. (1982)). The result is interpreted to imply an uniaxial compression driving the commensurate (here a 2x2 structure) incommensurate in one direction only. The hexagons shown to the left in the figure illustrate the uniaxial deformation of the reciprocal lattice and that the powder diffraction response becomes a doublet with intensity ratio 1:2. The full curves through the measured diffraction groups are calculated using this model and show detailed agreement with the data points.

The C-I ransition has been studied in most details for Kr-films (Chinn et al. (1977)), (Stephens et al. (1979), Nielsen et al. (1981)). An important characteristic observed in this system at moderate and high temperatures is that the weakly incommensurate phase is a fluid phase, although with an unusual long correlation length. The most dramatic demonstration of this fluid character has however been observed in H_2 and He monolayers. They also form the commensurate $\sqrt{3} \times \sqrt{3}$, $30°$ structure and in the weakly incommensurate phase at a layer density slightly above commensuration their melting temperature drops by a factor of 4 or more (Nielsen et al. (1980)).

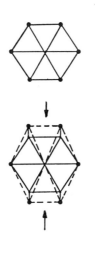
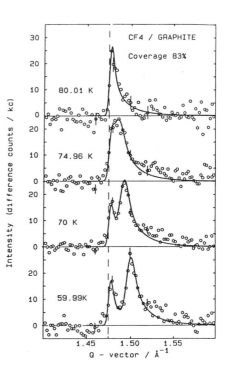

Fig. 5. Diffraction profiles from the commensurate-incommensurate transition of CF_4. The figures to the left show how the triangular reciprocal lattice of the (2x2) phase is transformed by an uniaxial compression in real space. The measured groups show the corresponding splitting of the powder diffraxtion pattern.

4. Conclusion

The application of synchrotron X-ray diffraxtion to the study of physisorbed monolayers on graphite has opened possibilities for studying the 2-d behaviour of these films in great details and many new systems will undoubtedly be studied in the near future.

Very recently Bak et al. (1982) have studied theoretically the uniaxial incommensurate structure of CF_4 and predicted that also in this case a fluid phase, now only melted in one direction, should appear at sufficiently small incommensuration.

References

P. Bak (1982), Rep. Progress Phys., June.
P. Bak and T. Bohr, 1982, to be published in Phys. Rev.
R.J. Birgeneau, E.M. Hammond, P. Heiney, and P.W. Stephens (1980), in Ordering in Two Dimensions, ed. S.K. Sinha (Elsevier North Holland, 1980) p. 29.
S.C. Fain, Jr., M.D. Chinn, and R.D. Diehl (1980), Phys. Rev. B 21, 4170.
B.I. Halperin and D.R. Nelson (1978), Phys. Rev. 41, 121
P.A. Heiney, R.J. Birgeneau, G.S. Brown, P.M. Horn, D.E. Moncton, and P.W. Stephens (1982), Phys. Rev.Lett. 48 104.
J. Kjems, L. Passell, H. Taub, J.G. Dash, and A.D. Novaco (1976), Phys. Rev. B 13 1446.
H.J. Lauter, H. Wiechert, and R. Feile (1980) in Ordering in Two Dimensions, ed. S.K. Sinha (Elsevier North Holland, 1980) p. 291.
J.P. McTague, J. Als-Nielsen, J. Bohr, and M. Nielsen (1982) to be published in Phys. Rev.
D. Moncton, J.P. McTague, J. Als-Nielsen, and M. Nielsen (1982), to be published.
M. Nielsen, J.P. McTague, and L. Passell (1980). Proceedings of Nato Advanced Study Institute on Phase Transitions in Surface Films, held in Erice 1979, Plenum Press 1980, p. 127.
H. Taub, K. Carneiro, J.K. Kjems, L. Passell, and J.P. McTague (1977), Phys. Rev. B 16 4551.
P. Thorel, B. Croset, C. Marti, and J.P. Colomb (1976), Proceedings of Conference on Nuetron Scattering, Gatlinburg, Tenn., ed. R.M. Moon, p. 85 (Oak Ridge Nat. Lab. Report No. Conf.-760601) (unpublished).

Molecular reorientation and tunnelling motions

S Clough
Department of Physics, University of Nottingham, U.K.

Abstract

The quantum tunnelling rotational motion of small symmetric molecular groups like CH_4, CH_3, NH_4, is revealed in solids at low temperatures by inelastic peaks in the high resolution INS spectrum. Detailed information has been obtained on the symmetry and magnitude of rotational potentials in many different molecular environments. The temperature dependence of the INS tunnel spectrum has a wider significance. The rotational motion is analagous to that of a particle moving in a periodic potential and coupled to lattice phonons. It provides new insights into this important problem.

1. Introduction

High resolution inelastic neutron scattering has proved a very important tool in the study of the coherent quantum mechanical tunnelling rotational motions of small molecular groups like H_2, CH_3, CH_4 and NH_4. The motion of these systems has two different kinds of interest. Since they can occur in different environments e.g. in molecular crystals, in clathrates on surfaces etc, the tunnelling motion is informative about the local crystal potential. This is a topic which can be called tunnelling spectroscopy and its objectives are essentially structural. The second kind of interest is rather more fundamental. It relates to the general study of the motion of a particle moving in a periodic potential and simultaneously coupled to lattice phonons - a common and difficult problem in solid state physics. The rotational motion of small symmetrical groups turns out to be a close analogue of this problem, and uniquely able to provide answers to many important general questions. This study may be described as rotational tunnelling dynamics and its objectives are to provide a deeper understanding of the phenomenon crudely described as 'hopping' in classical models and to relate measured hopping rates to the crystal potential.

The key role of neutron scattering in this field arises from (a) a wavelength comparable with the size of the molecular groups and (b) the spin dependent interaction between the neutron and the hydrogen nuclei. The second is of crucial importance because the neutron scattering transitions which reveal the molecular tunnelling are necessarily ones in which a change in the nuclear spin state occurs. This fact makes the transitions less accessible to other forms of spectroscopy, though nuclear magnetic resonance relaxation spectroscopy, which exploits inter-nuclear magnetic dipole-dipole interactions as the necessary short range spin-dependent interaction, is also of importance in this field. The combination of both techniques has been very fruitful.

2. Spin Symmetry Species of Rotating Symmetric Groups

The necessity for a spin-dependent interaction to reveal the tunnelling motion is due to the effect of the exclusion principle on a system whose rotations interchange particles distinguishable only by their nuclear spins. The methyl group provides a simple example. In the absence of a hindering potential, it is a simple one-dimensional free rotor with wavefunctions $(2\pi)^{-1/2}\exp(im\phi)$ with eigenvalues $\hbar^2 m^2/2I$ where I is the moment of inertia. Since the addition of $2\pi/3$ to ϕ (the rotational coordinate in $\exp(im\phi)$) has the effect of cyclically permuting the three hydrogen atoms, we can identify three species of wavefunction, namely

A(m=o, ±3, ±6 etc), E^a (m=7, -4,-1,2,5 etc), E^b (m=7,4,1,-2,-5 etc)

which under cyclic permutation are multipled by 1, ε and ε^* respectively where $\varepsilon = \exp(2\pi i/3)$. The nuclear spin function which multiplies a free rotor wavefunction must have the property that cyclic permutation multiplies it by 1, $\varepsilon^*, \varepsilon$ respectively so that the total function can be invariant under cyclic permutation of both space and spin functions, as required by the exclusion principle. (It is unnecessary in this case to antisymmetrise for single pair exchanges since the practical consequences of this step are not detectable). The A quartet of spin functions are $\alpha\alpha\alpha, (1/3)^{1/2}(\alpha\alpha\beta + \alpha\beta\alpha + \beta\alpha\alpha)$ and a pair obtained from these by interchanging α and β. The E^a doublet is similarly obtained from $(1/3)^{1/2}(\alpha\alpha\beta + \varepsilon\alpha\beta\alpha + \varepsilon^*\beta\alpha\alpha)$ and the E^b doublet is the complex conjugate of E^a. These spin functions are also eigenfunctions of total nuclear spin with quantum numbers 3/2 (A species) and 1/2 (E).

When a hindering potential is present, this introduces terms in the Hamiltonian of the form

$$\frac{1}{2}\sum_{n=1}^{\infty} V_{3n} \cos(3n\phi) + \sum_{n=2}^{\infty} W_{3n} \sin(3n\phi)$$

where the 3-fold symmetry is due to the structure of the group. In a symmetrical environment some terms in this series are zero, but in almost all cases studied, it turns out that the environment is not symmetrical and the potential is strongly dominated by the leading term $(V_3/2)\cos 3\phi$. This means that measurements in different materials can be compared very easily and the differences attributed simply to different barrier heights V_3.

The energy levels for the CH_3 group as a function of V_3 converge with increasing V_3 towards three-fold degeneracy as the free rotor states are changed towards torsional oscillator states localized in the wells of the potential. The residual splitting is then the tunnelling splitting and is related to the rate at which a wavefunction localized in a single well leaks into the wells on either side. The tunnel splitting measured at low temperatures is that characteristic of the ground torsional state.

For CH_3 the Hamiltonian $\mathcal{H} = (-\hbar^2/2I)\partial^2/\partial^2\phi + (V_3/2)\cos(3\phi)$ is clearly analagous to that of a particle moving in a one dimensional periodic potential. The cyclic boundary conditions appropriate to rotational motion mean that instead of energy bands, only a few discrete levels occur, with the concept of tunnel splitting replacing that of bandwith. The motion of a tetrahedral molecule (NH_4 or CH_4) may be modelled in a similar way. For example, if two of the protons have spin state α and two β, then there are six distinguishable orientations of the tetrahedron. This can be represented by a particle moving in a four-dimensional space between potential wells located at (11-1-1), (1-1-1 1), (1-11-1), (-11-11), (-1-1 11), (-111-1). From any well, four of the remaining five are equidistant and represent orientations of the tetrahedron accessible from the first by rotations about the four three-fold axes. The fifth is more remote, and represents the orientation accessible only by a single rotation about a two-fold axis. One may think in terms of vibrational states of a particle localized in each of the wells (the pocket states) and the splitting of a six-fold degenerate ground vibrational state due to tunnelling between the wells. The six pocket ground states combine into Bloch wave functions A + 3T + 2E with tunnelling energy splittings dependent on the potential barriers between the wells. If we restrict attention to the ground state, it is a simple matter to write an effective tunnelling Hamiltonian

$$\mathcal{H} = \sum_{i=1}^{6} \sum_{j<i} a_{ij} T_{ij} \quad (1)$$

where T_{ij} are operators which interconvert pocket states i and j, $<\psi_i|T_{ij}|\psi_j> = 1$, and a_{ij} depends on the overlap of the two pocket state wavefunctions and the potential barrier between the potential wells. Of the fifteen a_{ij} only seven are independent, corresponding to rotations about the four three-fold and three-two fold axes. Then (1) relates the observable energy splittings between A, T and E levels to the tunnelling parameters a_{ij} which in turn may be deduced from the potential.

Nuclear spin may be introduced into this description by using a spin Hamiltonian $\mathcal{H}_S$ operating, not on six pocket functions, but on a basis of sixteen (2^4) nuclear spin functions, each a product of four terms α, or β.

$$\mathcal{H} = \sum_{i=1}^{7} a_i (R_i + R_i^{-1}) \quad (2)$$

The operators R_i permute the spin functions in the same way that the corners of a tetrahedron are permuted by rotations, and the seven parameters a_i are simply related to the seven independent a_{ij} of (1). In a symmetric environment, rotation about all 3-fold axes is equally hindered and the same is true for 2-fold rotation, so (2) contains only two different parameters. The three T species are then degenerate, though in a lower symmetry field this degeneracy is lifted.

The spin parts of the matrix elements for neutron scattering transitions can be obtained from the eigenfunctions of (2) and the spatial parts are combinations of functions like $\exp(i\underline{q}.\underline{r}_{ij})$ where $\underline{r}_{ij}$ is a side of the tetrahedron. Thus tunnel spectra can be computed from a rotational potential for comparison with experiment.

3. Torsional Spectroscopy

The tunnel peaks in the INS spectrum occur below about 300μeV, a spectral region otherwise empty. Transitions to excited torsional states occur at a few meV where they may be confused with other modes. Two features have been used to identify inelastic peaks due to torsion of small symmetrical groups: (a) they may be split by tunnelling in the excited state, (b) they show an unusual degree of broadening with increasing temperature. The combination of the data from the tunnelling and

torsion regions of the spectrum is capable of yielding considerable
detail on the shape of the potential.

A recent review of tunnelling spectroscopy by Press (1981) presents a
comprehensive range of examples of applications. Probably the most
interesting spectroscopic information relates to CH_4 and CD_4 in their
several phases, CH_4 in rare-gas matrices and CH_4 adsorbed on grafoil.
Hindering potentials have been found for CH_3 and NH_4 in many molecular
solids. In the case of CH_3, two or three different potentials can sometimes be observed, corresponding to different chemical or crystallographic
environments. A number of pressure dependence studies of tunnelling have
also been carried out. The reader is referred to the review by Press for
an extensive survey of the literature.

4. Molecular Rotational Dynamics

While it may be claimed that low temperature tunnelling spectroscopy is
thoroughly understood and uncontroversial, the same cannot be said of its
temperature dependence. Because of its structural and dynamical
simplicity, the methyl group has been the focus of most attention. The
main experimental features are clear. They are (1) a progressive
reduction of the observed tunnel splittings with increasing temperature
and (2) an associated progressive broadening of all the peaks, inelastic
and quasi-elastic, until the structure merges into a single broad peak
which continues to broaden as the temperature rises. In this high
temperature regime the width can be identified with a hopping rate in the
classical model of the molecular motion. This classical regime has
features common to many hopping problems. What makes molecular rotational
motion specially interesting is the appearance of the well understood
structure at low temperatures. This provokes the question whether a
dynamical theory of the temperature dependence, taking account of the
lattice vibrations, can be built from the simple time-independent
Hamiltonians which describe the low temperature spectroscopy. This may
then be expected to lead to a deeper understanding of the concept of
hopping.

There have been two very different approaches to this problem. The first
(Haupt 1971) has been to attempt to describe the interaction of the
group with the lattice phonons through perturbation theory. In this case
a hop from one potential well to another is viewed as occurring through
the absorption and subsequent emission of phonons, with the hop occurring
during the excitation. This gives a hopping rate which depends on the
density of phonons resonant with the torsion frequencies and on their
degree of excitation. This approach leads to rather complicated
expressions for the hopping rate which depend very much on the phonon
spectrum of the lattice. The reduction of tunnel splitting occurs on
this model due to the coupling of molecular torsion and phonon modes,
with a consequent reduction in the overlap of wavefunctions localized in
different wells of the hindering potential with the degree of excitation
of the coupled phonons (Huller 1980). The broadening of the motional
spectrum associated with this temperature-dependent reduction in tunnel
frequency has been calculated by Hewson (1982).

The second approach to the hopping problems is in strong contrast. It
takes the view that the phonons may be regarded simply as a thermal
reservoir. By modulating the amplitude or phase of the hindering
potential, this has the effect of maintaining the torsional motion in

an agitated state which, on average, is a thermal mixture of eigenstates of the time independent Hamiltonian. The hopping rate is simply calculated from a thermal average of the angular velocity (Clough et al 1982a) or even more simply, as a thermal average of the tunnel frequencies of all the torsional states (Stejskal and Gutowsky 1958). This leads to the conclusion that the hopping rate is a unique function (and a relatively simple one) of only two parameters, the potential barrier height V_3 and the lattice temperature. Essentially the methyl group, through its hopping rate, behaves simply as a thermometer, with a calibration curve determined by V_3. The phonon spectrum plays no role, because the equilibration between phonons and molecular torsion is much faster than the hopping rate. This approach therefore, claims to take account of those many phonon scattering processes which do not cause hops as well as those which do, while the perturbation approach focusses on those phonon scattering processes which do cause hops.

Methyl group motion offers the unique possibility of testing this second model very thoroughly since the potential barrier height and shape can be inferred from the tunnelling and torsion splittings observed by neutron scattering at low temperatures and so a detailed prediction can then be made of the hopping rate at high temperatures. The hopping rate can then be measured as the width of the broadened INS spectrum, or by nmr from the proton spin lattice relaxation times. This has been done now for a large number of methyl containing samples (Clough et al, 1981b) and the experimental values agree well with the predictions (which do not involve adjustable parameters) in all cases. There is no evidence of a strong dependence on the phonon spectrum. The model works equally well for the methyl pyridines which are liquids at room temperature and for the metal acetates which have high melting (or decomposition) temperatures. This evidence seems to point strongly in favour of the simple thermal average treatment. It also indicates that the classical concept of hopping may be misleading.

The model suggests that the dominant process over a wide temperature range is tunnelling between excited states, and that only at high temperatures does transport by levels above the potential barrier become most important. An intriguing feature of tunnelling is that a function localized in one potential well leaks symmetrically into the two wells on either side, and so does not have a direction, clockwise or anti-clockwise, such as the hopping model assumes.

The concept of a thermal average was invoked by Allen (1974) to account for the observed reduction in CH_3 tunnel splittings with increasing temperature. It arises simply because the tunnel splittings alternate in size, and increase in magnitude in each successive torsional state. Again no adjustable parameters are involved and the model usually works very successfully. Another satisfactory experimental feature is the very similar values which are normally observed for the temperature dependent shift and width of CH_3 inelastic tunnelling peaks (Clough et al 1981a), since this emerges naturally from the thermal averaging procedure. A key fact which is not fully accounted for, is the rough equality of the widths of the inelastic and quasi-elastic peaks. The former arise due to transitions between A and E species and the latter to transitions between the two E species. If the width is due to a thermal average of the tunnel splittings as suggested above, then this accounts for the widths of the inelastic peaks, but as there is no tunnel splitting between the E species, it does not explain the quasi-elastic width. If on the other hand the width is mainly a result of fluctuations due to excitations between different torsional levels, then equality of the two widths is expected (Clough 1981). It is then not clear though why the width should correlate so well with a thermal average of the tunnel splittings. In the hope of answering puzzling questions of this kind, computer simulations of the evolution of the motional state of a CH_3 group in a fluctuating potential are being carried out by Dr A J Horsewill and the writer.

The widths of CH_3 and NH_4 torsional peaks in the INS spectrum have not been extensively investigated, though they are also a means of testing theoretical models of the motion. The broadening of torsion peaks with increasing temperature is a measure of the increasing rate of thermalization between torsional oscillations and lattice vibrations. The thermal averaging procedure is justified on the basis that this process is very fast, so it is satisfactory that the width of torsion peaks is always much greater than the width and shift of related tunnel peaks. Like the widths of the tunnel peaks, torsional broadening is strongly correlated with the barrier height V_3. There has so far been little theoretical discussion of torsional broadening.

5. Spin symmetry conversion and level crossing spectroscopy

At low temperatures, transitions between spin symmetry species occur very slowly, and it is possible to establish non equilibrium distributions by rapid temperature change. The approach to equilibrium may then be followed using INS or nuclear magnetic resonance. Interesting phenomena connected with resonant symmetry conversion occur when an applied magnetic field makes the nuclear Zeeman splitting equal to a molecular tunnel splitting. A second type of resonant conversion process uses paramagnetic impurities, with the electronic Zeeman splitting equal to the molecular tunnel splitting (Glattli et al 1972). An associated dynamic polarization of the nuclei (Clough et al 1982b) serves as a clear signature of this process. Since the conversion only occurs at impurity sites, a concentration gradient of spin symmetry species is set up and the diffusion of tunnelling energy can be studied (Clough et al 1981c) From INS spectra, the conditions for successful nmr level crossing experiments can be predicted and then the nmr experiments provide high resolution for study of widths of tunnel peaks and the mechanisms of resonant spin symmetry conversion.

References

Allen P S 1974 J Phys C: Solid State Phys 7 L22
Clough S 1981 J Phys C: 14 1009
Clough S, Heidemann A and Paley M N J 1981a J Phys C: 14 1001
Clough S, Heidemann A, Horsewill A J, Lewis J D and Paley M N J 1981b
    J Phys C: 14 L525, 1982a J Phys C: 15 2495
Clough S, Horsewill A J and Paley M N J 1981c Phys Rev Lett 46 71, 1982b
    J Phys C: 15 3803
Glattli H, Sentz A and Eisenkremer M 1972 Phys Rev Lett 28 871
Haupt J 1971, Z Naturf a 26 1578
Hewson A C 1982 J Phys C: 15 3841, 3855
Huller A 1980 Z Phys B 36 215
Press W 1981 'Single-particle rotations in molecular crystals', Springer
    Tracts in Modern Physics Vol 92. Springer-Verlag.
Stejskal E O and Gutowsky H S 1958 J Chem Phys 28 388

Neutron small angle and diffuse scattering

W. Schmatz

Kernforschungszentrum Karlsruhe, Institut für Angewandte Kernphysik,
D-7500 Karlsruhe, P.O.B. 3640, Federal Republic of Germany

Abstract. The history of neutron small angle and diffuse scattering is described with emphasis on the principal ideas, the technical development and first applications in different scientific disciplines.

1. Introduction

In the Science Abstracts Index the major keyword for neutron scattering research is "Neutron Diffraction Examination of Materials". In 1981, about 1000 publications were cited. The major increase in the number of annual publications occured in the period 1960 - 72. The contribution of neutron scattering to Materials Science - in its most general sense - is documented with about 20 000 publications since 1932.

In this lecture I report on the history of two special scattering techniques - neutron small angle scattering (NSAS) and diffuse elastic neutron scattering (DENS) -, emphasizing the principal ideas, the technical development and first applications in different scientific disciplines. NSAS and DENS give information on the disorder of materials. Especially by NSAS many contributions are given to disciplines in Materials Science which are closely related to applied research, as for instance Metal Physics and Polymer Science.

2. Neutron Small Angle Scattering (NSAS)

By scattering of neutrons at small angles the scattering law close to the origin of reciprocal space is determined within a Q-range of a few 10^{-4}Å^{-1} to $0,3 \text{Å}^{-1}$. In other words, long-wavelength fluctuations in density, chemical composition and in magnetization are investigated. Besides many long-wavelength fluctuations of fundamental interest - as for instance density and magnetization near the critical point - there are also many large scaled inhomogenieties in the range from a few tens up to a few thousands of Angstroms which are of interest in the understanding of materials parameters as for instance precipitates, voids, cracks and dislocations.

A short summary of NSAS-work up to about 1960 - the **so-called** "heroic age" - is given in the book "Neutron Diffraction" (Bacon, 1962). In this book - based on about 10 original papers - there is reported on critical magnetic scattering, multiple refraction by magnetic domains, scattering by pores of graphite, small-angle scattering by spin-waves, determination of the coherent scattering length by NSAS and scattering from dislocations in cold-worked metals. For the last-mentioned application the use of long-wavelength neutrons in order to avoid double Bragg-scattering was essential. At this time already three of the basic advantages of NSAS compared to X-ray small-angle scattering (XSAS) have been exploited in experimental work, namely magnetic interaction, penetration of thick materials and the possibility of avoiding double Bragg-scattering by use of neutrons with wavelengths beyond the Bragg-cut-off. For XSAS in case of weak scattering phenomena double Bragg-scattering was recognized as the major source of background contribution by Webb and Beeman, 1959. The draw back for neutrons at this time was the low intensity of cold neutrons ($\lambda > 4$ Å approximately). The situation changed completely in the period from 1960 to 1973 in course of the development of better neutron sources and scattering techniques.

Following the idea of Maier-Leibnitz and Springer it was demonstrated by Christ and Springer, 1962, that high subthermal neutron fluxes can be obtained with neutron guide tubes. Based on this a major effort was started at the research reactors in Munich and Jülich in the period from 1961 - 67 to study dislocations, radiation damage zones, magnetic precipitates and critical magnetic scattering by NSAS. Also the first work on biological substances - namely hemoglobin - was performed. Another important impact was the development of cold sources at the research reactors in Harwell and Saclay, which demonstrated the possibility of getting even higher fluxes of subthermal neutrons for NSAS. In addition to further work on critical magnetic scattering at Saclay, NSAS was used for the first observation of small angle diffraction from the flux-line lattice in type II-superconductors by the Saclay group, 1966. The experience obtained from the above-mentioned studies was the basis for the development of the high-performance instruments at Jülich and at Grenoble (D11 and D17), which have been in operation since 1961, 1971 and 1974 respectively. Closely related also at Saclay the technique was continuously improved. The major steps for 1964 - 1972 can be characterized as follows.

Immediately after Néel and Maier-Leibnitz initiated the German-French plans for a high flux reactor in 1964 according to the study of Ageron, there was an intensive discussion on the design of a new generation of neutron scattering instruments promoted by Maier-Leibnitz. As a part of this, the essential design criteria for a high performance NSAS-instrument was already clearly recognized in early 1967: i) At a reactor with a cold source the optimum wavelength for a general purpose NSAS, instrument ranges from 4 to 20 Å. ii) The instrument can be installed far from the reactor by use of a neutron guide tube. iii) Considerations on the relation between information and resolution favoured a point-collimation system, which nevertheless allows to relax towards slit-geometry for suitable problems. iv) Considerations on resolution also showed that monochromatisation with mechanical velocity selectors is sufficient in most cases. v) In order to illuminate sample areas from about 1 to 10 cm^2 for small Q-values very long instruments (up to 100 m) are required. vi) With a fixed sample position (for practical reasons) the reduction of length - in order to go to higher Q-values - the detector has to be moved towards the sample while the collimator is shortened by replacing part of it by guide tubes. vii) By careful design and construction of the collimator, the beam-stop and the shielding the instrumental background must be and can be reduced to a level far below the inherent sample background in nearly all cases of interest. viii) Two-dimensional multi-detectors have to be used. ix) Fast data collection and processing with experimental computers is necessary.

The presentation of the 40m-Jülich NSAS-instrument at the Small Angle Scattering Conference in Graz by Schelten, 1970, attracted the interest of the X-Ray Small Angle Scattering Community. Based on this, and following the general policy of the Institute Laue-Langevin, Ibel interested many scientists in the use of the 80m-instrument D11 at the ILL built under his responsibility. It was also at this time that the inherent possibilities of the H/D-contrast for NSAS-work in biology and polymer science interested a broader community. Guinier - the XSAS-pioneer - encouraged the NSAS-community and the next SAS-conference was held in Grenoble in 1973. At this conference, already a broad review on NSAS was presented by Ibel and Schelten covering the successful instrumental development and the broad spectrum of applications. Especially in polymer science and biology promising results were obtained, mostly by use of the H/D-contrast. Stuhrmann and Higgins report at this conference on the value of NSAS for these research fields. A recent review on Metal Physics and other disciplines in materials science is given by Kostorz (1979). At present, about 60 original papers concerned with NSAS are published annually.

NSAS is now an established technique at many reactors in the world with a dozen high-performance instruments especially in Western Europe, North America and Japan. At Oak Ridge, a high-performance NSAS instrument together with a high-performance point collimation XSAS instrument is the instrumental basis for a National Small Angle Scattering Center. These instruments and also special set-ups (double crystal spectrometers and direct-beam isntruments) will allow many investigations for the next decades. Thinking about the possible developments until the 100th birthday of the neutron one may raise the question of further impacts for NSAS. Ultracold neutrons (UCN) can be used for NSAS. They would allow the study of thin films and possibly even surface phenomena. A polarized beam instrument would be desirable. For this, the supermirror-technique could be used. The combination of a supermirror-analyzer with a multidetector would allow a polarization analysis instrument, certainly of interest for many problems in magnetism. A basic assumption for the instruments existing today was the small sample size (1 - 10 cm^2). In case large sample areas (up to a few m^2) could be provided by focussing the neutron beam with large mirror systems a factor up to 100 could be gained in intensity. Pulsed sources with sufficient medium flux open new possibilities. A very promising new field is inelastic small angle scattering especially for biological substances and polymers. A series of results has already been obtained by the neutron spin-echo method. Also UCN-techniques may enter into this field.

3. Diffuse Elastic Neutron Scattering (DENS)

The elastic coherent neutron scattering of "perfect" crystalline material vanishes for $\underline{Q}$-vectors not equal to reciprocal lattice vectors ($\underline{\tau}_{hkl}$). The "background" is the incoherent elastic neutron scattering. For $\underline{Q} \neq \underline{\tau}_{hkl}$) therefore the elastic neutron scattering is a good measure for local fluctuations in atomic arrangements in so far as the incoherent elastic background is sufficiently small. In addition, local fluctuations of static magnetic moments can be investigated by DENS. Information is desirable for a rather extended $\underline{Q}$-space (up to 7 $Å^{-1}$ approximately) in general and especially for displacive disorder. On the other hand, the DENS-cross-sections for many disordered materials are small. Therefore most DENS-experiments with high sensitivity have used, up to now, mainly cold neutrons resulting in a restricted $\underline{Q}$-range.

The history of DENS is rather "diffuse". In the "heroic" age (up to 1960) there were efforts to determine small concentrations of point defects by transmission experiments for materials with small incoherent scattering cross-sections as Be, graphite and quartz. Also some scattering experiments on reactor-irradiated MgO and graphite were performed. A rather well-developed discipline was diffuse scattering from magnetic alloys.

In the period from 1960 to 1970, magnetic DENS - mostly performed by the Harwell and the Oak Ridge group - was much more dominant than nuclear DENS. However it was also within this period that the use of cold neutrons allowed the initiation of a very ambitious experiment: a scattering study of low-temperature reactor irradiated aluminium at the Munich research reactor. A Frenkel-pair concentration of 2×10^{-3} was undoubtedly measurable. This experiment and also the next one on a Pb-Bi alloy (demonstrating again the sensitivity) were performed without energy analysis of the scattered beam. Energy-gain scattering was avoided by cooling the sample to about 10 K and energy-loss scattering by use of sufficiently long neutron wavelengths. Of course the latter condition restricted the Q-range. Our colleagues concerned with inelastic neutron scattering looked suspiciously at these special studies, admiring the sensitivity but expressing considerable doubt for broader use. DENS was easily measurable for them with TOF- and TA-spectrometers in a much larger Q-range. They got it as a by-product and for many materials the limit was anyhow set by the incoherent elastic scattering.

We reconsidered the situation around 1968 for further experiments in Jülich and eventually at the HFR in Grenoble. Evidently the high luminosity of cold sources was an encouraging advantage. In order to allow higher sample temperatures and a larger Q-range, a special TOF-spectrometer with very low resolution and carefully designed with respect to background and double-scattering processes seemed to be the best solution. Such an instrument was installed by Bauer at the FRJ2 in Jülich and by Just at the HFR in Grenoble in 1970 and 1973, respectively. At 1979 with these special instruments a high level for investigation with low defect concentrations has been obtained complementary to experimental work with other instruments (Bauer, 1979) and to the continuation of magnetic scattering by disordered materials (Hicks, 1979). Further progress to be expected in nuclear and magnetic DENS will depend mainly on the development of research interest, which is not so stream-lined as in NSAS. In experimental techniques polarization analysis and extension of DENS to higher Q-values are of interest.

Acknowledgement

Since the discovery of the neutron, many scientists have contributed to the success of NSAS and DENS. I personally thank the many colleagues with whom I have had contact over the years and especially for long-time cooperation Prof. Springer, Dr. Schelten and Dr. Bauer.

References

Bacon GE   1962 Neutron Diffraction (London: Oxford Univ. Press)
Bauer GS   1979 in Treatise on Materials Science and Technology, volume 15,
                  ed. by G. Kostorz (New York: Academic Press) pp 291-336
Hicks TJ   1979 as above, pp 337-380
Kostorz G   1979 as above, pp 227-290

Inst. Phys. Conf. Ser. No. 64: Section 4
Paper presented at Conf. on Neutron and its Applications, Cambridge, 1982

Neutron diffraction studies of amorphous solids

Adrian C Wright

J J Thomson Physical Laboratory, Whiteknights, Reading RG6 2AF, UK

Abstract. A brief survey is presented of the role of neutron diffraction in structural studies of amorphous solids. The inherent limitations of the diffraction technique are discussed, together with modern instrumentation and methods for separating individual component correlation functions. An introduction is given to the use of modelling and the extraction of structural parameters from experimental data.

1. Introduction

The science of amorphography is concerned with the structures of amorphous solids and their systemmatic classification. As with crystalline solids, neutron diffraction techniques provide an important structural probe. There is, however, an important difference between diffraction studies of crystalline solids and the corresponding investigation of amorphous materials. A crystal structure is specified in terms of a unit cell and translational symmetry, the latter leading to characteristic sharp Bragg diffraction peaks. Only a limited number of parameters are required to define the atomic positions within the unit cell and, given a single crystal sample and good diffraction data over a reasonable region of reciprocal space, it is in practice possible to determine the structure of a simple crystalline solid absolutely.

The same is not true for amorphous materials. Their complete lack of symmetry, periodicity and long range order, together with the fact that they are normally isotropic on a macroscopic scale, leads to a diffraction pattern which is a continuous function of the magnitude of the scattering vector Q (not its direction) and means that the maximum that can be obtained from a diffraction experiment is a one-dimensional correlation function, from which the regeneration of the underlying three-dimensional structure can never be unique. It is for this reason that modelling plays such an important role in structural studies of amorphous solids and why the choice between structural models involves not only diffraction data but a wide range of other techniques such as vibrational and magnetic resonance spectroscopy.

2. Experimental Techniques

A neutron diffraction experiment comprises a measurement of the scattered intensity as a function of the scattering vector

$$Q = (4\pi/\lambda) \sin \theta \qquad (1)$$

where λ is the incident wavelength and 2θ the scattering angle. It is conventional to perform a total diffraction experiment in which the

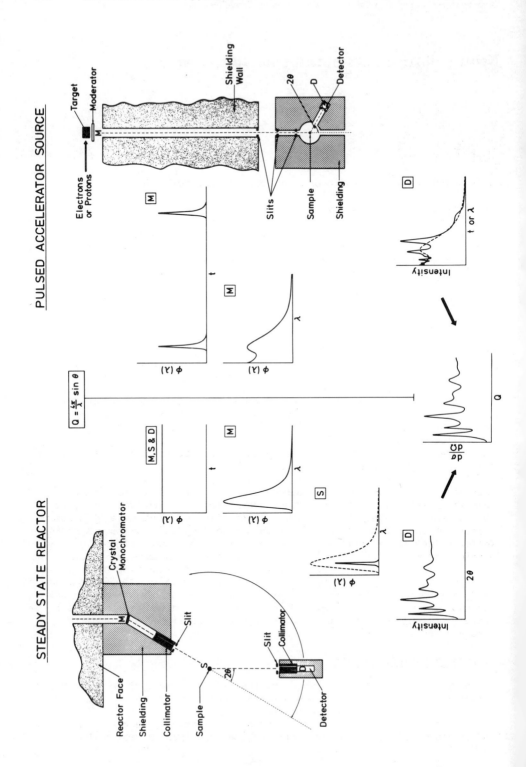

detector records both the elastic and phonon inelastic scattering and in order to achieve the necessary variation in Q it is possible to scan 2θ at a fixed incident wavelength (conventional technique) or to make measurements as a function of λ at constant scattering angle (time-of flight technique).

The two techniques are compared opposite. In the conventional steady state reactor twin-axis experiment the flux $\phi(\lambda)$ of thermal neutrons extracted from the moderator reflector of a steady-state reactor has a Maxwellian distribution of velocities and is time independent. A beam of wavelength λ, selected by the monochromator crystal, is incident on the sample and scattered into the detector through a variable angle 2θ. The variable λ time-of-flight technique is usually employed with a pulsed accelerator source. Electrons or protons strike a heavy metal target producing pulses of fast neutrons which are partially moderated before being incident on the sample and scattered into detectors situated at fixed angles 2θ. Each detector records the scattered intensity as a function of time-of-flight for the distance from the moderator via the sample to the detector. For any neutron the time-of-flight is simply related to λ through the velocity and De Broglie's relationship and the diffraction pattern $I(Q)$ may be extracted by dividing the measured intensity by the incident neutron spectrum shape. The great advantage of a pulsed accelerator source over a steady-state reactor is that the former is under-moderated which gives rise to a strong epithermal component in $\phi(\lambda)$ and allows data to be collected to much higher values of Q.

The relationship between the corrected diffraction pattern $I(Q)$ and the real space correlation function $T(r)$ is shown diagrammatically overleaf. The data are for vitreous silica. For a polyatomic material the correlation function

$$T(r) = \sum_j \sum_k \bar{b}_j . \bar{b}_k t'_{jk}(r) \qquad (2)$$

comprises a weighted sum of the individual component correlation functions $t'_{jk}(r)$ each of which is a convolution of the true component correlation function $t_{jk}(r)$ with the peak function $P(r)$, which defines the experimental resolution in real space and is the Fourier cosine transform of the modification function $M(Q)$. (See for example reviews by Wright (1974) and Wright and Leadbetter (1976).) $\bar{b}$ is the neutron scattering length, the j summation is taken over the atoms in one unit of composition and that for k over atom types (elements). The width of the central maximum in $P(r)$ is inversely proportional to Q_{max}, the maximum scattering vector to which reliable data are available.

For a few favourable samples it is possible to obtain a partial or complete separation of the components $t_{jk}(r)$ through a variation of the scattering length $\bar{b}$ for one or more of the constituent elements (c.f.Equ.2). This may be achieved either by isotopic substitution ($\bar{b}$ changes for different isotopes of the same element) or by anomalous dispersion (the wavelength dependence of $\bar{b}$ near an absorption resonance). Very few elements have absorption resonances in the thermal region, however, and suitable isotopes are not available for most common glass-formers. In other cases, as with many liquid studies, systems are ill-conditioned with the result that the data obtained are likely to be of very dubious quality. The anomalous dispersion technique (Krogh-Moe, 1966) has recently been successfully employed to study the environment of Sm in vitreous $Sm_2O_3-Al_2O_3-GeO_2$ (Wright et al., 1982). In addition to probing the magnetic structure of

T(r) from a Neutron Diffraction Experiment

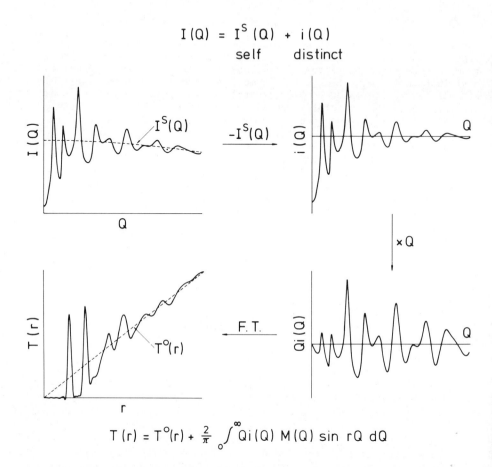

$$I(Q) = I^S(Q) + i(Q)$$
$$\quad\quad\quad\text{self}\quad\quad\text{distinct}$$

$$T(r) = T^°(r) + \frac{2}{\pi} \int_0^\infty Qi(Q) M(Q) \sin rQ \, dQ$$

amorphous solids, neutron magnetic diffraction can also be used to obtain information concerning the distribution of the magnetic ions within a sample. A review of neutron magnetic scattering studies of amorphous solids is given by Wright (1980).

3. Methods of Interpretation

The data obtained from an experimental investigation of an amorphous solid may range from a single total correlation function $T(r)$ to a complete determination of all the individual components $t'_{jk}(r)$. In each case the object is the same, namely to extract the maximum structural information for the material in question. Most amorphous solids contain a well defined structural unit, which gives rise to one or more relatively sharp peaks in the correlation function at low r and may be investigated by means of peak fitting techniques. In order to obtain accurate values for co-ordination numbers, peak positions and r.m.s. bond length variations, it is necessary to have data to high Q (good real space resolution), especially since typical peak assymmetries are only detectable for $Q \gtrsim 15\text{-}20 \text{Å}^{-1}$ (Hayes and Wright, 1983). The primary peak parameters may be used to extract information on bond angle distributions and also in favourable cases the distribution of bond torsion angles.

Currently the greatest interest for covalent amorphous solids is in the intermediate range order or network topology. This is usually investigated through modelling studies. The interference function $Qi(Q)$ and correlation function $T(r)$ emphasise different aspects of the structure and it is therefore most important to compare model and experiment in both real and intensity space and in the former to correctly incorporate the peak function $P(r)$. Similarly, no matter how good the agreement obtained, the model in question is only one possible fit to the data and hence it is essential to consider, and if possible eliminate, all alternative models. A variety of structural models and modelling techniques has been employed in the study of amorphous solids including:
(i) <u>Random Network Models</u>. The Random Network Theory has been the most successful in describing the structure of traditional oxide glasses and amorphous semiconductors. Random network models may be built either by hand or using a computer.
(ii) <u>Crystal Based Models</u>. A great deal of information on the structure of amorphous solids can be obtained from a simple comparison with the corresponding crystalline polymorphs. More formal crystal based models start with an appropriate crystal structure and restrict the longer range order to interatomic spacings less than some characteristic correlation length.
(iii) <u>Chain and Ring Models</u>. This type of model is appropriate for materials containing only divalent atoms (e.g. Se) or larger two-connected groupings of atoms such as found in organic polymers ($-CH_2-$ etc.).
(iv) <u>Random Sphere Packing Models</u>. Random sphere packings provide a useful starting point for models of amorphous metallic alloys and materials whose structures are governed by predominantly ionic forces such as vitreous $ZnCl_2$.
(v) <u>Molecular Models</u>. Models based on a random packing of approximately spherical molecules have been used to describe the structure of some vapour deposited thin films (e.g. CCl_4 and $As_{1-x}S_x$). The formalism employed closely follows that for molecular liquids.
(vi) <u>Amorphous Cluster Models</u>. Amorphous clusters comprise atoms in a regular but non-crystallographic configuration frequently based on pentagonal dodecahedra or fivefold symmetry. Cluster size is limited due to a slight mismatch between adjacent structural units, which leads to increasing distortion as the cluster grows.
(vii) <u>Monte Carlo and Molecular Dynamics Simulations</u>. The structures of a range of simple inorganic glasses have been simulated using purely ionic potentials. Again the methods employed closely follow those for related liquids.

4. Accuracy and Limitations

A discussion of accuracy and limitations can be divided into two parts: first the accuracy of the experimental data themselves, and that of the resulting transform, and second the extent of the ambiguity of any interpretation of these data. The accuracy of the experimental data can most easily be assessed by the behaviour of the transform at low r, below the first true peak. If the transform is well behaved in this region the data are of reasonable quality. Great care is needed in making this assessment, however, as some authors either plot the radial distribution function $rT(r)$, which has the effect of reducing the amplitude of error ripples at low r, or use these false oscillations as a criterion for tidying their data before publication and do not include the original unadulterated transform. Similar techniques are sometimes used to "remove" the effects of terminating the data at finite Q_{max}. Information theory, however, indicates that it is impossible to replace the unmeasured data without making some assumption about the material under investigation so that the resulting correlation function merely becomes one possible model which fits the results rather than an unbiased Fourier transform. It should also be noted that in principle both $Qi(Q)$ and $T(r)$ contain the same information expressed in a different form, although in practice the information content of $T(r)$ is slightly reduced by the use of a modification function. The finite upper limit in Q means that in $T(r)$ each component $t_{jk}(r)$ is convoluted with the appropriate peak function $P(r)$, but the information for a particular distance is still concentrated around that distance whereas it is spread throughout reciprocal space.

The uniqueness of any structural information extracted from the data is much more difficult to judge. As already stressed, the one-dimensional nature of the correlation function means that it is impossible to unambiguously determine the structure of an amorphous solid. The reliability of any interpretation therefore depends not only on the quality of the fit to the experimental data in both real and intensity space, but also on the number of other possible models which have been tried and satisfactorily rejected.

5. Conclusion.

At their best, neutron diffraction techniques, particularly on modern pulsed sources, are capable of yielding accurate high resolution data. The greatest barrier to progress thus lies not so much with the technique itself but in the generation of structural models which can predict these data within their known uncertainties. This feature is unique to amorphous materials and has no parallel in corresponding studies of the crystalline state.

References.

Hayes T M and Wright A C 1983 The Structure of Non-Crystalline Materials II eds P H Gaskell, E A Davis and J M Parker (London : Taylor and Francis), in press.
Krogh-Moe J 1966 Acta Chem. Scand. 20 2890
Wright A C 1974 Adv. Struct. Res. Diffr. Meth. 5 1
Wright A C and Leadbetter A J 1976 Phys. Chem. Glasses 17 122
Wright A C 1980 J. Non-Cryst. Solids 40 325
Wright A C, Etherington G, Desa J A E and Sinclair R N 1982 J. de Phys., in press.

Contribution of neutron scattering experiments to the understanding of conformation and dynamics of polymer molecules

J.S. Higgins

Department of Chemical Engineering, Imperial College, London, S.W.7.

Abstract. Neutron scattering techniques are widely applied in polymer science to the study of both conformation and dynamics. The two most important areas have been small angle scattering and quasi-elastic scattering. In this article a number of general reviews are cited and a few experiments highlighted to demonstrate the range of interests covered.

1. Introduction

The basic chemical structures of synthetic polymers are usually very simple (unlike many naturally-occurring biological macromolecules). Small groups formed mainly from carbon and hydrogen are repeated many thousands or even millions of times. It is these large molecular weights and the consequent problems for organisation and motion of long, entangled molecules in solutions, melts, rubbers or glasses which generally interest the physicist and physical chemist, rather than the local chemical structure.

Most of the currently available neutron scattering techniques have at some time been applied to polymeric systems. Neutron diffraction, inelastic and quasielastic scattering measurements have all made contributions to our understanding of how high molecular weight molecules organise themselves. These experiments largely involved neutron scattering specialists choosing to study polymeric systems. The advent of large area detectors and, in particular, their application in small angle scattering brought the direct involvement of the polymer specialists themselves. The use of deuterium labelling to pick out single molecules in concentrated solutions and melt samples has uniquely allowed questions about, for example, lamellar crystal structures, deformed and flowing samples, copolymers, blends and composite structures such as microemulsions and colloids to be addressed at the molecular level. With the spin-echo technique the molecular motion in these systems is now being studied on a time scale which reaches up to 10^{-7} s overlapping and complementing other techniques such as NMR and photon correlation spectroscopy.

In the short space available it would be inappropriate to attempt a general review of such an expanding field of study. Many thorough and detailed review articles have appeared in recent years. Among the many available, we may quote Maconnachie and Richards (1978) for a general review, Higgins and Stein (1978) for small angle scattering, and Higgins (1982a, b) for inelastic and quasielastic scattering. In this present article only a few examples will be presented which illustrate the various problems amenable to study with neutron techniques. Very little will be said about crystalline and semicrystalline polymers, not because these are unimportant - many commercially important polymers are semicrystalline - but because this is not the area of expertise of this author, while it is that of Professor Keller, whose review follows this.

2. Organisation of Macromolecules - Small Angle Scattering

Although, because of the dimensions covered (from 10 $\sim$ 1000 Å) and the differing hydrogen deuterium cross-sections, it was potentially the most widely applicable neutron technique as far as polymer problems are concerned, small angle scattering was in fact not in general use until relatively recently. For intensity reasons, the technique really only got under way with the advent in the early seventies of large area position sensitive detectors.

The problem that was actually 'waiting' for the technique was the question of the Gaussian conformation of a macromolecule in a melt. This had long been the subject of speculation and argument. Use of deuterium labelling quickly led to resolution of the question for amorphous samples, although, as Professor Keller will show, for crystalline samples the neutron data somewhat refuelled the fires of argument which are only now beginning to subside, and a generally agreed picture emerge.

The molecule in the melt was shown to be Gaussian, while light scattering experiments had long indicated that in dilute solution, interactions with the solvent led generally to chain expansion (except under certain so-called theta-conditions, where Gaussian statistics were regained). The question immediately arises as to how the molecular conformation changes with concentration between these two extremes. This is one of several areas where initial neutron results led to an explosion of theoretical interest. Phase diagrams were developed for molecular dimensions as a function of temperature and concentration based on the random phase approximation and from mean field theory and predictions compared with experiment. Fig. 1, for example, shows the variation of molecular dimensions with temperature for labelled chains in a semi-dilute solution. The plateau value is that which would be observed for a melt or dilute solution under theta conditions.

Many of the interesting problems now under investigation are even more complex examples of the effect of interactions on the molecular conformation. Among these may be mentioned three component systems with, for example, two polymers dissolved in one solvent or a copolymer in a solvent, polymer blends where the solvent itself is polymeric, systems involving electrostatic interactions such as polyelectrolytes and ionomers, and multiphase systems such as microemulsions, dispersions, block copolymers and latices. In every case the deuterium labelling techique has allowed unique information

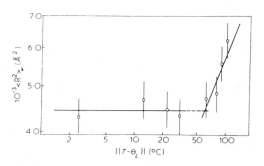

Fig. 1. Molecular dimensions R as a function of temperature for a semi-dilute (C = 0.179 g ml^{-1}) solution of polystyrene in cyclohexane.
θ_L = 213°C. Richards et al. (1981).

to be obtained about single molecule conformation in a many-molecule environment.

Having said this, it is perhaps worth pointing out that despite the assumptions inherent in all these experiments, deuterated and hydrogenous polymers are not identical. Phase boundaries are shifted and crystallisation temperatures changed by deuteration. At present fortunately, in most cases, these seem to be minor perturbations if care is taken, but it is necessary to be alert for such effects, or spurious results may be obtained.

A different type of experiment has been involved with observing the effect of an external perturbation on the molecular shape - for example stretching, shearing or in some other way deforming the sample, and, by inference, the molecule. Dr. Oberthur will be describing such an experiment on flowing polymers. Fig. 2 shows what happens when the sample - a high molecular weight polymer above its glass transition temperature - is stretched and held at constant length. The force necessary to stretch the sample reduces with time and the experiment shows that this occurs in parallel with the molecules, whose shapes initially deformed affinely with the external deformation, gradually resuming their normal isotropic shape distribution. The entanglements of the long molecules first hold the sample together so that it can be stretched (unlike a normal liquid!) and then slowly allow slippage back to equilibrium.

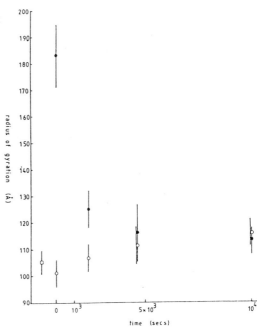

Fig. 2. Molecular dimensions R parallel ● and perpendicular O to the direction of stretch for a sample of polystyrene as a function of lapsed time after stretch. □ - unstretched sample. Maconnachie et al. (1981).

3. Dynamics of Macromolecules - Quasielastic Scattering

If a polymeric sample is subjected to a mechanical oscillation of varying frequency, then loss peaks are observed at certain frequencies corresponding to vibrations and torsions of side groups, to rotations of whole short sections of the main chain, and finally to freeing of the whole backbone into a wriggling motion which is effectively the glass to rubber transition. If the sample is semicrystalline there will be a further loss associated with melting of the crystalline regions.

Techniques for observing these motions include mechanical and dielectric relaxation, NMR, Raman and infrared spectroscopy, fluorescence spectroscopy and light scattering. Neutron scattering has been used to observe each type.

For crystalline samples, in principle, observation of phonon dispersion curves can give important information on the force constants and hence the ultimate material strength to be obtained in, for example, oriented fibres. The inherent semicrystalline nature of polymeric samples has, however, severely limited this area of application after a promising start in some of the earliest neutron experiments on polymers.

In certain cases the side chain motions are difficult to pick out by other techniques and incoherent scattering from partially deuterated samples, either via the inelastic spectrum or, more cleanly, via quasielastic scattering has helped to give more precise values for the barriers to rotation of methyl and phenyl groups in the same way as for small molecules.

The motion of the main backbone has important implications for the truly 'polymeric' property of a polymer, its viscoelasticity. Quasielastic neutron scattering, even with the recent welcome addition of spin-echo is confined to a relatively high frequency range. Nevertheless, by choosing the system studied it is possible to explore the main features of the motion of a single chain in dilute solution.

When distances longer than the polymer dimensions (R) or shorter than a basic segment (σ) are explored, i.e. at very low or high Q values, then an essentially Brownian motion is observed, characterised by a Q^2 dependence of the inverse correlation time Ω (or in frequency domain the half width parameter, $\Delta\omega$). In the intervening Q-range the connectivity of the chain leads to a higher powered Q-dependence. This is Q^3 for a system where hydrodynamic effects of the solvent are important and Q^4 when hydrodynamic effects are screened out (e.g. in a melt). A plot of Ω/Q^2 is then 's'-shaped with two plateaus at low Q ($\leq R^{-1}$) and high Q ($\geq \sigma^{-1}$). Both these plateaus have been observed in neutron spin-echo experiments on different samples. R is adjusted by changing the molecular weight of the polymer, σ by changing the chemistry to obtain a 'stiff' backbone. Results from high molecular weight polymers, some of them relatively stiff, show the upper turn-over $R^{-1} < Q \approx \sigma^{-1}$ (Nicholson et al. (1981)). Results for flexible but low molecular weight polydimethylsiloxane samples give the lower turn-over $R^{-1} \sim Q < \sigma^{-1}$ (Higgins et al. (1982)). In each case the Q^3 region is visible.

In a polymer melt matters are complicated by the molecular entanglements though, in general, the short observation scale of the quasielastic measurements ($Q^{-1} < 30$ Å) and fast time scales seem to preclude direct observation of the entanglement slippage or reptation which has recently been the subject of a large theoretical effort.

Incoherent quasielastic measurements have also given information on the rates of the main chain motion in polymeric melt samples. Fig. 3 shows the half width $\Delta\omega$ as a function of Q for some network samples.

The question posed was whether the chemical cross-links in such a system move more slowly than the free sections away from these cross-links. A model network was prepared with cross-links every 30 repeat units. The free centres were deuterated but hydrogen was retained around the cross-links. Incoherent scattering then allowed separation of the motion of these two areas. The results in Fig. 3 do indeed show that the cross-links are slowed down, by about a factor of 2, but the free chain centres move just as though they were part of long, uncross-linked molecules.

This brief resumé, highlighted with examples from our own experimental programme can only given an indication of the breadth of interest covered by neutron scattering applications to synthetic polymers. The interested reader is warmly recommended to spend some time with the literature, starting perhaps with the review articles cited at the beginning.

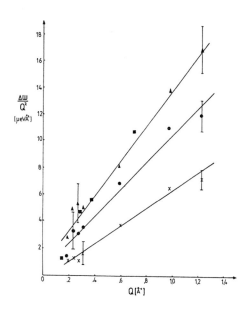

Fig. 3. Energy width $\Delta\omega$ as a function of Q for ● hydrogenous network (1), X network (2) with hydrogen only around the cross-links, ▲ subtraction of (2) from (1) and ■ high molecular weight uncross-linked polymer. Higgins et al. (1981).

References

Higgins J S and Stein R S 1978 App. Cryst. 11 346
Higgins J S, Ma K and Hall R H 1981 J. Phys. C Solid State Physics 14 4995
Higgins J S 1982a in Developments in Polymer Characterisation 4 ed J V Dawkins (London: Applied Science Publishers Ltd.)
Higgins J S 1982b in Static and Dynamic Properties of the Polymeric Solid State ed R Pethrick and R W Richards (Netherlands: D Reidel Publishing Company)
Higgins J S, Ma K, Hayter J B, Dodgson K and Semlyen J A 1982 Polymer in press
Maconnachie A and Richards R W 1978 Polymer 19 739
Maconnachie A, Allen G and Richards R W 1981 Polymer 22 1157
Myers W, Summerfield G C and King J S 1966 J. Chem. Phys. 44 189
Nicholson L K, Higgins J S and Hayter J B 1981 Macromolecules 14 836
Richards R W, Maconnachie A and Allen G 1981 Polymer 22 153

Application of neutron scattering to crystalline polymers

A. Keller, H.H. Wills Physics Laboratory, University of Bristol, BS8 1TL.

This is a brief account on the application of small angle neutron scattering (SANS) to the study of crystalline structure of polymers. The scattering is of the coherent elastic variety and the polymers we are concerned with consist of regularly repeating sequences of units such as result from the polymerization of a single monomeric species. The regular sequence within the chains ensures crystallizability and the so called 'crystal structure', the arrangement of the atoms within the lattice formed by the repeating units, is usually known from X-ray crystal structure analyses traditionally performed on fibres. The crystal structure, thus determined, however says nothing about the full trajectory of the chain, only of those portions which are part of the lattice. It is information on the full trajectory, that is the path of a chain within its own environment, which is provided to varying levels of certainty and resolution by neutron scattering. Knowledge about this trajectory is vital for the understanding of how macromolecules crystallize, and beyond that provides information about the ability of macromolecules to form organized entities in the widest generality. It follows that in the case of crystalline polymers the SANS technique has a particularly central part to play, in many respect even more so than in the case of traditional crystals.

The main issue turns around chain folding. It has emerged exactly 25 years ago that long chain molecules form well defined crystalline lamellae. These lamellae are a few hundred Å-s thick within which the chain direction is perpendicular to the lamellar surface. In molecular terms this is only conceivable if the chains fold up and down repeatedly within the confines of the lamellar thickness, hence the picture of chain folding (1). This was totally unexpected in the light of traditional conceptions, according to which long chain macromolecules are so entangled with each other and possess such limited mobility that at their best they are expected to be capable of forming only small regions of order envisaged as fringed micelles. The discovery of chain folded lamellae has basically modified this structural picture but the problem of how good the order is within the lamellae, and in particular how regularly the chains are folded remains an issue of acute controversy (2). The advent of neutron scattering through isotopically labelled molecules was expected to bring evidence in this matter. Expectations were high as regards the importance of information to be forthcoming and this was fully borne out by the results. Nevertheless as regards the expectation that SANS can produce a straightforward decision between the opposing views the results have been disappointing. This is not the fault of the SANS technique; it is because the variety of trajectories a chain can take up can vary according to the polymer used and the conditions of crystallization, all, however, within the broader framework of chain folding.

It is not possible to describe even the principle results, not to speak of
the divergences of opinion which exist in such a brief account, only the
most salient points can be mentioned.

Most works were concentrated on polyethylene where the crystal morphology
is best known. Unfortunately, however, the deuterated species used as
the labelled molecule tended to segregate confining works to rapidly crystallized material, thus limiting the generality of the scope of the investigations. In spite of this there are numerous highly informative studies with this material including from our own laboratory. Work also
exists on other polymers, isotactic polystyrene (i-PS) and polypropylene,
which do not show isotopic segregation, however, in the former case the
level of achievable crystallinity is low, while in the latter the lamellar
morphology is little explored and complicated. At present we all have to
contend ourselves with these limitations set by the materials, the only
ones available for such studies so far in both deuterated and protonated
forms required by such studies.

An important variable of course is the angular range of scatter examined.
Scattering at the smallest angles provide information on the global conformation, radius of gyration in particular. In this respect the chains in
melt crystallized material often (but not always) retain the same global
configuration they possessed in the melt apparently lending support to
the view that on crystallization the chains merely freeze-in, only with
some local ordering so as to produce the lattice (e.g. 2,3,4). However,
this much publicized finding is essentially confined (by necessity) to
rapidly crystallized polyethylene and even here local groupings of
straight chain stems into rows consistent with at least a limited amount
of adjacently reentrant folds could be verified (5). In solution grown
crystals however, the radius of gyration is much reduced (6) and is consistent with sheet like deposition of folded molecules along the crystal
faces (see below), an inference which for i-PS is reported to hold also
for the melt crystallized material at least under specified crystallization conditions and molecular weights.

At larger angles (still well within crystallographic, i.e. Bragg reflections) we should have more intimate information on the nature of chain
folding itself.

From scattering in the range of $q \sim 0.1 - 0.4 \text{Å}^{-1}$ (q being $\frac{4\pi \sin\theta}{\lambda}$, where
θ is the scattering angle and λ the wave length) the sheet like nature of
the volume circumscribing the labelled molecules could be definitely
identified (7) for solution grown crystals, which in the case of polyethylene could consist of a double or multiple row of stems, the number of
rows increasing with chain length (6). This we interpreted as the chain
folded sheets doubling up on themselves with increasing chain length, an
effect we termed superfolding. However, the absolute level of the scattered intensity was too low for each stem position to be occupied by the
same molecule, which means that the sheet cannot consist of the isotopically labelled species alone. This dilution of the setms (by about a factor of 2X) of a given isotopic species within the chain folded sheet means
that strictly adjacent reentry is not uniformly realized, an issue which
has acquired a central position in the controversies relating to the
regularity or otherwise of chain folding, and beyond that to the issue of
the organizational ability of macromolecules in general. Nevertheless in
i-PS, a polymer which owing to the absence of isotopic segregation could
be crystallized at much lower supercooling, adjacent stem reentry along

the planes of folding appears to pertain (8).

Most recent work has extended the range of measurements up to q values of 1.0 and beyond (9,10). At these higher angles it is becoming possible to obtain information about the near neighbour arrangement of labelled fold stems. In the case of solution grown crystals, of polyethylene in particular, where according to the foregoing overall stem adjacency is not realized and in average only half of the stems in a chain folded layer belong to the same molecule, the nature of this 'stem dilution' is receiving detailed attention. The trend in the process of emergence indicates that stems from given molecules tend to cluster, so that there are runs of adjacently reentrant stems separated by gaps filled by stems belonging to other molecules. It follows that we must have a preponderance of sharp folds intermixed with fewer long loops. Such structure features, together with many others, are not mere details in the structural characterization of polymer crystals; they are providing us with vital clues for the way long chain molecules tend to crystallize.

Another separate issue relates to what happens to individual molecules when a solid crystalline polymer is being stretched, an important process in the production of technological fibres and films. In this instance the anisotropy of the scattering pattern which develops on stretching is being mapped and correlated with the macroscopic deformation. Work on this topic of both fundamental and technological importance is in progress in several laboratories (e.g. 11). In particular, it should provide information on the way the crystals break up and the chains straighten out by unfolding to give rise to oriented structures, a process intrinsic to macromolecular matter.

1. Keller, A., Phil. Mag. 2, 1171 (1957).
2. Faraday Discussions, 68, 288 (1979).
3. Schelten, J., Ballard, D.G.H. and Wignall, G.D., Polymer, 15, 685 (1974).
4. Ballard, D.G.H., Cheshire, P., Longman, G.W. and Schelten, J., Polymer, 17, 751 (1976).
5. Sadler, D.M. and Harris, R., J. Polymer Sci. Phys. Ed. 20, 561 (1982).
6. Sadler, D.M. and Keller, A., Science, 203, 263 (1979).
7. Sadler, D.M. and Keller, A., Macromolecules, 10, 1128 (1977).
8. Guenet, J.M., Macromolecules, 13, 387 (1980).
9. Stamm, M., Fischer, E.W. and Dettenmaier, M. Faraday Discussion, 68 263 (1979).
10. Sadler, D.M. and S.J. Spells, to be published.
11. Sadler, D.M. and P.H. Barham, Polymer, in the press.

Neutron scattering of flowing polymers

R.C. Oberthür

Institut Laue-Langevin, 156X Centre de Tri, 38042 Grenoble, France

Abstract. The use of neutrons for the investigation of polymer conformation in a shear gradient by scattering experiments offers a lot of advantages in comparison with X-rays and light.

1. Introduction

Flowing polymers are polymer melts or solutions subjected to a velocity gradient. A velocity gradient appears due to the combined action of the cohesional forces of the polymer system and some external force, which causes the system to flow. Essentially two different types of a velocity gradient can be distinguished:

1) a longitudinal (or elongational) gradient G_L.

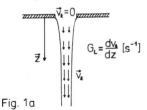

Fig. 1a

It occurs, e.g., when a polymer fluid flows from a hole in the bottom of a vessel due to the action of the gravitational force of the earth (Fig. 1a). As a result of the increasing velocity of the flowing system, the diameter of the fluid stream decreases.

2) a transverse (or shear) gradient G_T.

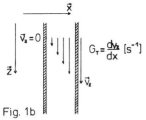

Fig. 1b

It occurs, e.g., when a polymer fluid is confined between two plates, one at rest and the other plate moving at constant velocity $\vec{v}_z$ due to the action of a constant external force (Fig. 1b).

In pure form or in combinations both types are encountered in almost all technical processes concerned with the manufacture of plastic materials, like moulding or injection moulding of shaped parts, spinning or stretching of synthetic fibres, extrusion of tubes and foils, blowing of bottles and the expansion of polymeric foams, with gradients up to $G \approx 10^5$ s^{-1}. Similar shear gradients are encountered in lubricating oils, which often contain polymer additives to control their viscosity. Most polymeric systems, in contrast to normal low molecular liquids, show the peculiar behaviour that their flow velocity does not increase proportional to the applied force (non-Newtonian flow behaviour), which means that part of the applied energy could be consumed by some constant structural changes within the flowing system. Finally, the presence of only a small amount of high-molecular-weight polymers dissolved in a Newtonian liquid can shift its transition from laminar to turbulent flow (formation of vortices in the flowing system with increasing external force) to higher values of the applied force (Toms effect), a phenomenon, which if exploited, could save much pumping energy.

All these macroscopic phenomena have been described by polymer rheology [1], however, only little is known about the underlying microscopic phenomena at a molecular level.

2. Theoretical Predictions and Experimental Approaches

The influence of one of the basic velocity gradients on polymer conformation has been considered theoretically by several authors. For a polymer in an increasing shear gradient a gradual transition towards an anisotropic intersegmental distance distribution is expected [2], whereas in an elongational gradient a sharp coil-stretch transition has been predicted [3] for $G_L \approx 1/\tau$, where τ is the largest relaxation time of the intramolecular chain motion.

Experimentally this latter phenomenon has been observed [4] by flow birefrigerence with polarized light (cf. Fig. 2) which gives an integrated information on the orientation of the coil segments in the flow field.

In an extensive way this latter technique has been used to investigate the orientation of rod shaped particles and of polymer coils subjected to a constant <u>shear</u> gradient, however, no <u>detailed</u> information on the statistical conformation of the deformed coil can be obtained [14].

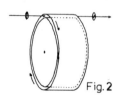
Fig. 2

A much more detailed information could be obtained if the polymer molecules in solution or melt could be observed in a velocity gradient by <u>scattering methods</u>.

3. Scattering Experiments

Scattering methods give a reciprocal image of the distance distribution between the constituant parts of the polymer coil, eventually down to its atoms, depending on the resolution attainable with the applied radiation.

This reciprocal image is experimentally given by the scattered intensity at the detector Y, which is proportional to $d\Sigma/d\Omega [cm^{-1}]$, the probability of a quantum or particle from the primary beam with intensity Y_0 to be scattered by a volume element of the investigated sample into a solid angle element $d\Omega$ of the detector (differential scattering cross-section per unit sample volume), as a function of $\hbar\vec{Q}$ the momentum transfer during the scattering process. These quantities are related by:

$$Y(Q) = \frac{Y_0}{D^2} \cdot \Delta V \cdot T \cdot \frac{d\Sigma}{d\Omega}(Q) \quad \text{where} \quad \frac{d\Sigma}{d\Omega}(Q) = \rho_2 \left(\frac{d\tilde{e}}{d\rho_2}\right)^2 \cdot M_2 \cdot P(Q)$$

with D sample-detector distance
 ΔV scattering volume
 T sample transmission (T < 1 due to absorption and scattering)
 ρ_2 mass concentration of the polymer

M_2 molecular mass of the polymer

$(d\tilde{\rho}_2/d\rho_2)$ scattering length density increment = scattering contrast of the polymer

P(Q) product of form factor of the polymer coil and structure factor of the coil distribution normalized to P = 1 for $\rho_2 \to 0$ and $Q \to 0$

$Q = (4\pi n/\lambda).\sin(\theta/2)$, where θ is the angle between incident and scattered beam, λ the wavelength of the applied radiation in vaccuo and n the refractive index of the sample.

In the electromagnetic spectrum only few windows exist where the absorption of radiation in the sample is small enough to allow the scattering from a sample of reasonable thickness to be observed. It is fulfilled for visible light and somewhat less for X-rays with wavelengths around 0.1 nm. Neutrons, though, can easily pass through a sample, being only absorbed by several rather exotic elements like B, Cd or U, which do not occur in normal polymer systems.

The scattering contrast between the polymer to be investigated and its surrounding (solvent) is given by its scattering length density increment $d\tilde{\rho}/d\rho_2$ [cm/g] which is the change in scattering length density $\tilde{\rho}$ when 1 gram of polymer is added to a large volume of the solvent. This contrast can be enhanced for neutrons by tagging the chain molecule under investigation with an isotope of one of its constituent atoms, which only made possible the investigation of individual polymer molecules in the pure molten state [5]. For light and X-rays, however, the contrast is given and cannot be changed without essentially changing the whole system.

Furtheron, the distance range which can be explored by the different scattering techniques and the related angular range, where the scattering has to be recorded, are of interest.

Table I : Comparison of different types of radiation suitable for scattering experiments with polymer melts and solutions.

	λ	θ	n	Q
vis-light	0.36 µm - 0.78 nm	2° - 180°	1.3 - 1.6	0.5 µm^{-1} - 50 µm^{-1}
X-rays	0.05 nm - 0.2 nm	0.15° - 8°	≈1	0.1 nm^{-1} - 15 nm^{-1}
neutrons	0.05 nm - 2 nm	0.15° - 8°	≈1	10 µm^{-1} - 15 nm^{-1}

Table II : Typical size scale for vinyl polymer chains with different number of monomers units N

	$N = 10^5$	$N = 10^3$
root mean square coil radius	≈0.2 µm	≈10 nm
statistical chain element		≈3 nm
monomer distance		≈0.25 nm

Comparison of Tab. I with Tab. II shows that only large polymer molecules can be investigated with light scattering but the large θ-range to be used together with a high reflectivity (due to a refraction index difference) at interfaces limits the construction and the use of a flow apparatus designed for light scattering [6].

X-rays and neutrons, instead, cover exactly the Q-range to investigate the interesting transitions from the scattering of the total coil to the scattering of the flexible segment and the monomer unit [7]. Moreover, most of the scattered radiation of interest can be collected close to the primary beam (small angle scattering, SAS), which facilitates the construction of a flow apparatus. However, most materials (except those with light atoms, like H, Be, B, C) have a high absorption coefficient for X-rays. Nevertheless, X-ray SAS has already successfully been used to investigate flowing polymers in a Debye-Scherrer glass tube of 1 mm diameter and 10 μm wall thickness [8] and in a Couette-type set-up with inner rotating cylinder (cf. Fig. 2), the X-ray beam passing through polyimide-foil windows [9]. Both set-ups have the disadvantage that the X-ray beam does not encounter all molecules in the same constant gradient, even in the Couette-system in only 75 % of the scattering volume the gradient is constant.

3. Flow-Apparatus for SAS of Neutrons

Most of the disadvantages encountered with electromagnetic radiation can be avoided with an apparatus constructed for the observation of polymer systems in a constant shear gradient by neutron SAS [10]. The apparatus consists of an inner fixed thermostatable quartz piston (stator) and an outer rotating quartz beaker (rotor), leaving a cylindrical gap of 0.5 mm (or 1 mm) thickness, 50 mm diameter and 70 mm height to be filled by the polymer system. Quartz is highly transparent for neutrons and has a very low SAS background. The neutron beams of 1.5 cm^2 cross-section hits the sample twice, perpendicular to the axis of rotation (Fig. 3).

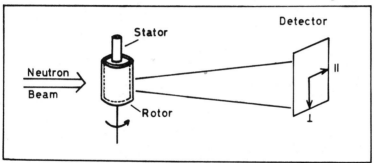

Fig. 3 Flow-apparatus for SAS of neutrons with coordinate axes on the detector parallel ($\parallel$) and perpendicular ($\perp$) to the flow direction.

The sum of the deviations in space and time from a constant shear gradient in the scattering volume due to thickness fluctuations of the gap with rotations, limited motor stability, the curvature of the cylinders and their limited height should be less than 1 %. The lowest attainable shear gradient ($G_T \approx 2$ s^{-1}) is limited by the motor stability and the highest attainable shear gradients are limited (a) by the onset of turbulence at low viscosities, (b) by the rise of the fluid towards the brim of the open beaker due to centrifugal forces, (c) by the energy dissipation in the fluid during rotation, and (d) by the maximum torque of the motor at high viscosities (Fig. 4). Temperature control is assured in the range from 5°C to 80°C.

The apparatus is essentially designed for the SANS-camera D11 at the ILL in Grenoble and can here be operated in the Q-range from 10 μm^{-1} to 2 nm^{-1}.

4. Experimental results

4.1. Isolated polymer coil

Polymer coils, when subjected to a shear gradient should be able to restore their equilibrium conformations distorted by the shear gradient according to the relaxation times τ_{seg} characteristic for their segmental diffusion (Fig. 5). Only if $G_T \gtrsim 1/\tau_{seg}$ a permanent distortion of the polymer coil at a given intersegmental distance can be expected.

Information on the intersegmental mobility as a function of intersegmental distance is provided by quasielastic neutron scattering (QENS) as performed on IN11, the spin echo spectrometer at the ILL. In Fig. 6 the experimentally determined reciprocal values of the quasi-elastic line broadening $\Delta\omega^{-1}$, which should be proportional to τ_{seg}, as a function of Q are given for polydimethylsiloxane (PDMS), polyoxyethylene (POE), polymethylmethacrylate (PMMA), and polystyrene (PS), all in benzene at 72°C [11]. Using the relation

$$\Delta\omega = \text{const.} \cdot \frac{kT \cdot Q^3}{\eta}$$

where the constant is determined to (28.6 ± 0.5) mrad for PDMS and POE and to (16.7 ± 0.5) mrad for PMMA and PS, conditions can be found, where a shear deformation of a polymer coil can be observed. The overall dimensions of the macromolecular coil limit the intramolecular relaxation at $Q \approx R^{-1}$ where R is the root mean square coil radius.

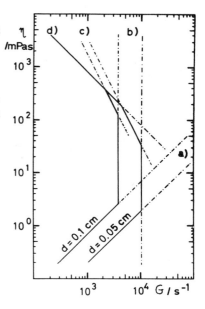

Fig. 4 Viscosity/shear gradient diagram for the range where the flow-apparatus can be used (see text).

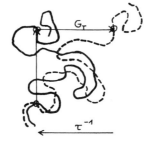

Fig. 5 Influence of G_T and τ on a polymer chain segment

Fig. 6 Intersegmental mobility of polymer coils as a function of Q ($\sim$ reciprocal intersegmental distance) in solvents of different viscosity, together with the experimental range of the spin-echo spectrometer IN11 and the flow-apparatus mounted on D11.

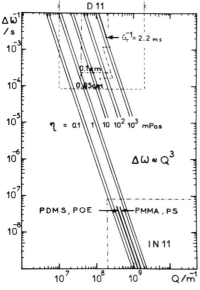

To enable the investigation of the shear deformation of PS of M_w = 279000 g/mole (M_w/M_n = 1.13) with N_w = 2490 monomer units in solution, the viscosity of the solvent has been increased to $\eta \approx 1$ Pa.s by mixing 20 parts of toluene with 80 parts of oligostyrene (N_w = 7.7 ; N_w/N_n < 1.3). To overcome the loss in scattering contrast a deuterated polymer in a protonated solvent has been chosen. The shear gradient where the scattering measurements were done in the given conditions was G_T = 450 s^{-1}.

The experimental result is shown for a sample at rest and in the given shear gradient (Fig. 7 a,b). Solvent subtraction and extrapolation

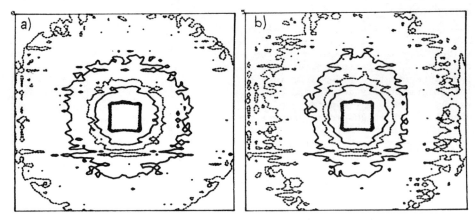

Fig. 7 Uncorrected scattering pattern of a PS-solution a) at rest b) in the shear gradient (cf. text)

of the two-dimensional data to the axes parallel and perpendicular to the flow direction gives the scattering curves of Fig. 8. After extrapolation to zero concentration ($0 < \rho_2 < 8$ mg/cm^3) and zero angle it turns out that the root mean square half axis perpendicular to the flow remains unchanged with respect to the coil at rest, whereas the value parallel to the flow has increased by 19 % from 14.8 nm to 17.6 nm. This can be pictured in the following way (Fig. 9)

In the given set-up the neutrons have "seen" the coil from the top, whereas flow birifringence in general "sees" the molecules from the side (cf. Figs. 2 and 3).

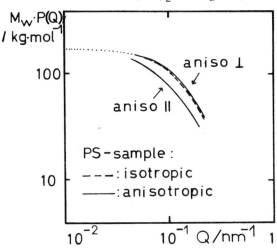

Fig. 8 Scattering curves of d$_8$-PS at a mass concentration ρ_2 = 6 mg/cm^3 at rest (isotropic) and in the shear gradient (anisotropic).

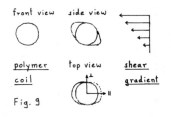

Fig. 9

4.2. Interacting polyelectrolyte systems

Little is known about the dynamics of polyelectrolytes in unscreened solution (no added salt) [12]. So it was a surprising result that an anisotropy could be observed (Fig. 12) when PTMA-Cl poly-(trimethylammonium-ethyl)-methacrylate-chloride (PTMA-Cl) of $M_w \approx 1.4 \cdot 10^6$ g/mol, $M_w/M_n \approx 2$, at ρ_2 = 4 mg/cm^3 dissolved in pure D$_2$O ($\eta \approx 1$ mPa.s) was subjected to a shear gradient of $G_T = 3000$ s^{-1}. This anisotropy is probably due to a slight disentanglement of the flexible polyelectrolyte chains, which are kept at maximum distance from each other by electrostatic repulsion (ring of maximum intensity in the scattering pattern), with a slight tendency to form a two-dimensional lattice of repelling rods.

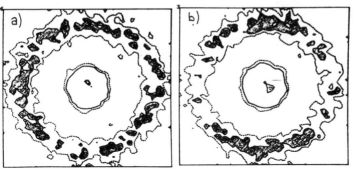

Fig. 12 Scattering pattern of PTMA.Cl in D_2O a) at rest b) in the shear gradient, corrected for solvent scattering and detector efficiency.

A similar, but much more pronounced result (Fig. 13) was obtained with polymeric vanadium oxide (V_2O_5) at $\rho_2 = 12.5$ mg/cm^3 in D_2O, an inorganic polyelectrolyte with a stiff ribbon-like structure [13]. With this system a saturation of the orientation was already achieved at $G \simeq 500$ s^{-1} in agreement with the results of Sjöberg [8]. He interpreted his findings with a rotational diffusion coefficient $D_{rot} \simeq 10$ s^{-1} for a single rod from the onset of anisotropy at $G_T \simeq D_{rot}$. However, from the disappearance

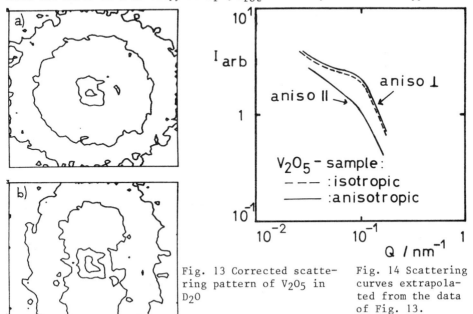

Fig. 13 Corrected scattering pattern of V_2O_5 in D_2O

Fig. 14 Scattering curves extrapolated from the data of Fig. 13.

of the shoulder in the scattering curve perpendicular to the flow it seems possible that the liquid has just undergone a transition from randomly oriented stiff

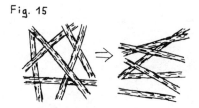

Fig. 15

ribbons with preferred maximum distance at the given concentrations to more or less parallel oriented ribbons with a similar maximum distance (Fig. 15). For this transition the intermolecular relaxation would then be of similar importance.

5. Outlook

Only a few examples have been given to illustrate the many possibilities for investigations with the newly designed flow-apparatus.

Future investigations could show, down to which intersegmental distance the shear gradient can distort the equilibrium statistical chain conformation. In concentrated solution the influence of local solvent viscosity and macroscopic solution viscosity on the distance dependent chain dynamics could be studied and finally in the melt chain dynamics could be measured in a time and space scale which is inaccessible by the neutron spin echo technique (cf. Fig. 6).

Literature

[1] a) F.R. Eirich, Ed., "Rheology, Theory and Applications", 5 Vols., Academic Press, New York 1956-1969
   b) R.B. Bird, R.C. Armstrong, O. Hassager, "Dynamics of Polymeric Liquids", 2 Vols., John Wiley & Sons, New York 1977
[2] A. Peterlin, W. Heller, M. Nakagaki, J. Chem. Phys. 28, 470 (1958)
[3] P.G. de Gennes, J. Chem. Phys. 60, 5030 (1974)
[4] cf. M.J. Miles, K. Tanaka, A. Keller, Polymer Preprints, Vol. 22 No. 1 p. 118 (1981)
[5] R.G. Kirste, B.R. Lehnen, Makromol. Chem. 177, 1137 (1976)
[6] W. Heller, R. Tubibian, M. Nakagaki, L. Papazian, J. Chem. Phys. 52, 4294 (1970)
[7] R.G. Kirste, R.C. Oberthür, in "Small Angle X-ray Scattering", ed. by O. Glatter and O. Kratky, Academic Press, London 1982, p. 387
[8] B. Sjöberg, J. Appl. Cryst. 13, 154 (1980)
[9] H. Tjakraatmadja, R. Hosemann, J. Springer, Colloid & Polymer Sci. 258, 1015 (1980)
[10] R.C. Oberthür, P. Lindner (to be published)
[11] D. Richter, J.B. Hayter, F. Mezei, B. Ewen, Phys. Rev. Letters 41, 1484 (1978), and B. Ewen, pers. Communication
[12] J. Hayter, G. Jannink, F. Brochart-Wyart, P.G. de Gennes, J. Physique Lettres 41, L-451 (1980)
[13] P. Aldebert, pers. communication
[14] H. Janeschitz-Kriegl, Fortschr. Hochpolymer-Forschung 6, 170 (1969)

Problems involving H and D polymer mixtures

S.F. Edwards

Theory of Condensed Matter, Cavendish Laboratory,
Madingley Road, Cambridge, CB3 OHE, U.K.

Introduction

The ideal situation is when the forces H - H, H - D, D - D are the same. But they are not. See [Buckingham, Hentschel J. Poly. Sci 18 853 (1980)]. Does this cause problems? Can we exploit it? Let us look at a series of cases of increasing complexity.

1. Case 1

A single species in concentrated solution
Let the density $\rho(r)$ have fourier components ρ_k. Internal energy $= \sum_k w \rho_k \rho_k^*$

$$\left(\sum_{ij} W(r_i - r_j) = \sum_k \rho_k \rho_k^* w \qquad \rho_k = \sum_i e^{ik \cdot r_i} \right)$$

If we look at the distribution of ρ_k we need the Jacobian to change variables from coordinates r_j to ρ_k. This is

$$\exp\left[- \sum_k{'} \rho_k \rho_k^* a^{-1}(k) \right]$$

where $a(k)$ is the correlation function in the absence of interaction (roughly speaking the entropy term).

$$a(k) = \left\langle \sum{'} e^{ik(r_\alpha - r_\beta)} \right\rangle \quad \alpha, \beta \text{ molecules on polymer}$$

$$= N \int_0^L \int_0^L e^{-k^2 \ell |s-s'|/6} \, ds \, ds' \qquad \ell \text{ step length of polymer}$$

$$= N \frac{6L}{k^2 \ell^3}\left(1 - \frac{6}{k^2 L \ell}\left(1 - e^{-k^2 L \ell/6}\right)\right)$$

$$= \frac{6L}{k^2 \ell^3} N \qquad \text{when } k \text{ large } \text{of } L^{-1}$$

$$= L^2 \qquad \text{when } k \text{ small } \text{of } L^{-1}.$$

So: $P(\rho) = \mathcal{N} \exp\left(- \sum_k \rho_k \rho_k^* \frac{k^2 \ell}{\rho_0}^{-1} - \sum_k w \rho_k \rho_k^* \right), \rho_0 = \frac{NL}{V\ell}$

and $\langle \rho_k \rho_{k'} \rangle = \delta(k+k') \left(\frac{k^2 \ell^{-1}}{\rho_0} + w \right)^{-1}$

$$= \delta(k+k') \frac{\rho_0/\ell}{k^2 + \xi^{-2}} .$$

ξ screening length

2. Case 2

Two Species with identical interactions but different scattering power (Idealised H : D mixture)

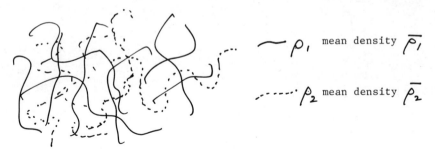

ρ_1 mean density $\bar{\rho}_1$

ρ_2 mean density $\bar{\rho}_2$

The distribution is

$$\exp\left[-\left(\sum_k \rho_{1k}\rho_{1k}^* \frac{k^2 l_1}{\bar{\rho}_1} + \rho_{2k}\rho_{2k}^* \frac{k^2 l_2}{\bar{\rho}_2}\right) - w\sum_k (\rho_1+\rho_2)_k (\rho_1+\rho_2)_k^*\right]$$

The scattering matrix is (for long polymers and taking $l_1 = l_2$)

$$\langle \rho\rho \rangle = \begin{pmatrix} \frac{k^2 l_1}{\bar{\rho}_1} + w & w \\ w & \frac{k^2 l_2}{\bar{\rho}_2} + w \end{pmatrix}^{-1}$$

$$= \frac{\begin{pmatrix} \frac{k^2 l}{\bar{\rho}_2} + w & -w \\ -w & \frac{k^2 l}{\bar{\rho}_1} + w \end{pmatrix}}{\frac{k^4 l^2}{\bar{\rho}_1 \bar{\rho}_2} + w k^2 l (\bar{\rho}_1^{-1} + \bar{\rho}_2^{-1})}$$

There are now two denominators

$$\frac{1}{k^2} \text{ and } \frac{1}{k^2 + w l (\bar{\rho}_1 + \bar{\rho}_2)} = \frac{1}{k^2 + w l \bar{\rho}} = \frac{1}{k^2 + \xi^{-2}}.$$

Thus if only one species is visible one gets a mixture of single polymer k^{-2} and screened polymer $(k^2 + \xi^{-2})^{-1}$.

The coefficients reflect the physical conditions; if D is the species (2), then

$$\langle \rho_2\rho_2^* \rangle = \frac{1}{k^2 l} \frac{\bar{\rho}_1 \bar{\rho}_2}{(\bar{\rho}_1 + \bar{\rho}_2)} + \frac{\bar{\rho}_2^2 (\bar{\rho}_1+\bar{\rho}_2)^{-1}}{k^2 + \xi^{-2}}$$

If $\bar{\rho}_1 = 0$ all one species

If $\bar{\rho}_2 \sim 0$ Dominated by $\frac{\bar{\rho}_2}{k^2}$, unscreened.

3. Case 3

Different visibilities and different interactions. Now we have $w_{11}, w_{12}=w_{21}, w_{22}$

$$\langle PP^* \rangle = \frac{\begin{pmatrix} \frac{k^2 \ell}{\bar{\rho}_2} + w_{22} & -w_{12} \\ -w_{12} & \frac{k^2 \ell}{\bar{\rho}_1} + w_{11} \end{pmatrix}}{k^4 \ell^2 (\bar{\rho}_1^{-1} + \bar{\rho}_2^{-1}) + k^2 \ell \left(\frac{w_{11}}{\bar{\rho}_2} + \frac{w_{22}}{\bar{\rho}_1}\right) + Det(\underline{w})}$$

This clearly now has two screening lengths

$$(k^2 + \xi_1^{-2}) \qquad (k^2 + \xi_2^{-2})$$

where $\xi_1 \to \xi \quad \xi_2 \to \infty \quad$ as $\underline{w} \to \begin{pmatrix} w & w \\ w & w \end{pmatrix}$.

Thus one only gets the k^{-2} term if all the forces are the same and in general one gets screening albeit weak in one term.

4. Case 4

Semi-dilute i.e. very long polymers, but also very dilute, so there is enough of <u>one</u> polymer between neighbours to have excluded volume effect.

Then we must have for Case (1), $f(k\xi)$ but this function is not known. It must look something like $k^2 + \xi^{-1}$ and k^2.

5. Case 5

High density i.e. incompressible material. Modify w to

$$\left(w - \frac{k^2 \ell}{\bar{\rho}_0}\right)$$

so that the polymer effect vanishes when $\bar{\rho}_1 + \bar{\rho}_2 = \bar{\rho}_0$

This leaves two terms

$$k^{-2} \quad \text{as before and} \quad \left(k^2 \left[(\bar{\rho}_1 + \bar{\rho}_2)^{-1} - \bar{\rho}_0^{-1}\right] + w\right),$$

Where w^{-1} is now to be identified as the melt scattering i.e. as $k \to o$ the compressibility.

6. <u>Now return to different forces.</u> For simplicity put two species to have equal densities. Then also for simplicity take the matrix $\underline{\underline{w}}$ to be $\begin{smallmatrix} w(\rho_1+\rho_2)^2 \\ -u(\rho_1-\rho_2)^2 \end{smallmatrix}$. Note the signs. The interesting case is to take the material incompressible $\rho_1 + \rho_2 = 1$ (w.l.g) but beware, there are self energy problems. Then since $k=0$ is a fixed mean, if we write

$$\varphi_k = \rho_{1k} - \rho_{2k} \qquad \varphi_0 = \phi = 0 .$$ But leave ϕ in the formulae

and the distribution is

$$exp\left[-\sum \rho_1 \rho_1^* \frac{k^2 \ell}{\bar\rho_1} - \sum \rho_2 \rho_2^* \frac{k^2 \ell}{\bar\rho_2} - w \sum (\rho_1+\rho_2)^2 + u \sum (\rho_1-\rho_2)^2\right].$$

Using
$$\rho_{1k} + \rho_{2k} = \delta(k)$$
$$\rho_{1k} - \rho_{2k} = \varphi_k$$

we can rapidly transform the probability into

$$exp\left[-\sum \frac{\varphi_k \varphi_k^* k^2 \ell}{1-\phi^2} + u \sum \varphi_k \varphi_k^*\right]$$

where ϕ is the mean φ_0 and is zero. $a(k)$ is a function with a minimum L^{-1} (we have neglected L^{-1} relative to k^2, but now we must be careful) $a^{-1} = k^2\ell + L^{-1}$ is a good model. Thus

$$\langle \varphi_k \varphi_{k'}^* \rangle = \frac{\delta(k+k')}{a^{-1}(k) - u}$$

provided $a^{-1} > u$. Otherwise it is physically undefined: the system then phase separates.

At high temperatures u contains $\frac{1}{T}$ and is small, but there is evidence that at lower temperatures the system does break up into H rich and D rich regions for polyethelene. Not apparently for polypropylene etc.

But the effect is small as the driving force is small and hence slow.

It is interesting to complete a quadratic calculation:

$$exp\left[-\sum_k \frac{\varphi_k \varphi_k^* (k^2\ell + L^{-1})}{1-\phi^2} + u \sum \varphi_k \varphi_k^*\right]$$

Remembering normalisations, if we take ϕ to be slowly varying, we get

$$exp\left[-\frac{V}{(2\pi)^3} \int d^3k \, log\left(1 - \frac{u(1-\phi^2)}{k^2\ell + L^{-1}}\right)\right]$$

The model for a^{-1} is $k^2\ell + L^{-1}$ and then the integral can be performed assuming that u is truly $u(k)$ and there is a high k cut off. The integral gives

$$\frac{V}{6\pi\ell}\left[(L^{-1} - u(1-\phi^2))^{3/2} - L^{-3/2} + \frac{3}{2} u(1-\phi^2) \int^{a^{-1}} dk\right]$$

a the cut off

The problem of this cut off associated with the assumption of strict incompressibility. If this is softened in a way which is the normal mathematical method of studying short distances, then the integral becomes

$$\frac{V}{6\pi}\left[\left(L^{-1}-u(1-\phi^2)\right)^{3/2} - L^{-3/2} + \frac{3}{2}u(1-\phi^2)\right]$$

(i.e. linear term in u is removed.)
This discussion is for strict equilibrium where ϕ is a constant zero. In fact it will have a spatial variation and it is clear physically that when u is an attraction the sample will ultimately separate into part with $\phi = 1$ and the rest $\phi = -1$. If ϕ is taken to be slowly varying, one must add in the term $u\int \phi^2 d^3r$ from the interaction and

$$\int \frac{(L\bar{L})^{-1}\phi^2 + (\nabla\phi)^2}{(1-\phi^2)} d^3r$$

from the entropy. The final form is now

$$\exp\left[-\int \frac{(\ell L)^{-1}\phi^2 + (\nabla\phi)^2}{(1-\phi^2)} d^3r + u\int \phi^2 d^3r + \frac{1}{6\pi}\int \left(L^{-1} - u(1-\phi^2)\right)^{3/2} d^3r \right.$$

(b from the cut off). $\left. - \frac{3}{2}\frac{ub}{6\pi}\int (1-\phi^2) d^3r \right].$

We are now assuming that there are variations which are slow enough to integrate our fluctuations leaving an effective free energy depending on a mean $\phi(r)$ the difference in density of H polymer and D polymer.

The driving force of the phase separation is u and if this is small the process is slow. Although the field minimizing the free energy is ± 1, dynamically we can start with $\phi = 0$ at $u = 0$, change u to be an attraction (by varying temperature for example). If the free energy functional is $F(\phi)$, the Smoluchowski equation will be

$$\left(\frac{\partial}{\partial t} - \int d^3r \frac{\partial}{\partial \phi} D \nabla^2 \left(\frac{kT}{\partial \phi} + \frac{\partial F}{\partial \phi}\right)\right) P([\phi], t) = 0$$

where D is known from polymer theory.

The point to make here is that phenomenological forms for such problems have been developed since the first paper on alloy nucleation by Cahn and Hilliard (J. Chem. Phys. 31, 688 (1959) and notably in papers by Langer. (Langer J.S. Bar-on M, Miller H.D. Phys. Rev. A 11, 1417 [1975]), but we are here using the long chain nature of polymers able to write a derived expression.

Acknowledgement This work arose from the author reading the Ph.D thesis of Nigel Goldenfeld (1982)

The contribution of neutron scattering to molecular biology

Heinrich B. Stuhrmann

Institut für Physikalische Chemie der Universität Mainz
D6500 Mainz, Germany

Abstract. About half of the atoms of living cells are hydrogens, and nearly all biological applications of neutron scattering rely on the well-known difference in the scattering lengths of the proton and the deuteron. This intruduces us to a wide variety of biological problems, which are related with hydrogen in water, proteins, nucleic acids and lipids. Neutron scattering gives an answer to both structural and dynamical aspects of the system in question. With deuterium labelled samples unambiguous information about molecular structure and motion becomes accessible. The architecture of viruses, cell membranes and gene expressing molecules has become a lot clearer with neutron scattering.

1. Introduction.

Of many physical techniques that have been used to study biological structures, the use of thermal neutrons has been one of the more recently applied. Within the last ten years since 1972, neutron beams have been shown to provide useful information about the properties of subcellular structures with the following types of measurements:

a. Diffraction from protein crystals and most recently from virus crystals

b. Small angle diffraction from lipid bilayers of biological membranes and connective tissue

c. Small angle scattering from solutions of biological macromolecules

d. Quasielastic and inelastic scattering measurements of dynamical properties.

The emphasis is predominantly on small angle scattering experiments. More than 80% of the biological projects submitted to the Institut Max von Laue - Paul Langevin at Grenoble fall into the categories b) and c). The majority of the small angle experiments has been carried out at the ILL instrument D11, which started to operate in September 1972, just ten years ago. Let us first look into the specifications

of this instrument (Ibel 1976), which is attracting many
biologists and biophysicists not only from the member states
of the ILL but also scientists from all over the world.

2. The Neutron Small Angle Camera D11.

This instrument (Fig.1) at the high flux reactor in Grenoble
is the realization of a concept which has been developed at
the FRJ2 reactor in Jülich (Schmatz and Schelten 1971) and
further optimized by Ibel, Schmatz and Springer (1971).

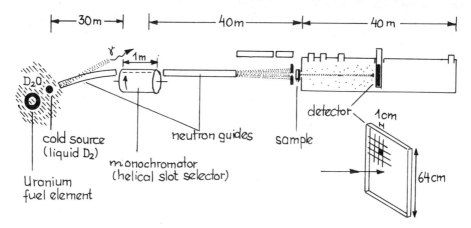

Fig. 1 Small angle camera D11 at the high flux reactor
of the Institute Max von Laue - Paul Langevin in Grenoble.

Neutrons released from the fuel element are thermalized in
heavy water. They are further cooled down by liquid deuterium
and reflected into the small angle camera by a 28 m long bent
guide tube (horizontal width: 3 cm, radius: 2700 m) at
grazing incidence, which eliminates γ-radiation and fast
neutrons. The resulting wavelength distribution of the cold
neutrons is centered at 6 Å wavelengths approximately. The
spectral distribution is further narrowed to 9% fhw by a
helical slot selector, which can be tuned to any wavelength
between 4 and 25 Å just by changing its rotatory speed.

Then a further limitation of the directional spread of the
neutron flight paths has to occur. This is done in a 40 m long
collimation path which can be shortened by the introduction
of straight neutron guide sections. The typical beam size
accepted by the specimen will be 15 mm wide 25 mm high.
The detector is a boron trifluoride containing multielectrode
chamber with gas amplification. Its sensitive area consists
of 64x64 channels, the cross section of one element being
1 cm^2. The direct beam is caught by a rectangular beam stop.
(3x5cm) in front of the center of the area detector. At 11 cm
distance from the center, the corrections for the wavelength
spread on the one side and beam divergence and spatial resolution of the detector on the other side are equal. On the
whole, these corrections are small, except very near the

beam stop where substantial collimation errors do occur.
The scattering process is described by the conservation of
momentum and energy

$$\vec{Q} = \vec{k}_1 - \vec{k}_o \qquad (1)$$

and

$$\hbar\omega = \hbar^2(k_o^2 - k_1^2)/2m \qquad (2)$$

$\vec{k}_o$ and $\vec{k}_1$ are the wave vectors of the incident and scattered
neutron respectively. In the case of inelastic scattering the
magnitudes of $\vec{k}_o$ and $\vec{k}_1$ are no longer equal. There has been
a change in the neutron energy and a phonon of frequency ω has
been generated. This is also the way, that fast neutrons from
the fuel element are slowed down to speeds of some km/s by
collisions with the D_2O molecules of the moderator. Small
angle scattering involves nearly no energy transfer. For
elastic scattering the magnitude of the transferred momentum
is

$$Q = \frac{4\pi}{\lambda} \sin\theta \qquad (3)$$

λ = wavelength, 2θ = scattering angle

D11 is an extremely versatile instrument. The scattering intensity may be measured over a wide range of Q, from 0.0005 $Å^{-1}$ to 0.5 $Å^{-1}$. According to the Bragg equation

$$n\lambda = 2d\sin\theta \qquad (4)$$

spacings of periodic structures from d = 12000 Å to 12 Å can
be resolved. This range of structural resolution matches the
characteristic overall dimensions of proteins (30 - 60 Å),
ribosomes (200 - 300 Å), viruses and phages (200 - 2000 Å)
and smaller living cells.

3. Basic Features of Neutron Scattering.

The central objective of scattering studies on biological
macromolecules is to determine molecular conformations and
arrangements, particularly those which may contribute to
specific function. Usually this must be done from a very limited amount of scattering data and so independent methods of
analysis are valuable. The principles of neutron scattering
are very similar to those of X-rays, but some important differences exist which are due to the differences in the two
radiations.

If the scattering specimen is a single macromolecule with no
regular separations between neighbouring structures, then the
scattering is continuous. The amplitude of the scattered
radiation is determined by the scattering density distribution $\rho(\vec{r})$ of the macromolecule according to the Fourier
relationship :

$$A(\vec{s}) = \int \rho(\vec{r}) e^{-2\pi i \vec{s}\cdot\vec{r}} d^3r \qquad (5)$$

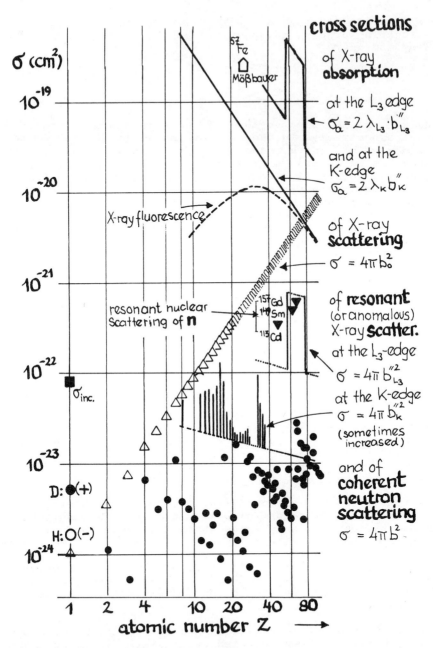

Fig.2 Cross sections of scattering absorption and fluorescence as a function of the atomic number Z. The neutron coherent cross sections compare to the anomalous X-ray scattering cross sections at the K edges. The resonant neutron scattering of ^{113}Cd, ^{149}Sm and ^{157}Gd equals that of X-ray anomalous scattering at L_3-edges. Mößbauer scattering of ^{57}Fe excels any other coherent scattering. The measurements have to be corrected for absorption and incoherent fluorescence.

The Fourier inverse of this expression is

$$\rho(\vec{r}) = \int A(\vec{s}) e^{2\pi i \vec{s} \cdot \vec{r}} d^3s \tag{6}$$

where $\vec{s} = \vec{Q}/2\pi$

It gives a three-dimensional coherent scattering density distribution from the values $A(\vec{s})$, which must be determined in the scattering experiment. The problem of phase determination results from only being able to measure the magnitude of $A(\vec{s})$, since the measured intensity is proportional to $|A(\vec{s})|^2$.

In writing the Fourier relationships, the structure was expressed as the density of the coherent scattering lengths. For X-rays, the scattering centers are the electrons of the atoms. Every electron has the same X-ray coherent scattering length b_e ($=0.28 \cdot 10^{-12}$ cm for $\theta=0$). Consequently, $\rho(\vec{r})$ is just the familiar electron density of the structure. The X-ray scattering cross section of an atom with Z electrons then is $4\pi Z^2 b_e^2$ (Fig.2). For neutrons, the scattering centers are nuclei of atoms and each nucleus has a characteristic neutron coherent scattering length b. From Fig.2 it is seen that there is hardly any regularity of the neutron scattering cross sections, as there is with X-rays. This is because neutron scattering is the result of nuclear forces and is dominated by a process known as resonance scattering. Some few nuclei (e.g. ^{149}Sm, ^{113}Cd, ^{157}Gd) exhibit a strong wavelength dependance (dispersion) at thermal neutron energies. As is shown in Fig.3 there is a change of the magnitude and the phase of the scattering length b near the resonance energy which is taken into account by adding the dispersion of the real b' and the imaginary b" to the constant b_o.

$$b = b_o + b' + i b" \tag{7}$$

b' is an order of magnitude larger than b, and it assumes both negative and positive values. A consequence of this is that the scattering length of nuclei may be negative even at energies far away from the resonance energy, as for hydrogen.

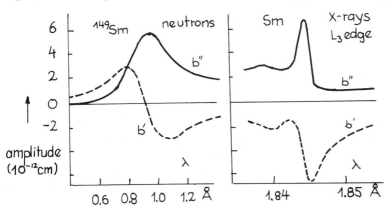

Fig.3 Dispersion of the resonant contributions to b' and b"

The dispersion of resonant (or anomalous) X-ray scattering, especially at many L_3-absorption edges, compares to that of resonant neutron scattering. However, the constant part b_0 is no longer relatively small, but it is comparable or larger than the X-ray resonant terms. In principle, also with X-rays negative scattering lengths might be encountered.

Because neutrons are scattered by nuclei of atoms, the scattering lengths are different for different isotopes of the same element. Isotopic substitutions provide a direct means of isomorphous changes of the scattering amplitudes. The large difference between the coherent scattering lengths of hydrogen ($b=-0.374 \cdot 10^{-12}$ cm) and of deuterium ($b=+0.667 \cdot 10^{-12}$ cm) is particularly useful in neutron scattering studies of biological problems. It corresponds to a change of 3.7 electrons in X-ray scattering, which is in fact slightly smaller than the change of the X-ray scattering lengths at the K-absorption edges of other elements. However it is much bigger than the X-ray scattering length of the single electron of the hydrogen atom. Furthermore, hydrogen is the most abundant atom in all biomolecules. The techniques of specific substitution of hydrogen by deuterium have made considerable progress in the last years, as similar requirements exist in other fields, like nuclear magnetic resonance spectroscopy.

The coherent scattering density $\rho(\vec{r})$ may be calculated for biological substances by summing the coherent amplitudes of the constituent atoms and dividing by the volume occupied by these atoms (Table 1). It is noted that the scattering density of lipid, protein and ribonucleic acids fall between the very low value for H_2O and the very high value for D_2O. This property has considerable use in neutron beam studies of many subcellular structures.

Table 1. Coherent scattering densities (in units of 10^{10} cm^{-2})

	X-ray	neutron
H_2O	9.35	-0.56
D_2O	9.35	6.38
$(-CH_2)_n$	7.5 to 8.2	-0.28 to -0.34
protein	11. to 12.	1.5 to 2.3
rRNA	14.	4.2

4. Contrast variation.

The easiest way of isomorphous replacement in a biological system is to diffuse D_2O into its aqueous phase. The effect on the neutron scattering pattern will be tremendous. It is particularly informative in the case of those macromolecules which consist of two different components, like lipids and proteins in lipoproteins. As the scattering density of lipids nearly equals that of water, neutron scattering of lipoprotein in H_2O shows predominantly the protein arrangement. The H_2O/D_2O mixture containing 40% D_2O matches the density of proteins and polar groups of lipids, and only the aliphatic fatty acid chains are seen (Stuhrmann et al. 1975).

A more rigorous approach starts from the concept of contrast. In dilute solutions the neutron scattering intensity originates from the difference between the scattering density distribution $\rho(\vec{r})$ of the dissolved macromolecule and the density of the solvent:

$$\rho(\vec{r}) = \rho(\vec{r})_{solute} - \rho_{solvent} \qquad (8)$$

The solvent density has been assumed to appear uniform at low momentum transfer. A similar equation holds for the mean values:

$$\bar{\rho} = \rho_{solute} - \rho_{solvent} \equiv contrast \qquad (9)$$

At vanishing contrast, the scattering density distribution $\rho_s(\vec{r})$ merits special attention. $\rho_s(\vec{r})$ describes the positive and negative deviations of the density distribution of the dissolved particle from its mean value. The integral over $\rho_s(\vec{r})$ is zero. Consequently, there is no coherent scattering of the particle at zero angle, and only relatively weak scattering is encountered at wider angles. It is due to the internal structure, e.g. the mutual arrangement of lipids and proteins in lipoprotein.

On decreasing the scattering density of the solvent, the excess scattering density of the particle with respect to the solvent increases everywhere inside the particle boundaries. The added contrast is taken into account by:

$$\rho(\vec{r}) = \rho_s(\vec{r}) + \bar{\rho}\,\rho_c(\vec{r}) \qquad (10)$$

where $\rho_c(\vec{r}) = 1$ everywhere inside the particle volume and it is zero elsewhere. It is the product $\bar{\rho}\,\rho_c(\vec{r})$ which describes the uniform increase of the excess scattering density $\rho(\vec{r})$. The idea of contrast variation was first developed in crystallographic studies on hemoglobin by Bragg and Perutz (1952).

With neutron scattering in H_2O/D_2O mixtures, $\rho_s(\vec{r})$ and $\rho_c(\vec{r})$ need a slightly different interpretation. About one quarter of the hydrogen atoms in proteins may dissociate and are therefore readily exchanged by deuterium. The variation of the contrast therefore is somewhat smaller than the change in density of the H_2O/D_2O mixture. This is taken into account by $\rho_c(\vec{r}) \leq 1$. In fatty acid domains no hydrogen exchange will occur ($\rho_c(\vec{r})=1$) whereas polar regions are described by $\rho_c(\vec{r}) \sim .7$. $\rho_s(\vec{r})$ represents the internal structure with the hydrogen atoms partly exchanged by deuterium according to the H_2O/D_2O ratio of the solvent at bouyancy.

With this in mind we can start to analyze neutron scattering. As the intensity is the square of the amplitude, its contrast dependence always assumes the following form (Stuhrmann and Kirste 1965):

$$I(\vec{s}) = I_s(\vec{s}) + \bar{\rho}\,I_{cs}(\vec{s}) + \bar{\rho}^2 I_c(\vec{s}) \qquad (11)$$

For randomly oriented particles the intensity only depends on the magnitude of $\vec{s}$. One of the basic scattering functions, $I_s(s)$, has already been discussed, as it is the scattering function of $\varrho_s(\vec{r})$. $I_c(s)$ is the scattering function of the volume excluded to the solvent and $I_{cs}(s)$ is a cross term originating from the convolution of the macromolecular shape $\varrho_c(\vec{r})$ with the internal structure $\varrho_s(\vec{r})$. The measurement of neutron scattering in at least three different H_2O/D_2O mixtures provides the three basic scattering functions.

The analysis of $I_c(s)$ is most promising. The known family of possible structures belonging to a one-particle scattering curve can be greatly reduced, if the assumption of a shape model can be made. Using a multipole expansion and restriction to spherical harmonics $Y_{\ell,m}$ with $l < 4$, a rather detailed model has been found in the case of the large ribosomal subunit of E.coli ribosomes (Stuhrmann et al. 1977). It agrees remarkably well with the model from observations in an electron microscope (Fig.4).

The analysis of the internal structure relies on the evaluation of the quadratic terms of the power series of the basic scattering functions. It tells us whether the macromolecule has a core of relatively high or low density with respect to its peripheric domains. Many complex macromolecules have been investigated by this method. Ribosomes have

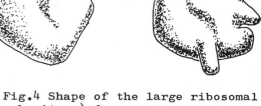

Fig.4 Shape of the large ribosomal subunit. a) from neutron scattering, b) from electron microscopy (Lake 1976).

a relatively dense core surrounded by ribosomal proteins of lower density (Serdyuk 1976). With chromatin the inverse model is true: the histones are surrounded by a DNA coil (Hjelm et al. 1977). - An exhaustive analysis of the scattering functions has been proposed recently by Svergun et al.(1982).

With spherical structures the analysis of the basic scattering functions can easily go beyond the quadratic approximation. The radial density distribution is directly calculated from the square root of $I(s)$ (see Eq.6). As an example, we mention the investigation of tobacco bushy stunt (TBS) virus (Chauvin, Witz and Jacrot 1978). The radial density distribution unambiguously shows the spherical protein shell and the RNA core penetrated by unfolded protein (Fig.5).

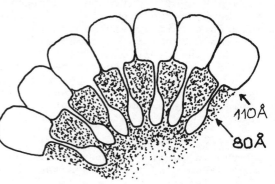

Fig 5. Model for the internal structure of TBS virus. Shaded area corresponds to RNA. Proteins are shown schematically. The interior of the virus is disordered

5. Three-dimensional Models.

These can be obtained from continuous neutron scattering only in some favourable cases. More rigorous attempts to resolve the detailed structure of complex macromolecules rely on deuterium labelling techniques. Engelman and Moore (1972) and Hoppe (1972) proposed a new method for the determination of quaternary structures which became a most powerful tool for the investigation of the structure of ribosomes.

Each living cell contains ribosomes. This cell organelle translates the genetic information, which is coded as a sequence of four different nucleotides in nucleic acids (DNA or mRNA) into the sequence of twenty different amino acids of proteins. Each triplet of nucleotides (codon) corresponds to either an amino acid or to a processing signal for the ribosome. Most of the investigations on protein biosynthesis are performed on the ribosome of E.coli bacteria. This ribosome with a relative molecular mass of about 3 million can be separated into two unequal subunits. The small subunit consists of one ribosomal RNA and 21 different proteins (S1-S21), the large subunit of two (a short and a very long) rRNA molecules and 32 different proteins (L1-L34). The 55 components of the ribosome can be separated and reconstituted again to fully active ribosomes (Traub and Nomura 1968, Nierhaus and Dohme 1974).

The multicomponent character of the ribosome on the one hand and the lack of suitable crystals together with the difficulties that can be forseen for a crystallographic study on the other hand has led to the method for the reconstruction of the three-dimensional arrangement of the ribosomal proteins within

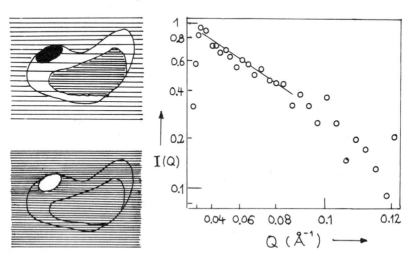

Fig.6 Schematic density distribution of the ribosome. a) native, one protein deuterated. b) "invisible" ribosome, one protein native.

Fig.7 The small angle scattering curve of the L4 protein in situ. Mol. weight: 22000, radius of gyration 20 Å, axial ratio 3:1 (Nierhaus et al.1982)

the subunits. The idea of this method consists in labelling two individual proteins and then to measure the distance between them. From a large number of distances between the proteins the relative co-ordinates of the proteins can be determined by triangulation. The feasability of such distance measurements had first been reported by Engelman, Moore and Schoenborn (1975) and Hoppe et al.(1975). The study of the small ribosomal subunit has progressed considerably since then (Ramakrishnan et al.1981). Meanwhile, the position of 14 proteins has been determined by triangulation. Unfortunately, the proteins have to be represented as spheres in this model. Although the experiments provide the pair distribution, which contains additional information on the shape and the orientation of the mostly anisotropic proteins, this information is not easily disentangled.

Nierhaus (1982) therefore proposed a new strategy, which aims at the structure determination of the large ribosomal subunit, and which is particularly suitable for the structure determination of each ribosomal component <u>in situ</u>. The labelling technique is reversed, as it starts from hydrogen labelled protein in deuterated ribosomes. More precisely speaking, the degree of deuteration of rRNA and the proteins is chosen in such a way that its scattering densities both match the scattering density of heavy water (Fig.6). This not only renders the ribosome invisible to neutron small angle scattering, but it also eliminates nearly all of the otherwise large incoherent scattering from protons. Another sample is prepared from the same deuterated material, but with one or two native proteins, which have been added in excess during the reconstitution process.

The weight fraction of one labelled protein in the ribosome molecule is of the order of 0.01. In a 2% solution its weight fraction in the irradiated volume is 0.0002. Consequently, coherent scattering is small. Experiments have shown that it is only some percent of the total scattering (mainly incoherent scattering of deuterons and residual protons). A more delicate question is to what extent the reference particle is really invisible, as any inhomogeneity of the matrix would lead to irrecoverable errors in the shape determination of the labelled protein (Moore 1981). Enormous efforts of cell biologists and biochemists were necessary to adjust the density of the residual proteins and rRNA within the reqired limits. Fig.7 shows the small angle scattering curve of the L4 protein <u>in situ</u> (Nierhaus et al. 1982, May et al. 1982).

6. Periodic Structures (Membranes).

If the scattering specimen consists of layered structures, e.g. membranes, then the scattered radiation experiences interference, which is constructive in directions given by the Bragg equation (Eq.4). The amplitude of each diffraction peak is determined by the Fourier relationship (Eq.5). For most lamellar structures, that give Bragg diffraction, the structure factors F_n are calcultated from the diffraction intensities according to

$$F_n = (\sin 2 \theta_n \ I_n)^{\frac{1}{2}} \tag{12}$$

If the structure is centrosymmetric, the Fourier expression for
the coherent scattering density becomes a simple Fourier cosine
series

$$\varrho(x) = \sum_n (\pm) |F_n| \cos 2\pi nx/d \qquad (13)$$

A plus or minus sign is used in the summation depending on
whether the phase of reflection is 0 or π. A central problem in
the structure determination is that of determining the phase
assignments which correspond to the physical structure.
A method of phase assigment, which is unique to neutron diffraction, involves changing the scattering amplitude of the water
in the structure by H_2O/D_2O exchange. For centrosymmetric structures the structure factors of the Bragg reflections are linear
functions of the isotopic composition of the water (Fig.8).
With only simple assumptions about the distribution of water in
the bilayer, these functions may be interpreted in terms of
phase assignments (Worcester 1976).

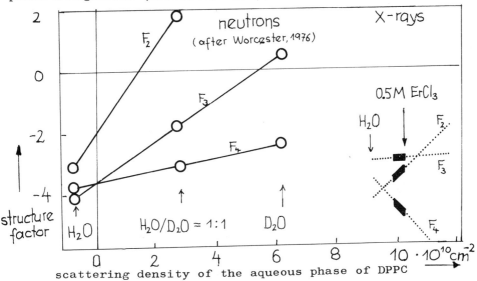

Fig.8. Structure factors of dipalmitoyl phosphatidylcholin bilayer containing 20% water. For anomalous X-ray scattering
0.5M $ErCl_3$ solution has been used. The decrease of the density
by b' of erbium at the L_3-absorption edge compares to a change
by 3% D_2O in neutron scattering (Stuhrmann and Büldt 1982).

A most spectacular result from neutron scattering has been obtained in the case of purple membrane of halobacterium halobi-
um. The importance of the structure of this membrane is that
it is by far the simplest known example of a specialized biological energy transducer. Light is absorbed by retinal bound
to 27000 molecular weight protein which then pumps the protons
across the membrane against the electrochemical potential gradient, and the proton gradient is then used as source of stored energy for the organism. - Starting from the low resolution
density map of electron microscopy and the amino acid sequence
neutron scattering offers a unique way of positioning the amino acid chain in the 7 α-helices of the protein. The amino

acids valine and phenylalanine were deuterated by biosynthetical incorporation. Fourier difference techniques showed clearly that valine is distributed towards the periphery of a single bacteriorhodopsin molecule, whereas phenylalanine is distributed towards the center. The consequence of this is that the charged polar groups of the protein tend to lie at the molecular interior, while the non-polar surfaces make contact with the lipid regions. This observation may help in the understanding of many membrane proteins.

References

Bragg W.L. and Perutz M.F. 1952 Acta Cryst 5 277-283
Chauvin C.,Witz J. and Jacrot B. 1978 J.Mol.Biol. 124 641-651
Engelman D.M. and Moore P.B. 1972 Proc. Nat. Acad. Sci. USA 69 1997-1999
Engelman D.M., Moore P.B. and Schoenborn B.P. 1976 Brookhaven Symp. Biol. 27 IV 20-27
Engelman D.M. and Zaccai G. 1980 Proc. Nat Acad. Sci. USA 77 5894-5898
Hoppe W. 1972 Isr. J. Chem. 10 321-333
Hoppe W., May R., Stöckel P., Lorenz S., Erdmann V.A., Wittmann H.G., Crespi H.L., Kratz J.J., and Ibel K. 1976 Brookhaven Symp. Biol. 27 IV 38-48
Hjem R.P., Kneale G.G., Suau P., Baldwin J.P., Bradbury E.M. and Ibel K. 1977 Cell 10 139-151
Ibel K., Schmatz W. and Springer T. 1971 Atomkernenergie 17 15-18
Ibel K. 1976 J.Appl.Cryst. 9 296-309
Lake J. 1976 J.Mol.Biol. 165 131-159
May R.P., Stuhrmann H.B. and Nierhaus K.H. 1982 Brookhaven Symp. Biol (in press)
Moore P.B. 1981 J.Appl.Cryst. 14 237-240
Nierhaus K.H. and Dohme F. 1974 Proc.Nat.Acad.Sci.USA 4713-7
Nierhaus K.H., Lietzke R., May R.P., Nowotny V., Schulze H., Simpson K., Wurmbach P. and Stuhrmann H.B. 1982 Proc. Nat Acad. Sci. USA (submitted)
Ramakrishnan V.R., Yakubi S., Sillers I.Y., Schindler D.G., Engelman D.M. and Moore P.B. 1981 J.Mol.Biol. 153 739-760
Serdyuk I.N. 1976 Brookhaven Symp. Biol. 27 IV 49-59
Svergun D.I., Feigin L.A. and Schedrin B.M. 1982 Acta Cryst.A in press
Stuhrmann H.B. and Kirste R.G. 1965 Z.f.physik.Chem. Neue Folge 46 247-250
Stuhrmann H.B., Tardieu A., Mateu L., Sardet C., Luzzati V., Aggerbeck L. and Scanu A.M. 1975 Proc. Nat.Acad.Sci. USA 72 2270-2273
Stuhrmann H.B. and Büldt G. (manuscript in preparation)
Traub P. and Nomura M. 1968 Proc. Nat.Acad.Sci. USA 59 777-784
Worcester D.L. 1976 Biol. Membranes 3 1-48

Label triangulation

Roland P. May

Institut Laue-Langevin, B.P. 156X, 38042 Grenoble Cedex, France

Abstract. Label Triangulation (LT) with neutrons allows the investigation of the quaternary structure of biological multicomponent complexes under native conditions. Provided that the complex can be fully separated into and reconstituted from its single - protonated and deuterated - components, small angle neutron scattering (SANS) can give selective information on shapes and pair distances of these components. Following basic geometrical rules, the spatial arrangement of the components can be reconstructed from these data. LT has so far been successfully applied to the small and large ribosomal subunits and the transcriptase of E. coli.

1. Introduction

X-ray, and recently also neutron crystallography have proven to be the most powerful tools for the structural investigation of biological macromolecules. Yet there are two main obstacles for their application to all interesting biological objects: a) Not all of them can be readily crystallised, or the crystals are not sufficiently well ordered. b) In the case of large asymmetric objects, the numerical requirements might exceed the capacity of computers available to date. We may add that in certain cases the crystallisation conditions deviate extremely from the native conditions necessary for the _function_ of the molecule. The same is certainly true for electron microscopy, which has its merits mainly at the level of larger structural elements.

Conventional small angle scattering with its advantage of maintaining the natural environment gives only coarse structural information, even if advanced methods are used like in the SANS study of the large (50S) ribosomal subunit (Stuhrmann et al., 1976).

An alternative approach - LT - was proposed in 1972 by Hoppe and by Engelman and Moore on the basis of a paper by Kratky and Worthmann (1947), who had applied the idea to organic compounds.

2. Theoretical background

The basic idea of LT is to reconstruct the positions of the components of a multicomponent complex from the distances between the components. The distances can be determined by labelling pairs of individual components in the complex and separating their scattering from all other contributions in a SANS experiment. Thus limited information is obtained more accurately at the expense of an increased biochemical effort: labelling takes advantage of the difference in coherent scattering length between hydrogen ($b=-0.374 \times 10^{-12}$ cm) and deuterium ($b=+0.667 \times 10^{-12}$ cm). The three main requirements for LT are: 1) The complex (e.g. ribosome) can be separated into its components (proteins and RNA). 2) The components reassemble to form the original complex using appropriate biochemical conditions (reconstitution). 3) The components are available in labelled form, e.g. by growing micro-organisms in culture media with high contents of D_2O.

The distance measurement requires the preparation of four different equimolar samples of the complex (Fig. 1, top): Solution 1 contains complexes where both components are labelled, solution 2 and 3 complexes where one of the two is labelled, and solution 4 complexes where none of the two components are labelled. Subtracting the scattering from samples 2 plus 3 from the scattering of samples 1 plus 4 yields a pair distance interference term arising exclusively from the vectors connecting volume elements in one component with those in the other.

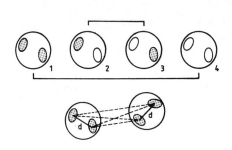

Fig. 1

Interparticle interferences (dashed lines in Fig. 1, bottom), which can appear due to high concentrations or aggregates, are also cancelled out if mixtures of 2 plus 3 and 1 plus 4 are used instead of homogeneous solutions (Hoppe, 1973).

In order to minimize scattering contributions from the unlabelled part of the complex, the buffer solution is usually chosen to match the average scattering length density of the unlabelled complex (Fig. 2, top). This is simply achieved by replacing a part of the water in the buffer solution with heavy water (contrast variation, see contribution of Stuhrmann in this volume).

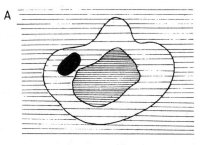

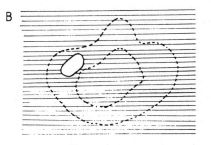

Fig. 2

3. Applications of LT

LT can be applied to a variety of multicomponent complexes, such as ribosomes, DNA dependent RNA polymerase, complexes of messenger RNA or ribosomal RNA with proteins, aminoacyl tRNA synthetases, tryptophan synthase and nucleosomes. In any case, the limiting factor is the biochemical expense involved. Three successful examples are described here.

3.1 30S ribosomal subunit of E. coli

Ribosomes are small cell organelles present in all living cells. They translate the genetic information coded in nucleic acids into the amino acid sequence of proteins and peptides (for a review see Nierhaus, 1982). All ribosomes can be subdivided into two unequal subunits. In the case of the bacterium E. coli, the small subunit consists of 21 different proteins and a ribonucleic acid (RNA) of 1541 nucleotides, which makes up two thirds of the total molecular mass of about 1 million.

Engelman, Moore and coworkers are undertaking a programme to build a model of the spatial arrangement of the proteins within the small ribosomal subunit of E. coli. Their procedure uses the second moments of the pair distance distribution functions obtained by a sine Fourier transform of the pair distance interference curve for the calculation of protein position coordinates and rough values of the protein radii of gyration (Ramakrishnan and Moore, 1981). At present, 62 protein pair separations have been determined and the positions of 14 proteins are known in this most advanced LT study (Ramakrishnan et al. 1981, Moore 1982).

3.2 DNA-dependent RNA polymerase of E. coli

RNA polymerase (transcriptase) is a multi-subunit enzyme which catalyses the first basic step in gene expression, the transcription of DNA into mRNA. The enzyme can be separated into five protein components two of which are identical and form a dimer. In contrast to ribosomes, polymerase consists only of proteins. Matching the coherent scattering length density of the protonated proteins in a polymerase reconstituted with one deuterated protein by a buffer solution containing about 40% D_2O, shape parameters of the single components could be obtained in addition to the protein positions in a LT study of RNA polymerase (Stoeckel et al. 1979, Stoeckel et al. 1981). This was facilitated by the fact that the single components of polymerase are relatively large (40000 to 150000) with respect to the total molecular mass (580000).

3.3 50S ribosomal subunit of E. coli

The large ribosomal subunit (50S) of E. coli consists of 32 different proteins, and a small and a large RNA (120 and 2904 nucleotides). The RNA again accounts for two thirds of the total molecular mass of about 2 million. The hope to obtain direct information on the shape of a protein in situ led Nierhaus et al. (1982) to try a special procedure to prepare a 50S subunit which is invisible in a buffer solution of nearly 100% D_2O (see Fig. 2, bottom).

The new strategy drastically reduces the incoherent background level as compared to the matching conditions of a normal ribosome. The gain in signal/noise ratio allows work with dilute solutions under conditions where ribosomes are active. Also, interparticle contributions become negligible, so that the four LT samples can be measured separately, leading to a reduction of the reconstitution effort by a factor of about four, since identical samples no longer have to be prepared repeatedly for mixing. The single-label samples can be used for in situ shape determination. Finally, the low-yield isolation procedure for single proteins is done with the cheaper to grow cells with natural abundance hydrogen.

50S subunits were reconstituted from the RNA fraction, a fraction of all proteins of the 50S subunit (TP50), and single proteins isolated from E. coli cells grown in culture media containing three different levels of D_2O (76%, 84% and 0%, respectively).

The difference of the scattering curves of a particle containing labelled (protonated) protein L4 and of a reference particle containing no labelled proteins, calculated for the contrast matching point of the reference particle from a contrast variation series of both samples, is shown in the Guinier plot (ln intensity vs. Q^2, $Q = (4\pi \sin \theta)/\lambda$, 2θ = scattering angle, λ = wavelength) in Fig. 3 (Nierhaus et al., 1982; May et al., 1982). The same technique has been applied to all five proteins reported on so far (L1, L2, L3, L4 and L23). They all appear to be moderately elongated in situ (axial ratios 1:2 to 1:5). In the two cases where solution SANS data of the same isolated proteins have been obtained, the radii of gyration were identical within error. The molecular weights determined from the scattering intensity at zero angle are in very good agreement with the sequence data.

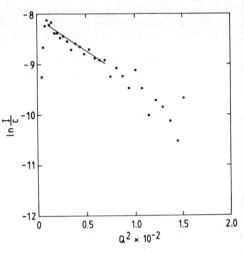

Fig. 3

The straight forward LT study on the 50S subunit, which does not depend on the direct shape determination, has already provided three protein-protein distances.

References
Engelman D M and Moore P B 1972 Proc. Natl. Acad. Sci. USA 69 1997-1999
Hoppe W 1972 Isr. J. Chem. 10 321-333
Hoppe W 1973 J. Mol. Biol. 78 581-585
Kratky O and Worthmann W 1947 Monatsh. Chem. 76 263-281
May R P, Stuhrmann H B and Nierhaus K H 1982 Brookhaven Symposia in Biology 32 submitted
Moore P B 1982 Brookhaven Symposia in Biology 32
Nierhaus K H 1982 Curr. Topics Microbiol. Immunol. 97 81-155
Nierhaus K H, Lietzke R, May R P, Nowotny V, Schulze H, Simpson K, Wurmbach P and Stuhrmann H B 1982 Proc. Natl. Acad. Sci. USA submitted
Ramakrishnan V R, Yabuki S, Sillers I-Y, Schindler D G, Engelman D M and Moore P B 1981 J. Mol. Biol. 153 739-760
Stoeckel P, May R, Strell I, Cejka Z, Hoppe W, Heumann H, Zillig W, Crespi H L, Katz J J and Ibel K 1979 J. Appl. Cryst. 12 176-185
Stoeckel P, May R, Strell I, Cejka Z, Hoppe W, Heumann H, Zillig W and Crespi H L 1980 Eur. J. Biochem. 112 419-423
Stuhrmann H B, Haas J, Ibel K, deWolf B, Koch M H J, Parfait R and Crichton R R 1976 Proc. Natl. Acad. Sci. USA 73 2379-2383

Neutron scattering studies of virus structure

Stephen Cusack

European Molecular Biology Laboratory
Grenoble Outstation, c/o ILL, 156X, 38042 Grenoble, France

1) Why study viruses ?

Viruses are the simplest of lifeforms. They need to infect the cells of higher organisms in order to reproduce and thereby frequently cause disease. In the extracellular state they generally consist of the viral genetic material (DNA or RNA) packaged in a protective capsid composed of protein molecules and sometimes lipid. Viruses vary enormously in size and complexity from simple plant viruses to complicated human pathogens such as vaccinia virus. Knowledge of the structure of viruses is important to the understanding of the mechanism of viral infection and can ultimately contribute to the production of improved vaccines and other anti-viral agents. More generally viruses provide well defined model systems to study how biological macromolecules assemble into more complicated functional units.

2) Why use neutrons to study virus structure ?

Over the last ten years neutron scattering has emerged as an important new technique in the study of virus structure. The usefulness of neutrons for studying multicomponent particles such as viruses is largely due to the possibility of exploiting the H_2O/D_2O contrast variation method. Briefly this can be explained as follows. In a typical small angle neutron scattering experiment, a purified sample of virus in buffered solution is exposed to a monochromatic beam of neutrons (wavelength typically 5-10Å) and the small angle scattering recorded on a two dimensional detector which is at a variable distance from the sample. A series of different scattering curves are usually measured for each virus by varying the percentage of heavy water (D_2O) in the buffer. Since the scattering length densities of H_2O and D_2O are very different the effect of this is to vary the contrast (i.e. scattering length density difference) between the buffer and the various components of the virus (whose scattering densities vary only slightly when immersed in different H_2O/D_2O mixtures, due to H/D exchange). The crucial point is that different biological macromolecules (e.g. lipid, protein and nucleic acid) have very different mean scattering densities and are therefore contrast matched at different percentages of D_2O. For instance protein has a mean scattering density equivalent to 41% D_2O. In buffers of this D_2O content therefore the contribution to the forward scattering from the protein component of the virus is effectively zero. The scattering is dominated by contributions

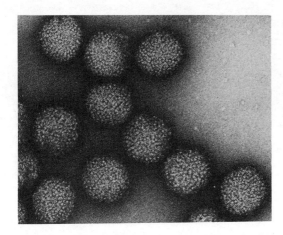

Figure 1
Electron micrograph of Influenza virus. The particles are about 1200Å in diameter.

from the lipid (if present) or nucleic acid (e.g. RNA), which have contrast match-points of respectively 12% and 69% D_2O. By analysing the scattering curves obtained at different contrasts it is possible to interpret the data in terms of the distribution within the virus of its different chemical constituents. Although solution neutron scattering is only capable of leading to low resolution models of virus structures, even this can be useful, particularly in the case of large viruses for which it is very difficult to obtain the same information by other techniques.

Many plant and animal viruses have now been investigated by neutron scattering and I shall illustrate the methods used by referring to recent work on Influenza virus (Mellema et al. 1981). Influenza virus is a quasispherical particle having a diameter of 1200Å and a total molecular weight of 200×10^6 daltons (see Figure 1). It is an enveloped virus, the envelope consisting of a lipid membrane into which are embedded externally projecting glycoprotein spikes of two kinds, hemagglutinin and neuraminidase. Interior to the viral membrane are found two other structural proteins, the M-protein and the ribonucleoprotein which is complexed with the viral genome. The latter consists of eight distinct segments of RNA and constitutes only about 2% of the viral mass.

3) Analysis of neutron scattering curves

Figure 2 shows the neutron scattering curves of Influenza virus obtained in 12,22,41,68 and 100% D_2O to a resolution of $1/150Å^{-1}$. The measurements were made on instrument D11 at the Institut Laue-Langevin, Grenoble, France. The changes of the scattering curves with contrast are clearly evident. However all show behaviour consistent with the assumption that at low resolution the particle is spherically symmetrical. At higher scattering angles one might expect to see deviations from spherically symmetrical scattering due to for instance the arrangement of spikes on the surface of the virus.

It is well known that for any particle, the innermost part of the

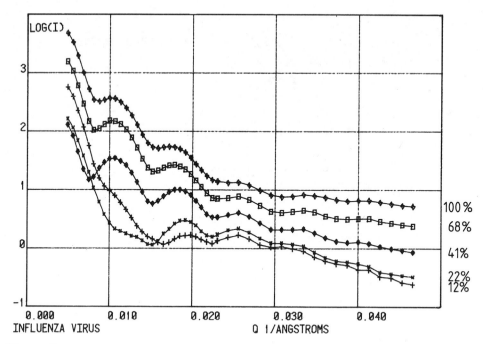

Figure 2
Neutron scattering curves of Influenza virus measured in solutions of the indicated D_2O content.

scattering curve follows the Guinier approximation enabling the model independent parameters, the radius of gyration and the zero-angle scattering I(0) to be experimentally derived. In the case of neutron scattering, it must be remembered that both these quantities are dependent on the contrast. Provided intensity measurements are made on an absolute scale I(0) can be used to deduce the molecular weight of the virus particle (Jacrot and Zaccai, 1981). This parameter, not easy to obtain with precision by other techniqes, can be of immense value in the determination of the structure of large viruses. This is because it constrains the number of protein subunits within each particle which in combination with symmetry principles can suggest how the virus is constructed.

The most widely used method for the determination of the radial structure from the spherically symmetrical domain of scattering is the model fitting technique. In this method a model for the virus is proposed and the parameters defining the model obtained by calculating the corresponding scattering curves and least squares fitting to the data. For spherical viruses, a model consisting of a small number of concentric shells is taken. Use of a full contrast variation set of data enables the amount and chemical nature of the material in each shell to be determined. Models derived by this method cannot be shown to be unique, but must be demonstrated to be internally consistent (i.e. in their behaviour with contrast) and agree with information obtained by independent methods. For a more detailed review of the methods used and results obtained with neutrons in the study of virus structure, the reader is refered to Jacrot (1981), Timmins and Jacrot (1982) and Cusack (1982).

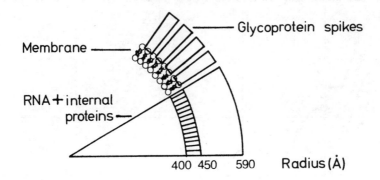

Figure 3
A low resolution model of the structure of Influenza virus as deduced from small angle neutron scattering.

4) The structure of Influenza virus

The results of such an analysis for Influenza virus are shown in Figure 3 (Mellema et al. 1981, Cusack 1982). The analysis is less straightforward for this complex virus than for simpler plant viruses (e.g. Tomato Bushy Stunt Virus, Chauvin et al. 1978). However some interesting results have emerged concerning the virus composition which is found to be significantly different from that hitherto reported. The outer shell, from a radius of 450 to 590Å, represents the glycoprotein spikes and constitutes some 45% of the viral mass, much higher than hitherto reported. The lipid bilayer is located between 400 and 450Å. Interior to the bilayer it has not been possible to unambiguously determine the positions of the internal proteins but is is clear that there must be less M-protein than had previously been thought. An interesting extension of this work, currently in progress, is to elucidate the structural changes in Influenza virus which occur below pH5.5. These have recently been shown to be important in the mechanism of entry of the virus into the cytoplasm of infected cells (Matlin et al. 1981).

5) Conclusions

Small angle neutron scattering is a powerful technique for the determination of the structure of viruses, particularly very large, complex animal viruses that are not readily studied by other scattering techniques. The contrast variation method however can be equally useful in low resolution crystallography of viruses. In the near future more detailed information than can be obtained from solution scattering should become available from crystallographic studies on the smaller plant viruses.

6) References

Chauvin C , Witz J and Jacrot B 1978 J.Mol.Biol. 124 641
Cusack S 1982 To be published in the proceedings of the Brookhaven
   Symposium on Neutrons in Biology ed B P Schoenborn (Plenum)
Jacrot B 1981 Comp. Virol. 17 129
Jacrot B and Zaccai G 1981 Biopol. 20 2413

Matlin K S, Reggio H, Helenius A and Simons K 1981 J.Cell.Biol. $\underline{91}$ 601
Mellema J E, Andree P J, Krijgsman P C J, Kroon C, Ruigrok R, Cusack S, Miller A and Zulauf M 1981 J.Mol.Biol. $\underline{151}$ 329
Timmins P A and Jacrot B 1982 To be published in 'Neutron Scattering in Molecular Biology' ed D L Worcester (Elsevier/North Holland)

Neutron studies of connective tissue

A. Miller, Laboratory of Molecular Biophysics, Zoology Department, Oxford University.

Much of our information on molecular structure in biological materials came from X-ray diffraction analysis. This is well illustrated in the case of the structure of collagen in connective tissue (Miller 1976, 1982). If molecular arrangement as distinct from molecular structure is under scrutiny, then electron microscopy is a powerful complementing technique for producing general models ; however X-ray diffraction, since it can be carried out on tissues in their native state, is preferable to determine the final, functional model. Here I summarise how the special characteristics of neutron scattering have complemented electron microscopic and X-ray methods and extended our understanding of the structure and biological properties of collagenous connective tissue. The three special characteristics described are (a) the availability of long wavelength ($\lambda \sim 18$Å) with low absorption (b) isotopic substitution leading to contrast variation and (c) inelastic neutron scattering leading to information about molecular dynamics.

(a) The characteristic structures of collagenous connective tissue are fibrils. These are very long, cylindrical in cross-section and of variable diameter in the range 8 - 550 nm. as determined by electron microscopy (Parry and Craig, 1978). Fibrils are the units into which the collagen molecules are packed. Fibrils then pack together in parallel bundles within tendons. A neutron scattering pattern of an array of parallel tendons was taken with $\lambda = 18$Å and 40 m. between specimen and detector on the D11 instrument at I.L.L. Grenoble. This diffraction pattern contained an intensity maximum at a spacing corresponding to 400 nm. in a direction perpendicular to the fibril axes. Theoretical calculations have shown that this observation is due to inter fibrillar interference effects and leads to a value for the packing fraction of the fibrils of greater than 0.8 (Finney, Hulmes, Cusack & Miller, unpublished).

(b) The 300 nm. long collagen molecules are staggered longitudinally within the fibrils by 67 nm. with respect to nearest neighbours. This yields a set of meridional xray and neutron reflections. The intensities of these at a variety of contrasts provided by D_2O/H_2O mixtures have been interpreted in terms of the known amino-acid sequence of the molecules and have led to a determination of the axial molecular conformation about the telopeptides on the collagen molecules. These are sites of the intermolecular crosslinks which stabilise connective tissue and change on ageing of organisms. Hulmes et al. (1981).

A similar contrast variation study was made of mineralised and unmineralised turkey leg tendons. This allowed a determination of the axial location of the calcium hydroxyapatite crystallites within the collagen fibrils. (White et al. 1977).

2.
In cartilagenous connective tissue, the collagen fibrils are embedded in a proteoglycan matrix. Contrast variation studies on solutions of purified components of this matrix have led to a low-resolution model for protein-polysaccharide complexes and to an indication that the contrast match point of proteoglycans is usefully different from that of collagen. (Perkins et al. 1981).

(c) The biological function of connective tissue is mechanical. Hence it is of interest to relate the molecular structure of these tissues to their mechanical properties. Since many animal connective tissues are bicomposite, consisting of collagen plus another type of molecule, studies on the microscopic mechanical properties have been performed. Measurements of inelastically scattered radiation can yield information about the energies of phonons propagating in biological fibres (Harley et al. 1977, Cusack & Miller, 1979, Randall & Vaughan 1979). These experiments involved Brillouin scattering of visible light. Inelastic neutron scattering has the advantage of a range of energy transfers and the possibility of making measurements at high momentum transfer. Here I will briefly describe one type of such experiment - observation of quasi-elastic broadening of incident neutron energies due to diffusion of water molecules in collagen (see Miller et al. I.L.L. Report 08-04-004, 1976; White, S., 1977).

Experiments were carried out on the IN5 instrument at I.L.L. using an energy resolution of 25μeV and energy window of I±300μeV. The specimen was an array of parallel collagen fibres dissected from rat tail tendons in 0.15 M Na Cℓ to preserve the native structure. Measurements of the quasi-elastic broadening were made with the momentum transfer vector Q perpendicular and parallel to the fibre axis. Different experiments were carried out with the tendon fibres in atmospheres of relative humidity 0%, 47%, 60%, 90% and 95%. As an internal calibration of the effect of the relative humidity on the molecular packing, an equational trace was made through the intensity peak due to lateral intermolecular interactions.

For the case of Q perpendicular to the fibre axis, the plot of ΔE against Q^2 is shown in Figure 1. for relative humidities 95%, 90% and 60%. At R.H. values less than 60%, no measurable energy broadening occurs. The diffusion coefficient calculated from the slope of a line drawn through the ΔE vs Q^2 points is 5×10^{-6} cm^2 sec^{-1}. This may be compared with a value of 2×10^{-5} cm^2 sec^{-1} for bulk water. Hence the diffusion of water is reduced in the lateral direction. Measurements with Q parallel to the fibre axis showed a diffusion coefficient similar to that in bulk water. It is possible that this anisotropy of water mobility is imposed by the 1.5 nm. diameter, 35 nm long gaps in the collagen fibrils, the long axis of these gaps is parallel to the fibril axis. The reason for the non-zero intercept of ΔE vs Q^2 at $\Delta E \sim$ 70μeV is not obvious. A jump reorientational model of the water motion would predict a line of constant ΔE vs Q^2.

Preliminary experiments on the instrument IN10 at I.L.L. were also carried out on similar samples. This had an energy resolution of around 1μeV and energy window of ±20μeV. Very slight broadening in the region 0.1 - 0.2μeV, just detectable with this instrument, was recorded on hydration of dry tendons, and at RH of 98%, a hint of a free water component.

Figure 1. Variation of ΔE with Q^2

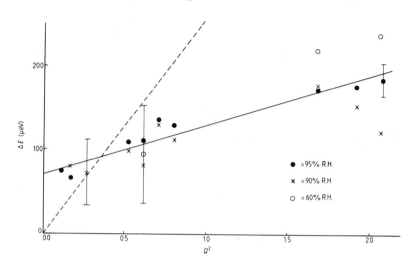

References

Cusack, S. and Miller, A. 1979. J. Mol. Biol. <u>135</u>, 39-51.

Harley, R., James, D., Miller, A. and White J.W. 1977. Nature <u>267</u>. 285-7.

Hulmes, D.J.S., Miller, A., White, S.W., Timmins, P.T. and Berthet, C. 1981. Int. J. Biol. Macromol. <u>2</u>, 338-346.

Miller, A. 1976. In 'Biochemistry of Collagen' (G.N. Ramachandram and A.H. Reddi, eds.). pp. 85-136. Plenum Press.

Miller, A. 1982. Trends in Biochem. Sciences. <u>7</u>, 13-18.

Miller, A., Timmins, P.T., Jenkin, G.T., Ghosh, R., White, J.W. 1976. I.L.L. Annaul Report on experiment no. 08-04-004.

Parry, D.A.D. and Craig, A.S. 1979. Nature <u>282</u>, 213-215.

Perkins, S.J., Miller, A., Hardingham, T.E. and Muir, I.H. 1981. J. Mol. Biol. <u>150</u>, 69-95.

Randall, J.T. and Vaughan, J.M. 1979. Phil. Trans. Roy. Soc. Ser. A. <u>293</u>, 133-140.

White, S. 1977. D.Phil. Thesis., Oxford University.

White, S.W., Hulmes, D.J.S., Miller, A. and Timmins, P.T. 1977. Nature <u>266</u>, 421-425.

Neutrons and proteins

B. P. Schoenborn and N. V. Raghavan

Brookhaven National Laboratory, Upton, NY, USA 11973

1. Introduction:

Cambridge not only is the birthplace of the neutron but also saw the birth of protein crystallography. A few years after Chadwick discovered the neutron, Max Perutz, a young Austrian, joined the Cavendish Laboratory and commenced his heroic development of protein crystallography. His efforts culminated in the elucidation of the structures of myoglobin and hemoglobin (Perutz et al., 1954 and Kendrew, 1962).

These and subsequent studies of protein structures by x-ray crystallographic techniques demonstrated the inherent physical order of proteins and gave insight into the relationship between their architecture, mechanism, specificity, and evolutionary differences, (Blow and Steitz, 1970).

The three-dimensional structure of proteins is determined largely by the behavior of hydrogen atoms, which are responsible for many catalytic reactions and control conformational fluctuations. The general role of hydrogens as structural and functional determinants is well understood, but their role in specific instances is seldom known. Unfortunately, hydrogen atoms are weak x-ray scatterers and therefore cannot be directly located in x-ray Fourier maps.

Neutrons, however, provide a unique probe for localizing hydrogen atoms and distinguishing between hydrogen and deuterium. This distinction is particularly important in the study of enzymatic sites that include charge relay systems and in studies that probe protein dynamics by hydrogen exchange. Neutron diffraction studies also provide the ability to distinguish N from C and O and allow correct orientation of groups such as histidine and glutamine (Schoenborn, 1969).

2. Experimental techniques:

Data collection from protein crystals is no trivial matter. Neutron beams from even the most modern high flux beam reactors have a relatively low flux and are, at best, less intense than common x-ray sources by about 5 orders of magnitude. In addition, protein samples produce a large background due to the incoherent scattering by the unpaired nuclear neutron of hydrogen atoms. Increasing specimen size and deuteration of the mother liquor can partially overcome these handicaps. Fortunately, protein crystals do not show any radiation damage in the neutron beam, and all data can be collected from one crystal, thereby eliminating scaling problems. To improve the utilization of neutrons and permit data collection in a reasonable time (about 2 months), position-sensitive detectors were developed by Alberi et al. (1975). Such devices have, however, a limited resolution (~3mm FWHM) producing considerable spot smearing. To allow collection of statistically accurate intensities, novel spot integration schemes have been developed that use position-sensitive detectors linked to modern data acquisition systems. Instead of dealing with conventional scans like the θ 2θ scan, which provides an integrated intensity as a function of one rotational parameter (Cain et al. 1976), the computer-linked counter can be used to produce a three-dimensional reflection profile. As the crystal steps through a reflection, the observed data for each step are stored in an external memory as a function of extent in 2θ and height (y) of a reflection. In this space, the reflection will be a three-dimensional distribution with dimensions determined by such basic geometrical constants as $\Delta\lambda$, crystal size, mosaic spread, counter resolution, and beam collimation parameters. The interactions of these basic parameters have been analyzed (Schoenborn, 1982a + b) to determine optimal conditions for a protein crystallographic station and to calculate shape parameters that will permit the delineation of reflections to allow accurate background subtraction. Another approach, which uses a statistical noise filtering technique to determine the spot shape from the observed data, has been developed by Spencer and Kossiakoff (1980) and modified by Sjolin and Wlodawer (1981). These procedures produce statistically well determined reflection intensities, reducing the effect of the high background as much as possible.

3. Protonation states of active sites:

To date, seven proteins and protein derivatives have been studied to elucidate particular functional aspects of the specific protonation of active groups such as histidine and glutamic and aspartic acids.

Norvell et al. (1975) showed that in carbon monoxide myoglobin the distal histidine linking the functional CO ligand to the iron of the heme group is not protonated whereas in oxy and met myoglobin that histidine is indeed protonated (Phillips and Schoenborn, 1981, Schoenborn 71). Two other studies also involved the determination of the protonation states of histidine groups. Wlodawer and Sjolin (1981) studied the role of imidazole groups in the enzymatic activity of ribonuclease, and Kossiakoff and Spencer (1981) determined the charge relay system in trypsin. In triclinic lysozyme, Mason (1982) has demonstrated the involvement of protons in the catalytic activities of glutamic and aspartic acids.

4. Hydrogen exchange:

Neutron diffraction offers a unique way to study hydrogen-deuterium exchange in proteins (Schoenborn, 1971) to probe the structural integrity of local regions. The exchange rate of hydrogens on the peptide amide group is a measure of the local stability versus structural fluctuations (Englander et al. 1972) and can be determined from a detailed analysis of the neutron Fourier maps. The observed scattering density P at amide proton positions is proportional to the relative exchange x, with $x = \frac{P + 3.7}{+ 10.3}$. The scattering length of a proton is -3.7 F and that of a deuteron + 6.6 F. Such H/D exchange data must, however, be interpreted with some caution. They are obviously subject to experimental conditions and to errors in phasing, incomplete refinement, structural disorder, and degree of accuracy of the measured structure factors. The accuracy of the observed structure factors is of particular concern. The atomic density observed in Fourier maps is strongly dependent on the high-order reflections, which are often weak. Because of the high background associated with hydrogen-containing crystals, the weak higher-order reflections have a particularly large counting statistical error. In most cases the reflections with a Bragg spacing of less than 2Å are observed with an average intensity less than 3 σ (σ is the standard deviation). An analysis of the observed scattering length of the stable and unexchangeable hydrogen on the peptide Cα positions gives a good indication of the minimal error to be expected. Such an error estimate based on a histogram of the scattering length of Hα gives a width of 1.2 F units (FWHM) or nearly a third of a hydrogen atom. Interpretations of the scattering length of exchangeable H atoms have to be viewed with an error of this magnitude in mind. The experimental conditions of crystal growth and soaking time in H_2O or D_2O will affect the observed H/D exchange rate, as illustrated in the table below.

This table lists the experimental conditions and the observed H exchange for four different myoglobin data sets. It should be noted that myoglobin forms large crystals (25 mm^3) and is an ideal case for neutron studies, producing good diffraction intensities. All four data sets have been refined by reciprocal least-squares techniques, but they differ in treatment of solvent and structural water (Raghavan and Schoenborn, 1982). The level of refinement varies for the different data sets, as indicated by the listed R factor.

Exchange in myoglobin of main-chain amide hydrogens.

Derivative	Met	Met	CO	O_2[1])
H	5.6	5.6	5.6	5.4
Growth	H_2O	D_2O	H_2O	H_2O
D_2O soaking time	few wk	10 yr	6 mo	4 mo
R in %	27	17	18	15
Observed water molecules used in refinement	0	50	40	120
Unexchanged (± 20%)	16	2	5	17
Partial exchange	115	110	70	49
Full exchange (± 20%)	18	37	74	83

1) Data from Phillips (1982), this crystal was soaked at pH 8.4.

Preliminary exchange results have now also been published for ribonuclease A by Wlodawer and Sjolin (1982), for trypsin by Kossiakoff (1982), for lysozyme by mason (1982), and for crambin by Teeter and Kossiakoff (1982). It is still too early to describe definitive exchange patterns, but generally it seems that β-sheet structures show less exchange than α-helical regions.

Neutron diffraction studies of proteins have made important contributions to the understanding of the mechanism of protein functions, and with the recent development of new experimental techniques at ILL and BNL the use of neutrons to elucidate structural features of enzymatic function will rapidly increase.

This research was carried out at Brookhaven National Laboratory under the auspices of the United States Department of Energy with partial support from the National Science Foundation.

5. References:

Alberi J, Fischer J, Radeka V, Rogers L C and Schoenborn B P 1975 IEEE Trans. Nucl. Sci. NS-22 $\underline{1}$ 255
Blow D M and Steitz T A 1970 Ann. Rev. Biochem. $\underline{39}$ 63
Cain J E, Norvell J C and Schoenborn B P 1976 Brookhaven Symp. Biol $\underline{27}$ VIII-43
Englander S W, Downer N and Teitelbaum H 1972 Ann. Rev. Biochem. $\underline{41}$ 903
Kendrew J C 1962 Brookhaven Symp. Biol. $\underline{15}$ 216
Kossiakoff A A and Spencer S 1981 Biochemistry 20 6462
Kossiakoff A A 1982 Nature $\underline{296}$ 713
Mason S A 1982 Brookhaven Symp. Biol. 32, Plenum Press, New York ed B P Schoenborn (in press)
Norvell J C, Nunes A C and Schoenborn B P 1975 Science $\underline{190}$ 568
Perutz M F, Green D W and Ingram V M 1954 Proc. Roy. Soc., London, A $\underline{225}$ 287
Phillips S E V and Schoenborn B P 1981 Nature $\underline{292}$ 81
Phillips S E V 1982 Brookhaven Symp. Biol. 32 Plenum Press, New York ed B P Schoenborn (in press)
Raghavan N and Schoenborn B. P 1982 Brookhaven Symp. Biol. 32 Plenum Press, New York ed B P Schoenborn (in press)
Schoenborn B P 1969 Nature $\underline{224}$ 143
Schoenborn B P 1971 Cold Spring Harbor Symp. Quant. Biol. $\underline{36}$ 569
Schoenborn B P 1982a Acta Crystallogr. (in press)
Schoenborn B. P. 1982b Brookhaven Symp. Biol. 32. Plenum Press, New York ed B P Schoenborn (in press)
Sjolin L and Wlodawer A 1981 Acta Crystallogr. A $\underline{37}$ 594
Spencer S A and Kossiakoff A A 1980 J. Appl. Crystallogr. $\underline{13}$ 563
Teeter M and Kossiakoff A A 1982 Brookhaven Symp. Biol. 32 (in press)
Wlodawer A and Sjolin L 1981 Proc. Natl. acad. Sci. USA 78 2853

Physics design methods for pressurized water reactors

R. L. Hellens

Combustion Engineering, Inc. Windsor, CT, U.S.A.

Abstract. An account is given of some recent improvements in PWR physics and their relation to earlier design methods. Comparison of predictions with data from operating power reactors is discussed.

The physics of pressurized water reactor (PWR) cores has advanced during the last thirty years through a number of stages, first dealing with small nearly homogeneous cores, then uniform slightly enriched uranium lattices, and, most recently, large non-uniform power reactor cores. Evolution of the theory has been guided both by design requirements and by experimental data from critical assemblies and operating plants, but it has also been significantly shaped and limited by computer capabilities as well. In this paper, some of the major areas of current interest are outlined in the context of the historical development of the theory. These comments will be limited to that small area of reactor physics which plays a role in the PWR design methods used today, and thus will overlook many interesting topics in favor of practical aspects.

For nuclear design, reliable predictions are required for reactivity and power distributions throughout core life, control rod worths and the reactivity coefficients important to accident response. Because of the large changes in PWR characteristics between laboratory and operating reactor conditions, it was recognized at an early stage that even elaborate mockup experiments were of very limited practical value, and that reliable design could only be based on accurate theoretical methods. By a happy chance, digital computers with steadily growing capability became available just as work on PWR analysis began.

Early PWR cores were small and reactor physics in the early 1950s naturally emphasized neutron leakage. Multi-group diffusion theory was unsuitable because it neglects neutron energy loss in anisotropic scattering by hydrogen. Direct multi-group approximations of the transport equation were considered and codes written, but computers of the day could not deal with this level of complexity for routine calculations, particularly in multi-dimensional situations. The approach eventually chosen for reactivity and global power distribution calculations was inherently an attempt to generalize two-group theory and to relate it to the transport equation. The first step was to solve the multi-group transport equation with linear anisotropic scattering for a single Fourier component of the slowing-down spectrum with a buckling or wave-number roughly that of the actual reactor. Few-group cross sections were then obtained by integration of reaction rates, fluxes and currents over the broad energy bands corresponding to the few-group scheme desired. Effective few-group diffusion coefficients were defined as the ratio of the integrated current to integrated flux gradient from the multi-group calculation. This procedure ensured, at least for a given slab reactor, that the multi-group transport theory and the few-group diffusion equations would give exactly the same results for reactivity, gross leakage, fluxes and currents.

Few-group theory was initially formulated in this way at Bettis for homogeneous media, with the expectation that heterogeneous or lattice cell effects could subsequently be introduced into the multi-group calculations by some form of flux weighting of the cross sections of the various materials. Analysis of these lattice effects proved to be difficult, however, even in uniform lattices where reflecting boundary conditions could be employed to isolate the cells. It was not until about the mid-1960s that good representations of resonance capture and neutron thermalization in lattice cells were developed and incorporated in US design codes.

During this same period (1950-1965), a large experimental program had been carried out in a number of laboratories to develop buckling and cell parameter data for uniform lattices of slightly enriched uranium metal and oxide fuel rods. This data played an important role in judging the adequacy of both the theory and the isotopic cross section data. The agreement attained by 1965 for uniform lattices seemed adequate for design purposes. The picture soon changed, however, when new experiments were performed in several systematic studies of non-uniform lattices including the water holes, poison shims and control fingers characteristic of large power reactors. In addition, the analysis of the reactivity effect of fuel depletion in a number of operating power reactors revealed significant errors which could introduce unacceptable errors in the prediction of both fuel enrichment and power distributions. This new data led to a reexamination of design methods about 1970, and to the development of new lattice design codes during the next ten years.

The cause of these discrepancies seemed to lie in the ambiguities inherent in applying the single-cell methods already outlined to non-uniform reactor lattices. The practice had been to perform most reactivity and power distribution calculations, particularly for quasi-separable first cores, using fine-mesh two-dimensional diffusion calculations with one mesh point for each fuel pin. Although the fuel pin-cells could be reasonably approximated by the uniform lattice model, proper definition of similar cell spectrum calculations for water-holes, burnable poison shims and control fingers was more difficult since they require some estimate of an incident current spectrum. This was usually provided by surrounding the cell of interest with a homogeneous annulus of fuel and moderator to act as the neutron source. Once these single-cell cross sections were generated, some further spectrum interaction occurred in the diffusion calculation between the disparate cells, but this information was not readily fed back to improve the single-cell spectra. The approximation scheme clearly needed to be reformulated to account for the strong spectrum interaction effects between the various components of a fuel assembly.

This problem had already been encountered for different reasons in heavy-water reactor design, and methods of treating "double heterogeneity" had been developed in Sweden and the U.K. The adaptation of these methods to PWR problems has been actively pursued during the last ten years, including modifications required in the highly-loaded fuel assemblies designed for extended burnup cycles. The forerunner of such lattice assembly spectrum codes, called WIMS-D (Askew et al 1966) was developed at Winfrith for HWR geometries. The basic concept of a two-stage calculation was introduced here, i.e., a fine-group spectrum calculation retaining enough of the double two-dimensionality of the fuel geometry to produce reasonably accurate space-dependent spectra was followed by a few-group assembly calculation with more accurate geometrical representation. This concept and the original method for performing the spectrum calculation were due to Leslie (1963).

In the early 1960s, Carlvik (1964) provided the basic tools for accurate assembly calculations by developing essentially exact numerical methods for producing one- and two-dimensional collision probability matrices in complex geometries. In principle, these matrices provided the means for direct integration of the transport equation including anisotropic scattering. If explicit pin geometry is preserved, the two-dimensional solutions required in a PWR fuel assembly are, however, still excessively time consuming even on today's fast computers using exact methods, and further approximations are required. Jonsson et al (1974) have shown that Carlvik's one-dimensional numerical methods could be combined with the so-called "interface current" technique to reduce the computational effort significantly. In this method, the full angular distribution of the neutron flux is retained from integral transport theory in each pin cell, but the coupling at pin cell boundaries is limited to a double spherical harmonics expansion. For most practical purposes DP-1, or even DP-0, is sufficient for multi-group cell coupling. These boundary conditions eliminate the need for costly calculations of inter-cell collision probabilities while permitting retention of the explicit interior pin cell geometry.

The foregoing concepts and approximations have been gathered together into a number of lattice spectrum codes, such as EPRI-CPM (Eich 1979) and the C-E code called DIT (Jonsson 1979). Some features of the latter are worth describing briefly since it represents an advanced example of this type of integrated design code. Four of the code blocks deal with spectral features:

1. 85-group calculations are performed in representative spectrum geometries abstracted from the assembly, such as single fuel cells (as earlier), or clusters of fuel-rod, poison-rod, and water-hole cells. The clusters of cells are coupled by the interface current method, and initial multi-group incurrent guesses are applied to the cluster periphery.

2. The energy spectrum is condensed in each geometrical subregion to yield 4-12 group average cross sections including transport-corrected isotropic scattering.

3. Few-group quarter assembly or checkerboard calculations are performed for the explicit components using interface currents to link the cells and zero net-current boundary conditions. From these calculations, improved few-group incurrents on the cluster boundaries are recycled to the first block without changing the incurrent energy distribution within a broad group.

4. The leakage spectrum calculation is performed with cross sections

weighted by both volume and the piece-wise discontinuous flux spectrum in each region of the assembly.

As can be seen from this brief sketch, an enormous amount of information is generated by codes of this sort to describe the flux in space and energy throughout an entire fuel assembly or checkerboard segment. One important by-product of the assembly spectrum concept is the capability to produce directly from the lattice calculations few-group cross sections for both fine-mesh and coarse-mesh spatial calculations on a consistent basis.

The need for three-dimensional reactor design calculations had been recognized at an early stage for isotopic depletion, changes in control rod worth, thermal analysis and for the reactivity feedback effects of xenon, fuel temperature and moderator density. This has proved to be particularly true in the shuffled later cores of large power reactors. Although three-dimensional diffusion codes, such as PDQ, have been available for some time, they have remained too slow and costly for design use, and a number of alternatives have evolved. One of the simplest, called the "window-shade" model, stitched together several radial flux shapes with a single axial calculation using cross sections inferred from the radial calculations. Elaboration of this crude model led to a variety of "synthesis" models based on the classical notion of compounding a solution from a set of carefully-chosen trial functions with weighting coefficients determined in some way to represent the problem in hand. The approach attracted a lot of serious analysis effort during the 1960s, and some aspects are implemented in later versions of PDQ. However, the method has found little application in large power reactor design, possibly due to the known sensitivity of the result to choice of trial functions, and to the lack of any generally valid measures of the accuracy of the final solution.

In the early 70s, interest turned to nodal methods for generating coarse-grained, spatial flux distributions by relating the neutron current between two mesh blocks or nodes to the average fluxes in the nodes. Early attempts to exploit this approach had produced rapid 3-D solutions, but the reliability and potential for generating detailed information seemed limited. During the past ten years, many nodal or coarse-mesh schemes have been developed, differing principally in the way the ratio of edge current to average flux is estimated. The method chosen at C-E (Robinson and Eckard 1972), which can serve as an example, derives higher order difference equations from an expansion of the flux in a Taylor's series at the center of each mesh block or node. The nodes are then coupled by requiring continuity of flux and normal current at the nodal interfaces, truncating the Taylor's series with fourth-order terms. Approximations are made for the second and higher order terms, and a set of equations with nearest-neighbor coupling almost identical to the ordinary finite difference equations is obtained. Two neutron groups are normally used with four x-y nodes per fuel assembly and 20-30 axial planes. Core edge effects of reflector and core shroud are represented by albedo boundary conditions usually generated by S_n multi-group calculations.

Both analyses of operating reactors and comparisons of fine-mesh PDQ benchmark calculations with this coarse-mesh code, called ROCS, have been surprisingly successful. As a result, efforts have been made to extract more detailed power distributions beyond the node average power from the information implicit in the nodal calculation. One promising approach uses the nodal interface currents as sources in fine-mesh imbedded diffusion calculations, which serve to introduce detailed fuel assembly descriptions. This provides a means of tracking explicit fuel depletion particularly in limiting assemblies, but the feedback of such detailed information into the coarse-mesh cross sections would not necessarily be required. Further improvement in these assembly power distributions can be made by correcting the tendency of diffusion theory to underestimate fuel power peaking near water holes. These adjustments are now made through the cross sections of the imbedded calculations, but it may be preferable to imbed transport calculations instead.

Experimental verification of physics design methods initially came from small critical assemblies in which measurements of buckling and the relative reaction rates of various important isotopes were made. During the past decade, data from operating plants has become dominant as a basis for appraising the methods since only here are the conditions described by the analysis actually realized. Important core characteristics can be inferred from soluble boron concentration, control rod worth, power and moderator temperature defects and coefficients. PWR cores now usually contain either fixed or movable miniature neutron detectors in about 20% of the fuel assemblies so that gross flux levels can be inferred repeatedly during cycle life at several hundred radial and axial points. The major problem in exploiting plant data is the effort and expense involved, since, except for the initial startup measurements, all subsequent characteristics become influenced by prior operating history which must be represented in the analysis.

Space permits only a brief summary of the type of agreement observed between a sample of operating plant data and 3-D analyses using the new methods outlined.

Fuel Depletion: An analysis of the rods-out soluble boron rundown done twelve years ago (Hellens 1971) using 2-D PDQ, and the old lattice methods showed an error for individual plants increasing by about $0.5\%\Delta k$ for each 10 MWd/kgU burnup; the scatter from one plant to another was of comparable size. Similar calculations with present codes, DIT-ROCS in 3D, yield the scatter pattern shown in the Figure for nine different plant cycles. First cycles with burnable poison shims tend to dip at midcycle, but return at end of cycle (10-14 MWd/kgU) to reactivity errors similar to those at the end of later cycles, namely, $-0.2\%\Delta k$ with a standard deviation of $0.1\%\Delta k$.

Boron Worth: Soluble boron worth influences the apparent reactivity depletion errors just mentioned and the measurement of the worth indicates the adequacy of the theoretical description of flux peaking in the coolant. The early calculations underestimated the worth by 3-13%, but recent calculations have reduced this deviation to about 3% with a standard deviation of 4% among 12 measurements. This tends to confirm the improvement in the assembly calculation expected from the new calculations.

Moderator Temperature Coefficient: Both measured and computed zero power moderator temperature coefficients of reactivity are found to correlate equally well with soluble boron concentration in similar plants for various cycles. The difference between individual pairs of calculations and measurements shows a standard deviation of 1.6 pcm/$^\circ$C independent of soluble boron concentration. However, the average slope of this difference versus boron concentration is one-third of the slope of the moderator coefficient itself. The size of this discrepancy is not understood as yet.

Global Power Distribution: A direct measure of the accuracy of global 3-D flux shapes can be obtained from the differences between miniature in-core detector signals and their prediction at various points in cycle life. The largest deviation observed for any planar set of about 40 detector signals shows an rms deviation of 3.5% relative to the largest detector signal in the set.

New methods, such as those outlined above, appear to provide a significant improvement in the prediction of PWR core characteristics and form the basis for procedures of increased reliability, computer efficiency, and ease of use for the designer.

References

Askew J R, Fayers F J, Kemshell P B 1966 J. Brit. Nucl. En. Soc. 5 p 564
Carlvik I 1964 ICPUAE U.N. Geneva 2 pp 225-231
Eich W J 1979 Symp. Proc. on Nucl. Data Problems for Thermal Reactors EPRI NP-1098
Hellens R L 1971 Proc. Am. Nucl. Soc. Conf. on New Developments in Reactor Physics, Conf 720901, pp 3-22
Jonsson A, Goldstein R, Mockel A J 1974 Atomkernenergie 24 pp 74-84
Jonsson A, Rec J R, Singh U N 1979 Symp. Proc. on Nucl. Data Problems for Thermal Reactors EPRI NP-1098
Leslie D C 1963 J. Nucl. Energy 17 p 293
Robinson C P, Eckard J D 1972 Trans. Am. Nucl. Soc 15 p 297

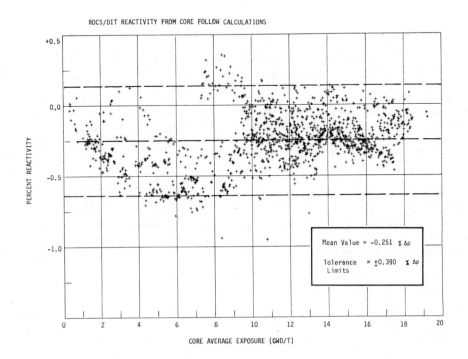

ps. Conf. Ser. No. 64: Section 5*
Paper presented at Conf. on Neutron and its Applications, Cambridge, 1982

50 years of neutron transport

Noel R. Corngold

California Institute of Technology, Pasadena, California 91125 U.S.A.

Abstract. The process of neutron transport is fundamental to all neutronic systems. The mathematical theory of neutron transport, an offshoot of the linearized, semi-classical kinetic theory of gases, has been developed to a considerable degree in recent decades. In this lecture we shall describe some of the development, indicating how it has drawn from, and enriched other branches of mathematics, science, and technology. We shall also indicate where the "frontiers" are, at present, both in the mathematical challenges, and in application of the theory to unusual systems.

1. Introduction

It is appropriate, at this Anniversary Celebration, to survey briefly a half-century's development of neutron transport theory. The subject had a modest birth in Rome in the 1930's as Italian scientists sought to describe the diffusion and slowing-down of neutrons in matter. It matured rapidly during World War II and now plays many roles. Encrusted with computational algorithms, it is the language of the nuclear power industry; scrubbed clean, it is being treated surgically by mathematicians. However, the middle ground, where direct and perspicuous mathematics is linked with good physics appears, today, a little thin.

2. Early Development

The theory developed by the Roman School, under Enrico Fermi (1934), is what one calls "age-diffusion theory" today. Supplemented by the particularly simple case of slowing-down in hydrogen, it provides an effective description of the neutron field in large, simple, and weakly absorbing systems. Indeed, it was adequate for the design of a generation of graphite-moderated reactors. Nevertheless, the limitations of diffusion theory were appreciated early in the comparison of theory and albedo-experiment (Flügge 1938). A proper transport theory, with characteristic Boltzmannian form (Boltzmann 1910, Sommerfeld 1955) was required.

At the time, Boltzmann's non-linear equation was hardly unknown to physicists. Many of its implications for irreversible flows had been worked through (Hilbert 1912, Chapman and Cowling 1952). In fact, a linear version of Boltzmann's equation, more appropriate to neutrons, had been described quite early by Lorentz (1905). His was a two-component system, electrons and ions. The ions act as passive, Maxwellian host, while the electrons drift under the combined action of electric field and collisions with the ions. The transcription to "neutrons and moderator" is immediate, but seems to have been overlooked, perhaps because the theory focussed upon the thermal regime. The first statement of the "modern" form of the one-speed neutron transport equation that I have been able to find occurs in a paper by Bothe (1941). (Earlier, Flügge (1938) used an integral equation that resembles what we today call "Peierls' Equation".) In Bothe's paper aspects of neutron transport are discussed, which now have achieved textbook status. For example, one finds there the famous transcendental equation for diffusion length in terms of mean free path and number of secondaries per collision.

The theory of radiative transport, which had its origin in the radiation theory of the 19th century, provided a second source for neutron transport theory. This "kinetic theory of radiation" was developed energetically by astrophysicists. The work of Eddington (1921) and of Hilbert (1912) was particularly important; the literature quickly assumed its modern aspect, with references to source and streaming terms, to expansions in Legendre polynomials. Chandrasekhar's text (1950) summarized much, and enriched a generation of transport theorists. One great gift was the solution, by Wiener and Hopf (1931) of Milne's problem, the escape of radiation from a semi-infinite star. A mutually beneficial relation between the two subjects continues to exist (Hunt 1971).

3. Recent Development

Soon after World War II ended, two works appeared which described much of the progress in transport theory that had been achieved in wartime. Marshak's report (1947) described the theory of slowing-down, while Davison's (1957) classic surveyed the entire field. As neutron physics and engineering flourished subsequently research concentrated upon limited aspects of the subject. The complex geometries and irregular cross-sections encountered in reactor engineering demand a mix of physical and mathematical approximations, dependent upon computer algorithms, to describe the neutron field (Sanchez and McCormick 1982). Though the approximations are usually uncontrolled, the process of design has turned out to be, overall, quite effective. The design and analysis of both thermal reactors and fast critical assemblies agree well. On the other hand, scholarly work in transport theory has developed by emphasizing either the spatial, or the velocity variable. The former might be called "pure" transport theory. It concerns the equation of one-speed transport with precise boundary conditions and its extension beyond plane geometry and isotropic scattering. This subject has become more and more a branch of mathematics in recent years, being concerned greatly with proofs of existence and uniqueness. The latter branch, evolving from the theory of slowing-down, is concerned with the exchange of energy between neutron gas and moderator. A principal aim is to understand the distribution-in-energy, the neutron spectrum, that obtains in various situations, both transient and steady. These studies have remained close to neutron physics and the physics of condensed matter. To a degree, this division of the subject into two disjoint branches is artificial, and scholars strive to bridge the gap, but only in special cases (Wick 1949; Bednarz and Mika 1963) can one progress with an even-handed and fairly rigorous treatment of spatial and velocity coordinates.

3.1 Pure Transport Theory

In the classic case of a steady distribution function with plane-symmetry, we use the variables x, $v_x = v \cos \theta = v\mu$, $v = |\underline{v}|$ and write

$$\frac{\partial}{\partial x} f(x,\mu,v) = \frac{1}{\mu v} \left(\frac{\delta f}{\delta t} \right)_{coll} = Lf \qquad (1)$$

where the operator L operates upon functions of v and µ. The equation, being linear and partially separable, would appear to yield to an old technique, the superposition of separable solutions. The Ansatz $f(x,\mu,v)= e^{-x/\upsilon}\phi(\upsilon,\mu,v)$ leads to difficulty, however, in that one is forced to consider the spectral theory of L, a singular operator. In fact, L is the product of an unbounded operator and, in simple--but interesting--models of scattering, an operator of finite rank. L has a point spectrum, with regular eigenfunctions φ, and a continuous spectrum with associated singular eigenfunctions. The superposition is, in effect, an expansion in a set of (eigen) functions one hopes is complete. Boundary conditions at planar bounding surfaces or interfaces complicate the analysis, for often they refer to only a portion of the angular range ("no neutrons enter the system from vacuum").

In spite of its complexity, the method of eigen-function expansion has had considerable success. It was introduced into neutron physics by Case (1960) in a classic paper. (Similar considerations appear in an earlier paper by van Kampen (1955), which dealt with plasma oscillations and, even earlier, Davison, in unpublished work (1945), had begun to grapple with the singular solutions.) Its triumphs are described in two works which cover the period 1960-71 (Case and Zweifel 1967, Kuščer and McCormick 1973). The latter contains several hundred references to research papers, the earlier ones dealing with transport at fixed energy.

Equation 1, being linear, may be attacked in many ways. As Williams (1973) points out in his review, Fourier-transform techniques, combined with the idea of Wiener-Hopf decomposition, compete well with the method of eigenfunctions. Other methods, closely linked to plane geometry, are also effective. Chief among these are the methods of invariant imbedding (Pahor and Zweifel 1964), borrowed from astrophysics (Ambartzumian 1943), and the method of transfer matrices (Aronson and Yarmush 1966). The trend towards increased mathematical rigor may be seen in the pages of the journal "Transport Theory and Statistical Physics", whose editors have written a review of the movement (Greenberg and Zweifel 1976). Most interesting here has been the realization (Larsen and Habetler 1973) that analysis of the resolvent operator associated with Eqn. (1) yields the spectral resolution (expansion theorems) with less effort than does "Case's Method".

3.2 Transport Theory Emphasizing Inelastic Processes

With the slowing-down and thermalization of neutrons (Williams 1966, Parks et al. 1970) inelastic scattering processes rather than precise treatment of boundary conditions are our principal concern. The simplest case has neutrons thermalizing in a large, homogeneous system. The spectrum is a distorted Maxwellian, the distortion controlled by the interplay of the energy-dependence of the capture cross-section (nuclear physics) and the inelastic scattering processes in condensed matter. In small systems, where neutron leakage plays a role, elastic processes (mean-free paths) enter, and the analysis becomes more difficult.

Transient processes in the neutron field are particularly challenging. A neutron pulse n(v,t), as it slows down in a non-multiplying system, will scan the energy scale. It is, in effect, a slowing-down spectrometer. When fission and nuclear inelastic scattering are present, the process becomes richer. In the passive materials one asks whether, after a sufficient temporal interval, a stable distribution $n(v,t) \sim e^{-\lambda t}\phi(v,\lambda)$ will be formed, displaying a balance of thermalizing against absorbing processes. The answer, "it depends", summarizes a rich topic for both theoretical and experimental investigation (Corngold 1975). One interesting result: when stabilization is impossible, the waning distribution becomes progressively cooler until, to describe it properly, one needs to deal with sub-Bragg interactions in crystalline materials.

The $n(v,t)$ experiments are designed to have simple spatial variation; a complementary class has simple harmonic temporal variation, and spatial variation that is interesting. One asks whether stable distributions $n(v,x) \sim e^{-\kappa x}\psi(v,\kappa)$ exist for a particular source-frequency. Since it is easier to wait many e-folding lengths, the $n(v,x)$ experiments are somewhat more difficult to analyze. But discussion of these intriguing experiments and their potential is moot for, East and West, they have been discontinued as research funding has disappeared.

4. New Frontiers

While neutron transport theory is quite adequate for reactor design, the theory has its own life. New and elegant results have appeared in one-speed transport in materials with spatially varying properties (Larsen 1981). Then, there is the largely unexplored field of multi-velocity ("thermal") transport in media that support gradients (of temperature, density,...) flows, significant order, and random structures (Stepanov 1965). Random also describes the number of neutrons emitted in fission, and this feature introduces the subject of stochastic transport, or neutron 'noise' (Bell 1965, Lewins 1978). Finally, an exotic application, the transport of ultra-cold neutrons, those with energies of 10^{-7}eV, and wavelengths of 10^3 Å (Steyerl 1977, Golub and Pendlebury 1979). In their interaction with matter, these neutrons have lost their particle-aspect, but on the other hand, in their motion through guides and bottles they act as a Knudsen gas with a complex surface interaction.

5. Connexions

Neutron transport theory has drawn upon other fields and, in turn, enriched them. Conferences (Hunt 1971, Inönü and Zweifel 1967) have been held to freshen the relationship with radiative transfer; some of the models used in both fields are identical, and the interested reader can trace the interaction through the literature. The connexion with kinetic theory of gases languished until one noticed the relation between Boltzmann's linearized equation and a particular model of neutron thermalization (Kuščer and Williams 1967). An immediate consequence was insight into the manner in which a classical gas approaches equilibrium. Later the nature of high frequency sound waves in gases was elucidated; then, the gas-surface interaction. One may trace the contribution of neutron physicists in the proceedings of a series of conferences on rarefied gas dynamics. See, for example, Potter (1977).

But these are established fields. That insight into neutron transport might energize a developing subject has been demonstrated by Williams' impressive work on the theory of sputtering (Williams 1979). A more exotic application is to the transport of neutrinos in a collapsing star (Bethe et al. 1979). The stellar dynamics is influenced strongly by the pressure exerted by its (degenerate) electrons. They, in turn, are involved in the reaction $e + p \leftrightarrow n + \nu$. In the early stages of collapse the neutrinos simply escape, but, as the star's density increases they tend to be trapped, and the electron loss rate is modified. Thus, the epoch before super-nova is controlled by neutrino transport theory, a subject similar to ours (Bludman et al. 1977).

Finally, one should acknowledge the contribution that neutron transport theory has made to the training of scientists, and to making them keen for the more difficult challenges of kinetic theory (Williams 1971, Duderstadt and Martin 1979). In the early 1960's, a conjunction occurred when a decent model for the scattering law for thermal neutrons upon liquids was needed. van Hove (1954) had emphasized that the scattering was controlled by fluctuation and transport processes in the fluid. To the struggle came those skilled in neutronics, and their contributions were impressive (Nelkin and Ghatak 1964, Akcasu et al. 1970, Desai and Yip 1968). As Waterloo was won on the playing fields of Eton, so many a puzzle in kinetic theory has been illuminated by insight gained from neutron transport. It is a nice subject.

6. References

Akcasu A Z, Corngold N, and Duderstadt J J 1970 Phys. Flds. __13__ 2213
Ambartzumian V A 1943 Dokl. Akad. Sci. USSR __38__ 229
Aronson R and Yarmush D L 1966 J. Math. Phys. __7__ 221
Bednarz R and Mika J 1953 J. Math. Phys. __4__ 1285
Bell G 1965 Nucl. Sci. Eng. __21__ 390
Bethe H A, Brown G E, Applegate J, and Lattimer J M 1979 Nucl. Phys. __A324__ 487
Bludman S A, Lichtenstadt I, Ron A, Sack N, and Wagschal J J 1977 Nucl. Sci. Eng. __64__ 294
Boltzmann L 1910 Vorlesungen über Gastheorie (Leipzig: Barth) Transl. by Brush S G 1964 (Berkeley: U Cal)
Bothe W 1941 Zeits. für Physik __118__ 401
─────── 1942 Zeits. für Physik __119__ 493
Case K M 1960 Ann. Phys. (NY) __9__ 1
─────── and Zweifel P F 1967 Linear Transport Theory (Reading: Addison-Wesley)
Chandrasekhar S 1950 Radiative Transfer (Oxford: Oxford UP)
Chapman S and Cowling T G 1952 Mathematical Theory of Non-Uniform Gases (Cambridge: Cambridge)
Corngold N 1975 Advan. Nucl. Sci. Technol. __8__ 1
Davison B 1945 Natl. Res. Council of Canada Rept. MT-112 (unpublished)
─────── 1957 Neutron Transport Theory (London: Oxford UP)
Desai R and Yip S 1968 Phys. Rev. __166__ 129
Duderstadt J J and Martin W R 1979 Transport Theory (New York: Wiley)
Eddington A S 1921 Zeits. für Physik __7__ 351
Fermi E, Amaldi E, Pontecorvo B, Rasetti F, and Segre E 1934 Ric. Scient. __5__ 282; transl. in collected papers (EF) 1962 (Chicago: U Chic)
Flügge S 1938 Zeits. für Physik __111__ 109
Golub R and Pendlebury J M 1979 Repts. Prog. Phys. __42__ 441
Greenberg W and Zweifel P F 1976 Trans. Thy. and Stat. Mech. __5__ 219
Hilbert D 1912a Math. Ann. __72__ 562
─────── 1912b Phys. Zeits. __13__ 1056
Hunt G E 1971 (ed) Proc. Symp. Interdisciplinary Appl. Trans. Thy., J. Quant. Spectr. Rad. Transfer __11__ 511
Inönü E and Zweifel P F 1967 Developments in Transport Theory (New York: Academic)

Kuščer I and Williams M M R 1967 Phys. Flds. 10 1922
────── and McCormick N 1973 Advan. Nucl. Sci. Technol. 7 181
Larsen E and Habetler G 1973 Comm. Pure Appl. Math. 26 525
────── 1981 Prog. Nucl. Ener. 8 203
Lewins J 1978 Proc. R. Soc. Lond. A362 537
Lorentz H A 1905 Arch. Néerl. 10, 336; also 1936 Coll. papers V.III Hague: Nijhoff)
Marshak R 1947 Rev. Mod. Phys. 19 187
Nelkin M and Ghatak A 1964 Phys. Rev. 135 A4
Pahor S and Zweifel P F 1964 J. Math. Phys. 10 581
Parks D E, Nelkin M S, Beyster J R, and Wikner N F 1970 Slow Neutron Scattering and Thermalisation (New York: Benjamin)
Potter J L 1977 (ed) 10th Intl. Symp. Raref. Gas Dyn. (New York: AIAA)
Sanchez R and McCormick N J 1982 Nucl. Sci. Eng. 80 481
Sommerfeld A 1955 Thermodynamics and Statistical Mechanics (New York: Academic)
Stepanov A 1965 in Pulsed Neutron Research (Vienna: IAEA)
Steyerl A 1977 Springer Tracts in Mod. Phys. 80 57
van Hove L 1954 Phys. Rev. 95 249
van Kampen N 1955 Physica 21 949
Wick G-C 1949 Phys. Rev. 75 738
Wiener N and Hopf E 1931 Berliner Ber. Math. Phys. Kl. 696
Williams M M R 1966 The Slowing Down and Thermalisation of Neutrons (Amsterdam: North-Holland)
Williams M M R 1971 Math. Methods in Particle Trans. Theory (London: Butterworth)
────── 1973 Advan. Nucl. Sci. Technol. 7 283
────── 1979 Prog. Nucl. Ener. 3 1
Yip S and Boon J P 1980 Molecular Hydrodynamics (New York: McGraw Hill)

Inst. Phys. Conf. Ser. No. 64 © 1983: Section 5
Paper presented at Conf. on Neutron and its Applications, Cambridge, 1982

Hilbert space method for the numerical solution of reactor physics problems

R T Ackroyd

Central Technical Services, UKAEA, Northern Division, Risley, Warrington, Cheshire WA3 6AT

Abstract. A Hilbert space approach is used to give a unified treatment of neutron transport by finite element methods. Global solutions can be found by least squares, variational and weighted residual methods stemming from an identity. Bounds for local characteristics of solutions are found by a bi-variational method.

1. Introduction

Hilbert space methods for the numerical solution of some partial differential equations arising in electrostatics and elasticity were initiated by Prager and Synge (1947). Their method of the hypercircle, Synge (1957), using pyramid function approximations to solve second and fourth order partial differential equations for regions with irregular boundaries, is in effect a finite element method. This idea was extended by Ackroyd (1960), (1962), to solve the first order Boltzmann equation by introducing the concept of a relaxed equation so that the apparatus of the method of the hypercircle could be applied to an equation which lacked an extremum principle. Apart from the elementary code GRAB, Black (1960), there was no immediate development.

By the early 70's the undoubted success of finite element methods for problems in solid and fluid mechanics attracted the attention of reactor physicists. By that time the intuitive basis of the pioneer work in finite element methods had been replaced by treatments using either variational principles or weighted residual techniques. In applying the finite element idea to problems of neutron transport a Hilbert space presentation provides a unified approach to a variety of methods. One can use either the method of the hypercircle or extremum principles for the second order form of the Boltzmann equation, Ackroyd (1978); alternatively variational and weighted residual methods for the first and second order forms of the transport equation can be regarded as variants of a generalised least squares method, Ackroyd (1982a). However for those who prefer to partake of their analysis neat there is no need for a dash of Hilbert space, but for others geometrical intuition makes the analysis more palatable. A fundamental problem for neutron transport is to solve approximately the Boltzmann equation

$$\underline{\Omega}\cdot\nabla\phi_0(\underline{r},\underline{\Omega}) + \sigma(\underline{r})\,\phi_0(\underline{r},\underline{\Omega}) - \int_{\Omega'} \sigma_s(\underline{r},\underline{\Omega}\cdot\underline{\Omega}')\,\phi_0(\underline{r},\underline{\Omega}')\,d\Omega' = S(\underline{r},\underline{\Omega}) \quad \ldots (1)$$

for the angular flux ϕ_0, at $\underline{r}$ for the direction $\underline{\Omega}$, which is induced in an arbitrary region V by a distributed source S and a surface source T. The boundary of V has a bare part S_b, a perfect reflector S_{pr} and the part S_s

with source T. The total cross-section σ and the differential scattering cross-section satisfy

$$\sigma(\underline{r}) > \int_{\Omega'} \sigma_s(\underline{r},\underline{\Omega}\cdot\underline{\Omega}')\, d\Omega'$$

The Boltzmann equation (1) is equivalent to the mixed parity equations

$$\begin{aligned}\underline{\Omega}\cdot\nabla\phi_0^- + C\phi_0^+ &= S^+ \\ \underline{\Omega}\cdot\nabla\phi_0^+ + G^{-1}\phi_0^- &= S^-\end{aligned} \quad \ldots (2)$$

Here the even-parity angular flux ϕ_0^+ and the odd-parity angular flux ϕ_0^- are defined by

$$\phi_0^\pm = \frac{1}{2}\left(\phi_0(\underline{r},\underline{\Omega}) \pm \phi_0(\underline{r},-\underline{\Omega})\right)$$

and G, G^{-1}, C and C^{-1} are positive definite self-adjoint operators defined in terms of σ and σ_s, Ackroyd (1981). The parity components $S^\pm$ are defined similarly to $\phi_0^\pm$. The mixed parity equations are used to derive the second order forms of the Boltzmann equation

$$\begin{aligned}-\underline{\Omega}\cdot\nabla\left[G\underline{\Omega}\cdot\nabla\phi_0^+\right] + C\phi_0^+ &= S^+ - \underline{\Omega}\cdot\nabla G\, S^- \\ -\underline{\Omega}\cdot\nabla\left[C^{-1}\underline{\Omega}\cdot\nabla\phi_0^-\right] + G^{-1}\phi_0^- &= S^- - \underline{\Omega}\cdot\nabla C^{-1} S^+\end{aligned} \quad \ldots (3)$$

for which it is not difficult to establish very useful complementary extremum principles. The boundary conditions for the first and second order equations are given in Table 1.

In developing a finite element method for neutron transport it is convenient to use a finite element representation for the spatial dependence of the angular flux with some form of expansion for the directional dependence, usually either a discrete ordinate or spherical harmonic expansion.

2. Least Squares Formulation of a Finite Element Method for Neutron Transport

A least squares formulation of a finite element method for neutron transport has been given by Ackroyd (1981), (1982a), (1982b). It embraces classical variational methods and some weighted residual methods for the second order forms of the Boltzmann equation. These methods are well-behaved numerically, for example they are free from the well-known ray effect. The formulation which is based on the mixed parity equations (2) also gives a maximum principle and weighted residual methods for the first order Boltzmann equation. The parent K^{+-} maximum principle gives rise to the maximum and minimum principles listed in Table 2, and described in the above references. If ϕ^+ and ϕ^- are trial functions for ϕ_0^+ and ϕ_0^- respectively then one has an identity

$$K^{+-}(\phi^+,\phi^-) + W^{+-}(\phi^+,\phi^-) = \int_V \left\{\langle C^{-1} S^+, S^+\rangle + \langle G S^-, S^-\rangle\right\} dV$$

$$+ 2 \int_{S_s} \int_{\underline{\Omega}\cdot\underline{n}<0} |\underline{\Omega}\cdot\underline{n}|\, T^2\, d\Omega\, dS \quad \ldots (4)$$

Table 1 Comparison of Boundary Conditions for Transport Equations

Surface / Equation	Bare S_b	S_s with source	Perfect reflector S_{pr}	Interface $S_i \cap S_j$
		$\underline{n}$ outward normal for exterior surfaces		$\underline{n}_i$ outward normal to S_i
1st order	$\phi_0 = 0$ for $\underline{\Omega}\cdot\underline{n} < 0$	$\phi_0 = T(\underline{r},\underline{\Omega})$ for $\underline{\Omega}\cdot\underline{n} < 0$	$\phi_0(\underline{r},\underline{\Omega}) = \phi_0(\underline{r},\underline{\Omega}^x)$ for all $\underline{\Omega}\cdot\underline{n} = -\underline{\Omega}^x\cdot\underline{n} = 0$ where $\underline{\Omega}^x$ is the reflected direction to $\underline{\Omega}$	ϕ_0 continuous for $\underline{\Omega}\cdot\underline{n}_i \neq 0$
2nd order Even-parity	$\phi_0^+ + G[S^- - \underline{\Omega}\cdot\nabla\phi_0^+] = 0$ for $\underline{\Omega}\cdot\underline{n} < 0$. $\phi_0^+ - G[S^- - \underline{\Omega}\cdot\nabla\phi_0^+] = 0$ for $\underline{\Omega}\cdot\underline{n} > 0$.	$\phi_0^+ + G[S^- - \underline{\Omega}\cdot\nabla\phi_0^+] = T(\underline{r},\underline{\Omega})$ for $\underline{\Omega}\cdot\underline{n} < 0$. $\phi_0^+ - G[S^- - \underline{\Omega}\cdot\nabla\phi_0^+] = T(\underline{r},-\underline{\Omega})$ for $\underline{\Omega}\cdot\underline{n} > 0$.	$\phi_0^+(\underline{r},\underline{\Omega}) = \phi_0^+(\underline{r},\underline{\Omega}^x)$ $G[S^-(\underline{r},\underline{\Omega}^x) - \underline{\Omega}\cdot\nabla\phi_0^+(\underline{r},\underline{\Omega})$ $- S^-(\underline{r},\underline{\Omega}^x) + \underline{\Omega}^x\cdot\nabla\phi_0^+(\underline{r},\underline{\Omega}^x)] = 0$	ϕ_0^+ and $G[\underline{\Omega}\cdot\nabla\phi_0^+ - S^-]$ continuous for $\underline{\Omega}\cdot\underline{n}_i \neq 0$
Odd-parity	$\phi_0^- + c^{-1}[S^+ - \underline{\Omega}\cdot\nabla\phi_0^-] = 0$ for $\underline{\Omega}\cdot\underline{n} < 0$. $\phi_0^- - c^{-1}[S^+ - \underline{\Omega}\cdot\nabla\phi_0^-] = 0$ for $\underline{\Omega}\cdot\underline{n} > 0$.	$\phi_0^- + c^{-1}[S^+ - \underline{\Omega}\cdot\nabla\phi_0^-] = T(\underline{r},\underline{\Omega})$ for $\underline{\Omega}\cdot\underline{n} < 0$. $\phi_0^- - c^{-1}[S^+ - \underline{\Omega}\cdot\nabla\phi_0^-] = -T(\underline{r},-\underline{\Omega})$ for $\underline{\Omega}\cdot\underline{n} > 0$.	$\phi_0^-(\underline{r},\underline{\Omega}) = \phi_0^-(\underline{r},\underline{\Omega}^x)$ $c^{-1}[S^+(\underline{r},\underline{\Omega}) - \underline{\Omega}\cdot\nabla\phi_0^-(\underline{r},\underline{\Omega})$ $- S^+(\underline{r},\underline{\Omega}^x) + \underline{\Omega}^x\cdot\nabla\phi_0^-(\underline{r},\underline{\Omega}^x)] = 0$	ϕ_0^- and $c^{-1}[\underline{\Omega}\cdot\nabla\phi_0^- - S^+]$ continuous for $\underline{\Omega}\cdot\underline{n}_i \neq 0$

where $\langle u, v \rangle$ denotes $\int_\Omega u(\underline{r},\underline{\Omega}) v(\underline{r},\underline{\Omega}) d\Omega$. The positive definite functional $W^{+-}(\phi^+, \phi^-)$ vanishes iff $\phi^+ = \phi_0^+$ and $\phi^- = \phi_0^-$. This functional has the volume term

$$\int_V \left\{ \begin{array}{l} \langle C^{-1}[\underline{\Omega}.\nabla\phi^- + C\phi^+ - S^+], \underline{\Omega}.\nabla\phi^- + C\phi^+ - S^+\rangle \\ + \langle G[\underline{\Omega}.\nabla\phi^+ + G^{-1}\phi^- - S^-], \underline{\Omega}.\nabla\phi^+ + G^{-1}\phi^- - S^-\rangle \end{array} \right\} dV$$

surface terms involving positive weights $w^\pm(\underline{r},\underline{\Omega})$ and surface terms without weights. The surface terms allow $\phi^\pm$ to be free from all boundary conditions on both external and internal surfaces. The K^{+-} maximum principle is obtained by noting that the minimisation of $W^{+-}(\phi^+, \phi^-)$ implies the maximisation of $K^{+-}(\phi^+, \phi^-)$. The treatment ensures that the Euler-Lagrange equations for $W^{+-}(\phi^+, \phi^-)$ are the parity transport equations and not some higher order equations, a defect which sometimes can occur with the method of least squares, Mitchell and Wait (1977).

The parent principle K^{+-} has the merit of permitting the use of trial functions which are discontinuous across boundaries. Consequently it is possible to use different expansions for the directional dependence of ϕ^+ and ϕ^- in different parts of a system. For example, it is possible to use a low order expansion where neutrons are effectively diffusing and a higher order expansion where neutrons are streaming. If continuous trial functions are used the K^{+-} principle reduces to the classical principles K^+, K^- and E^+ for the second order Boltzmann equations, and gives a new principle $\tilde{K}$ for the first order equation.

3. Weighted Residual Methods

Weighted residual methods were developed for the first order Boltzmann equation in the first place because of a lack of a variational principle. A good account of this method has been given by Martin et al (1981). However, care has to be exercised with this method to eliminate oscillations in the flux profile when a discrete ordinate representation is used for directions. Although mesh refinement ameliorates the problem, the cure lies in the use of trial functions which are discontinuous in both space and directions along the edges of the finite elements. This freedom in the choice of the trial function is made possible by employing the 'weak' or integral form of the transport equation, Duderstadt and Martin (1979). In the 'weak' form the approximate solution ϕ of the first order transport equation

$$L \phi_0 = S \qquad \qquad \ldots (5)$$

is sought in principle by equating the scalar products of a suitable Hilbert space

$$(L \phi, \psi) = (S, \psi) \qquad \qquad \ldots (6)$$

for all test functions ψ belonging to that Hilbert space. In practice a weak form is obtained from the above weak form by an integration by parts to eliminate any continuity requirements on ϕ. By making appropriate choices of the test functions various 'weighted' residual methods can be obtained, but in practice it is difficult to ascertain good choices for the test functions.

The weighted residual equations that can be derived from the K^{+-} maximum principle are a little more restricted than the above weighted residual equations, because the weights are automatically given by the variational principle. However the trial functions $\phi^{\pm}$ in general need not satisfy any boundary conditions. Variationally derived weighted residual equations have been obtained from the K^+ principle and its derivatives. These weighted residual equations are well-behaved because they are derived from an extremum principle as follows, Ackroyd (1982a), (1982b).

The solution $\phi_0 = \phi_0^+ + \phi_0^-$ is represented by a vector $\underline{\Phi}_0$ in a function space with a suitable metric, which ensures that the right-hand side of the identity (4) is $\underline{\Phi}_0^2$, ie the magnitude of the solution vector is known. A suitable metric is suggested by the form of the functional W^{+-} associated with the K^{+-} maximum principle. A connection between suitable metrics and functionals of extremum principles has been noted by McConnell (1951). The approximation $\phi = \phi^+ + \phi^-$ is given by a vector $\underline{\Phi}$ and

$$(\underline{\Phi} - \underline{\Phi}_0)^2 = W^{+-}(\phi^+, \phi^-) \qquad \ldots (7)$$

The use of the maximum principle is equivalent to minimising the distance of $\underline{\Phi}$ from $\underline{\Phi}_0$, and by the projection theorem this implies that $\underline{\Phi} - \underline{\Phi}_0$ is orthogonal to every vector $\underline{\Phi}_1$ of the basis used to represent $\underline{\Phi}$, ie

$$(\underline{\Phi} - \underline{\Phi}_0) \cdot \underline{\Phi}_1 = 0 \qquad \ldots (8)$$

where

$$\underline{\Phi} = \sum_{l=1}^{N} a_l \underline{\Phi}_l \qquad \ldots (9)$$

Here the expression (9) is the geometrical analogue of the finite element approximation

$$\phi = \sum_{l=1}^{N} a_l (\phi_l^+ + \phi_l^-)$$

The orthogonality condition (8) leads by an integration by parts to a variety of weighted residual methods when some natural choices are made for ϕ_1^+ and ϕ_1^-.

4. Bounds for Local Characteristics

The variational and weighted residual methods sketched above give rise to global approximations to the solution of the Boltzmann equation in the sense that the mismatch between the approximation and the solution is minimised over the whole system space and for all directions.

Having obtained a global approximate solution one may ask how accurate it is in a local region of specific interest. To assess the local accuracy of an approximate solution Ackroyd and Splawski (1982) have defined local characteristics of a solution and have obtained rigorous bounds for them. A typical local characteristic is the capture rate in a small region of a system. The upper and lower bounds for a local characteristic are obtained by applying a bi-variational technique to complementary principles for the second order Boltzmann equations. In a tricky cell problem the capture rate in the clad of the fuel was bracketted to within $\frac{1}{2}\%$, but much remains to be done to achieve reasonable accuracy for local characteristics for deep penetrations in shields.

5. Concluding Remarks

In this snapshot of a unified treatment of finite element methods for neutron transport emphasis has been placed on principles and detailed applications neglected. The current status of finite element methods for neutron transport is well documented in the volume edited by Goddard and Williams (1981).

References

Ackroyd R T 1960 Internal UKAEA document
Ackroyd R T 1962 J. Math. Mech. 11, pp 811-850
Ackroyd R T 1978 Ann. nucl. Energy 5, pp 75-94
Ackroyd R T 1981 Ann. nucl. Energy 8, pp 539-566
Ackroyd R T 1982a Ann. nucl. Energy 9, pp 95-124
Ackroyd R T 1982b To appear in Ann. nucl. Energy
Ackroyd R T and Splawski B A 1982 Ann. nucl. Energy 9, pp 315-330
Black G 1960 Internal UKAEA document
Duderstadt J J and Martin R M 1979 Transport Theory (New York: Wiley)
Goddard A J H and Williams M M R (editors) 1981 Ann. nucl. Energy 8, 11/12
McConnell A J 1951 Proc. R. Ir. Acad., A54, pp 263-290
Martin W R 1981 Ann. nucl. Energy 8, pp 633-646
Mitchell A R and Wait R 1977 The Finite Element Method in Partial Differential Equations (New York: Wiley)
Prager W and Synge J L 1947 Q. Appl. Math. 5(3), pp 241-269
Synge J L 1957 The Hypercircle in Mathematical Physics (Cambridge Univ. Press)

Table 2 K^+ Principle Family Tree

Principle	Type	Derivation	Trial Functions	Equation	Boundary Conditions on Trial Functions
K^+	Maximum	Least squares	ϕ^+ and ϕ^- for ϕ_0^+ and ϕ_0^-	Transport parity pair	None anywhere, but weights $w^+(\underline{r},\underline{\Omega})$ and $w^-(\underline{r},\underline{\Omega})$ assigned to $S_{perfect}$ and interfaces reflector
$\bar{K}$	"	From K^+	ϕ for ϕ_0	First order	None anywhere. Weight constraint $w^+ = w^-$
$\tilde{K}$	"	Least squares or $\bar{K}$	ϕ	"	Continuity of ϕ across interfaces Reflection condition for S_{pr}
K^+	Maximum Classical	$\phi^- = \phi_0^-$ in K^+	ϕ^+	Second order for ϕ_0^+	Continuity of ϕ^+ across interfaces Reflection condition for S_{pr}
K^-	"	$\phi^+ = \phi_0^+$ in K^+	ϕ^-	Second order for ϕ_0^-	Continuity of ϕ^- across interfaces Reflection condition on S
R^+	Minimum Classical	$\phi^- = G(S^- - \underline{\Omega}.\nabla\phi^+)$ in K^+	ϕ^+	Second order for ϕ_0^+	Continuity of ϕ^+ and $G(S^- - \underline{\Omega}.\nabla\phi^+)$ across interfaces Reflection condition for ϕ^+
$\pi - K^-$	"	$\phi^+ = \phi_0^+$ in K^+	ϕ^-	Second order for ϕ_0^-	As for K^- $2\sigma = \int_V \left[\langle G^{-1}s^+, s^+ \rangle + \langle Gs^-, s^- \rangle \right] dV$
$K^+-\sigma$	Maximum Classical	$\phi^- = \phi_0^-$ in K^+	ϕ^+	Second order for ϕ_0^+	As for K^+ $+ 2 \int_{S_a} \int_{\underline{\Omega}.\underline{n}<0} |\underline{\Omega}.\underline{n}| T^2 d\Omega\, dS$

Inst. Phys. Conf. Ser. No. 64: Section 5
Paper presented at Conf. on Neutron and its Applications, Cambridge, 1982

Neutron spectra

Reginald John Brissenden

Atomic Energy Establishment, Winfrith, Dorchester, Dorset DT2 8DH

Abstract. The calculation of neutron spectra has always been a major theoretical issue in support of the Nuclear Power Programme. Current interest centres on the design of safe and reliable plant for reprocessing and tackling the problems of transport and storage of used fuel. Improving the economic performance of fuel manufacturing plant and the disposal of waste are also areas of current interest. These new calculational requirements have led to a resurgence of interest in point energy Monte Carlo. This paper deals with neutron thermalisation by bound light atoms in Monte Carlo codes.

1. Introduction

The need to predict the multiplication of systems containing large quantities of fissile materials in complicated geometrical configurations has led to the development of very powerful Monte Carlo computer codes able to do realistic simulations. In the United Kingdom the standard code for this purpose is MONK (1). It has been developed continuously for 20 years and is recognised and used internationally.

The complete nuclear power cycle includes mining, ore extraction and refining; enrichment by gaseous diffusion or centrifuge; blending, manufacture of fuel pellets and fuel element fabrication; storage and transport of new and used fuel and waste disposal; decommissioning of obsolete plant and equipment. In all these areas criticality is a major design constraint and a safety issue. Modern plant and process design demands constantly-improving calculational support and the increasing speed of modern computers has made this possible in principle.

Calculating the neutron spectra has always been a major source of error, and this paper describes new work aimed at improving the thermal spectrum treatment in MONK.

2. Summary of the physics of thermal scattering

Thermal agitation increases the neutron scattering cross section and reduces the angular anisotropy of scattered neutrons. The mean energy loss per collision for neutrons just above thermal energies too is sensitive to the finer details of this thermal agitation.

Nuclides with atomic weights greater than 12 can be treated as free gases because they have small velocities compared with the thermal neutron and exchange little energy with it per collision. This leaves certain light nuclides, notably hydrogen, deuterium and carbon which appear in various chemically bound forms in large quantities and their correct treatment is important.

The increase in scattering cross section is caused by the atomic motions producing collisions even with very slow-moving neutrons. With free hydrogen gas the increase is shown in Table 1.

Table 1

Mean Incident Energy (ev)	Free Gas Broadening Factor	Nelkin Broadening Factor
.008	3.37	4.32
.032	1.83	3.18
.073	1.40	2.58
.120	1.20	1.70
.200	1.15	1.60

Because the atomic motions are randomly orientated, the collisions with low energy neutrons will also tend to be isotropic thus reducing $\bar{\mu}$, the mean cosine of the angle of scatter, below the high-energy asymptotic value of 2/3. Table 2 shows $\bar{\mu}$ for free hydrogen as a function of incident neutron energy.

Table 2

Energy (eV)	Free gas $\bar{\mu}$	Nelkin $\bar{\mu}$
.008	.145	.044
.032	.280	.127
.073	.364	.202
.120	.512	.355
.200	.520	.364

Finally the mean energy of scattered neutrons, due to the injection of thermal energy, rises above the asymptotic value as the neutron energy decreases, as shown in Table 3.

Table 3

| Energy (eV) | Mean energy of scattered neutrons | |
	Free gas	Nelkin
.008	.028	.016
.032	.038	.037
.073	.057	.065
.120	.078	.093
.200	.118	.141

3. Scattering by bound atoms

We consider hydrogen bound in light water which has been modelled by Nelkin (2) using 4 Planck oscillators in a molecule of mass 18 which is assumed to have free translational motion at ambient temperature.

Table 1 compares the free hydrogen broadening factor with values from the Nelkin model. There is much more broadening for the bound hydrogen due to the internal molecular oscillations which increase the target area of the hydrogen in the molecule. The oscillations also effect the angle of scatter; Table 2 compares $\bar{\mu}$ for free hydrogen with that of bound hydrogen and one can see that the bound hydrogen tends towards angular isotropy more quickly with decreasing neutron energy. Finally, in Table 3 the mean energy of emergent neutrons is shown for bound hydrogen compared with free hydrogen. The effects of binding are seen to be to decrease both the slowing down and speeding up of neutrons compared with the free gas. This is because at high energy the neutrons can 'see' the internal energy but at low energy it gets 'frozen out'.

4 Treatment of Thermal Events

Monte Carlo sampling must be very fast as well as accurate. Free gas scattering is relatively easy to cope with, but even Nelkin's model (which is a considerable simplification in itself) is too complicated for fast direct simulation. Fortunately there are, for criticality calculations, 6 factors needing accurate representation and for the remaining details only a token representation is necessary. The factors are in decreasing order of importance.

(a) Correctly-broadened mean free-paths

(b) Correct $\bar{\mu}$ for all incident energies

(c) Correct thermal equilibrium for neutron flux in absence of absorption and leakage

(d) Correct asymptotic behaviour at high neutron energies

(e) Correct mean emergent energy for all incident energies

(f) Correct angle/energy correlation

The first step in a sampling algorithm is to select a value for the relative velocity S between the incident neutron $\underline{V}_n$ and the water molecule $\underline{V}_a$. The relevant bivariate distribution for $\underline{V}_a$ is

$$P(V_a,\rho) \propto \sqrt{V_a^2 + V_n^2 - V_n V_a \rho} \cdot V_a^2 \exp(-V_a^2/V_o^2) \qquad 4.1$$

V_o is the most probable velocity and ρ is the cosine of the angle between $\underline{V}_n$ and $\underline{V}_a$. To draw from this we use a rejection technique with the covering distribution

$$Q(V_a,\rho) \propto \sqrt{V_a^2 + V_n^2 + 2V_a V_n} \cdot V_a^2 \exp(-V_a^2/V_o^2) \qquad 4.2$$

Sampling from P may be done by sampling from Q and rejecting with probability

$$1 - CP/Q, \quad C = \text{MAX }\{Q/P\} \qquad 4.3$$

The maximum rejection rate is 32%; the average is much less, so that Q is an acceptably close covering distribution which is easy to sample.

The equation for conservation of energy and momentum (using a unit neutron mass) may be written as:

$$q = \tfrac{1}{2}P^2(1+1/A) + SP\cos\phi \qquad 4.4$$

where $\underline{S} = \underline{V}_n - \underline{V}_a$

$q = E_i - E_f; \underline{P} = \underline{V}_n' - \underline{V}_n$

In this equation q is the internal energy change of molecule, which starts with Energy E_i and is left with Energy E_f. P is the momentum blow which is registered[1] in the change in the neutron's velocity from $\underline{V}_n$ to $\underline{V}_n'$ in the lab frame; S is the relative velocity between the neutron and the centre of mass of the molecule; A = 18 and ϕ is the angle between P and S. Transforming to the CM frame, the neutron's velocity becomes:

$$\underline{W}_n = \frac{A}{A+1}\underline{S} \quad \text{and after the collision} \quad \underline{W}_n' = \underline{W}_n + \underline{P}$$

Squaring this equation and combining with 4.4 leads to $W_n'^2 = W_n^2 + 2q_o$

where $q_o = Aq/A+1$.

Now also, $\underline{P}^2 = (\underline{W}_n' - \underline{W}_n)^2$

which leads to

$$P^2 = 2W_n^2 + 2q_o - 2W_n\sqrt{W_n^2 + 2q_o}\cos\chi \qquad 4.5$$

where χ is the angle of scatter in the CM frame.

If q and χ are assumed known, P can be found from 4.5 (since S is obtained inter alia while sampling for $\underline{V}_a$).

Hence $\cos\phi$ is obtained from 4.4. The magnitude of P has now been found and the angle it makes with S. To find the final energy of the neutron in the Lab Frame, ψ, the angle between $\underline{V}_n$ and P is needed and this is easily related to ϕ.

If ζ is an arbitrary azimuth chosen at random in $(0,2\pi)$ to position P precisely about S, then $\underline{P} = \underline{s} + \underline{r}$; where s is the projection of P in the

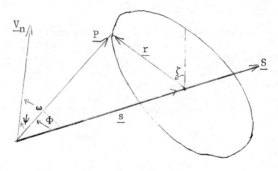

direction of $\underline{S}$ and $\underline{r}$ is the azimuthal radius vector. Projecting $\underline{P}$ with $\underline{V}_n$ we get $\underline{P}.\underline{V}_n = \underline{s}.\underline{V}_n + \underline{r}.\underline{V}_n = V_n s\sin\omega + V_n r\cos\omega\cos\zeta$; ω being the angle between $\underline{V}_n$ and $\underline{S}$.

Thus

$$\cos\psi = \cos\phi\sin\omega + \sin\phi\cos\zeta\cos\omega \text{ ; where } \sin\omega = (V_n - V_a\rho)/S; \quad 4.6$$

and $\cos\omega = \pm V_a \sqrt{1 - \rho^2}/S$

with the + and − signs equally probable.

Since $\underline{V}'_n = \underline{V}_n + \underline{P}$, we have on squaring,

$E' = E + \frac{1}{2} P^2 + PV_n\cos\psi$; and the angle of scatter θ is given by

$$\cos\theta = \frac{V_n + P\cos\psi}{\sqrt{2E}} \quad 4.7$$

Thus $(E', \cos\theta)$ the post-collision neutron parameters can be found once χ and q have been sampled and this is where the quantum-mechanical modelling is required. The value of E_i is taken from a discrete distribution of the possible eigenvalues of the hindered rotation oscillator.

$$E_i = (n + \tfrac{1}{2})\varepsilon_o \quad n = 0, 1, 2, \ldots$$
$$\varepsilon_o = .06 \text{ eV}$$

.06 eV being the point energy of the hindered rotation oscillator which is the only one switched on below .2 eV. A good choice for the distribution of these n-values is

$$P_n \propto (n + \tfrac{1}{2} + \alpha)\varepsilon_o \exp\{-(n + \tfrac{1}{2})\varepsilon_o/KT\} \quad 4.8$$

This is the planck time-state distribution suitably broadened by an energy factor to allow for the 'hot' molecules being bigger than 'cold' ones and therefore more likely to be hit; $\alpha\varepsilon_o$ represents the inert target area energy. A good fit to the Nelkin kernel at low energy is achieved by choosing $\alpha = 2.0$ and E_i then, is sampled from 4.8.

The results of the impact we assume follow the ding-dong model (3), which is a quasi-classical picture of events. On impact the neutron + molecule interaction energy $E_a = \tfrac{1}{2} A/A+1 \; S^2$

may excite the .06 eV oscillator to a state at most m levels above its initial state n (the ding):

$$m = \text{intpt } (E_a/\varepsilon_o) \quad 4.9$$

The hydrogen atom will then move very fast in a transient vibrational motion having a high probability of 'immediately' returning and hitting the neutron again (the dong). If E_a exactly equals $m\varepsilon_o$ the neutron comes to rest in the molecular frame and then it is 'bound' to be hit again. In practice the small residual speed of the neutron is $V_{nr} = \sqrt{E_a - m\varepsilon_o}$ so we choose a probability for this second collision.

$$P_c = 1 - V_{nr}/\sqrt{\varepsilon_o} \quad 4.10$$

which has reasonable limiting behaviour and is suitably concave. Thus P_c is the probability that the .06 eV oscillator will immediately reject E_a. The oscillator is now immediately left in a microstate with quantum number k where

$$k = n: \text{probability } P_c \text{ or } k = n + m: \text{otherwise}$$

This established microstate now decays in a longer time interval over the set of equiprobable states ℓ, $0 \leqslant \ell \leqslant k$ depositing energy in the neutron.

Thus

$$E_f = (\ell + \tfrac{1}{2})\varepsilon_o \quad 0 \leqslant \ell \leqslant k \text{ with probability } 1/k+1. \text{ Thus}$$

$$q = (E_i - E_f) = (n-j)\varepsilon_o \qquad 4.11$$

where j is a uniform random integer in the range (0,k), and n has been sampled from 4.8.

When $q \neq 0$ one expects the interaction details will cause the neutron to forget incident path so that a uniform angular distribution for χ is good enough. For scattering with $q = 0$, however, the neutron remembers its incident path and we fit a double P1 distribution for $\cos\mu = \chi$ using the exact Nelkin treatment. By these means we can reproduce the Nelkin $\bar{\mu}$ of Table 2 and the Nelkin energies of Table 3. The kernels too can be scored accurately using 20,000 collisions for the sample and these show close agreement with the accurate Nelkin model, reproducing the fine structure of the quantised energy exchange.

Since the Nelkin kernels can be altered by $\pm$ 10% by changing the orientation-averaging procedure we consider the quasi-classical fit to be sufficiently accurate for criticality purposes. It can be made to execute almost as rapidly as the free-gas algorithm.

Above .15 eV one has to start a smooth join of the thermal treatment into the free gas model at a few eV. Nelkin does this by replacing the .06 eV oscillator by its WICK asymptote (4) and introducing the .208 oscillator. The effect of this asymptotic replacement is to reduce the effective mass of the molecule and to increase the water temperature. This process is repeated several times incroducing a stiffer oscillator each time until a smooth join is effected at 1 eV with a free gas of mass unity, but a very high effective temperature.

5. References

1 Sherrifs V S W - SRD R86 (Jan 1978) MONK - A General Purpose Monte Carlo Neutronics Program
2 Nelkin M - Phys. Rev. Vol. 119, No. 2, (July 1975) Scattering of Slow Neutrons by Water
3 WICK G C - Phys. Rev. Vol. 194, No. 5, 1230 (1954)
4 WICK G C - Phys. Rev. Vol. 95 No. 5, (1954) The Scattering of Neutrons by Systems Containing Light Nucleii.

Nuclear data for fission reactors

A. Michaudon

Centre d'Etudes de Limeil, B.P. 27, 94190 Villeneuve-St-Georges (France)

Abstract. Nuclear reactions induced by neutrons play an important role in fission reactors and relevant cross-sections need to be known with great accuracy. After a brief history and a general presentation, the nuclear data situation is discussed for Pressurized-Water and Liquid-Metal Fast Breeder Reactors.

1. History and general background

The discovery of nuclear fission was immediately followed by intensive research to find out whether this process could be used for energy production. At that very early stage, nuclear data were extremely scarce, though some of them were crucial to answer this question of paramount importance. At that time microscopic measurements of such data were barely possible and scientists had to rely on what is now called integral experiments such as the ones carried out by F. Joliot and his team. Despite precarious conditions, this team was able to take out several patents on the production of nuclear energy. These predictions were rapidly confirmed : the first man-made reactor became divergent in 1942 at Chicago, and the first atomic bomb exploded in 1945 at Alamogordo.

During a long period, the work on fission energy uses, including relevant nuclear data, was carried out in great secrecy because of its importance for military as well as future commercial applications. This situation lasted even after the war until the 1955 Geneva Conference on Peaceful Uses of Atomic Energy where basic data pertinent to nuclear energy were released. This was the beginning of a new era for nuclear data because international exchange of information and cooperation became possible, yet with some exceptions. At that time also, a new type of reactor, the so-called "Fast Breeder" started to be studied and required large quantities of data. Lastly, the rapid improvement in reactor calculations rendered necessary the use of more accurate data. To meet all these requirements, new facilities were constructed. They produced (and are still producing) enormous quantities of data that must be handled, treated, stored, retrieved and used on a vast scale. For all this new methods were developed, largely based on computers. Four compilation centres were created in this respect : National Nuclear Data Centre (Brookhaven, USA), Nuclear Data Bank (OECD, Paris, France), Nuclear Data Section (IAEA, Vienna, Austria), Centre for Nuclear Data (Obninsk, USSR), each one covering a well-defined service area. Gradually, international cooperation developed after 1955 and relevant committees were set up : the NEANDC (Nuclear Energy Agency Nuclear Data Committee) covering the OECD area and the INDC (International Nuclear Data Committee) formed later to advise the IAEA. Requests for data are now merged at the IAEA level in a world-wide list called WRENDA (World

Request List for Nuclear Data). More specific lists also exist either at the national level or between a few countries. Cooperation among laboratories is encouraged and periodically experts meet to take stock of the work accomplished and stimulate new work. All this results in a closely-knit network covering all aspects of nuclear data.

The vast increase in nuclear data has obviously proved valuable as well as creating a few problems to the users. When data were scarce, they could be used as such. But when several, often incompatible results are available, the users want to know which one to choose, why and its associated uncertainty. To answer this question a new kind of scientist was created in the mid-1960s : the "evaluator". Relying on sound experimental experience and nuclear theory, evaluators contribute to bridge the gap between data producers and users, in coming out with unique and recommended values for the data, which then can be used directly by the reactor physicists. Several so-called "evaluated data files" have been obtained in this manner : ENDF/B (USA), ENDL (USA), UKNDL (UK), AWRE (UK), KEDAK (FRG), JENDL (Japan), SOKRATOR (USSR) ... Another important point was raised by the use of sensitivity methods which led to some data accuracies far beyond experimental capabilities and required developments of new ones, hence long delays. On the other hand, these data had to be available in time to be incorporated in the calculations. To overcome this serious difficulty, "integral" experiments were designed for obtaining rapidly global results in conditions similar to those of the reactor project. Such experiments are called "integral" because they involve global values of the data, as opposed to "microscopic" measurements of detailed data. These experiments put overall constraints on some sensitive microscopic data for which the accuracy requirements can be somewhat relaxed. Further steps involve i) calculations of what the results of those "integral" experiments should be, using evaluated data, ii) adjustment of these data within well-defined ranges to obtain agreement between measured and calculated values. When adjusting the cross-sections, care should be taken to preserve their own internal consistency, hence the use of correlation coefficients between cross-section changes. This method leads to "adjusted data files" having usually a restricted distribution. All these files are then processed before being used in reactor calculations.

With this general background information in mind, it is possible to discuss in more detail nuclear data for fission reactors in which nuclear reactions induced by neutrons play a dominant role. Since a full account of this subject is far beyond the scope of this paper, attention is focused here only on the present situation for two types of reactors already greatly developed : the "Pressurized-Water Reactors" (PWR) and the "Liquid-Metal Fast Breeder Reactors" (LMFBR).

2. Pressurized-Water Reactors (typically 900 to 1300 MW of electric power)

This type of reactor uses a fuel of uranium enriched to 3 % in ^{235}U and the fission neutrons are moderated in light water which also acts as a coolant. The spectrum there is composed of a Maxwellian part associated with a $1/E_n$ tail of undermoderated neutrons. Therefore, nuclear data are important in the thermal neutron region and that of the first low-energy resonances whose $1/E_n$ spectrum average contribution is called the "resonance integral".

The effective multiplication factor K_{eff} must be known to about 0.25 % for the fuel as a function of burnup from zero to a maximum of about 35000 MWD/T (Megawatt-day per metric ton). For an infinite medium, one has

$K_{eff} = K_\infty = \overline{\nu}\Sigma_f/\Sigma_f+\Sigma_\gamma)$ where $\overline{\nu}$ is the average number of fission neutrons and $\Sigma_f(\Sigma_\gamma)$ is the macroscopic fission (capture) cross-section. A 0.25 % uncertainty for K_∞ leads to 0.25 % uncertainty also for $\overline{\nu}$ but about 0.4 % for both Σ_f and Σ_γ. For a fresh fuel, Σ_f and Σ_γ come essentially from ^{235}U and ^{238}U respectively. Such accuracies are beyond present experimental capabilities. About 1 % accuracy is the lowest limit that can be achieved. This simple example illustrates the need for integral experiments to supplement microscopic measurements.

For an irradiated fuel, one has to take into account i) transuranic elements, essentially the Pu isotopes which have a positive overall effect on reactivity and for which both capture and fission (possibly also (n,2n)) cross-sections are needed, typically with 10 % accuracy and ii) fission products. Because the neutron spectrum is narrow, this latter effect is restricted to fission products formed with a substantial yield and having a large thermal-neutron capture cross-section caused by a strong low-energy resonance. Two of them dominate the effect : ^{135}Xe and ^{149}Sm. The other fission products have a smaller but growing effect with time in contrast to those for ^{135}Xe and ^{149}Sm which remain practically constant. Nuclear data for these two nuclei are fairly satisfactory.

The variation of K_{eff} with temperature plays an important role for safety. As far as the moderator is concerned, this involves the shapes of the cross-sections for ^{235}U and ^{238}U. Differences of about 3.10^{-5} in K_{eff} per °K between microscopic and integral measurements are observed but not understood. The temperature coefficient for the fuel itself is mainly determined by the resonance broadening, known as Doppler effect, for the 6.68 eV in ^{238}U. Thermal motion of the nuclei broadens the cross-section shape of the resonance in the laboratory frame without altering the area under the resonance. In contrast the area above the dip caused by the resonance in the spectrum ϕ increases with Doppler broadening because of self-shielding. Therefore the reaction rate $\sigma\phi$ varies with temperature, and the overall temperature coefficient must be known to about 10 %. Solid-state effects play a role also in Doppler broadening and their variation with temperature must be understood, especially when comparing zero power integral experiments (hence cold) with results from reactors.

The spatial power distribution must be known to 5 %. This is the most global parameter involving practically all nuclear data pertinent to the reactor.

The kinematics of the reactor require the knowledge of β_{eff} caused by delayed neutron emission to an accuracy of about 5 %. This involves by order of decreasing importance i) the total yield of these neutrons, ii) their repartition in groups of precursor half-lives and iii) their energy spectrum. An overall accuracy of only 10 % is presently achieved.

The decay heat of the irradiated fuel needs to be known to about 3 % for safety reasons. In case of accident, this parameter is critical for times comprised between a few seconds and a few hundreds of seconds. It can be determined both by microscopic and integral methods. In the former case, yields and decay properties of individual fission products are measured and used to compute the decay heat. In the latter method, the heat released by a sample of irradiated fuel is directly measured with a calorimeter. Both methods seem now to agree for ^{235}U fission but discrepancies are still observed for ^{239}Pu. The ANS (American Nuclear Society) decay heat standard is above these curves, especially at short times.

The handling of irradiated fuel after removal from the reactor necessitates the knowledge of heat production as well as neutron and hard γ-ray emission after several months of cooling time. Again, a detailed knowledge of secondary actinides (transactinides) and fission products at that time is necessary. Among the former let us quote ^{244}Cm which is responsible for most of neutron emission by spontaneous fission. Capture cross-sections of ^{242}Pu and ^{243}Am are the most important data for prediction of ^{244}Cm production.

Future trends include :

i) higher burnup from 35000 MWD/T to about 45000 MWD/T hence more ^{235}U depletion and greater importance of transactinides and fission products.

ii) compensation of ^{235}U depletion during the operation of the reactor by insertion of elements other than boron with high capture cross-sections. Hafnium and Gadolinium are considered in this respect.

iii) use of reprocessed fuel having the same ^{235}U enrichment but a higher amount of ^{236}U presenting a strong resonance at 5.4 eV.

iv) reconsideration of the shapes of $\sigma_{n\gamma}$ for ^{238}U and η_f for ^{235}U in the thermal neutron region in order to meet results from recent integral data.

3. Liquid-Metal Fast Breeder Reactors (such as Super-Phenix with 1200 MW of electric power).

The breeding performance of a reactor is usually expressed in terms of its breeding ratio BR or gain G with $G \simeq BR - 1 \simeq \eta_f - 2$ (G must be positive for breeding). The variation of η_f with E_n for the main fissile isotopes shows the interest of ^{239}Pu which has the highest η_f value at high energy ($\eta_f > 2$ above about 20 keV). This explains the development of the ^{238}U - ^{239}Pu LMFBR cooled by liquid sodium and of course without moderator. Yet, the primary fission neutron spectrum is degraded by fast neutron scattering and a typical energy spectrum in the reactor covers a wide range from several MeV down to hundreds of eV. This includes the resonance region and the continuum. Therefore, resonance properties are far more important in LMFBR than in PWR.

The main high accuracy requirements on the data come from K_{eff} and BR. The effective multiplication factor must be known to about 0.25 % for the fresh fuel. The uncertainty calculated with available microscopic data is about ten times greater but can be reduced to the required value with the help of integral measurements. For an irradiated fuel, the accuracy on K_{eff} can be tolerated to 0.5 % but integral experiments are more difficult. The Breeding ratio needs to be known to about 2 %. Taking into account integral measurements, these target accuracies have the following consequences for nuclear data uncertainties.

The capture and fission cross-sections for the main actinides : ^{235}U, ^{238}U, and ^{239}Pu are requested with a typical uncertainty of 2 to 3 % from 100 eV to 5 MeV, somewhat greater at higher energy. These requests are approximately met for their fission cross-sections ; that of ^{235}U is of good quality and is now used as a standard. The capture cross-sections of ^{235}U and ^{239}Pu are obtained through measurements of $\alpha = \sigma_{n\gamma}/\sigma_{nf}$. Discrepancies up to 15 % are reported for ^{235}U. For ^{239}Pu, the situation is better but the α-value for this nucleus played an important role around 1967 when measurements gave results having about twice the accepted value, therefore decreasing the η_f value. Though this decrease proved later to be smaller, it was yet sufficiently important to cause the abandonment of the steam-cooled fast reactor project. This unexpected behaviour of α was interpreted as an intermediate structure effect in the fission cross-section, a consequence of the double-humped fission barrier. Direct measurements of the ^{238}U capture cross-section can be carried out but the associated uncertainty is greater than the request by at least a factor of 2.

The inelastic scattering cross-sections must also be known for ^{238}U (to 5-10 %) and ^{239}Pu (to 20 %) but are difficult to measure because of the fission neutrons. The requests seem to be satisfied for the total inelastic cross-sections but large discrepancies exist as to their components.

The fission products (FP) contribute to about 5 % in reactivity loss at full burnup. Their properties are also important for the fuel cycle. An overall 10 % accuracy is required for their capture cross-sections. The effect of fission products in LMFBR is entirely different from that in PWR because of large differences in neutron spectrum. Whereas a few prominent low-energy resonances play a dominant role in PWR fission products, the situation is quite different in LMFBR for which the spectrum covers a large number of resonances. A kind of smooth spectrum average is obtained and many fission products significantly contribute to the loss in reactivity, 10 (or 20) of the most important ones for about 62 % (84 %) of their overall effect. The required 10 % uncertainty is not presently met. Measurements are incomplete because samples are scarce and, when results are obtained, discrepancies of 15 % - 25 % are commonly observed, sometimes amounting to a factor 2. Theoretical calculations are therefore essential in supplementing measurements. But, they must rely on systematics of nuclear physics parameters and differences of a factor 2, sometimes up to 5 or 10 can be noticed.

The requests for structural materials depend on the composition of stainless steel used in the reactor. Typical constituents are : Fe (60 - 70 %), Cr (18 - 20 %), Ni (10 - 13 %), Mo, Mn (2 %). Their overall effect is comparable to that of the fission products. This results in about 5 % required accuracy for the capture cross-sections of Fe, Cr and Ni and 5 % to 10 % for Mo and Mn. Those requests are still not met by measurements. Experimental difficulties come from i) changes from resonance to resonance in the γ-ray spectrum dominated by a few strong γ-ray transitions and ii) the large contribution of elastic scattering compared to capture. The capture in the coolant (Na) is small and does not present a serious problem.

Secondary actinides are produced in large number. Only Pu isotopes contribute to the reactivity change, depending on the initial composition of the fresh fuel. The other secondary actinides do not contribute much to the neutronics but are important for fuel handling and specific production of a few of them. Typical 10 to 20 % accuracy is requested for their cross-sections. They are still more difficult to measure than for thermal neutrons. Fission cross-sections have been obtained only for some of them because of the lack of availability of samples and of their high radioactivity. Capture cross-sections are even more difficult to carry out because they need larger samples. Therefore, the set of requests is far from being met by measurements. One has to rely also on calculations using nuclear theory, based itself on parameter systematics of known data. It is difficult to assess the quality of such calculations. They seem to meet the requirements for fission cross-sections of some nuclei near those having known cross-sections. The uncertainty gets worse away from this zone, especially for lower-Z actinides (Th region) presenting a triple-humped fission barrier. For capture cross-sections, calculations are less reliable because of the still more limited data base for parameter adjustment.

The <u>Doppler effect</u> is mainly due to ^{238}U, that for ^{239}Pu being smaller because fission and capture components, though important, are of opposite signs and compensate each other. This effect must be known to about 10 % implying a similar accuracy on parameters for a very large number of resonances. This request seems satisfied at low energy where the resonance shapes are well determined by experiments. At the high energy end of the resolved-resonance region, this accuracy cannot be met because experimental resonance broadening is too wide. In the unresolved region, use is often made of computer-simulated cross-sections calculated by extrapolating resonance properties determined in the resolved-resonance region.

The <u>sodium-void coefficient</u> must be known to about 15 % for safety reasons but is consequences for nuclear data depend on the fuel being used.

The <u>decay heat</u> for short times and long times must be known for safety reasons and for fuel handling with similar requirements as for PWR. Activation of structural materials (mainly due to ^{59}Co) must be known to 10 %. But the main uncertainty comes from flux calculations at very low energy where activation is the most important.

<u>Neutron transport</u> calculations outside the core and the U blanket must be known for the activation of secondary sodium coolant. Very large calculational uncertainties exist. Accurate knowledge of the minima in the total cross-sections (so-called neutron windows), around 25 keV for Fe and 300 keV for Na, is requested.

4. Conclusion

This broad survey shows the importance of neutrons in nuclear energy and the evolution of nuclear data since the discovery of fission. Crucial at this very early stage, kept secret for a long time, the nuclear data field has greatly expanded afterwards as a consequence of the development of Fast-Neutron Reactors but has now levelled off somewhat lower. No major surprise is expected any longer from improved nuclear data but they can greatly help understanding many aspects of the study and operation of nuclear reactors now used on a very large scale. Several points are not treated here, because of lack of space but some of them have been illustrated in the oral presentation. The basic interest of Neutron Physics is treated elsewhere in this Conference. The economical aspects of nuclear data are very controversial and could not be dealt with here. Many references should be mentioned but could not be inserted here also because of lack of space. The reader may find them as well as further details in many Proceedings such as those of i) Conferences : Geneva (1955 and 1958) and, more recently, Antwerp (1982), Kiev (1980), Knoxville (1979), etc... and ii) Specialists'Meetings : Fast Neutron Capture (Argonne, 1982), Fast Neutron Scattering on Actinides (Paris, 1981), Fission Products (Bologna, 1979, and Petten, 1977), Transactinium Isotope Nuclear Data (Cadarache, 1979, and Karlsruhe, 1975)... Also books about relevant fundamental and applied aspects are being published : Fission and Neutron-Induced Fission Cross-Sections (1981, Ed. A. Michaudon), Neutron Sources (1982, Ed. S. Cierjacks) and Neutron Radiative Capture (Ed. R.E. Chrien, 1983).

The author is very grateful to MM. Bouchard and Ravier from C.E.N. Cadarache for many helpful discussions and to Mrs Gouget for typing this manuscript.

The role of the neutron in controlled thermonuclear fusion

V.S. CROCKER≠, T.D. Beynon*, L.J. Baker≠

≠Materials Physics Division, AERE Harwell, Didcot, Oxon. OX11 ORA, UK
*Department of Physics, University of Birmingham, Birmingham B15 2TT, UK

1. Introduction

Nuclear fusion predates nuclear fission by many years and has been the subject of research since the 1930's. The first fusion reaction, in which deuterium was accelerated onto a deuterium target and interacted to form helium, was achieved by Oliphant et al (1934). Today interest is largely concentrated on the deuterium-tritium fusion reaction, which releases a relatively large amount of energy (17.6 MeV) and which has the largest cross-section at low energies ($\simeq$ 20 keV). The goal is the utilisation of energy from this reaction to produce power efficiently. This paper briefly outlines the progress made in the research towards a fusion power reactor and more particularly looks at the role of the neutron in some conceptual systems.

For thermonuclear fusion to work as a practical source of electricity, physics requires that two conditions be achieved simultaneously. For a deuterium-tritium mixture these are:

(a) The hot plasma must be adequately confined; that is, the product of the ion density and the confinement time, $n\tau$, should be close to 10^{20} m^{-3} sec (Lawson 1957).

(b) The temperature should be in the region of kT = 20 keV (T=2.4.10^8 °K).

A necessary step in the development of fusion is 'scientific breakeven' wherein the burning plasma liberates energy equal to the amount required to heat itself. This can be achieved for an $n\tau$ value an order of magnitude lower and a temperature a factor of 2 less than the figures quoted above. 'Engineering breakeven', in which the released energy equals the total energy required for the whole system and which involves the thermal-to electrical conversion efficiency, is an intermediate step between scientific breakeven and commercial exploitation.

In tokamaks, progress in the technology of magnetic confinement has yet to reach the scientific breakeven conditions although they have been attained singly by ALCATOR at MIT ($n\tau$ = 3.10^{19}) and with the Princeton Large Torus (PLT) attaining a temperature of 7.3 keV. The generation of machines currently under construction - JET by EURATOM, TFTR in the USA, JT-60 in Japan, and T-15 in the Soviet Union, have performance predictions which achieve the scientific breakeven conditions simultaneously, and will, if successful, justify the construction of a device designed to demonstrate engineering breakeven.

Whilst magnetically confined systems aim at confinement times in excess of 10-20 sec, inertial confinement fusion (ICF) must achieve values of a few tens of nanoseconds coupled with high particle densities. Various ICF systems have been proposed employing laser, electron, and ion beams to implode and heat target pellets. For example, the confinement conditions for deuterium-tritium fusion might be achieved using beams of 10 GeV singly-ionized uranium ions, in pulse lengths of 30 ns, with a beam current of 30 Ka focussed onto a deuterium-tritium pellet of 2.5 mm radius. For an energy delivered onto the pellet of about 3 MJ per pulse, an energy release of 300 MJ should be achieved. A number of types of accelerator have been proposed for this application.

2. The Interactions of Neutrons in Fusion Reactors

The first generation of fusion reactors would employ the most readily achieved reaction, which is that between the second and third isotopes of hydrogen:

$$D + T \rightarrow {}^4He + n + 17.6 \text{ MeV}$$

Natural hydrogen contains 0.015% deuterium giving a total terrestial inventory of $\approx 3.10^{13}$ tonnes. However tritium has a half-life of 12.3 years and thus must be manufactured as required. (A 1000 MW(e) fusion reactor would consume $\approx$400 gm of tritium per day). The only practical nuclear reactions which can be utilised for this purpose are the neutron-induced reactions with the two naturally occurring isotopes of lithium:

$$^7Li + n \rightarrow T + {}^4He + n - 2.5 \text{ MeV}$$
$$^6Li + n \rightarrow T + {}^4He + 4.8 \text{ MeV}$$

There are adequate known resources of lithium to support a large power generation programme. The D-T source neutron has a reaction energy of 14.06 MeV, but since the plasma is thermonuclear there is some thermal broadening, 99.6% lying within 0.5% of the nominal value. Clearly it is possible for source neutrons to interact in lithium-7 producing a triton and a less energetic neutron which may generate a further triton from a lithium-6 reaction. This, in principle, is the proposed fuel cycle. There are some possible variations. The source energy is sufficient to realise neutron multiplication in materials such as beryllium or lead, so it is possible to dispense with the lithium-7 reaction. Alternatively in the 'hybrid' reactor concept neutrons could interact with fertile isotopes such as thorium-232 or uranium-238 producing fission reactor fuel as well as sufficient neutrons for tritium breeding.

The essential feature of a breeding blanket for a tokamak, therefore, is a hollow toroidal-shaped layer of lithium surrounding the plasma vessel, in which the neutron economy produces sufficient tritium to allow the injection into the torus of one triton for each primary fusion reaction. A number of effects and constraints conspire to make this difficult. The theoretical maximum of just under two tritons per source neutron is reduced due to inevitable effects of scattering, leakage and parasitic absorption, and due to constraints imposed by other features of the reactor. The most important considerations are:

(a) The breeding blanket is separated from the plasma by a first wall, which performs the functions of a vacuum vessel in most reactor designs. The effects of sputtering, neutron damage and cyclic thermal loading combine to create a severe environment for this component, which must necessarily be of high integrity and heavily built (2 cm stainless steel). This first wall scatters a proportion of the source neutrons below the ^{7}Li(n,n'α)T threshold and absorbs back-scattered neutrons.

(b) As much as 10% of the volume of the blanket will be structural material, which is similar to the first wall in its effect on the neutron economy.

(c) In the case of a tokamak, some of the space around the torus is required for fuel injectors, divertors and plasma heating ducts. This reduces the blanket coverage to 70-80% of the total area, implying that local tritium breeding ratios of at least 1.25-1.45 must be achieved. Similar effects apply to the blankets for ICF chambers, where allowance must be made for beam ports and for the pellet injection system.

(d) Some of the tritium produced will decay to helium-3 before it can be used; also excess tritium would be needed to supply the initial inventories of other reactors. For these reasons a breeding ratio in excess of the minimum is required.

In ICF reactor concepts, various types of blanket have been proposed in order to absorb the fusion reaction products (X-rays, alpha particles, neutrons and pellet debris) which will occur explosively. A 300 MJ output per pulse corresponds to the energy release of about 70 kG of high explosive. An important feature of ICF reactors is that confinement of the D-T fusion is decoupled from the various functions of the first wall and blanket. The absence of the toroidal field coils of the tokamak concept and the simple spherical or cylindrical geometry can be exploited by placing fluids or magnetic deflection fields inside the reaction chamber to protect the first wall from the fusion products.

A simple example of a liquid-wall concept is the BLASCON reactor proposed by Fraas (1971). This features a cavity formed by a vortex in a pool of rotating liquid lithium. The fusion pellets and the driver beams enter the vortex from the top of the chamber, and the vortex surface defines the first wall of the reactor. The lithium layer performs the functions of tritium breeding and of heat transfer from the reactor vessel, and is continuously injected with an inert gas to improve its performance as a shock absorber.

A variation on this approach is the lithium 'waterfall' concept (Meier et al 1977), which features a thick continuous annular flowing curtain of liquid lithium. An alternative approach is the 'falling pellet' concept which substitutes lithium oxide spheres for the metallic lithium of the waterfall blanket (Meier et al 1977). Among their other advantages, the falling pellets do not respond to the peaked energy release as would a continuum, and thus the amplitude of the shock waves transmitted to the reactor structure is reduced.

In addition to their role in the fuel cycle, the source neutrons carry 80% (14.06 MeV) of the D-T reaction energy which must be converted to high-temperature steam to drive a conventional turbogenerator. It is usual to express the operating reactor power density in terms of the energy carried across the first wall by neutrons, values of up to 10 MW/m^2 having been proposed although most current reactor concepts operate close to 4MW/m^2. A high proportion ($\approx$99%) of this energy is deposited in the breeding blanket via charged particle, scattering, and absorption events. The non-elastic interactions produced gamma-rays which are also largely absorbed in the blanket; up to 40% of the energy may appear in this form, so the prediction of photon as well as neutron fluxes is important. Exothermic reactions such as ^{6}Li(n,α)T predominate over endothermic events, so the blanket energy deposition is in the range 16-18 MeV/source neutron depending on the choice of materials. An important requirement is that the cooling circuit should have a high outlet temperature allowing an efficient thermodynamic efficiency, a condition which is not easily met since structural materials need to be kept cool ($\approx$300°C) to minimise damage swelling and embrittlement effects, and many of the possible lithium compounds have limited operating temperatures (500°C - 600°C) because of their chemical reactivity. One of the possible solutions to this dilemma is the inclusion of a graphite moderator which can be operated at a higher temperature than the rest of the blanket, but the need to obtain high energy deposition in a graphite layer imposes a severe constraint on the optimisation for tritium breeding.

The reactor physicist seeking a solution to the various requirements of a breeding blanket design is faced with a different and probably more difficult task than that presented by a fission reactor, thus:

(a) The considerable flux gradient in the blanket region requires a good representation of anisotropic scattering processes. Multigroup finite difference solutions to the Boltzmann transport equation provide an adequate means of calculation in simple geometric approximations.

(b) For energy deposition calculations it is necesary to link neutron and photon transport methods. The use of polynomial expansions of the scattering functions allows both particle types to be represented in a single calculation, since the physics of scattering is then decoupled from the transport equation.

(c) The range of neutron energies is large due to the 14 MeV source energy, although thermalisation effects may be ignored. The neutron spectrum in the first wall of a tokamak reactor is compared with the core centre spectrum for the prototype fast reactor (PFR) in fig. 1. Above 8 MeV the fast reactor spectrum is insignificant, for which reason the development of fission reactors has not given rise to a satisfactory nuclear data base for fusion reactor neutronics. The paucity of cross-section measurements in the 8-14 MeV range causes considerable uncertainty in the calculation of integral parameters. Cross-section data is also inadequate in the evaluation of secondary neutron distributions

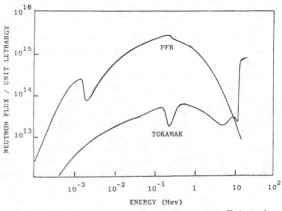

Fig. 1   Neutron Spectrum at a Tokamak First Wall Compared with the PFR Core Spectrum.

in energy and space for (n, 2n) processes, and in the energy distribution of photon production data. A notable example of such uncertainties is the ^{7}Li(n,n'α)T cross-section. Swinhoe et al (1979) have reported measurements giving values ≈26% lower than the accepted standard, and Baker (1980) has shown that this change would reduce breeding ratios by as much as 8%. In an attempt to resolve such uncertainties, a number of integral benchmark experiments have been made. In essence, these consist of an assembly of blanket materials arranged in a simple geometry and irradiated by a 14 MeV neutron source. Measurements of neutron flux distributions and reaction rates can then be compared with computations using nuclear data bases which can be modified accordingly.

(d) Toroidal geometry represents a severe test for calculation techniques by comparison with the cylindrical form of fission reactor cores. Currently much study is directed at parametric surveys for which one-dimensional representations are adequate, but as detailed reactor designs emerge there will be an increasing demand for accurate three-dimensional models. At present, monte carlo techniques are the only possible approach to this requirement, although finite element methods may be applied eventually. As indicated, this difficulty is significantly lessened for the simpler geometry of ICF and linear magnetic confinement reactors.

Many combinations of materials and geometric configurations have been proposed for breeder blankets. Early designs utilised metallic lithium contained in refractory structural materials such as niobium, usually with a graphite or steel neutron reflector. The disadvantages of using lithium are that the problems of liquid metal technology are introduced into the blanket and that breeding layers have to be thick (≈1 m). A possible advantage was seen in the circulation of the lithium, thus obviating the need for a separate coolant and removing tritium rapidly from the blanket, but it was realised that the energy requirements for pumping liquid metal in the presence of high magnetic fields were unacceptably high. For this reason the use of static lithium in a gas-cooled blanket was proposed, and a further variation employed to minimise the lithium inventory was the use of a beryllium multiplier backed by a graphite moderator, this combination producing a high neutron flux at intermediate energies which could be absorbed in a thin layer of enriched lithium. The latter concept suffers from enhanced parasitic neutron absorption by structural materials. A different approach is made possible by the use of compounds of lithium. Ceramics such as the silicates, aluminates and oxides, intermetallic liquid phases of lithium, lead and bismuth, and alloys of lithium with aluminium or lead have all been proposed as candidate materials. Some of these have advantages in leading to more compact designs, in having greater chemical stability, or in having large thermal capacities which allow smoothing of the reactor power output. No one combination of materials has emerged as the preferred choice for reactor blankets. This reflects the uncertainty which still exists in reactor design concepts and predictions of materials performance under irradiation, and also the lack of experimental information. The next generation of tokamak experiments (eg the INTOR concept) should provide space for the irradiation of blanket modules and therefore provide an opportunity to begin the development of the necessary reactor blanket technology.

Zone	Thickness (cm)	Materials		Density Factor
First Wall	2.5	Niobium	30%	1.0
		Helium	70%	
Blanket 1 (Natural Lithium)	14.0	Inconel	4%	0.8854
		Graphite	20%	
		Lithium-lead	60%	
		Helium	16%	
Blanket 2 (90% enriched Lithium)	6.0	Inconel	4%	0.8554
		Graphite	60%	
		Lithium Oxide	25%	
		Helium	11%	
Reflector	25.0	Stainless Steel	95%	0.8144
		Helium	5%	
Fixed Shield	80.0	Stainless Steel	80%	1.0
		Water	20%	

TABLE 1. The Configuration of Materials in a Minimum-size Tokamak Blanket

An example of the configuration of materials required for a reactor blanket is given in table 1. In this case the blanket has been optimised for minimum size (Baker 1978). The first breeding layer multiplies and moderates the neutron population to provide a large $^6Li(n,\alpha)T$ reaction rate both in itself and also in the second layer which is enriched to 90% ^6Li. The principal nuclear reaction rates and energy depositions for this blanket are shown in table 2, which indicates a tritium breeding ratio of 1.16. Higher tritium breeding ratios can be obtained from this configuration by increasing its size, the maximum value being 1.4.

Zone	Reaction Rates per Fusion Neutron				Energy Deposition (MeV/neutron)
	$^6Li(n,\alpha)T$	$^7Li(n,n'\alpha)T$	$Pb(n,2n)$	Parasitic Capture	
First wall	–	–	–	0.095	1.50
Blanket 1	0.433	0.193	0.291	0.062	11.20
Blanket 2	0.530	0.001	–	0.029	3.49
Reflector	–	–	–	0.160	2.88
TOTAL	0.963	0.194	0.291	0.346	19.07

TABLE 2. Principal Nuclear Reaction Rates in a Tokamak Blanket

No description of fusion neutronics would be complete without mention of the deleterious effects of neutron interactions. These effects are grouped under the general headings of damage and activation, the former including atomic displacements, helium and hydrogen production, and transmutation, and the latter implying problems of afterheat, active handling, and safety. The 14 MeV source energy is in excess of the (n,p) and (n,α) thresholds of most blanket materials, and the production of hydrogen and helium in structural materials reaches several 1000 appm. Also the high neutron energy in combination with the required power densities produces an atomic displacement rate in and near the

first wall which is considerably higher than that found in the permanent
structure of a fission reactor. This combination of effects leads to
changes in the tensile, creep, fatigue, fracture and swelling properties
of materials which present formidable problems to the metallurgist, and
to the designer of neutron sources. No neutron source exists which can
simulate the fusion reactor neutron spectra at the required intensities,
so that materials investigations are currently progressing with the use
of accelerator-based methods of simulation and low fluence fission
reactor irradiations (Gold et al 1981). Present indications from the
materials development programmes are that first wall and blanket
structures will have a lifetime of 10-20 $MW.yr/m^2$, corresponding to
reactor operating periods of 2-4 yrs.

The problems arising from the activation of materials are less severe,
and in most ways present a favourable picture when compared with
fission. In particular there are few long-lived isotopes comparable to
those produced by fission, so that on long timescales activities decay
to very low levels. This is a relatively benign situation from the
point of view of waste disposal, although it should be remembered that
this activity will be distributed through a very large mass of material.
The large mass of activated blanket is also a problem from the point of
view of reactor maintenance, when segments of several 100 tonnes must be
removed, repositioned, welded, etc, using remote handling equipment.
Afterheat in breeding blankets is a somewhat less severe problem than
for fission reactors, but the possibility of melting and release of
activated materials is still present and must be considered.

3. Hybrid Fission-Fusion Reactors

A possible role for D-T thermonuclear reactors has emerged in the
context of hybrid fission-fusion devices. The basic principle is that
14 MeV D-T neutrons can be employed to breed fission fuels from the
fertile isotopes thorium-232 or uranium-238, while creating a fission-
enhanced neutron population which is capable of breeding sufficient
tritium to complete the fusion fuel cycle. Several variations on this
theme are possible. The relative energies associated with fission and
fusion reactions are such that fission plays the dominant role in terms
of power production when the complete fuel cycles are evaluated. The
utility of hybrids therefore depends on the suppositions that the future
build-up of fast fission reactor installation will exceed the growth
rate of the fuel inventory achievable in conventional breeders, or that
the use of fast breeder reactors is to be avoided by employing hybrid
breeders to fuel thermal fission reactors.

An obvious mechanism for generating the necessary neutron population in
a hybrid blanket is fast fission in the fertile isotopes, but this
approach brings some of the problems of the fast reactor with it. An
alternative approach is the 'fission-suppressed' system in which the
neutrons are moderated below the 1.5 MeV fission threshold energies of
the fertile isotopes before they enter the regions of the blanket
containing them. Bethe (1978) has shown how such blankets can maximise
the total energy available from fission power reactors using fuel from a
specified fusion fuel breeder.

An example of a suppressed fission blanket is the relatively simple
design concept proposed by Lee et al (1981). It consists of an inner
region which is a conventional tritium breeder, and an outer zone
optimised to breed fission fuel. The inner region, adjacent to the
plasma, contains a thick beryllium neutron multiplier and a breeding

compound such as lithium oxide or lithium-beryllium oxide. The Be(n,2n) reaction multiplies and moderates the neutron population thus minimising the fast fission rate in the outer zone. The fertile material is in the form of a thorium or uranium salt in pellet form and coated in graphite to enhance the moderation and subsequent capture. As the quantity of bred uranium-233 or plutonium-239 increases the fission rate becomes significant, at which point the pellets would be removed and processed to separate the fertile and fissile isotopes.

In an optimised design of this type, a small number of suppressed fission hybrids could reduce the dependence of a larger number of thermal reactors (LWRs, HTGRs, or CANDUs) on new uranium resources. For example, Lee et al (1981) show that a breeding factory based on the scheme outlined above can supply the fuel for 25 thermal LWRs or 50 CANDUs, assuming that the fusion breeder has the same power rating as each of the fission systems supplied. The effective cost of an LWR in this scheme would be increased by about 20% by the existence of the breeder. Alternatively, or additionally, the initial fuel inventories for a fast fission breeder programme could be provided.

4. Summary

The neutron has been shown to play a vital role in conceptual fusion reactors based on the D-T reaction. It not only provides the major source of energy production but also the tritium for maintaining the fuel cycle. However, a by-product is the radiation damage it produces, particularly in the first wall. A thorough understanding of the interaction of neutrons with the first wall and blanket is a key feature in the development of fusion technology and will attract increasing attention as the prospect of ignition experiments and demonstration electricity producing reactors draws nearer.

REFERENCES

Baker L.J. (1978) Proc. 10th Symp. on Fusion Technology (Padova: Pergamon Press) pp 661-666

Baker L.J. (1980) Proc. 11th Symp. on Fusion Technology (Oxford: Pergamon Press) pp 289-294

Bethe H.E. (1979) Phys. Today (May 1979)

Fraas A.P. (1971) ORNL report TM-3231

Gold R.E., Bloom E.E., Clinard F.W., Stevenson R.D. and Wolfer W.G. (1981) Nuclear Technology/Fusion, 1,169

Lawson J.D. (1957) Proc. Phys. Soc. London, 70B, 6

Lee J.D. and Moir R.W. (1981) J. of Fusion Energy, 1, pp 299-303

Meier W.R. and Maniscalco J.A. (1977) Lawrence Livermore Laboratory report UCRL-79694

Oliphant M., Harteck P. and Rutherford E. (1934) Proc. Roy. Soc., A,144,692

Swinhoe M.T. and Uttley C.A. (1979) Proc. Int. Conf. on Nuclear Cross-Sections and Technology, Knoxville.

Inst. Phys. Conf. Ser. No. 64: Section 6
Paper presented at Conf. on Neutron and its Applications, Cambridge, 1982

Radiation damage

M.W. Thompson

University of East Anglia, Norwich, NR4 7TJ, England

Abstract. After the successful operation of the first chain reacting pile, and the prospect of developing it for energy and isotope production, the effects of neutron irradiation on reactor materials set limits for reactor designers to work within. Great advances have been made in understanding the underlying phenomena of radiation damage, many of which have intrinsic scientific importance. Some of these are described.

1. Engineering Background

There is a mistaken belief that fundamental discoveries only follow from the pursuit of fundamental questions and precede inventions. Some fields of science have indeed progressed in this way, exemplified by Cosmology or Elementary Particle Physics. But there are others that spring up as sideshoots from an engineering problem. Thermodynamics is the classic example, coming long after the invention of the heat engines that first posed the problems. Radiation damage is another, whose phenomena have forced enquiries in several branches of fundamental science leading on to important discoveries.

Although it was radiation damage in photographic emulsion that formed a major tool for the early discoveries of Nuclear Physics, and one can read occasional speculations of the Nuclear patriarchs concerning the effects on the solid of the radiation, rather than vice versa, the subject was hardly touched until the invention of nuclear reactors. These, with their enormous neutron fluxes, caused strange effects in reactor materials, at first seeming so severe as to prejudice the commercial development of nuclear power.

The most sensitive substances were the organic polymers used for electrical insulation which rapidly degenerated and these materials had to be replaced by ceramic oxides. The uranium metal of the earlier reactors at first seemed little affected but as the irradiations became more prolonged in attempts to extract more energy, or to breed more plutonium, from each tonne of fuel, the metal rods became distorted. In some cases they ruptured their cladding sheath, there to protect them from corrosion and to retain their fission products. Some of this distortion was a predictable volume increase due to the fission process producing two new atoms for each uranium atom lost. But the irradiation of single crystals of uranium showed weird distortions: elongation by many hundreds of percent along one axis and shrinkage along another (Paine and Kittel, 1955). To avoid the distortion of fuel elements,

a metallurgical process was developed which produced uranium with very
small crystallites orientated at random in order that growth of one would
be compensated by shrinkage of another. The economic consequences for
the cost of nuclear electricity were enormous. In the earlier fuels we
were lucky to burn up 0.1% of the uranium atoms before trouble started;
but in the current fuel elements of the Magnox reactors ten times this
percentage burn-up is achieved. Since much of the operating cost of a
power station is in processing and replacing the fuel, the cost of
electricity was reduced far below the level originally thought possible.

Another engineering problem was thrown up by the graphite which forms
the moderator in many reactor types, including Magnox. In doing its job
it is irradiated with very high doses of neutrons, sufficient in the
reactor's lifetime to cause every atom to have been thrown violently out
of position hundreds of times. The internal structure of the graphite
crystallites becomes defective and metastable. If their temperature is
raised a partial reversion to the undamaged structure occurs, accompanied
by the release of energy as heat (Kinchin, 1955; Woods et al, 1956). In
some irradiated specimens the rate of release of heat can be more than
enough to compensate for the rate of heat input necessary to raise the
temperature. The temperature then goes on rising without external
heating until much of the stored energy has been released. That is not
a situation which could be tolerated in an operating reactor.

All the reactors in British power stations are designed to run with the
graphite at a high enough temperature that such levels of stored energy
are not approached. But in some earlier reactors it was, and special
heating procedures had to be designed to remove the stored energy
controllably. An historic account of the British team's experience in
devising this procedure has recently been published by Cottrell
(1981). (I was a member of his team.) Again the economic consequences
are considerable, not only in the production lost in releasing the
stored energy but also in limiting the operational life of the reactor
and increasing the amortisation costs.

These are just two examples of neutron irradiation effects. They were
quite outside the range of known phenomena and demanded a study at the
most fundamental levels of solid state science. A massive programme of
research was launched in the nineteen fifties, first in the atomic energy
research establishments, then in the universities as the more fundamental
scientific implications became apparent. In the remainder of this lecture
I shall highlight a few of the scientific advances that have sprung from
the study of radiation damage by neutrons; and end by pointing to some
problems for the future. I do not believe that much of this would have
happened by now had the neutron not been discovered in 1932.

2. Elementary Processes

In the solid exposed to fast neutrons there will be occasional collisions
which scatter the neutron and cause the scattering nucleus to recoil. In
a collision with a 1MeV neutron the nucleus of medium mass recoils with
an average kinetic energy of order 10keV, at least a thousand times
greater than the potential energy that binds an atom to its near neighbours in the solid. As the recoiling atom strikes its neighbours a
cascade of interatomic collisions follows, sharing out the energy of
recoil until the kinetic energies become too small to overcome the binding
energy (Kinchin and Pease, 1955). The structure of the solid will then be

riddled with holes (vacancies) and peppered with extra atoms in abnormal positions (interstitials). These are the primary defects caused by radiation damage.

Neutrons may also induce nuclear reactions, fission being the most obvious to consider. Here the two fragment atoms, each with a mass of order 100 and a kinetic energy of order 100MeV, recoil apart and generate a dense cylindrical collision cascade. Each of these events is estimated to generate between 10^4 and 10^5 primary defects (see Thompson, 1969).

Another common neutron induced reaction is the (n,γ) type. Here the momentum of the emitted gamma ray must be subtracted from the recoil and can cause an energy transfer of order 10 to 100 eV even when thermal neutrons with a very small kinetic energy are being captured. Such events produce a very small collision cascade and a very simple form of primary damage.

If the temperature of irradiation is low enough, the primary defects may be immobile, but in the power reactor the temperatures will often be high enough to make them highly mobile. Some will then combine, either like with like to form clusters of vacancies or interstitials, or in unlike pairs to annihilate each other. The clusters are the simplest form of secondary defect. More complex forms might involve impurity atoms or, in a compound, atoms of several types forming defect molecules. It is the accumulation of defects that causes the radiation effect. The form of accumulation, hence the effect, is highly sensitive to the rate, the dose and the temperature of the irradiation. It will also depend on the neutron energy distribution, hence on the type of reactor and the irradiation position.

Going back to describe our two engineering problems in these terms: the graphite irradiated at about $100°C$ accumulates vacancies, as single defects, and interstitials in small clusters. If the temperature of the irradiated graphite is then raised to $150°C$ some interstitial clusters break up, the interstitials migrate to annihilate vacancies with the release of stored energy which appears as vibrational energy (heat) in the graphite.

In my second example, both types of primary defect in the uranium are highly mobile at the fuel temperature and they are produced at very high density in the core of the fission fragment's track (vacancies) and around it (interstitials). The conditions favour the formation of many clusters of each type but the uranium crystal, being non-cubic, has the property that interstitial clusters take on a plate-like form parallel to one set of crystal planes, whilst vacancies cluster on another set of planes. The result is an anisotropic distortion of the crystal with the interstitial clusters tending to elongate it along one axis and the vacancy clusters tending to shrink it along others (Buckley, 1961).

3. Some Important Discoveries

Clearly these descriptions of events in the fuel nd moderator draw upon an enormous body of knowledge gained through the efforts of many scientists in the last fifty years. Their research has covered the statistics and structure of collision cascades, the properties of point defects, their diffusion and clustering processes. All that I can do in the time available here is to highlight a few of the discoveries that underlie our ability to understand radiation effects, hence to design reactors where they are minimised.

The lowest temperature at which a particular type of defect is able to migrate through the solid may be determined by annealing experiments. In these a specimen of the solid is irradiated at very low temperature, often 4K, at which none of the primary defects is mobile. After the irradiation the concentration of the defects is monitored by measuring some defect-dependent property, like electrical resistivity in the case of a metal, as the temperature is progressively raised. Sharp changes in concentration are found to occur in narrow bands of temperature, known as recovery stages, and by comparing the annealing behaviour for several types of property change it has often been possible to identify a particular stage with the initial movement of a specific defect. In metals the interstitials generally become mobile far below room temperature (see review: Schilling, 1978), the vacancies at higher temperatures: roughly one-fifth of the melting temperature (Thompson, 1969).

Important results emerge from a comparison of the annealing after fast neutron irradiation (Blewitt, 1962), with that after MeV electron irradiation (Walker, 1962). In the latter case the recoil atoms from electron scattering have maximum energies in the range 10 to 100 eV, consequently the primary defects are mostly isolated interstitial-vacancy pairs. In the annealing, the low-temperature recovery stage associated with interstitial movement then splits into several distinct sub-stages. Several of these can be attributed to annihilation of the vacancy by the interstitial making a single jump from a specific neighbouring site in the crystal structure. From comparing such results with the behaviour of computational models of the damaged crystal (Gibson et al, 1960; Erginsoy, 1964) we know that the nearest neighbouring sites are usually unstable. By varying the direction of the irradiating electrons to favour particular directions of atomic recoil the shape of the unstable zone can in principle be determined (Bauer and Sosin, 1964).

In the case of fast neutron irradiation the sub-stages of interstitial recovery are not seen. It is thought that in the dense grouping of primary defects near a neutron collision site, the overlapping distortions of the crystal around each interstitial-vacancy pair so broaden the sub-stages that they cannot be resolved.

Another fundamental property of a crystal is the so-called anisotropic displacement energy. This is the function which gives the minimum recoil energy needed to displace the atom, and create a stable interstitial-vacancy pair, in relation to the direction of recoil in the crystal. Such functions have been measured in one or two cases (e.g. Jung, 1975) and again computational models have been of great assistance in interpreting the experiments.

Computer modelling has also shown that the symmetry of a crystal lattice greatly influences the propagation of collision cascades (Gibson et al, 1960; Robinson and Torrens, 1974). For example, focused sequences of collisions frequently occur along rows of near-neighbour atoms, the momentum being transported with remarkably little attenuation because the atoms' trajectories are focused into the rows by the potential energy contours of the crystal lattice.

These effects have been demonstrated by sputtering experiments in which collision cascades are generated near the surface of a crystal by irradiation with heavy ions having energy of order 10keV (see a review by Thompson, 1981). Such cascades are similar to those from fast neutrons but heavy ions are much more efficient at producing recoils. Whenever a cascade intersects the surface atoms are ejected, or sputtered, with momenta characteristic of the momentum distribution in the cascade. Ejection is found to be strongly anisotropic, with peak intensities along near-neighbour directions. The atoms' energy distribution along these directions has been measured by time-of-flight techniques and shows the predicted characteristics of the focused collision sequences, with a sharp cut-off above a limiting energy of order 100 eV.

A focused sequence starting above the displacement energy begins with a vacancy being created, the vacating atom collides with its neighbour and replaces it, the first neighbour then repeats the process with the second and so on until one of the atoms has too little energy to replace its neighbour: this atom then becomes the interstitial, not the one which creates the vacancy. This is thought to be an important process in separating the interstitial from the vacancy, hence favouring clustering in the recovery stages rather than annihilation.

A nice example of the focusing action of the crystal potential is provided by sputtering atoms from the {100} surface of a gold crystal observed along the <100> direction. Any atom ejected from the sub-surface layer has to pass through a ring of four atoms in the surface layer. The collective potential of the surface group acts like a lens to focus the possible trajectories of the ejected atom. Because the strength of the lens depends on atom's energy, there is one critical energy for which trajectories from all such lenses will be focused at a detector placed on the <100> axis. If this is an energy sensitive detector it will show its strongest signal when tuned to the critical focusing energy. This has indeed been seen (Reid et al, 1980) and the critical focusing energy for gold atoms ejected along <100> is 150 eV.

A collision effect of far reaching importance was discovered in the early nineteen sixties as a spin-off from radiation damage. Theoretical computation of the trajectories of primary recoils from neutron irradiation of a copper crystal showed that some became trapped into the natural channels of the crystal lattice, being gently steered by its potential energy contours, and were able to travel long distances with little energy loss (Robinson and Oen, 1963). This phenomena was named channeling and was quickly demonstrated by the discovery that a beam of fast ions would pass very easily through a thin crystal if they were directed along channel axes (Thompson and Nelson, 1963).

It was also shown that the backscattering of ions from a crystal was greatly reduced when the beam was aligned with channel axes (Thompson and Nelson, 1963). Because, once channeled, the ion has little chance of that close encounter with an atom needed for backscattering. Of course, if the crystal contains defects, say impurity atoms, which block a channel, backscattering will be seen. Some defects by their symmetry will block some classes of channel and not others. It is this principle that has led to an important new technique for the location of defects and the determination of their structure (see Chu, Mayer and Nicolet, 1978).

Another important class of experiment used to demonstrate channeling was heavy ion implantation (Piercy et al, 1963). These clearly showed that a substantial fraction of a channel-aligned beam penetrates much deeper into the crystal than in the non-aligned cases. This has been exploited in the manufacture of semiconductor devices by ion implantation (see Mayer, Eriksson and Davies, 1970).

My last set of examples are effects involving secondary defects. Their direct observation was made possible by developments in electron microscopy using diffraction contrast to reveal defect structures (Hirsch, Howie and Whelan, 1960). The studies were greatly assisted by the use of ion beams to simulate the effect of neutron irradiation. Because the collision generating efficiency of an ion is so much greater than a neutron, quite modest ion beams of a few microamperes can create in a few minutes a damage equivalent to many years in a reactor. Of necessity, sufficient point defects to produce observable secondary defects require such high levels of irradiation.

It was soon discovered that point defects often condense into flat platelets which then transform into dislocation loops. These loops may grow by absorbing further defects of the same kind, eventually joining up with others to form dense networks. Of course, if a loop that was formed from interstitials absorbs vacancies it will shrink. The conditions for growth and shrinkage is a complex problem in diffusion theory which must take account of all the competing sinks for the point defects and, possibly, attractive and repulsive forces associated with them. The problem has been solved for a number of cases and observable effects correctly predicted (Brailsford and Bullough, 1972, 1973a, b and c).

It is sometimes found that materials under mechanical stress sometimes deform gradually by creep during irradiation. This has been convincingly explained by the condensation of point defects onto certain segments of dislocation which then move to deform the solid in the direction of the applied stress. In essence the stress induces a preferential absorption of the point defects onto appropriate dislocations (Bullough and Willis, 1975).

Vacancies sometimes cluster to form polyhedral voids, usually when impurity atoms are present which, it is thought, prevent the transformation of the embryonic cluster to a dislocation loop. Under some conditions of irradiation, when the interstitials favour some more attractive sink, the vacancy voids grow numerous and large to fill an appreciable volume of the whole solid. The material then swells, because the interstitials in effect transport the material from the voids to push out the original boundaries (Cawthorne and Fulton, 1967). This effect can be an embarrassment to the reactor designer who needs to know that it is likely to happen in metals irradiated in a fast reactor for a year or so at a temperature roughly one-third the melting temperature.

A very pretty side effect was noticed: in some crystals the voids formed up on a regular lattice, often a magnified version of host crystal lattice (Evans, 1971). It has been shown theoretically that this could be due to weak interaction forces between the voids which move them as they grow into these stable configurations (Tewary and Bullough, 1972).

4. The Past and the Future

There can be no doubt that major scientific developments have taken place under the rather inadequate title of Radiation Damage. It draws together many different branches of science: nuclear physics, atomic collisions, point defects, dislocation theory, metallurgy and physical chemistry. All of these would have existed and prospered without the discovery of the neutron and the ensuing necessity to study radiation damage. But the rate of investment has concentrated the research into a short time span that has enabled direct interaction between scientists from these various disciplines. Had the research been spread out over a century, many of the scientists could not have met, their interactions would have been weaker or non-existent, and many ideas would have been lost. Each of the disciplines would have been the poorer for this.

For the future many of the ideas that grew within radiation damage will develop to enrich fields outside nuclear energy: channeling, ion implantation, sputtering, defect microscopy, radiolytic chemistry are just a few examples. But so long as we want nuclear power there will be a continuing need for new or more efficient materials for reactors. Resistance to radiation damage will be a central criterion for them to meet. The depth of existing knowledge at a fundamental level, and the great successes our nuclear metallurgists have had in improving the performance of the Magnox reactors over twenty years, gives us great confidence for the more advanced types.

There is still a need for further basic research. The fast reactor has brought us some surprises like void swelling and creep, both of which are now accommodated. In the newest designs there may be others. In fusion reactors the engineering problems are still paramount but we can see that if they are overcome there will be quite new irradiation effects at the first containment wall due to interaction with the hot plasma. Here our experience with sputtering and ion implantation should be of great assistance. The neutrons from the fusion reaction have a much higher energy than those from fission, and they are likely to come in pulses. These new factors will bring us into an unknown area of radiation damage which will need careful study if fusion is to become a practical reality.

5. References

Bauer W and Sosin A 1964 J. App. Phys. 35 703

Blewitt T H 1962 Rad. Dam. in Solids ed D S Billington (Academic Press)

Brailsford A D and Bullough R 1972 J. Nuc. Mat. 44 121

Brailsford A D and Bullough R 1973a Phil. Mag. 27 49

Brailsford A D and Bullough R 1973b J. Nuc. Mat. 48 87

Brailsford A D and Bullough R 1973c Nuc. Met. 18 493

Buckley S N 1961 Proc. Int. Conf. on Reactor Materials and Effects of Rad. Dam. (Butterworth) pp413

Bullough R and Willis J R 1975 Phil. Mag. 31 855

Cawthorne C and Fulton E J 1967 Nature 216 575

Chu W K, Mayer J W and Nicolet M A 1978 Backscattering Spectrometry (Academic Press)

Cottrell A H 1981 J. Nuc. Mat. 100 64

Erginsoy C 1964 Interaction of Radiation with Solids ed R. Strumane et al (North Holland)

Evans J H 1971 Nature 229 403

Gibson J B, Goland A N, Milgram M and Vineyard G H 1960 Phys. Rev. 120 1229

Hirsch P B, Howie A and Whelan M 1960 Phil. Trans. Roy. Soc. A252 61

Jung P 1975 Atomic Collisions in Solids ed Datz et al (Plenum Press) pp87

Kinchin G H 1956 Geneva Conference on Peaceful Uses of Atomic Energy Vol. 7 (UN Publication) pp472

Kinchin G H and Pease R A 1955 Rep. Prog. Phys. 18 1

Mayer J W, Eriksson L and Davies J A 1970 Ion Implantation in Semiconductors (Academic Press)

Paine S H and Kittel J H 1955 1st Geneva Conference on Peaceful Uses of Atomic Energy Paper 745 (UN Publication)

Piercy G R, Brown F, Davies J A and McCargo M 1963 Phys. Rev. Lett. 10 399

Reid I, Farmery B W and Thompson M W 1980 Symposium on Sputtering ed Viehbock (Techn. Univ. Wien)

Robinson M T and Oen O S 1963 Phys. Rev. 132 1385

Robinson M T and Torrens I M 1974 Phys. Rev. B9 5008

Schilling W 1978 J. Nuc. Mats. 70 465

Tewary V K and Bullough R 1972 J. Phys. F 2 L69

Thompson M W 1969 Defects and Radiation Damage in Metals (Cambridge University Press)

Thompson M W 1981 Physics Reports 65 4

Thompson M W and Nelson R S 1963 Phil. Mag. 8 1677

Walker R M 1962 Rad. Dam. in Solids ed D.S. Billington (Academic Press)

Woods W K, Bupp L P and Fletcher J F 1956 Geneva Conference on Peaceful Uses of Atomic Energy Vol. 7 (UN Publication) pp455

Atomic processes occurring during irradiation damage by neutrons

R S NELSON

Materials Development Division, AERE, Harwell, Oxford OX11 ORA

1. Introduction

Irradiation damage, which occurs during fast neutron irradiation, to both nuclear fuels and to cladding and structural materials, has proven to be an important design problem in the development of nuclear power. It was in 1942 that E P Wigner first recognised that energetic neutrons and fission fragments resulting from nuclear fission would cause lattice atoms to be displaced from their equilibrium atomic sites. Now after 40 years of research the fundamental understanding of radiation damage is fairly well advanced and most phenomena are at least qualitatively understood. There is now a wealth of data concerning the fundamental interatomic collision processes, point defect motion, the initial configuration of damage, the physical manifestation of damage, the effects on mechanical properties and the technological implications. However, it is the role of this paper to outline the basic atomic processes which occur as a result of fast neutron irradiation. We will concentrate our discussion on the irradiation of metals, particularly steel, as such materials are of major technological importance. The detailed evolution of microstructure, physical effects and the technological implications will be dealt with by other authors at this conference.

2. Atomic Displacement

Fast neutrons from a nuclear reactor (having an average energy of, say, 2MeV) interact essentially via elastic collisions with the atoms of a solid. The scattering is assumed to be isotropic, in as much as every possible energy transfer from zero to the maximum is equally probable, with a total collision cross-section of typically $5 \times 10^{-24} cm^2$. Thus fast neutrons interact with lattice atoms every few centimetres, and using the laws of conservation of energy and momentum, it is simple to calculate the maximum transferrable energy.

Providing the recoiling lattice atom receives an energy in excess of a minimum lattice binding energy, about 25eV, called the displacement energy E_d, it can leave its lattice site, so leaving a vacancy, and become permanently displaced, as an interstitial atom, within the solid. The exact magnitude of this energy not only depends on the solid in question but also on the recoil direction within the lattice, a discussion of the anisotropy of displacement energy can be found for example in Nelson (1968). In most cases of interest the lattice atoms recoil from primary collisions with kinetic energies of many tens of keV, and as such are capable of penetrating many atomic distances into the surrounding lattice. To a good approximation their slowing down in the solid results from a succession of uncorrellated collisions and is, therefore, amenable to calculation. Energetic atoms interact via Coulomb forces screened by their orbital electrons and as the mean energy falls, the scattering cross-section becomes so large that the mean free path between collisions is essentially equal to the distance between adjacent atoms. In such a situation it is quite clear that a cascade of secondary collisions will be initiated very close to the original primary neutron event, (See figure 1). It is the spreading of such a cascade which is therefore important to radiation damage, as those regions remaining in the wake of the cascade will contain many displaced atoms.

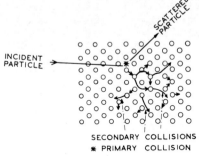

Fig.1 Schematic of a collision cascade

Originally, theories of the collision cascade assumed the solid to have no structure and the multiplication of collisions was considered to be completely random. However, it is now known that the regular nature of the crystal lattice plays a vital role in the spreading of the collision cascade. It was Silsbee (1957) who first pointed out that the ordered atomic array would impose a directional correlation between successive collisions and that energy and momentum would be focussed into those directions consisting of close-packed rows of atoms. A detailed review of focussed collision sequences may be found in the books by Thompson (1969) and Nelson (1968). The total number of displaced atoms is not thought to be significantly affected by the propagation of collision sequences. However, it should be pointed out that the initial spatial distribution of vacancies and interstitials remaining in the wake of the cascade, is in fact, largely determined by replacement collision sequences, and this has important consequences (see later).

The primary recoil atoms initiate collision cascades of many tens of keV, the energy spectrum of recoiling atoms within such cascades is proportional to E^{-2} and most collisions therefore occur with low energy transfers ($\lesssim 100$ eV) and as such can be assumed to be perfectly elastic (ie momentum is conserved). However, in the early stages of the cascade, where energies may be in excess of, say, 10keV, the nuclei of colliding atoms approach sufficiently close for their orbital electron shells to interpenetrate thus resulting in ionization and electron excitation. Such inelastic collisions therefore reduce the total energy available for atomic displacement. However, the work of Lindhard (1962) has shown that to a good approximation we can separate the contributions to elastic and inelastic collisions and hence estimate the fraction of total energy available for atomic displacement. The total number of displaced atoms can then be calculated from the well known Kinchin and Pease model (1955), refined later by Norget et al (1975) as $n_d = 0.8 \Delta E/2E_d$ where ΔE is the fraction of total primary recoil energy dissipated in elastic collisions (calculated using Lindhard, 1962) and E_d is the displacement energy (~ 25eV). The total damage created during a particular irradiation can then be simply calculated from a knowledge of the neutron flux, its energy spectrum and the elastic scattering cross-section.

In recent years it has become the accepted convention to relate radiation damage phenomena to the average number of displacements created during an irradiation normalised to one atomic volume of the solid, eg displacements per atom - dpa. Such a convention has proven very convenient, particularly in comparing irradiations from different reactors or relating neutron irradiation damage to that produced during charged particle irradiation.

We have of necessity discussed only the basics of the collision events responsible for radiation damage, more sophisticated treatments would include the importance of correlated collision such as channelling or focussing, non-linear energy loss processes due to the interpenetration of sub-cascades, thermal spikes, fission fragment damage and damage in insulators etc. Details may be found in the many books or reviews on the subject.

3. The Physical Configuration of Damage

In the preceeding section we outlined the mechanisms responsible for the production of irradiation damage in neutron irradiated solids. However, in any practical case - and particularly in the context of high doses at elevated temperatures - the individual defects created within cascades can migrate under the influence of thermal activation. On the other hand, even at zero degrees some defect rearrangement is possible due to the mutual elastic interaction between defects. For instance there is a critical separation between an interstitial and a vacancy which must be attained before the defects do not spontaneously recombine. This sets an upper limit to the number of defects that are free to move out from collision cascades, and sets a saturation density of about 0.1% for the total number of defects that can build up even at temperatures too low for thermally activated migration.

In the context of the present discussion we will restrict ourselves to practical situations where both the interstitials and vacancies are freely mobile, eg. say above 200°C for steel, but below temperatures where the thermally activated creation of vacancies results in the annealing of damage. After the dynamic collision events have ceased a substantial fraction of mutual recombination will occur within the cascade thus reducing significantly the number of defects from those originally created. However, as a consequence of the dynamic segregation of defects within cascades (due largely to focussed replacement sequences) a vacancy rich core remains at the centre of the cascade, thus leaving the interstitials and the residual vacancies free to migrate away from the damaged region. The vacancy rich core rearranges under the influence of elastic forces to form a compact cluster. The precise morphology of these clusters, ie wherever they form three dimensional voids, dislocation loops or stacking fault tetrahedra, depends on a variety of parimeters, such as lattice structure, stacking fault energy, and surface energy etc. In light materials, especially those associated with poor focussing such as aluminium or graphite, where the segregation of defects is somewhat limited, the heterogenous nucleation of defect clusters is less likely, with the result that a large fraction of those defects originally created are free to migrate throughout the lattice. However, in medium and heavy atomic weight metals, like Ni, Cu, Au and steels, there is now ample experimental evidence that individual collision cascades leave vacancy clusters (eg. loops) in their wake, see fig.2.

We must now discuss the general behaviour as the irradiation dose steadily builds up to beyond that where the individual cascades overlap. Whilst the individual cascades are well separated a large fraction of the freely migrating defects will be lost to the surface and to grain boundaries. However, as the dose increases the isolated vacancy agglomorates will

build up linearly until eventually overlap
and saturation prevails - in reality
this occurs after quite low doses eg
<0.01 dpa. In fact the dynamic
density of vacancy clusters will depend
on their life-time in the sea of freely
migrating interstitials which will
tend to destroy them. Thus as the
dose builds up the interstitial
concentration will also build up to a
sufficiently high level that homogeneous
nucleation of interstitial clusters
occurs - most probably in the form of
dislocation loops (see fig. 3). As
the dose builds up still further new
vacancies and interstitials which
escape recombination will primarily
be captured by the existing defect
clusters rather than sustain further nucleation. The clusters will
therefore steadily grow until they interact and eventually form a complicated dislocation entanglement as illustrated in fig. 4. This situation
represents an ultimate saturation in the configuration of damage as
further irradiation will simply result in re-arrangement of this entanglement by the processes of slip and climb. In a typical steel the saturation dislocation density will build up to about 10^{11} lines cm^{-2} after a
dose of about 10 - 20 dpa at 500°C. In the background, but generally
too small to "see", will be a very fine vacancy loop population which is
continually being destroyed and replenished.

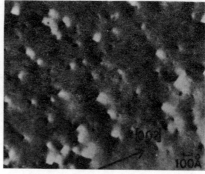

Fig.2 Individual vacancy loops in Cu (Courtesy M M Wilson)

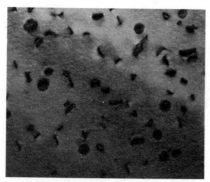

1000Å

Fig. 3 Faulted interstitial loops in neutron irradiated 316 steel, ∼1 dpa.

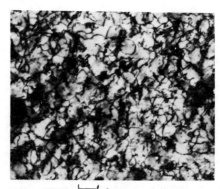

1000Å

Fig. 4 Dislocation network in neutron irradiated FV548 steel, 30 dpa.

(Courtesy T M Williams)

In reality, neutron irradiation damage, whether it be in nuclear fuels or
in structural materials, is always accompanied by nuclear transmutation.
In fuels, the inert gases Xe and Kr constitute 10% of all fission fragments, whilst (n,α) reactions occur in essentially all materials. Thus
for instance, in steel, He atoms are continuously created and steadily
build up as the damage increases. Such insoluble gases as well as other
partially soluble gases such as oxygen and nitrogen can interact with the
migrating defects, particularly vacancies, to form small gas bubble embrios.
We shall see later, and particularly in the following lectures, that such

three dimensional gaseous clusters can dominate the evolution of radiation damage effects in many materials.

4. Dynamic Defect Concentrations

The defect microstructure evolves in a sea of interstitials and vacancies which either recombine directly or after becoming tapped at impurity sites, or become lost at fixed sinks, such as dislocations. At high doses we are generally concerned with a "steady state" situation when dislocation nucleation and growth has saturated, and the microstructure is steadily changing by the absorption of point defects. Such a situation can be described by a simple rate theory and this will be the basis of the following paper by Bullough (1982). However, we will simply quote the basic results in order to provide a guide to the magnitude of the consequential effects. We assume a constant defect production rate (K) and that loss occurs either by direct recombination of loss to dislocations. In the simplest situation we will assume that the sink strength for both interstitials and vacancies is equal. However, in reality this is not so, as these differ by a few percent, and this is the fundamental driving force for the phenomenon known as "void swelling" (see Bullough, 1982). The vacancy is assumed to move more slowly than the interstitial and is therefore the rate controlling defect; we arrive at two simple approximations to the dynamic vacancy concentration depending on whether defect loss by recombination or fixed sinks dominate:

$c_v \sim (K/\nu_v)^{\frac{1}{2}}$ recombination dominant

$c_v \sim K/\rho_d \nu_v \lambda^2$ fixed sinks dominant

where ν_v is the vacancy jump rate, ρ_d is the dislocation density and λ is the vacancy jump distance. Recombination will dominate at lower temperatures when the vacancies move only slowly to sinks; however at higher temperature (eg say > 500°C in steel) vacancies move rapidly to sinks and recombination by the faster moving interstitials is less likely. A typical value for c_v for a steel irradiated at 500°C in a fast reactor would be $\sim 10^{-8}$, which at first sight might seem rather low. However, to put this into context this can be converted into an irradiation enhanced diffusion coefficient of $D' \sim 10^{-17}$ cm^{2}sec^{-1}, which would correspond to thermally activated diffusion at about 600 - 700°C in steel. Thus at temperatures below about, say, 700°C the radiation induced vacancies will be in super-saturation and will condense into clusters or onto gas bubble embrios to form voids, that is provided the corresponding interstitials are preferentially trapped at some other sink, eg. at dislocations.

5. Precipitation Effects

As a consequence of the supersaturated defect concentrations which exist during irradiation, precipitation of solute atoms into second phases is commonly seen in irradiated solids. In addition irradiation can lead to the disordering and dissolution of precipitates already existing. In general the net effect is a balance between enhanced precipitation and irradiation dissolution. Furthermore, precipitate nucleation during irradiation tends to occur uniformally throughout the solids, whereas preferential nucleation at grain boundaries often occurs under normal conditions.

Precipitate dissolution can be understood in terms of collision cascades intersecting the second phase particles resulting in solute atoms recoiling into the surrounding lattice. Models based on collision cascade

theory and radiation enhanced diffusion have been developed to describe
the phenomena, eg Nelson et al (1972). An interesting contrast to the
thermal situation is that theory predicts equilibrium conditions whereby
the precipitate size saturates (ie dissolution is balanced by growth).
Under thermal activation the system reduces its total free energy by the
growth of large precipitates at the expence of smaller ones.

6. The Future

The fundamentals of the atomic collision phenomena occuring during neutron
irradiation of solids are now fairly well understood. Macroscopic effects
which have technological implications can generally be explained and to
some extent can be predicted using simulation experiments (using both
neutrons and charged particles) coupled with a theoretical model. The
majority of design data still required for steels and other metals - for
both fission and fusion reactors - falls into the range of high doses
(> 50 dpa) at elevated temperatures (300 - 800°C). In addition to steady
irradiation environments, the effects of transients experienced during
operational conditions are becoming increasingly important - for example
temperature changes in fast reactors and pulsed irradiation in fusion
reactors. The synergistic effects of inert gas production and stress
during irradiation are also of prime concern - to date our knowledge of
how these interrelated variables influence the final microscopic or
macroscopic properties of irradiated solids is only very superficial.
Large fast neutron irradiation test facilities - particularly those
designed to study irradiation effects under well controlled and instru-
mented conditions providing the appropriate degree of flexibility, will
not be generally available. The experimentalists will therefore have to
rely on simulation studies using those neutron facilities which do exist
and studies using beams of charged particles. The interpretation of such
data in terms of the practical situation will therefore depend intimately
on relating one irradiation environment to another which in turn relies
on a continuing development of the theories of irradiation damage
production and the evolution of microstructure.

Nelson R S, 1968 "The Observation of Atomic Collisions in Crystalline
    Solids" (North Holland).
Thompson M W, 1969, "Defects and Radiation Damage in Metals" (Cambridge
    University Press).
Silsbee R M, 1957, J.Appl.Phys.$\underline{28}$, 1246.
Kinchin G H and R S Pease, 1955 Rep. Prog. Phys.$\underline{18}$,1.
Norget M J, M T Robinson and I M Torrens,1975, Nucl/^{r} and Engineering
    Design $\underline{33}$, 50.
Lindhard J, M Scharff and H Schiott, 1963, Kgl Danske Videnskab.
    Selskab, Mat.-Fys.Medd.$\underline{33}$ no. 4
Bullough R, 1982 (this conference)
Nelson R S, J A Hudson, D J Mazey, 1972 J.Nucl.Mat.$\underline{44}$, 318.

Radiation damage effects in fast reactor design and performance

J F W Bishop

Northern Division, UKAEA, Risley

Introduction

A large commercial fast reactor will typically have a sodium coolant entering the core at 370°C and leaving at a peak of 600°C. The hottest part of the reactor structure will be at ca 600°C; the fuel element cladding will be up to 670°C. Both the fuel and the reactor structure are partly in the high temperature regime where thermal creep effects must be taken account of in design. The design process is heavily conditioned by the rapid temperature changes which can occur in structural material due to the use of the high heat content and high thermal conductivity coolant, sodium, giving rise to substantial thermal stressing.

In the centre of the reactor core, a neutron flux of about 10^{16} n/cm^2/s is present with a modal energy of ca 0.15 MeV. This falls to less than 10^{13} n/cm^2/s in the restraint structure surrounding the core where the modal energy is about 10 eV. For design purposes the neutron energy is commonly tabulated over the range 0.1 eV to 16.5 MeV. The effects of interest to the Designer which arise from the neutron bombardment include

    i. Production of gaseous and solid fission products from transmutation of fuel atoms.
    ii. Activation of fuel cladding, and reactor structural materials so that they emit ionising radiations.
    iii. Production of helium from (n,α) reactions and hydrogen from (n,p) reactions.
    iv. Displacement of atoms in structural materials with associated changes in mechanical properties and dimensions.

This paper does not pursue the shielding requirement placed on the Designer by the production of highly radioactive fission products and structural material activation. It deals with the effects of dimensional changes and mechanical properties changes induced in the (replaceable) fuel elements, and the changes in mechanical properties occurring in the permanent reactor structure.

Part 1: Core

The term 'core' is used here in the broad sense of all the items - core and radial blanket fuel sub-assemblies, absorber rods and shield rods - which are mounted on the core support grid to generate nuclear heat in a controlled manner.

The energy required to displace a typical clad atom (iron, nickel, chromium) from its lattice site is about 25 eV. The neutron energy required to displace a clad atom (of atomic weight of about 60) is of order 1 KeV, so clad atoms in the core are easily displaced in the core flux. In fact in a typical dwell time of 2 years in the core centre, each clad atom will be displaced over 100 times from its lattice site by a total neutron fluence of over 10^{23} n/cm^2. These displacements per atom (dpa) will lead to three consequences as far as the designer is concerned, namely,

i. The structural material will undergo a volume increase due to aggregation of the voids left behind by displaced atoms.
ii. The material will deform under loads and at temperatures where deformation would not occur in the absence of the neutron flux.
iii. Such properties as yield strength, ductility and fracture toughness will be appreciably altered.

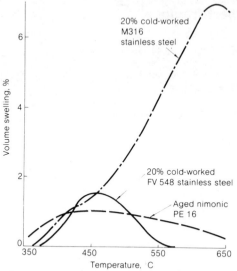

Fig.1 Temperature dependence of void swelling at a dose of 50 dpa

A paper by R Bullough (1982) to the present Neutron Conference indicates how the production, migration and recombination of interstitial atoms and vacancies produced in collisions leads to the first two phenomena 'neutron induced void swelling' – NIV, and irradiation creep. Fig 1 shows how swelling in the austenitic 316 type stainless steel and Nimonic PE16 used in UK fast reactor fuel elements occurs as a function of temperature. Fig 2 by Edmonds et al (1979) shows the displacement dose dependence of swelling found in irradiations in the Dounreay Fast Reactor in cold worked (C/W) and solution treated (S/T) materials of interest.

Structural engineers are familiar with dealing with low temperature dimensional changes due to loading (say 0.2% linear strain), high temperature thermal creep membrane strains of 1%, and thermal expansions of 1%. It is seen that the linear changes illustrated in Fig 2 can be substantially greater: this Void Swelling phenomenon is the dominant dimensional change which occurs in a fast reactor core. The illustrations portray the average changes occurring under nominally steady irradiation conditions: the Designer wishes to know what spread of data will occur and how the results will be affected by the changes to reactor conditions which will result

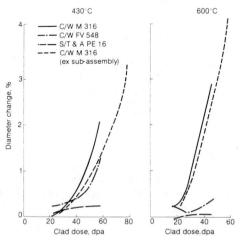

Fig. 2. Pin diameter changes at 430°C and 600°C as a function of clad displacement dose

from the operation of a power station, involving temperature cycling. Guidance is available on such queries but the topic is still in the active research phase.

Irradiation creep has been found to be broadly linearly dependent on displacement dose and to some extent on dose rate. The temperature dependence is not firmly established but has a trend to higher values at higher reactor temperatures. Mosedale et al (1977) amongst others have explored the phenomenon and some of their results for two stainless steels are shown in Fig 3: γ is shear strain and τ, shear stress. As a guide to the magnitude of the effect, it can be taken that a permanent strain equal to an elastic deflection occurs each 4 dpa: it has been noted above that over 100 dpa occur at the core centre in the fuel element dwell time.

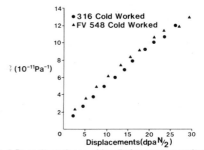

Fig. 3 Dose dependence of irradiation creep magnitude in two stainless steels irradiated at 280°C (after Mosedale et al., 1977)

Finally, irradiation can appreciably increase the lower temperature strength properties and reduce the elongation of stainless steels, and reduce the high temperature elongation. This is illustrated in Fig 4 by Bagley et al (1972). Changes to mechanical properties may be strongly influenced by neutron reactions which produce gases. About 0.1% of the clad atoms produce a hydrogen atom and 0.01% a helium atom at the core centre during the fuel element residence.

It is in the structural design of the sub-assembly that these neutron effects are of most consequence to the designers although they also have to be incorporated in individual fuel pin design. A sub-assembly for the proposed UK Commercial Demonstration Fast Reactor (CDFR) consists of 325 fuel 'pins', each 0.23 in. o.d., in a wrapper tube structure of 5.6 inches across flats. Generally this sub-assembly will be irradiated in a flux gradient which varies both axially and transversely. The net result is that the sub-assembly will elongate (by an inch or so on its length of 12 feet); it will bow sideways because the displacement dose (and therefore swelling elongation) are different on the two sides; and it will bulge in the middle due both to the higher swelling in the peak flux, and irradition creep under the excess coolant pressure inside the wrapper. These deformations are shown schematically in Fig 5, taken from a paper by Bishop (1982).

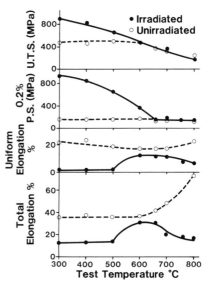

Fig.4 Tensile properties of solution treated M316 stainless steel irradiated in DFR to 10^{23} ncm^{-2}

In the Dounreay Prototype Fast Reactor, fuel closely similar to that which will be used in a Commercial Fast Reactor has been taken to nearly 9% burn-up of all the heavy atoms in the plutonium-uranium oxide fuel. In this reactor, there is little restraint on sub-assembly distortion and the following table gives an indication of the magnitude of distortion which has been seen in sub-assemblies and other core units.

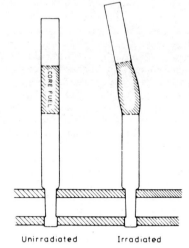

Sub-assembly elongation	8 mm
Sideways bow at top	15 mm
Dilation at wrapper centre	2.5 mm

Since the spacing between sub-assemblies is about 5 mm, it is seen that interference will be occurring between core units.

Two major considerations which the designer must deal with are the loading which arises from the distortions (affecting heat transfer in service, and charge/discharge operations), and the change to the reactivity worth of the core as the distortions change the location of the fuel by small amount. For the

Fig 5. Wrapper Distortions due to Void Swelling and In-Reactor Creep

CDFR, Lunt (1982) has set out the reasons why a core style using two re-straint planes located above the core fuel axial position has been chosen. This results in the bowing distortions occurring due to differential void swelling being largely compensated by irradiation creep in the reverse sense due to the developed interference loading. Further detail of the calculation routines evolved to undertake the design have been given in the review article by Bishop (1982).

Considerations of ductility and fracture toughness are important in the design both of fuel pins and the sub-assembly structure. Here the designer may be in an economic optimisation situation where reduction of stresses and strains can be undertaken at some cost penalty, or in a balancing situation where amelioration of one phenomenon will exacerbate another (as in thickening a wrapper to reduce dilation, but which in turn will result in an increased restraint loading).

Part 2: Reactor Structure

In proceeding outwards from the core centre to the high and low temperature permanent structural regions of the reactor, the flux level decreases and the neutron spectrum shifts to lower energy levels. This is illustrated in the following table for three locations of importance in CDFR: the author is indebted to Mr K Brindley for these data.

Location	Flux $n/cm^2/s$	Modal Energy
Core Centre	10^{16}	0.15 MeV
Above Core Structure (bottom)	2×10^{13}	30 KeV
Core Restraint Barrel	5×10^{12}	10 eV

The table below has been taken from a contribution by G J Lloyd to a paper by Harries et al (1982) showing other relevant end-of-life parameters for some important locations in CDFR with components made in 316 type stainless steel.

	Nominal Temp °C	Displacement Dose dpa (NRT)	Helium* Concentration appm
Top of Core Support Grid	370	0.1	5
Fuel Support Tubes	370	1.7	30
Core Restraint Structure	380-500	0.1	5
Lower end of Above Core Structure	550-600	1.0	30

* This statistic will depend strongly on boron impurity level in the steel and also on the steel composition.

It will be appreciated that the irradiation effects accrue over the 30 year design life of the reactor. Mechanical testing over such a period to produce design data is impracticable and the effects of radiation must be assessed from tests carried out over a substantially shorter period. The discussion which follows is based on unpublished work by Wood (1982).

In low temperature design (say up to $500°C$ in stainless steel) the Design Code requirements involve yield and ultimate stresses, and fatigue. In nuclear applications, specific consideration may be given to fracture toughness. In the context of CDFR design, the irradiation experienced at lower temperatures will increase the strengths and reduce elongation by some tens of percent. Fatigue resistance will vary somewhat. There is evidence that fracture toughness can decrease by a few tens of percent. Fig 6 taken from Harries et al (1982) shows this effect in terms of crack growth in cold worked 316 steel at 380°C irradiated to 2.03×10^{21} n/cm^2 (E $>$ 1 MeV): in this context, the total neutron flux in the reactor life is of order 10^{22} n/cm^2.

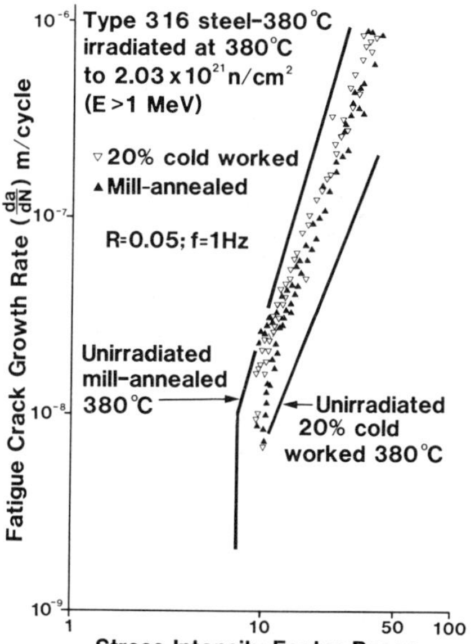

At temperatures where creep occurs, the main considerations relate to creep-fatigue effects, and especially the influence of helium on slow strain rate rupture ductility. Rupture life and creep-fatigue cycles to failure could be reduced by an order of magnitude by CDFR exposures relative to unirradiated material.

Fig. 6 Effects of irradiation on fatigue crack propagation in mill annealed and 20% cold worked type 316 steel at 380°C

The principal structures of concern to the designers are the core support structure, the above core structure and the core restraint structure. There is no problem in exposing material to the neutron fluence and temperature which occurs in service. Increasing difficulty is encountered in the post-irradiation testing of irradiated material (which requires shielded facilities) and yet again in carrying out testing during irradiation. It is then necessary to allow for the foreshortening of test time-scales relative to the reactor situation. Both safety and economic considerations combine to indicate that the appropriate design routes must contain substantial factors of conservatism and that design details chosen with structural redundancy and defence in depth against unacceptable failures are employed.

Concluding Remarks

It has been seen that both in Core and in Reactor design, the neutron environment of a fast reactor imposes dominant radiation damage considerations on designers. In spite of many of these effects only becoming apparent through the operation of fast reactors, the operation and development of fast reactors has proceeded safely and progressively for over 2 decades. This has been achieved through extensive monitoring and development programmes, and design with safety defences in depth. This process is now continuing towards the final target of safe, reliable, and economic commercial fast reactors. In the final stage there are still substantial advances in knowledge of radiation damage to be made and in an ability to select or tailor materials which will have specified properties under irradiation.

References

Bagley K Q, Barnaby J W and Fraser A S 1972 Proc. BNES Conf. on Irradiation Embrittlement and Creep in Fuel Cladding and Core Components, London pp 143-153
Bishop J F W 1982 Rodney Hill 60th Anniversary Volume (Pergamon; Oxford and New York) pp 1-12
Brindley K 1982 Unpublished work
Bullough R 1982 Proc. Neutron Conf. Cambridge
Edmonds et al. 1979 Proc. ANS Conf. Monterey pp 54-63
Harries D R, Standring J, Barnes W D and Lloyd G J 1982 Proc. 11th Conf. on Effects of Irradiation on Materials, ASTM
Lunt A R W 1982 To be published BNES Journal, October
Mosedale D, Harries D R, Hudson J, Lewthwaite G W and McElroy R J 1977 Proc. Int. Conf. on Radiation Effects in Breeder Reactor Structural Materials, Scottsdale (A.I.M.E., New York) pp 209-228
Wood D S 1982 Unpublished work.

Radiation effects in fuel materials for fission reactors

Hj. Matzke

Commission of the European Communities - Joint Research Centre - Karlsruhe Establishment - European Institute for Transuranium Elements - Postfach 2266 - D-7500 Karlsruhe - Federal Republic of Germany

Physical and chemical changes that occur in fuel materials during fission are described. Emphasis is placed on the fuels used today, or those foreseen for the future, hence oxides and carbides of uranium and plutonium. Examples are given to illustrate the most interesting neutron effects.

1. Introduction

The slowing down of fast, energetic ions (65 to 95 MeV fission products) is the dominant process to produce heat and thus finally electricity in nuclear fission reactors. This stopping process is connected with a variety of phenomena (electronic stopping, nuclear stopping, collision cascades, thermal spikes) that cause both changes in physical properties at a time-scale comparable to the life time of the slowing down of the primary particle ($\sim 10^{-11}$s) as well as changes that are stable for essentially longer periods of time. Both will be discussed. Also, the chemistry of the fuel changes due to transmutation and fission. In a modern fast breeder reactor, impurities (about 30 different elements) grow in at a rate of ~ 500 ppm per day.

The term "radiation effects" will therefore be used in a quite general sense. The fuel materials considered are metals, carbides and nitrides, or oxides of U, Th and/or Pu and cover thus a wide range in bonding properties and thermal conductivity. This causes the radiation effects to be essentially different in different fuel materials. Both fundamental processes as observed in specifically designed irradiation devices with well defined specimens of fissile fuels, performed under controlled irradiation conditions (temperature, stress etc.), as well as results from technological irradiations will be discussed. This implies that processes such as point defect formation, mobility and clustering, damage annealing, fission spikes, radiation enhanced diffusion etc. on the one hand will be treated, but bulk volume changes, swelling, fission gas bubble behavior etc. on the other hand will also be discussed thus yielding a broad picture on radiation effects in fissile fuel materials.

2. The slowing down of fission fragments in fuel materials

The partition of energy between atomic collisions (nuclear stopping power, S_n, and fraction ν of original energy E_0 transferred directly to lattice atoms) and of the electronic losses (S_e, and fraction η of the total energy given to electrons can be calculated using the methods of Lindhard et al. (1963) (LSS method). Extensions of this theory, originally established for monatomic targets, to diatomic targets, and in particular to UO_2 and UC, exist also. Fig. 1 shows the energy deposition along the path of a median heavy fission fragment, mhff, originating from the fission of a U-atom located at the origin. The median light fission fragment, mlff, is emitted to the other side. Table 1 summarises some of the properties of the fission fragments and Fig. 2 shows the fraction ν mentioned above.

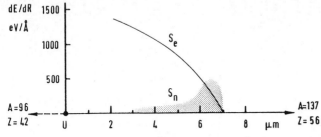

Fig. 1 Electronic (S_e) and nuclear stopping power (S_n) for the mhff in UO_2.

Table 1 : Some properties of the median heavy and median light fission fragments in UO_2

	mlff	mhff
atomic mass, A	96	137
atomic number, Z	42	56
original energy, MeV	95	67
velocity, v_o in 10^9 cm s^{-1}	1.4	1.0
most probable charge state	21	24

total length of track of ff (range) in UO_2	~ 8 μm
duration of collision cascade	$\sim 10^{-13}$ s
duration of temperature spike	$\sim 2 \times 10^{-11}$ s
volume for displacements, V_d	$\sim 6 \times 10^{-17}$ cm^3
temperature volume V_T	$\sim 140 \times$ "
pressure volume V_p	$\sim 225 \times$ "

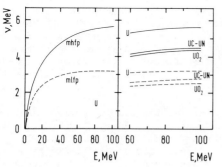

(see Blank (1972), Blank and Matzke (1973) and Matzke (1982) for details.

Fig. 2 Fraction ν of the total energy E_o of fission fragments, mhff and mlff, in U, UC or UN, and UO_2 (Lindhard and Thomsen (1962)).

At the start of the path (or fission track), a very high energy deposition indeed of ≥ 3 keV/atom layer occurs, but it is entirely due to S_e. In UO_2, a major part of S_e will be transferred into thermal exitations of the atoms. There is no theory available to exactly calculate the fraction μ_{th} of μ that is transformed into heat. We will therefore follow Blank's

(1972) estimate that only a small fraction (possibly 30 %) of μ is instantaneously carried away by X-rays and elastic waves whereas the majority is transformed into heat.

Table 2 : Displacements in UO_2

Particle	E(MeV)	Frenkel pairs formed	Frenkel pairs surviving	remark
electron	∿0.5	≲1	≲1	
α-particle	∿5	∿200	∿150	predominantly electronic energy loss
recoil atom	∿0.1	∿1500	∿150	predominantly nuclear energy loss
neutron	1	∿250	∿50	maximum pka-energy for U in UO_2 = 0.02 MeV
fission products mlff	95	∿60000	15000	∿ 3 % nuclear energy loss
fission products mhff	67	∿40000		∿ 6 % nuclear energy loss

(see Soullard et al. (1978), Weber (1981) and Matzke (1982) for details).

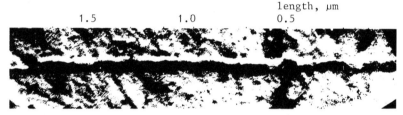

Fig. 3 : Fission track at the surface of UO_2. Replica electron micrograph (Ronchi, 1973) (Average width and height ∿ 500 Å as measured with stereo-techniques).

In U-metal, but also in UC and UN, the temperature effects will be much less pronounced than in UO_2 because of the much higher thermal and electrical conductivity of these materials. In contrast to thermal spike effects, the displacement damage will be quite similar for all the three classes of substances : U as a metal, UO_2 as a largely ionic crystal, and UC or UN with a mixed metallic-covalent bond. Strangely enough, though much work has been devoted to neutron damage and a satisfactory degree of knowledge is achieved on radiation damage produced by neutrons, particularly in metals, surprisingly little work has been devoted to the problem of stopping of fission fragments in nuclear fuel materials, though this process and the heat produced by it are the basis of generating nuclear energy and hence electricity, a power source that contributed e.g. with 12 % to the electricity production in the UK in 1982.

Previously, a number of concepts for locally damaged zones (spikes) along the path of charged particles was suggested, from the early concepts of displacement spikes (Brinkman, 1954) over Seitz and Koehler's (1956) thermal spikes to, more recently, ion explosion spikes (Fleischer et al. (1965)) or plasticity spikes, ionization spikes or energy spikes (see recent reviews by Thompson,(1981) and Matzke,(1982)). The concept of thermal spikes became unpopular about a decade ago though recently evidence in favor of it is increasingly reported. The argument against the concept is that the quenching times often are too short and the cascade dimensions too small for Maxwell-Boltzman statistics and normal thermodynamics to be applicable for this largely non-equilibrium situation. However, for a poor thermal conductor as UO_2, the quenching times can be long enough for a coupling between atomic and electronic excitation to occur. The conversion of the fraction η of the energy of the ff's into thermal excitation seems to be the only way to explain important features in UO_2 and to understand the difference between U, UC(UN) and UO_2.

Table 2 shows that a fission event in UO_2 produces about 10^5 displacements whereof 15 % survive. Table 1 gives the volume affected by displacements, V_d, as e.g. deduced from saturation measurements of changes with fission dose of a relevant physical property (electrical resistivity, fission gas release, etc.). Obviously, about 2 % of the atoms in V_d are displaced. However, V_T and V_P are much larger in UO_2 (Blank and Matzke, 1973). V_T in Table 1 is defined as that volume within which $T > T_m$ during the life time of the spike (for an operational temperature of 2000 °C). This hot volume causes a thermoelastic pressure pulse. V_P in Table 1 is the volume within which $P > 5 \times 10^3$ atm. The corresponding mean life-times for T in V_T and P in V_P are 1.4 and 4.4×10^{-11} s, respectively. These values are derived from estimates of T-t relations based on observed volumina within which phase changes occur in U_4O_9 needles in UO_2. This observation also provides the phase change that is often asked for before a thermal spike idea is accepted (Blank (1972)).

3. Contribution of different damage sources to the total damage

Table 2 summarizes also damage produced in events competing with fission. Obviously, electrons are not very effective because of their low mass. α-particles originating due to decay of e.g. Pu-239 can also be neglected. About 90 % of the displacement damage in α-decay is due to the recoiling daughter atoms (e.g. U in the decay of Pu), which, due to its high mass and low energy (∼ 100 keV) shows predominantly nuclear energy loss. However, the ratio of α-decay to fission events is $< 10^{-3}$, even in a Pu-containing fuel such that α-decays do not contribute essentially to the total damage. Fast neutrons can, of course, create collision cascades, as is very convincingly seen in structural materials in reactors. Also, the flux of neutrons through a given area of the fuel is >> than the flux of fission fragments. However, the maximum energy, a neutron of 1 MeV can transfer to a U-atom is ∼ 20 keV. Calculations easily show that the collision cascades produced by these primary knock-on atoms contribute to < 10 % to the total displacements even in a fuel irradiated with fast neutrons. The above damage sources will therefore not be treated here in detail.

4. Chemical effects of fission

In each fission event, two impurity atoms are created. In a modern fast breeder reactor, these grow in at a rate of about 500 ppm d^{-1}. Therefore, the original impurity content of the fuel of typically some 300 ppm ceases soon to determine the behavior of the fuel. A typical light water reactor fuel, at end of life at a burn-up of 33000 MWd/t fuel, contains, in weight-ppm : Xe (5420), Nd (3870), Zr (3647), Mo (3434), Cs (2720), Ru (2270) etc, the total being 34000 ppm. Some of these, as Xe (and Cs), are insoluble and precipitate to form gas-filled bubbles and cause thus the fuel to swell (see below), others are soluble (rare earth) and change the composition of the fuel since their valence is different from that of U, and others form solid precipitates (Pd, Ru etc). Therefore, all properties

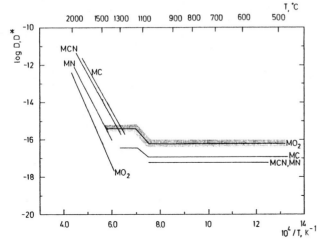

Fig. 4 Radiation enhanced (D*) and thermally activated (D) metal atom self-diffusion in MO_2, MC and MCN, MN (M = U and/or Pu), (Höh and Matzke (1973), Matzke (1976), Matzke and Bradbury (1978)).

depending on composition (e.g. O/U-ratio in UO_2) will change with burn-up but also the mechanical behavior will gradually change. Examples of properties that are largely affected are self-diffusion coefficients, elastic properties, creep rates etc. Before any change of a property during irradiation is attributed to radiation damage, a careful analysis is needed to decide whether the change is not possibly simply caused by the change in chemistry due to fission.

5. Evidence for thermal effects of fission spikes

The materials considered, U-metal, UN, UC and UO_2, show thermal conductivities, that decrease in the above order from 15^2 : 8 : 7 : 1, if normalized to UO_2 (λ = 0.03 W/cm deg at 1000 °C). The electrical conductivity decreases similarly. Thus, thermal effects of fission events are expected to dominate in UO_2 and to be largely absent in the more metallic fuels such as UC, UCN, UN, or U itself. Experimental evidence exists : i) fission tracks are visible on UO_2, but not on UC and UN (Fig. 3, Ronchi(1973)).

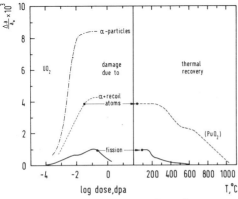

Fig. 5 Defect ingrowth (left) and thermal recovery (right) in UO_2 (or PuO_2) due to different damage sources (Nakae et al. (1978), Weber (1981)).

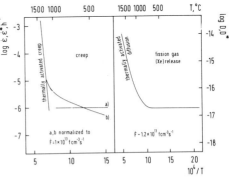

Fig. 6 Thermally activated (D,ε) and radiation enhanced (D*, ε̇*) fission gas diffusion and creep in UO_2 (Brucklacher and Dienst (1972), Dienst (1977), Turnbull et al. (1982)).

ii) The degree of lattice recovery during irradiation, as evidenced by lattice expansion following irradiation, decreases in the order UN > UC > UO_2. Also, the lattice expansion in a given material (UO_2, Fig. 5) increases whenever the damage becomes less overlapping. α-particles create largely isolated defects (Frenkel pairs). Thermal effects that cause defect annealing during irradiation, as observed during fission, are thus largely absent in irradiation with α-particles. iii) Fission fragments cause self-diffusion (Fig. 4) to be radiation-enhanced and largely athermal, temperature-independent at low temperatures ($\lesssim 1000$ °C). The corresponding diffusion coefficients, D*, decrease in the order UO_2 > UC > UN thereby reflecting the above ratios of conductivity. A similar dependence and a largely athermal behavior was also observed for radiation induced creep. Finally, fission gas diffusion is also radiation-enhanced and athermal at low temperatures (Fig. 6). D* and $\dot{\varepsilon}$* are proportional to fission rate and not to neutron flux confirming the argument that the damage during fission is mainly due to fission fragments. iv) Fission fragment induced re-solution of fission gas from gas bubbles can be explained in UC and UN with a displacement damage model (Nelson (1969)), whereas for UO_2 re-solution probabilities are much higher (Turnbull (1980)) and thermal and pressure effects (Blank and Matzke (1973)) must be invoked as explanation.

Physical properties that are affected by fission and reflect damage, surviving the life-time of the fission spikes are, in addition to lattice expansion and the mechanical properties mentioned above: thermal conductivity, electrical resistivity, magnetic properties, hardness etc. (see Tamaki et al.(1982) and Matzke (1982)).

References
Blank H 1972 phys. stat. sol. a10 465
Blank H and Matzke Hj 1973 Rad. Effects 17 57
Brinkman J A 1954 J. Appl. Phys. 25 96
Brucklacher D and Dienst W 1972 J. Nucl. Mater. 42 285
Dienst W 1977 J. Nucl. Mater 65 148 and other articles in this volume
Fleischer R L, Price P B and Walker R M 1965 J. Appl. Phys. 36 3645
Höh A and Matzke Hj 1973 J. Nucl. Mater. 48 157
Lindhard J and Thomsen R V 1962 in "Radiation Damage in Solids", IAEA Vienna 1 65, and private communication 1972
Lindhard J, Scharff M and Schiott H E 1963 Mat. Fys. Med. Dan. Vid. Selskap 33 nr. 10
Matzke Hj and Bradbury M H 1978 Euratom Report-5906 EN
Matzke Hj 1976 in "Plutonium 1975 and Other Actinides", North Holland Publ. Comp. Amsterdam pp. 801-31
Matzke Hj 1982 Proc. Conf. Radiation Damage in Insulators, Rad. Effects in print
Nakae N, Harada A, Kirihara T and Nasu S 1978 J. Nucl. Mater. 71 314
Nelson R S 1969 J. Nucl. Mater. 31 153
Ronchi C 1973 J. Appl. Phys. 44 3575
Seitz F and Koehler J S 1956 Solid State Physics (Academic Press, N.Y.) 2 360
Soullard J and Alamo A 1978 Rad. Effects 38 133
Soullard J, Leteurtre J, Genthon J P and Cance M 1978 Rad. Effects 38 119
Tamaki M, Matsumoto S, Ichimura K, Matsumoto G and Kirihara T 1982 J. Nucl. Mater. 108 & 109 671
Thompson D A 1981 Rad. Effects 56 105
Turnbull J A 1980, UK Report AERE-R 9733 (ed. Pugh S F) pp. 381-97
Turnbull J A, Friskney C A, Findlay J R, Johnson F H and Walter A J 1982 J. Nucl. Mater. 107 168
Weber W J 1981 J. Nucl. Mater 98 206

The use of sink strengths in the theory of radiation damage

R. Bullough

Theoretical Physics Division, AERE Harwell, Didcot, Oxfordshire OX11 ORA

1. Introduction

The choice of suitable materials for the structural components of the core of a fast fission reactor or for the first wall of a fusion reactor is constrained by the effects of fast neutron irradiation on a potential material's dimensional and mechanical stability. Fast neutron irradiation produces equal numbers of displacement defects (interstitials and vacancies) which, at the temperatures prevailing in the reactor environment, may be mobile and influence the materials stability by either interacting with existing microstructural sinks or aggregating to form new sinks; the overall effect involves an evolution of the total microstructure. In recent years, particularly since the discovery of the void-swelling phenomenon (Cawthorne and Fulton, 1967), a great deal of research has gone into observing and interpreting these microstructural changes and thereby assisting in the choice or development of suitable damage resistant alloys (Pugh et al, 1971; Corbett and Ianniello, 1972; Nelson, 1974; Robinson and Young, 1975). Of particular importance was the early recognition that the fast neutron damage could be usefully simulated, with a much higher damage rate (Hudson et al., 1971) by irradiating with ions, high energy protons or high energy electrons. However it is now apparent that the damage data obtained from such simulation experiments can only be used to predict accurately expected neutron damage when a comprehensive physical understanding of the damage evolution process exists (Bullough et al., 1977). Such a theoretical framework for understanding many aspects of this process is being developed by various groups with the aid of Chemical Rate Theory (Wiedersich, 1972; Brailsford and Bullough, 1972; Mansur, 1978; Brailsford and Bullough, 1981). The essential basis of such a rate theory is the allocation of sink strengths to all the sink types (dislocations, voids, grain boundaries, precipitates, traps, etc.) that make up the total microstructure; these sink strengths are then used in an effective lossy continuum (Brailsford and Bullough, 1972 and 1981) to deduce the simultaneous evolution of all the sink types in the presence of the diffusing displacement defects.

In this paper we review briefly the embedding procedure used to obtain the various sink strengths, indicate some recent improvements in the detailed models of some of the sink types and show how the rate theory can then be used to yield a quantitative prediction of the void swelling in a commercial 'M316' steel now being used as a core component material for the Fast Reactor and a proposed material for the first wall of a fusion reactor.

2. The Embedding Procedure and Rate Theory of Microstructure Evolution

The rigorous embedding procedure for obtaining the sink strengths of the various sink types that together constitute the total microstructure was introduced by Brailsford and Bullough (1972), developed further by Brailsford, Bullough and Hayns (1976) and Brailsford (1976) and has been recently reviewed in depth by Brailsford and Bullough (1981). The procedure involves considering each sink type in turn within an effective medium (lossy continuum) containing all the sink types identified by appropriate lossy terms (sink strengths); thus for voids, for example, a volume of the effective medium equal to that of a void is removed and replaced by a single spherical void. With a point defect generation term, equal to the displacement rate due to the radiation (see Nelson, 1982) the spatially varying diffusion equation (continuity equation) around the void can be solved for the flux of point defects into the void. The continuity equation contains a loss term for the loss of such defects into <u>all</u> the sink types deemed to be present in the effective medium including the voids. By equating the flux of defects into the geometically defined sink type (the void in this example) to the flux into a sink of the same type in the surrounding effective medium (an appropriate distance from the geometrically defined sink type) we obtain an implicit expression for the sink strength of the particular sink type as a function of the sink strengths of all the other sink types. This procedure is repeated for each sink type to yield a set of non-linear equations which can be solved numerically to yield the required sink strengths. The evolution of the microstructure during irradiation is determined by the rates at which all the various sink types (voids, dislocations, precipitates, traps, grainboundaries, etc) accumulate net interstitials or vacancies; thus, if there is a net accumulation of vacancies at the voids then the body will increase its volume and the phenomenon of void swelling will occur. Such processes can easily be calculated when we have obtained the above self consistent sink strengths for all the relevant sink types for both interstitials and vacancies.

Unfortunately the identification of the relevant sink types in a commercial alloy is very difficult and even when a sink type is identified for inclusion its physical properties are often only poorly understood. To emphasize this latter difficulty and, at the same time, to indicate some recent progress, some of these physical aspects will be itemised below.

1. The void

The point defect flux to the void may be diffusion or reaction controlled. The intrinsic point defects can cause minor (or major) alloying elements to segregate at the void surfaces and hence modify (continuously) the transfer velocity across the matrix-void interface. In general, however, it seems likely that the void is a relatively simple sink type and these complex segregation effects are probably not having a dominant effect; on the other hand, when the void density is high, or the voids are present within a high density microstructure, the interactive effects on the void sink strength can be large. However, for relatively low densities of the various sink types, we may regard the void as a diffusion controlled neutral sink for interstitials and vacancies (Quigley et al., 1981).

2. The Dislocation*

The interactions between dislocations and surrounding point defects are complex; there is a long range interaction which attracts more interstitials than vacancies to these sinks. This interaction is, by itself, too large and would, if it were not modulated by a compensating short range interaction, remove too many more interstitials than vacancies from solution; note, it is this preferential removal of the interstitials by the dislocations, that ensures an excess of vacancies are available to accumulate at the neutral voids. Such a large net loss of interstitials would lead to swelling values far in excess of those observed. The compensating short range interaction involves a preference for vacancies and arises from two sources:

(a) When a vacancy is very close to a dislocation it comes under the influence of the almost purely radial inhomogeneity short range attractive interaction; the long range size effect interaction, with its strong angular dependence, is too weak to dominate in the region very close to the core. This means that a vacancy can 'enter' the dislocation core from all the atom sites adjacent to the core. In contrast, the long range size effect interaction between the dislocation and an interstitial is much larger and will always dominate any short range inhomogeneity interaction; in consequence the interstitial is inhibited from entering the core over a certain fraction of the core perimeter because of the strong angular form of the size effect interaction. The vacancies that arrive at a dislocation can thus enter the core easier than the interstitials.

(b) In addition to this relative difficulty of entering a dislocation core, once the point defects enter the core this is evidence to suggest that the vacancy will migrate within the core, to the nearest jog, and cause climb, with a diffusivity greater than in the perfect crystal. On the other hand, the behaviour of an interstitial within a core is uncertain and its mobility there may well be relatively retarded[+]. Again, the effect of this relatively different pipe diffusion behaviour is to enhance the local preference for vacancies compared to interstitials at the dislocation sinks.

Both these long and short range effects have recently been included in a calculation of the dislocation sink strength (Bullough and Quigley, 1981, 1982). This analytic implicit sink strength has been included in the general rate theory of swelling to predict the expected swelling behaviour of solution treated M316 steel. The procedure is as follows (Bullough and Quigley, 1982).

The above short range interaction is first calibrated by comparison between the theoretical and observed swelling rate in this steel when irradiated with 1 MeV electrons over a wide range of temperature. The fitting process to the growth rate of the voids requires using the observed densities of the sink types: dislocations network and voids, that together make up the microstucture; in addition it is necessary to include the defect loss across the thin metal foil surfaces with the aid of an

---

*The term dislocation refers here only to the edge dislocations.

[+]Of course, if a point defect has a lower mobility within the core it will probably migrate along the dislocation outside the core and then 'enter' the core only adjacent to an appropriate recipiant jog.

appropriate sink strength for these foil surfaces (Bullough et al., 1981). It is, of course, no mean achievement of the model that a single calibration can be made to yield a swelling rate consistent with the electron data over a wide range of microstructure and temperature.

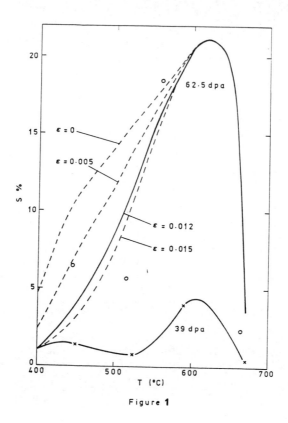

Figure 1

Fig. 1 The temperature dependence of the swelling in M316 steel irradiated with fast neutrons. The x and o indicate experimental swelling values, obtained by Kenfield et al (1978), at 39 dpa and 62.5 dpa respectively. The solid curve labelled $\varepsilon = 0.012$ is the predicted swelling at 62.5 dpa using the dislocation bias and cascade collapse efficiency parameters obtained by fitting the theory to electron and heavy ion data (Bullough and Quigley, 1982). The sensitivity to the cascade collapse efficiency ε, for fixed dislocation bias, is also indicated by the series of dashed curves.

Heavy ions and fast neutrons both create cascade damage with the resulting formation of small vacancy loops from the collapse of the inner depleted zones. Such vacancy loops are significant special dislocation sinks and their formation and sink effects must be included in the general rate theory (Bullough et al., 1975). The efficiency of this vacancy loop formation process must therefore also be calibrated and this was achieved by comparison with appropriate high damage rate heavy ion irradiation data. At this stage the rate theory is then completely calibrated and can be used to predict the expected swelling behaviour of the steel when irradiated with fast neutrons. The predicted temperature dependence of the swelling is shown in figure 1 together with the experimental observations obtained at 39 dpa and 62.5 dpa by Kenfield et al (1978). The physical parameters for M316 steel and details of the irradiation conditions etc., are given in the paper by Bullough and Quigley (1982). The parameter ε is a measure of the cascade collapse efficiency and the full curve in the figure, labelled $\varepsilon = 0.012$, is the predicted swelling with no change in any of the calibrated parameters; the other curves for ε values between $\varepsilon = 0$ and $\varepsilon = 0.015$ are included to indicate the sensitivity to the recoil spectra. An important conclusion from this study is that vacancy loop effects do not appear to be an _essential_ ingredient in the fast neutron situation. Thus, electron irradiation can provide a useful simulation technique for understanding and predicting the corresponding response to neutron irradiation. To do this correlation properly requires, we believe, a physically based theory of the evolution of the entire microstructure such as referred to in this paper.

Brailsford, A.D., 1976, J. Nucl. Mat. 60, 257.
Brailsford, A.D. and Bullough, R., 1972, J. Nucl. Mat. 44, 121.
Brailsford, A.D. and Bullough, R., 1981, Phil. Trans. Roy. Soc. 302, 87.
Bullough, R., Eyre, B.L. and Krishan, K., 1975, Proc. Roy. Soc. A346, 81.
Bullough, R., Eyre, B.L. and Kulcinski, G.L., 1977, J. Nucl. Mat. 68, 161.
Bullough, R. and Quigley, T.M., 1981, J. Nucl. Mat. 103 and 104, 1397.
Bullough, R. and Quigley, T.M., 1982, Harwell Res. Rep. AERE-TP.947, J. Nucl. Mat. (to be published).
Bullough, R., Wood, M.H. and Pierce, S.M., 1981, Metal Science 15, 342.
Cawthorne, C. and Fulton, E.J., 1967, Nature 216, 575.
Corbett, J.W. and Ianniello, L.C. (Eds), 1972, Proc. Int. Conf. Radiation - Induced Voids in Metals (Albany, U.S.A.).
Hudson, J.A., Mazey, D.J. and Nelson, R.S., 1971, J. Nucl. Mat. 41, 241.
Kenfield, T.A., Appleby, W.K., Busboom, H.J. and Bell, W.L., 1978, J. Nucl. Mat. 75, 85.
Mansur, L.K., 1978, Nucl. Technol., 40, 5.
Nelson, R.S. (Ed), 1974, Proc. Conf. Physics of Voids, Harwell Res. Rep. AERE-R.7934.
Pugh, S.F., Loretto, M.H. and Norris, D.I.R. (Eds), 1971, Proc. Brit. Nucl. Energy Soc. Conf., Voids Formed by Irradiation of Reactor Materials (Reading, U.K.)
Quigley, T.M., Murphy, S.M., Bullough, R. and Wood, M.H., 1981, Harwell Res. Rep. AERE-TP.916.
Robinson, M.T. and Young, F.W. (Eds), 1975, Proc. Int. Conf. Fundamental Aspects of Radiation Damage in Metals (Gatlinburg, U.S.A.).
Wiedersich, H., 1972, Rad. Effects, 12, 111.

Uses of neutrons in engineering and technology

John Walker

University of Birmingham, Birmingham B15 2TT, UK

Abstract. Applications of the neutron are considered to show how its properties make it a valuable engineering tool. They include nuclear safeguards, static and dynamic radiography, small angle scattering, neutron gauges, the transmutation doping of silicon, and the production and use of radioactivity in activation analysis and for industrial tracers.

1. Introduction

The discovery of the neutron led to the discovery of nuclear fission, and these two have created the whole field of nuclear reactor engineering, but nothing is said in this paper about the use of neutrons in the cores of research and power reactors in spite of their fundamental importance to the whole design, nor is there anything about the roles of neutrons in fusion reactors as transporters of energy and creators of new fuel. However, there should certainly be something from nuclear engineering, and I have chosen to give a little attention to nuclear safeguards before moving to other aspects of neutron technology.

2. Nuclear safeguards

The accounting and control of nuclear materials is obviously a very important aspect of nuclear engineering, and calls for the assaying of a variety of materials for fissile content; fuel elements before and after use, general waste, and the fissile material in reprocessing plants are examples of specimens to be analysed non-destructively. Neutrons can help in two ways: firstly, neutrons from spontaneous fission can act as indicators of fissile content, and secondly, externally-produced neutrons can interrogate specimens through the production of fission.

Ensslin et al (1981) and Caldwell (1981) have used the spontaneous fission of the even-mass isotopes for the assay of plutonium in solutions and in large packages of waste; the prompt neutrons were detected in coincidence by a high-efficiency counter consisting of a central source-cavity surrounded by ^{3}He proportional counters embedded in polyethylene. Coincidence counting (Swansen et al, 1980) enables multiple fission neutrons to be separated from single neutrons produced in other reactions. In these counters, the combination of moderation, diffusion, and absorption or leakage gives a neutron die-away time of about 50μs which can be reduced at the expense of counting efficiency by the addition of a cadmium liner to the moderator. A guide to detection sensitivity is given by Caldwell (1981) as approximately 10gm ^{240}Pu to produce a change in counting rate equivalent to 3 standard deviations in background rate. The isotopic composition of the specimens is required to convert coincidence rates to the concentration of individual isotopes such as ^{239}Pu. For the material from the Los Alamos plutonium processing facility used by Ensslin et al (1981), 1 litre of plutonium nitrate gave a coincidence rate of about 100cps for 10g l^{-1} (10kg m^{-3}) of plutonium, and as the plutonium concentration was increased the coincidence rate increased faster than linear because of neutron multiplication. As to accuracy, a concentration of 2g l^{-1} counted for 1000s gave a standard deviation of 1.5% and 20g l^{-1} gave 0.8%.

This procedure is often called passive assaying to distinguish it from active assaying in which there is neutron interrogation of specimens; the interrogating neutrons can come from either an isotope source such as ^{252}Cf or an accelerator. Passive assaying has the advantage of comparative simplicity but the active method detects the fissile isotopes directly through induced fission and its emitted neutrons. Fissile isotopes are usually emphasised by interrogation with near-thermal neutrons produced by moderation of fast neutrons from the source, but fast-neutron interrogation can be incorporated if information is required on fertile isotopes.

The ^{252}Cf shuttle provides an example of interrogation by an isotope neutron source; in the Harwell experimental version, a source emitting 5×10^8 n.s^{-1} is moved back and forth from its store to the interrogation station where it rests for times from seconds to minutes according to the delayed neutron precursors being emphasised. Delayed neutrons from fission, recorded after the return of the source to its store, are used to measure the fissile content of specimens. With the source intensity quoted, the shuttle method is sensitive to gram quantities of fissile material.

An alternative arrangement is to move the specimen instead of the source; this is suitable for small specimens and strong sources. With 100mg of ^{252}Cf (2.3×10^{11} n.s^{-1}), detection limits of 0.5ppm for natural uranium and 8ppb for ^{239}Pu in 15g samples of soil have been reported by MacMurdo and Bowman (1977).

For high sensitivity, and particularly with large samples, it is better to use a pulsed accelerator because it provides shorter pulses and allows the use of prompt neutrons from fission as indicators instead of having to rely solely on the lower-intensity delayed neutrons. Kunz (1981) has used a 14MeV neutron generator with a graphite-polythene moderator, and the associated neutron detectors were shielded with cadmium to restrict the counting to neutrons above about 1eV. With no fissile material, the duration of the detected pulse is settled by the accelerator burst and the neutron moderation time. In the presence of fissile material the detected neutrons extend over a longer time as a result of neutron multiplication. The differential die-away time depends on the quantity of fissile material, and can be used at present to measure amounts as low as about 1mg.

An entirely different method of using a pulsed accelerator involves its combination with a slowing-down spectrometer, a 1.8m cube (75 tons) of lead in which a 1μs burst of fast neutrons is moderated. The heavy moderator ensures that the neutron burst stays short during moderation so that time acts as an indicator of energy. The lead spectrometer has been applied to the fissile assay of reactor fuel elements by Bornt et al (1981). Threshold fission counters detected fast neutrons from fission at particular times, and therefore at known neutron energies characteristic of particular fissile nuclides.

3. Neutron Activation Analysis (NAA)

Neutron activation analysis (de Soete et al, 1972) has established a firm place in the whole range of analytical techniques. It is very valuable for standardisation purposes and is regularly used by the US National Bureau of Standards and Geological Service in the certification of reference materials. Thermal neutrons are particularly suitable for it because of the high cross sections of many elements, but 14MeV neutrons have grown in importance because of the ease of their production by low-voltage accelerators and because they detect elements such as N, O, Si, P, Tl and Pb which are difficult or impossible to detect with thermal neutrons. Neutrons with selected energies between thermal and 14MeV are also helpful for emphasising some elements in the presence of others, perhaps in greater abundance.

Neutron activation may be coupled with chemical separations (radiochemical neutron activation) to remove or reduce the interference between different activities. This can give the highest sensitivity at the expense of extra complexity, but obviously is only possible if the required induced activities have sufficiently long half-lives for the chemistry to be done. Radiochemical separations have the strong advantage over orthodox chemical separations that the activity gives a uniqueness to the atoms involved which clearly distinguishes them from all the other atoms present. Neutron activation analysis without chemical separation is often called instrumental neutron activation analysis (INAA).

Irradiations in a research reactor are appropriate for thermal neutron activation analysis, and the specimens are moved between the irradiation position and the counting position outside the reactor. It is often possible to irradiate many specimens simultaneously, and the method can provide high irradiation fluxes and high efficiency and low background in the detection of radiation. Guinn (1974) has given limits of detection for 71 elements when irradiated in a thermal neutron flux of $10^{13} \text{n cm}^{-2}\text{s}^{-1}$ for 1 hour and then counted with a NaI(Tl) detector; a Ge(Li) detector gives similar sensitivity but better resolution. A few examples from Guinn's data are:

Approx. detection limit: g	Element
$10^{-12} - 10^{-11}$	Mn, In, Eu
$10^{-11} - 10^{-10}$	Rh, Ag, Ir, Au
$10^{-10} - 10^{-9}$	Na, Al, Co, Cu, W, U
$10^{-9} - 10^{-8}$	Cl, Ti, Zn, Pt, Hg, Th
$10^{-8} - 10^{-7}$	F, Mg, Cr, Ni, Cd, Pb
$10^{-7} - 10^{-6}$	Ca, Zr
$10^{-5} - 10^{-4}$	Si, S, Fe

These limits are included here to indicate the high sensitivities possible with NAA, but the level reached in an actual case will depend on the matrix; practical interest is usually in the concentration of particular elements in mixtures.

A special form of INAA involves the use of prompt gamma rays from neutron capture (PGNAA or PGAA), and in this case the irradiation and counting have to be undertaken simultaneously. With a reactor, the alternative arrangements are external irradiations with neutron beam tubes or guides and with high-efficiency detectors close to the specimens, or internal irradiations with external detection of the gamma rays. Anderson et al (1981) have given approximate detection limits for the prompt gamma facility established by the University of Maryland and the National Bureau of Standards, which uses an external thermal neutron beam of $2 \times 10^8 \mathrm{n\ cm^{-2}\ s^{-1}}$ at the sample and a Ge(Li) detector. With an irradiation time of about 10 hours, the approximate detections limits range from $10^{-8} - 10^{-7}$g for B, Cd, Sm, Gd, through $10^{-5} - 10^{-7}$g for a group including Ni, Cu, Ag, Au, to $10^{-3} - 10^{-1}$g for a group including C, B, Sn, Pb. The limits should be considerably lower with the neutron guides at the ILL reactor, Grenoble. Comments on elemental sensitivities are also given by Glascock (1981). In practice, detection limits will depend on background levels and interference from other materials in the specimens. On the other hand, the simultaneous recording of information from many elements can often be a distinct advantage.

So far in this section nuclear reactors and, to a smaller extent, accelerators have received attention as neutron sources, but isotope sources (Ainsworth, 1975), although of lower intensity, are valuable when portable instruments are required for field work, such as borehole logging, or for mounting on industrial plant. ^{252}Cf, with neutrons from spontaneous fission, and ^{241}Am-Be, with neutrons from (α,n) reactions, are two important examples. The differences between their neutron spectra can help to discriminate between different elements in specimens undergoing analysis.

3.1 Applications of NAA in the coal industry

Coal is considered not only because of its importance as a fuel but also because the neutron techniques discussed can be used for on-stream analysis and control in other mineral and chemical industries. The increased demand on mineral resources and the greater care being taken to minimise industrial pollution have increased the need for improved efficiency in the processing of resources, and neutrons are valuable in this respect because they can penetrate considerable thicknesses of process streams and therefore reduce the effects of heterogeneity and particle sizes. The alternative approach in the case of coal involves selection of a sample which can be regarded as representative of the whole, crushing and mixing to give a smaller sample which is then analysed and, if necessary, burnt. NAA is also playing a very prominent role in direct measurements of atmospheric pollution.

C.G. Clayton and his colleagues at AERE, Harwell, have reported extensively on neutron (and other nuclear) techniques in the coal industry, and the topic received attention from a number of laboratories at the 1981 Toronto Conference on Modern Trends in Activation Analysis. One important aspect is the measurement of the ash content of coal to be used for electricity generation because it has to be less than 30% to avoid low generation efficiency and excessive time for maintenance of equipment; in the United Kingdom it is usually about 15-20% and needs to be known to about + 5%. Wormald et al (1979) have used the ^{27}Al(n,γ) reaction, which produces the gamma-emitting ^{28}Al of half-life 2.3m, as an indicator of the ash content of coal in moving railway wagons. Combustion and chemical analyses of a large number of coal samples had already shown the correlation between Al_2O_3 and the total ash content. A ^{252}Cf neutron source of about 10^8n s^{-1} irradiated the coal and a sodium iodide detector placed later along the line recorded the gamma rays. Monte Carlo methods were used to predict the neutron flux distributions in six energy groups and these gave the ^{28}Al production rate. These calculations, linked with transport calculations for gamma rays, gave a standard deviation of 3.4% on the ash content (e.g. 30 $\pm$ 1%).

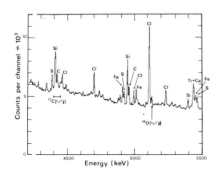

Fig.1 Gamma-rays from neutrons on coal (data from C.G.Clayton)

A knowledge of the composition of the combustible fraction of coal is important for the satisfactory operation of furnaces. As an example, the sulphur content must not exceed an amount acceptable for atmospheric pollution, and this calls for tests of efficient blending in coal preparation; in the case of chlorine, the amount must be low enough to avoid undue boiler corrosion. Neutron activation analysis based on high-resolution gamma-detectors is valuable for this elemental analysis: elements in coal which can be analysed by thermal neutrons are H, C, N, Na, K, Al, Si, Cl, Ca, Fe, Ti; and elements analysed by neutrons greater than 1MeV are O, C, Si, Al. A small part of a gamma spectrum produced by neutron activation of coal is shown in Fig.1.

3.2 Forensic applications of NAA

In the United Kingdom, NAA is used in cases of suspected poisoning to analyse for arsenic in hair and for elements in liver, kidney, blood or urine (Pounds, 1982), but it is not as prominent as it was in the late 1960's (Coleman, 1967). However, Pounds believes that the application to trace elements in hair, in particular, deserves a new approach with modern detectors. In North America, the technique has continued to be important due mainly to V.P. Guinn of the University of California and R.E. Jervis of the University of Toronto. Guinn (1974) has reviewed in detail the type of forensic help given by NAA, and he has discussed various cases more recently (1982). Features of particular importance are the high sensitivity for a large number of elements, the ability to detect many elements simultaneously, and the non-destructive character which means that specimens remain available for display in court. Very small specimens such as a single hair or a chip of paint can provide evidence which may be based on just one element (e.g. As in hair, or an element such as Ba, Sb or Pb from the primer of a bullet) or on as many trace elements as possible when a number of specimens are being compared. The investigation of sources of oil pollution is a modern example in which trace-element composition plays an important role.

3.3 Fission-track detection of uranium in very pure silicon

The α-particles from minute quantities of uranium and thorium in a silicon chip and its mounting materials can cause a loss of information in memory locations, an effect known as the 'soft error' which has become so important in high-density memory devices that the manufacturers require the α-particle emission rates to be less than 0.001 cm^{-2} h^{-1}. The corresponding concentration of uranium (and thorium by association) can be measured by using thermal neutron induced fission in ^{235}U, with the fission fragments being detected by track recording in polyimide film which is subsequently etched with sodium hydroxide. With a reactor neutron flux of $3 \times 10^{13} n$ cm^{-2} s^{-1} the detection limit for uranium is approximately 0.02 ppb (2×10^{-11}) (Mapper et al, 1981).

Obviously this method can be applied in any field in which uranium concentrations or distributions are required. For example, it can map the distribution of uranium in rock specimens, and it is particularly valuable when only a small amount of sample is available such as with single crystals of a particular mineral or with interstitial waters from marine sediments (Wilson, 1982). Application is also possible to other elements such as boron which give charged particles from neutron interactions.

4. Neutron sondes, moisture gauges, and level gauges

Neutron sondes used in oil and mineral exploration often include an isotope neutron source such as Am-Be or ^{252}Cf to irradiate the rock surrounding a borehole. These sources are robust and convenient, but compact sealed-tube accelerators can also be used to provide 14MeV neutrons, and are necessary if short pulses are required for measurements such as those on die-away times. A moderated neutron field is established in the inspected volume, the size of which depends on the composition of the rock but is likely to be about $1m^3$, and the stimulated gamma-rays from elements present are recorded in a germanium detector in the sonde. The full elemental analysis is complex because it depends on the detailed behaviour of both neutrons and gamma-rays in the surroundings of the probe, and Monte Carlo computations are often undertaken, as in the work on coal, to predict the neutron flux as a function of energy and space and to calculate gamma-ray transmission. A neutron detector is used instead of, or in addition to, a gamma-ray detector when the interest is in oil or water. The detector responds mainly to thermal or near-thermal neutrons and thus gives a strong signal in the presence of hydrogenous material which provides efficient moderation of the source neutrons. It is worth noting that the back-scattering of gamma-rays from a source in the probe can be used to supplement the neutron information because it depends on the density of the rock.

An interesting specialised form of sonde for uranium exploration involves the detection of delayed neutrons from neutron-induced fission. Smith and Rodriguez (1979) have used a pulsed accelerator source of 14MeV neutrons, with the delayed neutrons being detected after the accelerator pulse; Steinman et al (1982) have used a ^{252}Cf source well separated from the neutron detector in a long (5m) sonde which was moved in the borehole at an appropriate speed to give the delayed neutron response after the passage of the source, and they surveyed ore grades from 100 ppm to about 1% of U_3O_8.

Neutron-neutron and gamma-gamma measurements are also used in moisture/density gauges which can be applied in agriculture for irrigation control, in concrete technology for density and moisture measurements, and in the construction industry for soil foundation studies and to help detect and eliminate undesirable water intrusion in buildings. These gauges are designed to be used near the surface of the material involved.

Charlton, Heslop and Johnson (1981) of ICI consider that the detection of the level of hydrogenous material inside process vessels is one of the most useful applications of neutrons in the chemical industry; the ability of neutrons to penetrate a considerable thickness of metal, the strong moderation of neutron energies by hydrogenous materials, and the ease of detection of thermal neutrons are features favouring neutrons. They show how the combination of a fast neutron source (Am-Be) and a thermal neutron detector produces a change in the counting rate as the measuring head moves across a surface or a change in liquids. The instrument has also proved very useful for checking the potential hazard of incipient blockages in lines due, for example, to the freezing of condensed vapours.

5. Production and use of radioactivity other than in NAA

As with activation analysis, the range of applications of radioactive materials is very broad, and there is space for only a few examples of their use, but the value of neutrons in their production is indicated clearly by the different active nuclides marketed by Amersham International: of the total of 170, 10 have natural radioactivity, 30 are produced by cyclotrons, 10 are fission fragments, and 120 are produced by nuclear reactors. The last two groups, giving 130 out of 170, involve neutrons in their production.

The materials ^{60}Co, ^{192}Ir, ^{170}Tm, ^{169}Yb are important neutron-produced gamma sources for industrial radiography, with ^{192}Ir being particularly prominent. Intense sources of ^{60}Co are used for radiation sterilisation of pharmaceutical and other commercial products, and spent fuel elements from reactors can also be used for irradiations. ^{237}Np from processed fuel elements can absorb neutrons and lead by β-decay to ^{238}Pu, an α-emitter of use in portable electrical sources.

Radioactive tracers come quickly to mind as engineering tools because they can be detected with high sensitivity and used under conditions, such as high temperatures and pressures, which could be very difficult for other methods; the latter feature arises partly because the radiation detectors can be placed outside the containment vessel involved. The measurement of flow rates or residence times, the detection of leaks, and the observation of one fluid entrained in another (e.g. liquid droplets in gas streams) are examples of problems met in industrial plants which are suitable for solution by tracer methods; Charlton et al (1981) have described some actual cases in the chemical industry. Imperial Chemical Industries have found it worthwhile to install a small nuclear reactor to produce short-lived activities for use in their plants.

A variant on the method which can avoid radioactivity in parts of an industrial plant is to use tracer materials which can be activated by neutrons after passing through the sensitive region. Their subsequent path can be followed in the usual way, and their previous path can often also be deduced.

6. Production of changes other than radioactivity

6.1 Transmutation doping of silicon

Silicon is obviously important in electrical and electronic devices such as power rectifiers, thyristors, and integrated circuits, and its resistivity can be altered by adding various elements, including phosphorus in the case of n-type silicon. In conventional doping, phosphorus is added to the silicon melt but this produces variations in resistivity of about ± 20% which may be unacceptable. Neutron transmutation is a way of providing more homogeneous doping and accurate control of the concentration of phosphorus, and it is now important commercially. A comparison of the spreading resistance profiles for dislocation-free silicon crystals doped conventionally or by neutrons is given in Fig.2; the smoother curve in the latter case results from the more uniform concentration of phosphorus (Crick, 1982).

It is the isotope ^{30}Si (3.09%) which takes part in the transmutation to phosphorus through the radiative capture of thermal neutrons followed by β-emission to form ^{31}P:

$$^{30}Si(n,\gamma)^{31}Si \xrightarrow[2.6h]{\beta^-} {}^{31}P$$

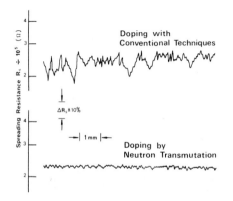

TABLE 1

Neutron doping of silicon: typical irradiation times

Resistivity Ω.cm	Total dopant conc. cm^{-3}	Dose n.cm^{-2}	Irradiation time ($\phi=8\times10^{12}$n.cm^{-2} s^{-1}) hours
150	.308x10^{14}	1.56x10^{17}	5.43
100	.462	2.41	8.37
70	.661	3.50	12.15
50	.925	4.95	17.19
30	1.541	8.34	28.96
10	4.624	25.28	87.77

Dopant concentration before irradiation			0.023 x 10^{14} atoms cm^{-3}
Resistivity	"	"	2000 Ω cm
Electron mobility			1350 cm^2 v^{-1} s^{-1}
^{30}Si cross section			0.118 x 10^{-24} cm^2

Fig.2 Doping of silicon

Table 1 shows typical irradiation times to produce different dopant concentrations and resistivities. During irradiation, the crystals suffer radiation damage in a number of ways. Thermal neutron capture produces recoiling atoms and, in the case of ^{31}Si, bombardment by β-particles; these processes cannot be avoided. Fast neutrons at the irradiation site produce atomic displacements and this effect can be reduced by having irradiations where the ratio of thermal to fast neutrons is as high as possible. The ability to select a low fast flux without unreasonable loss of thermal flux will depend on the type of reactor being used and on the detailed design of its irradiation facilities, but in any case annealing of the crystal is carried out after irradiation to re-establish atomic order.

6.2 Microfilter production

Nuclear track filters are made by exposing thin mica sheet or plastic film to a collimated beam of heavily ionising particles with sufficient energy to pass through the sheet or film. Carefully controlled chemical etching produces a hole along the track of each particle; the diameter of the holes is varied by changing the etching times, and their density is settled by the particle density. The technique is based mainly therefore on charged particles in track detectors, but neutrons have played a role in its development because fission fragments from neutron-induced fission have been used extensively as the bombarding particles. Polycarbonate film irradiated with fragments from ^{235}U in a nuclear reactor has formed the basis of the commercial production of nuclear microfilters with pore diameters ranging from $3 \times 10^{-2}\mu m$ to $8\mu m$. It is worth noting that energetic heavy ions up to uranium can now be provided directly by accelerators so that neutrons are no longer essential for the production of microfilters.

7. Information on materials from the scattering of low-energy neutrons

Thermal neutron diffraction (Bragg scattering) has become a standard tool in the study of crystalline materials in spite of suitable sources being more expensive and of lower intensity than X-ray sources which are also used extensively, of course, for diffraction measurements. Clearly neutron scattering must have special features, and in the present context we have to ask whether they are helpful in technological problems. Firstly, although electromagnetic coupling through the neutron's magnetic moment is important in magnetic studies, neutron scattering is predominantly due to short-range nuclear forces which can vary markedly from one element to another, and even from one isotope to another in the same element. This means that isotope substitution can be used to elucidate material behaviour, that light elements (especially hydrogen) can be studied, and that good contrast in behaviour is possible between adjacent elements; the ability to deal with hydrogenous and other light materials is worthy of special note for technological applications. The short range of the nuclear force results in low neutron absorption in many materials and allows the investigation of large specimens, clearly important in the non-destructive testing of engineering components. Another feature with neutrons is their low energy at a de Broglie wavelength which is suitable for structure measurements (e.g. 5meV for a wavelength of 4Å (0.4nm)); this low energy allows atomic mobilities to be measured comparatively easily by inelastic or quasi-elastic scattering, again a matter of importance for hydrogen in technical materials.

A cold source on a reactor or pulsed accelerator (Windsor, 1981) enhances the production of low-energy (long-wavelength) neutrons, and it is possible to avoid Bragg scattering by the selection of sufficiently long wavelengths (~ 10Å) for scattering measurements. Small-angle neutron scattering (SANS) can then be used to study heterogeneities up to about $1\mu m$ in size, i.e. much larger than atoms but smaller than features observed by neutron radiography. A SANS spectrometer usually incorporates a velocity selector and a two-dimensional detector array, although Allen and Ross (1981) have used photographic recording in SANS applied to magnetic materials; Galotto et al (1976) have described an instrument installed solely for technological applications.

A few examples of the use of various aspects of neutron scattering are:

Diffraction for texture in polycrystalline materials, minority phases in
  metals, and internal strains and stresses by high-resolution
  diffraction.
SANS for radii of metallic precipitates and changes in dislocation
  densities in turbine blades, with the aim of predicting residual life;
  effects of heat treatment in metals including welded zones; particle
  size distributions of catalysts, and pore sizes in cement pastes and

oil shales.

Quasi-elastic scattering for studies of free and bound water in concrete, and the mobility of hydrogen in metallic hydrides and clays.

8. Neutron radiography

The radiographic arrangement in which parts of a specimen remove thermal neutrons from a collimated beam and so cast a shadow on a suitable recorder is now well known, and there are a number of accounts which include descriptions of neutron sources and image recorders (Berger, 1965; Hawkesworth and Walker, 1969). The proceedings of the Birmingham conference, edited by Hawkesworth (1975), contain an excellent collection of radiographs of a range of technical subjects, and a more specialised group illustrating the behaviour of nuclear fuels has been presented by von der Hardt and Röttger (1981). An object being radiographed can remove neutrons by both scattering and absorption and, as with scattering alone, the combined effect can vary strongly from element to element and isotope to isotope. This makes possible the radiographing of some light elements when surrounded by much heavier ones; for example, to radiograph hydrogenous material inside steel or lead vessels is of particular value technically, and it is illustrated in Fig.3. Gadolinium foil and photo-sensitive film were used in this case but other combinations of converter screen and film can be selected to suit particular conditions. A track-etch plastic can replace the photosensitive film, but only the latter can autoradiograph appropriate converter foils away from the irradiation position. This so-called 'transfer technique' enables radiographs to be taken of highly active specimens such as fuel elements from power reactors, and it can also be adopted in other difficult environments; indium and dysprosium foils have radioactive lifetimes suitable for transfer radiography and are commonly used.

The ability to select different neutron energies is important in radiography for a number of reasons: a particular material can be emphasised by selecting an energy at which its neutron interaction is strong; on the other hand, changes in image density due to thickness variations in the sample can sometimes be enhanced by having an energy where the absorption is low - the use of epithermal rather than thermal neutrons for the inspection of uranium is a case in point; cold neutrons can increase the neutron penetration of materials surrounding a specimen and thus permit radiography in otherwise impossible situations.

A typical exposure for a high-resolution radiograph is 10^8n.cm^{-2}, and a thin gadolinium converter with a high-resolution photographic film can give a spatial resolution of about 10μm but up to about 200μm is characteristic of faster screen-film combinations.

8.1 Moving objects (dynamic radiography)

Moving objects can be radiographed by using the transmitted neutrons to activate a suitable fluorescent screen and then viewing the screen with some form of television system. In the United Kingdom, Rolls-Royce Aero Engine Division (Stewart, 1980) together with Burmah-Castrol (Stewart and Heritage, 1981) have used this approach to view the flow of lubricants and fuels in engines (Fig.5). A gadolinium oxysulphide screen, developed for X-ray detection but suitable for neutrons because of the presence of gadolinium, acted as the fluorescent neutron detector and formed the input to a Delcalix image intensifier, the output of which was passed to a TV

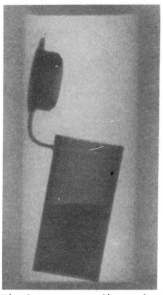

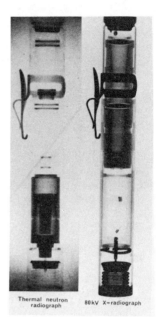

Fig.4............
Comparison of neutron radiograph and X-radiograph of dosimeter (15mm dia.) (after M.R. Hawkesworth)

Fig.3 Neutron radiograph of 12mm dia. polythene ampoule, half-filled with water, in lead pot with 25mm walls (after J.A. Izatt)

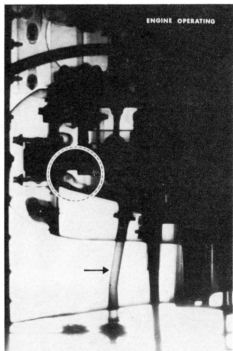

Fig.5 Dynamic neutron radiographs of aircraft engine. Arrows and circles indicate changes in flow of oil (Rolls-Royce/Burmah Castrol work at Harwell)

monitor. The recent work has used a Thomson-CSF neutron-sensitive image intensifier tube and has been based on the cold neutron facility (6H) of the DIDO reactor, Harwell (Baston and Harris, 1978), with radiography being performed in a blockhouse 25m from the reactor core; the cold neutron flux at this point is 3.3×10^5n cm^{-2} s^{-1} and the beam has a diameter of 300mm. Cocking (1982) is now using the same beam and an orthodox neutron fluorescent screen (NE426) which is viewed by a low-light TV camera linked to an image processor. Each frame is digitised into 512 x 512 elements at 256 levels of intensity, and frame summing is used to improve image quality.

Digitised images permit automatic measurement of shapes and intensities and the data can be processed to emphasise particular features; the presentation of only changes in images and not the whole of them is one example, and gating to select events in a short interval is another. Clearly, pulsed sources can also be used to provide a stroboscopic viewing of repetitive processes, and flash neutron radiography can be applied to single events. The advent of intense pulsed accelerator sources of neutrons should certainly enhance the value of dynamic radiography.

8.2 Radiography in three dimensions: tomography and holography

So far in this brief discussion only shadowgraphs have been considered, but a specific cross sectional view of a specimen can be obtained by tomography or related methods. In one approach, the specimen was rotated and about eight radiographs taken at different angles. The developed films were superimposed for viewing and moved relative to each other to bring different planes into focus. Zakaib, Harms and Vlachopoulos (1978) have used a neutron counter rather than film to measure the transmitted beam and then reconstructed tomographically the voids in test specimens as a preliminary to measurements of voids in two-phase flow.

Beynon and Pink (1980) have used cold neutrons scattered in a specimen to produce a non-coherent form of hologram on a foil-film image recorder. The hologram is in effect an infinite collection of shadowgraphs of a Fresnel zone plate with alternating aluminium (transmitting) and gadolinium (absorbing) zones which is placed between the scattering specimen and the image recorder. Reconstruction obviously requires an appropriate transformation of the hologram and this was done optically by illuminating a photographically-reduced version with converging light from a laser. A screen could be placed parallel to the original zone plate in a position corresponding to a plane in the specimen and movement of the screen thus effectively scanned the whole volume of the specimen.

References

Ainsworth, A., 1975, Radioisotopes 24, 794-804.
Allen, A.J. and Ross, D.K., 1981, Nucl. Instr. and Methods 186, 621-636.
Anderson, D.L. et al, 1981, Neutron-capture gamma-ray spectroscopy and related topics: von Egidy, T. et al (Eds.), (Bristol: Inst. of Phys.) pp. 655-68.
Baston, E.H. and Harris, D.H.C., 1978, AERE R9278 (HM Stationery Office).
Berger, H., 1965, Neutron Radiography (Amsterdam: Elsevier).
Beynon, T.D. and Pink, A.G., 1980, Nature 283, 749-751.
Bornt, F.W. et al, 1981, Trans. Am. Nucl. Soc. 39, 344-5.
Caldwell, J.T., 1981, Trans. Am. Nucl. Soc. 39, 339.
Charlton, J.S. et al, 1981, IAEA Conference on Industrial Application of Radioisotopes and Radiation Technology, Grenoble.
Cocking, R.J., 1982. Private communication.
Coleman, R.F., 1967, J. Brit. Nucl. Energy Soc. 6, 134-138.
Crick, N.W., 1982. Private communication.
de Soete, D., Gijbels, R. and Hoste, J., 1972, Neutron activation analysis (New York: Wiley-Interscience).
Ensslin, N. et al, 1981, Trans. Am. Nucl. Soc. 39, 335-6.
Galotto, C.P. et al, 1976, Nucl. Instr. and Methods 134, 369-378.
Glascock, M.D., 1981, Neutron-capture gamma-ray spectroscopy and related topics: von Egidy, T. et al (Eds.), (Bristol: Inst. of Phys.), pp.641-54.
Guinn, V.P., 1974, Ann. Rev. Nucl. Sci. 24, 561-591.
Guinn, V.P., 1982. To be published in J. Radioanal. Chem. 72.
Hawkesworth, M.R. and Walker, J., 1969, J. Mat. Sci. 4, 817-835.
Hawkesworth, M.R. (Ed.), Radiography with neutrons, 1975, (London: British Nucl. Energy Soc.).
Kunz, W.E., 1981, Trans. Am. Nucl. Soc. 39, 341-2.
MacMurdo, K.W. and Bowman, W.W., 1977, Nucl. Instr. and Methods 141, 299-306.
Mapper, D. et al, 1982, Solid State Nuclear Track Detectors (Proceedings of the 11th International Conference, Bristol, 7-12 September 1981), Fowler, P.H. and Clapham, V.M. (Eds.), (Oxford: Pergamon).
Pounds, C.A., 1982. Private communication.
Smith, R.C. and Rodriguez, R.G., 1979, Nucl. Instr. and Methods 158, 261-280.
Steinman, D.K., 1982. Private communication.
Stewart, P.A.E., 1980, ASTM STP716, Garrett, D.A. and Bracher, D.A. (Eds.), pp. 180-198.
Stewart, P.A.E. and Heritage, J., 1981, World Conf. on Neutron Radiography, San Diego. (Proceedings to be published: Barton, J.P. (Ed.). These proceedings will contain papers on various aspects of neutron radiography).
Swansen, J.E. et al, 1980, Nucl. Instr. and Methods 176, 555-65.
von der Hardt, P. and Rötter, H. (Eds.), 1981, Neutron Radiography Handbook (Dordrecht: Reidel).
Wilson, H.W., 1982. Private communication.
Windsor, C.G., 1981, Pulsed neutron scattering (London: Taylor and Francis).
Wormald, M.R. et al, Int. J. App. Rad. and Isotopes 30, 297-314.
Zakaib, G.D. et al, 1978, Nucl. Sci. and Eng. 65, 143-154.

… Section 7
Paper presented at Conf. on Neutron and its Applications, Cambridge, 1982

Neutron radiography — accomplishments and potential

John P. Barton

Neutron Radiography Consulting
P. O. Box 206, La Jolla, California 92038

Abstract. A review of current activities in neutron radiography shows the considerable diversity of the field, and the considerable potential for increased applications in the future. The nuclear industry is making considerable use of neutron radiography as one of the most important methods for inspection of irradiated fuel. Another major area concerns reactor-based service centers for general industry. Thirdly, there are now numerous custom-designed in-house facilities, facilities for in-house use, and finally there are special techniques developed for special applications.

1. Introduction

Of our five senses none is more valuable than our ability to see. Neutron radiography is a new way of seeing things. Visible light provides information of the surface of opaque objects. X-radiography allows penetration, and, since x-rays interact with the orbital electrons, provides a view of material density changes within the object. Neutrons penetrate and provide a picture dependent on differences in the nuclear properties of materials. For certain energy neutrons some materials (e.g., hydrogen, boron, cadmium, silver, gold, uranium) have neutron attenuation coefficients significantly higher than their neighbors in the periodic table. This feature of neutron radiography provides the broad range of potential applications. Other features also have importance in certain circumstances. The ability to neutron radiograph highly radioactive objects without interference from gamma rays, and the ability to distinguish isotopes of the same element.

Neutron radiography may appear at first to be slow in development. Within a year of the discovery of x-rays by Roentgen there were numerous centers taking x-radiographs. In 1962, thirty years after the discovery of the neutron, there was no one working to develop neutron radiography on this continent. The reason for the slow start is certainly not the lack of inherent capabilities of the neutron. It is perhaps related to limitations on available neutron sources, and some lack of understanding by sponsors or users who feel that x-rays have solved all our needs to see inside things.

During the past twenty years the momentum has increased significantly. At the First World Conference on Neutron Radiography, held a few months ago in California, sixty centers from twenty countries reported on their projects. The proven applications were remarkable not only for their number, but for their diversity. The diversity suggests that utilization will continue to grow rapidly. For example, if neutron radiography can be used for dentistry or forensic science in one country, it may not be long before it is used in most countries for these same purposes.

2. Nuclear Industry Applications

The single industry that most extensively exploits neutron radiography at the present time is the nuclear industry. In Europe alone there are over twenty reactor-based facilities in operation for neutron radiography of reactor fuel (Hardt, 1982). Some use a beam extracted through the concrete reactor shield to a separate containment designed to accept radioactive fuel elements. In others the radiography is performed below the reactor pool water, thus combining the reactor and specimen shields. In each case gamma-ray insensitive neutron imaging methods are used. The process of activating a foil (such as indium) in the transmitted beam, and then transferring the pattern to a film in a separate location is a standard method to avoid image fogging by the fuel radioactivity. However, recently the track etch method of imaging has become important, particularly if precise dimensional measurements are required from the image.

In the USA two TRIGA reactors (250 kW each) have recently been installed in nuclear laboratories specifically for neutron radiography (Tomlinson, 1982; Richards and McClellan, 1982) and a third is in an advanced stage of planning. Two of these reactors are part of hot cell complexes where complementary inspection methods are available.

In Japan and Germany steps have been taken to provide mobile neutron radiography systems that can operate in the spent fuel pool of power reactors (Matsumoto, 1982; Greim and Spalthoff, 1982).

For unirradiated fuel, neutron radiography is commonly used to provide qualitative information (assembly, foreign material, voids, rogue fuel pellets), and quantitative information (PuO_2 homogeneity, agglomeration size, average density).

For irradiated fuel, information is obtained concerning leaks, densification, swelling, migration, disintegration and redistribution. Comparisons of neutron radiographs taken before, during, and after irradiation are widely used.

Radiography of large arrays of fast reactor fuel pins inside special steel containers for exposure to simulated accident conditions present special problems. The solution has been use of epithermal rather than thermal neutrons to penetrate the enriched fuel, and use of multi-angle computerized tomography reconstruction to decipher the complex geometry (Rhodes, 1982).

Areas in nuclear industry currently requiring neutron radiography attention include: (1) standards for accurate interpretation of radiographs (Domanus, 1982); (2) methods for more precise dimensional measurements; (3) methods for penetration of fuel inside thick (10 cm) steel containments; and (4) methods for inspecting fuel or control rods at operating power stations.

3. General Service Neutron Radiography

To meet the increasing demand for neutron radiography by general industry reactor-based services have been set up in several industrial countries. In Britain and France it is the government centers (Harwell and Saclay) that provide this service. In the U.S.A. there are the privately owned companies (Aerotest and General Electric Company). The new facility on the ORPHEE reactor at Saclay is remarkable because it uses a cold neutron waveguide to extract the beam (Laporte, 1982). The U. S. services currently provide about 20,000 neutron radiographs per year, each on standard size film (14 x 17 inches) and costing from 30 to 85 dollars each, according to quantity (Newacheck, 1982).

Applications vary enormously, but chief among them are inspection of explosive actuated devices, ceramic capacitors, electronic devices, mechanical assemblies, and aircraft turbine blades.

4. Custom Designed Facilities

Demand for in-house systems to meet special needs is growing steadily with the proven value of the technique. They fall into two broad categories--stationary systems (for component inspection) and mobile systems (for inspection of aircraft, ships, etc.).

For stationary systems the choice of source is open to small reactors, subcritical multipliers, accelerators, and isotopic source. Each has sufficient advantages and disadvantages. The reactor is highest in flux and very well proven. However, there is some risk of delay in the license process in the U. S. The multiplier does not need licensing as a reactor in the U. S., only as special nuclear materials. However, the cost may be high in proportion to neutron yield. Accelerators cover a wide range of yields and costs; they have negligible residual radioactivity, but their performance and reliability for high-throughput radiography has not been widely demonstrated. Isotopic sources are, of course, reliable and relatively easy to work with safely. However the available neutron flux is generally limited.

Typical sources receiving serious consideration in the U. S. are the TRIGA reactor ($4 million), the high-yield D-T generator ($1.5 million), the 4 MeV Van de Graaff generator ($.7 million), and large sources of ^{252}Cf (available as a byproduct of other irradiations). Implementation of in-house systems in the past years has been sometimes at centers with unproven needs; they opted for low-budget sources, and achieved only limited capability. The future trend appears to favour much stronger neutron sources.

Mobile systems suitable for detecting corrosion in aircraft have been under development in the U. S. for several years using alternately ^{252}Cf and a D-T neutron generator. The performance levels demonstrated have, of course, been in proportion to the source neutron yields, and again the trend is to much higher yield sources in the future.

5. Special Applications

Cold neutrons have been used to significantly increase the contrast of hydrogenous or other materials behind metals. This has been employed for inspection of delicate pyrotechnic devices, and for inspection of thick metal ingots. Cold neutrons from a Harwell reactor have also been used for real-time neutron radiography of operating aircraft and automobile engines (Heritage and Stewart, 1982).

By making use of the very high neutron flux available when a TRIGA-type reactor is pulsed, it has been possible to obtain a high-speed motion film-radiograph showing details of the burnup cycle of propellant grains in a rifle shell inside the thick steel barrel. Rates of 8,000 frames per second have been demonstrated (Bossi, et al., 1982).

Intense pulsed neutron sources have been used in conjunction with time-of-flight techniques to take resonance neutron transmission radiographs. By recording information for different energy neutrons, corresponding to resonance cross sections in the object, it is possible to distinguish some different elements or isotopes in the object. A similar technique without geometric resolution has been demonstrated to have applications in nuclear fuels safeguards and allied programs (Bowman, et al., 1982).

These techniques, and others such as stop motion (flash source) neutron radiography, and high-resolution micro-neutron radiography, and even neutron holography (Beynon, et al., 1982), open numerous possibilities for use in engineering development diagnostics or materials research.

6. Conclusion

The neutron can be used as a radiation for everyday inspection or visualization, rather like light rays or x-rays. The development of neutron radiography is following behind x-radiography with a time lag of about 40 years, just as the discoveries were about 40 years apart. In both cases the widespread development is closely linked to the availability of convenient, high-yield, low-cost sources, and to the incorporation of this inspection requirement into industrial specifications. By looking at the progress over ten-year intervals one sees dramatic progress in both capability and utilization. Because the interactions of neutrons with matter are more complex than interactions of x-rays, one wonders whether the capabilities may also eventually be greater. No doubt the next 50 years will tell.

References

All references refer to the Proceedings of the First World Conference on Neutron Radiography, J. P. Barton, P. von der Hardt (Eds.) to be published by De Reidel Publishing Company, Holland.

# Neutron applications in the applied earth sciences

C.G. Clayton

Applied Nuclear Geophysics Group, AERE-Harwell.

1. Introduction

Neutron interaction techniques are important in the earth sciences because they are unique in possessing elemental specificity and a range which is significant compared with the local heterogeneity of minerals and large compared with the dimensions of particles normally encountered in minerals handling operations. In consequence, volumes interrogated provide better representative samples and reduce, or eliminate, the need for direct sampling procedures. The penetration of neutrons is also large compared with the thickness of containment vessels and is adequate to overcome borehole conditions and allow the examination of surrounding virgin formations.

Although, neutron techniques offer substantial potential for the solution of a number of problems in the earth sciences, there are several important factors which have to be resolved before operationally acceptable equipment can be designed. Variations in the composition and physical state of rocks and ores and in the environmental factors which have to be accommodated may strongly perturb the n-flux, energy spatial distributions and seriously modify measured intensities of neutrons and n-induced γ-ray spectra on which the methods depend. These perturbing effects must be understood and unfolded before an analysis of elemental compositions and evaluation of gross formation parameters can be accurately derived.

Because of the range of neutrons and the number of interacting parameters, an experimental attack to determine the influence of factors which affect neutron and γ-ray transport is often impossible and because of complex boundary conditions, calculation by purely deterministic methods is also unacceptable except for the simplest situations. A solution to understanding the influence of perturbing factors on n-transport, and hence to make allowance in equipment design and data interpretation, is beginning to lean heavily on n-tracking studies by Monte Carlo techniques and important successes are now being recorded.

Development of the full potential of n-techniques depends on the availability of suitable n-sources and neutron and γ-ray detectors, on 'fast' electronics and adequate interpretational and data processing techniques. Significant progress is now being made in all areas and this paper gives a brief comment on the state of some of the most important applications.

2. Analysis of coal

Because of the important role which coal is seen to have in satisfying future world energy demands, large efforts are now being made to develop better in-situ analytical equipment (inc. equipment based on n-interaction techniques) both for borehole logging and for on-line quality control.

A portion of a typical spectrum of n-induced prompt γ-rays from a sample of U.K. bituminous coal is shown in Fig. 1. If it is sufficient to derive the

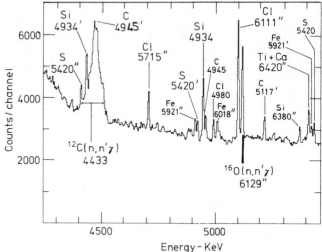

Fig. 1. Partial, n-induced γ-ray spectrum of U.K., bituminous coal. Neutron source: ^{241}Am-Be.

concentrations of a few elements only, interpretation of such a spectrum requires that, for acceptable accuracy, account be taken of the effects of variations in moisture content, bulk density and neutron poisons and in the disposition of the measuring system with respect to the coal to be interrogated. This is difficult in borehole logging but, on-line, presentation systems can be designed and independent measurements made which allow for the most important variables except moisture content. Microwave attenuation can give a satisfactory estimate of free moisture in coals of the same origin but different coals require different calibration constants.

Some relief to the problem of varying neutron and γ-ray flux can be obtained by considering the ratio of the γ-ray intensities from required analytes to the intensity from some reference element, such as H (Wormald et al 1979), but this is only valid if the n-interactions involved are of the same kind for all elements considered. It is worth noting that, at least for bituminous coals, if one or both of the principal ash-forming elements (Al, Si) are included, ash content can be inferred (Wormald et al 1979).

If all abundant elements can be identified in the spectra, even though of low analytical interest, and the volume interrogated is sufficient for the coal to be its own moderator, coal becomes a special material and a theory has been developed by Wormald and Clayton (1983) which shows that a total elemental analysis can be obtained without regard to perturbations in n-flux. It should be noted that uncertainties in the peak areas of the most abundant elements (e.g. C,O) will reflect on accuracies of determining concentrations of all other elements (Clayton et al 1983). An indication of the accuracy of the method is given in Table 1.

An accurate estimate of C content allows calorific value to be determined on-line, as indicated in Fig. 2 which shows the relationship between C.V. obtained from C and H contents with C.V. derived by incineration for a suite of samples supplied by the National Coal Board.

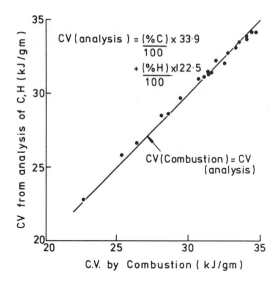

Fig. 2. Calorific value of coal from measurement of concentrations of C and H compared with calorific value by combustion.

Table 1. Analysis of a U.K. bituminous coal by n-induced prompt γ-rays compared with chemical assay.

Element	Wt % Prompt-γ analysis	Chem. assay
C	62.0 ±1.4	63.6
H	4.04±0.10	4.45
O	21.24±1.47	19.56
N	1.21±0.23	1.35
S	1.78±0.30	1.30
Cl	0.29±0.02	0.31
Si	4.25±0.21	4.28
Al	1.47±0.52	2.48
Fe	1.78±0.10	1.26
Ti	0.03±0.02	0.09
K	1.2 ±0.4	0.50
Ca	<.7(2σ)	0.35
Mg	0.45±0.48	0.21
Na	0.41±0.13	0.16

Equipments measuring S content, or ash, or total elemental composition are now operating or being installed at several sites in the U.S.A. (Gozani et al 1981, 1982).

3. Oil well logging

Logging is a method of obtaining direct information on the nature of oil-bearing strata in an oil reservoir. It is the only means of continuously monitoring the changing characteristics of a formation and its importance is increasing as more sophisticated methods of recovery are introduced. Neutron techniques dominate among the spectrum of techniques now employed. They have the greatest potential for further development.

The use of n-techniques in oil-well logging is aimed principally at determining formation porosity (ϕ) and hydrocarbon content by n-moderation since, in most formations, H is contained in the pore fluids (oil, gas, water) rather than in the rock matrix. Two thermal neutron detectors (^{3}He: pressure ∿15 atmos) are positioned at about 40 cm and 60 cm from an isotope n-source (^{241}Am-Be: 15-20 Ci). The ratio of countrates in near to far detectors is sensitive to the formation moderating power and provides some independence of borehole conditions. However, significant corrections have to be made for such effects on n-transport as borehole diameter, 'mud-cake' on the borehole wall, formation and borehole water salinity, formation lithology and, in a lined borehole, of casing and cement thicknesses.

The derivation of accurate correction factors is generally based on limited experimental data from carefully constructed test pits but there is significant interpolation and the effects of single parameter variables are virtually impossible to unfold. Monte Carlo techniques now allow n-flux space, energy distributions to be described for realistic formations and borehole situations and the changes in near/far count ratio to be determined for simple or compounded changes in important properties.

The results of computing variations in near/far count ratio with ϕ for a simulated probe are presented in Fig. 3. The probe response is equated with the ratio of n-fluxes in the detectors. For this particular simulation it is seen that the response is linear up to about $\phi=30\%$.

Partial saturation values are also included in Fig. 3 to demonstrate the degree of change in response when the pore space is filled with a gas having a hydrogen index (H density relative to water) of 0.437, corresponding to methane at about 4,000 psi. At $\phi=30\%$ for example, the probe response at partial saturation is the same as the saturation response at $\phi=9\%$. For a probe responding to H-index only the expected value would be $\phi=13\%$ (i.e. 30% x 0.437).

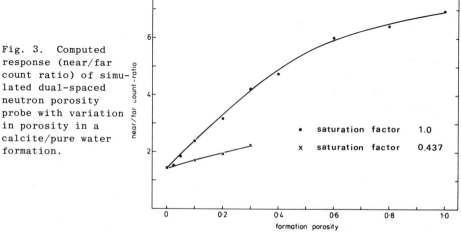

Fig. 3. Computed response (near/far count ratio) of simulated dual-spaced neutron porosity probe with variation in porosity in a calcite/pure water formation.

The difference is attributable to the fact that $\phi=30\%$ at 0.437 partial saturation is not equivalent to $\phi=30\%$ full water saturation as the former relates to a lower bulk density. The influence of the missing rock is known as the 'Excavation Effect'. The assessment of Excavation Effect is one way in which Monte Carlo calculations can be used to model a particular probe-formation system and to derive appropriate correction factors.

The n-die-away technique, using a 14 MeV pulsed n-source provides a measurement of the n_{th} decay time, τ, or its inverse the macro-absorption cross-section, Σ. If the n_{th} flux N_{t_1}, N_{t_2} refers to times t_1, t_2 after a pulse of fast neutrons where $t_2 > t_1$ and t_1 is greater than the n-slowing-down time

$$\frac{N_{t_2}}{N_{t_1}} = \exp{-\Sigma v(t_2-t_1)} \qquad \text{where } v \text{ is the } n_{th} \text{ velocity.}$$

Since the Σ of a complex material is the sum of the values of the cross-sections of the components, for a porous reservoir rock

$$\Sigma = [1-\phi]\Sigma_m + \phi[S_w \Sigma_w + (1-\phi)\Sigma_o]$$

where $\Sigma_m, \Sigma_w, \Sigma_o$ are the absorption cross-sections of the matrix rock, water and oil and S_w is the water saturation fraction of the pore space.

For saline reservoir waters Σ_w is proportional to salinity and is ~ 20 capture units (1 cu = 10^{-3}cm^{-1}) for pure water and $\sim$100 cu for water containing 220,000 ppm NaCl. For oil $\Sigma_o \sim$20 cu and for matrix rocks $\Sigma_m \sim$5-10 cu. Neutron-die-away techniques give estimates of S_w for salinities $>\sim$30,000ppm NaCl and for ϕ>15%.

Measurements of Σ are valuable for differentiating between oil, gas and water-bearing formations. They are particularly useful in detecting changes in S_w in cased boreholes during the producing life of a well where resistivity techniques are not applicable, for evaluating old wells and monitoring reservoir performances. However, for good quantitative interpretation, values of porosity and salinity are needed together with information on the lithology and on the nature of the hydrocarbons present.

Because n-die-away responds essentially to Cl content, the technique is not applicable to reservoir waters of low salinity. To overcome this problem the same basic equipment but with different timing modes and γ-ray energy windows has been developed to give carbon/oxygen ratios from the 4.43 MeV and 6.13 MeV γ-rays from (n,n'γ) reactions on C and O. In the most recent variant of this method (Westaway et al 1980), C, O, Cl, H, Si, Ca, Fe and S are observed by fast and thermal neutron capture reactions. Information on the extended range of elements allows a more comprehensive evaluation of formation factors.

4. Metalliferous and industrial minerals

Established applications in the metalliferous and industrial minerals industries are relatively few although several developments have been reported over the last decade. The most important deterrents to a wider acceptance of borehole logging are speed of response and the difficulty of providing a simple, reliable concentration measurement in the widely differing rock formations and borehole conditions encountered.

Development of borehole logging equipment for assaying U directly in roll front deposits in disequilibrium has probably received most attention. The delayed fission technique using a large ^{252}Cf source (Anon 1976) and the prompt fission technique using a pulsed n-tube have been developed. The prompt fission method has the advantage of low radiation risk. In this method a 10 μs pulse of 14 MeV neutrons ($\sim$2 x 10^6 n/pulse; 10 pps) is injected into the formation. With no U present the neutrons are thermalised in $\sim$200 μs but, with U present, fast neutrons are generated by fission of ^{235}U throughout the n_{th} lifetime ($\sim$2,500 μs). Thus, a detector sensitive to epithermal neutrons in the time window 200-2,500 μs gives a measure of U content. Corrections include a separate measurement of the macro n_{th} absorption cross-section of the formation (Humphreys et al 1983).

Nargolwalla et al (1977) using a prompt γ-ray technique quote limits of detection for Cu, Fe and Si of 0.2% in porphory Cu deposits and <0.1% for Ni in laterites. Neutron-induced γ-ray spectrometry has also been used in borehole logging to measure the concentration of S (Senftle et al 1978; Chrusciel et al 1977), Al (Senftle et al 1978), W (Fanger et al 1977) and Fe (Eisler et al 1977). Senftle et al (1983) report the use of a borehole sonde to determine the water content of a salt deposit being considered as a site for disposal of radioactive waste.

The possibility of overcoming interfering effects should be less on-line than in a borehole, but fewer applications have been reported. Basaru et al (1983) present results from measuring Fe, Al and Si in Fe-ores, Al and Si in bauxite and Mn in Mn-ores on-line. Equipment to sort lumps of rock by gold content at ~1 ppm by the $^{197}Au(n,n'\gamma)^{197m}Au$ reaction is being developed by Clayton and Spackman (1982). Several investigators have considered the possibility of measuring the Mn, Ni, Co, Cu contents of Mn-nodules in the deep ocean but operational acceptance is yet to be proven.

5. Hydrology, hydrogeology and sub-surface waters

Steady state and pulsed n-techniques have been considered for stratigraphic correlation, lithology identification, porosity, moisture content and location of water levels, for example, and there is widespread use to complement information from non-neutron methods.

Neutron logging is finding important application in the investigation and control of geothermal sites producing high pressure, high temperature steam for power generation and also at sites generating hot brines for use in space heating.

Soil moisture gauges based on n-moderation find routine use in civil engineering. The intensity of application varies in different countries.

6. Neutron sources and instrumentation

The use of ^{252}Cf and ^{241}Am-Be is widespread. ^{252}Cf is preferred when emphasis is on thermal neutron reactions, small size and low associated γ-radiation. ^{241}Am-Be finds application when range is important (oil well logging) and when fast neutron interactions (e.g. C, O) are inferred. Particle accelerators are currently being considered for plant control (Clayton and Spackman 1982) because of their ability to match n-energy to the preferred reaction cross-sections. Pulsed n-tubes have their main application in oil well and in U logging; but applications would be greater if lifetimes exceeding currently possible 100-200h could be achieved.

New scintillation detectors ($Bi_4 Ge_3 O_{12}$ and 'Polyscin') allow more efficient γ, n discrimination and higher detection efficiency and the new high purity n-type Ge allows high energy resolution and low sensitivity to n-damage. Instrumental cryogens are being introduced into field and on-site applications to give continuous operation and avoid the need for liquid N_2 supplies. New 'fast' electronics allow rapid data collection and on-line computers offer rapid data manipulation and display. All these developments are important to achieving a wide operational acceptance, more accurate analytical capability and more applications.

References

Anon 1976 Californium-252 Progress No. 20 25-9 (USERDA:South Carolina).
Bosaru M. Holmes RJ and Mathew PJ 1983 Int. J. Appd. Radn. and Isotopes 34 No.1.
Crusciel E, Niewodniczanski J and Palka KW 1977 Nuclear Techniques and Mineral
  Resources (Vienna:IAEA) p301-311.
Clayton CG and Spackman R 1982 (unpublished).
Clayton CG, Hassan AM and Wormald MR 1983 Int. J. Appd. Radn. and Isotopes 34 No.1.
Eisler PL, Huppert P, Mathew PJ, Wylie AW and Youl SF 1977 Nuclear Techniques and
  Mineral Resources (Vienna:IAEA) p215-228.
Fanger U, Pepelnik R and Michaelis W ibid 539-552.
Gozani T, Brown D and Bozorogmanesh H 1981 Proc. of Coal Technology 2 355.
Gozani T, Bernatowicz H and Brown D 1982 Proc. Met. Soc. of AIME, Dallas Feb. 1982.
Humphreys DR, Barnard RW, Bivens HM, Jensen DH, Stephenson WA and Winlein JH 1983,
  Int. J. Appd. Radn. and Isotopes 34 No.1.
Nargolwalla SS, Kung A, Legrady OJ, Strever J, Csillay A and Seigel HO 1977 Nuclear
  Techniques and Mineral Resources (Vienna:IAEA) p229-264.
Senftle FF, Tanner AB, Philbin PW, Bounton GR and Schram CW, 1978 Mining Eng. (AIME) 30 666.
Senftle FE, Mikesell HL, Doig NG and Brown FW 1983 Int. J. Appd. Radn. and Isotopes 34 No.1.
Westway P, Hertzog R and Plasek RE 1980. Proc. Conf. Soc. Petr. Engrs. Dallas Sept 21-24.
Wormald MR and Clayton CG 1983 Int. J. Appd. Radn. and Isotopes 34 No.1.
Wormald MR, Clayton CG, Boyce IS and Mortimer D 1979 ibid 30 297.

Laboratory neutrons — a breakthrough in non-nuclear disciplines

R.E. Jervis

SLOWPOKE Nuclear Reactor Facility and Dept. of Chemical Engineering and Applied Chemistry, University of Toronto, Toronto, Canada, M5S 1A4.

1. Introduction

Although in the early years, neutrons were readily available for the study of their behaviour and interactions in the various research programs of nuclear energy establishments of major industrialized nations, high intensities of neutrons were seldom accessible until the last two decades for utilization outside of such establishments or in non-nuclear disciplines, with the major exception of medicine. The first wave of research reactors: mainly, critical, pool-type assemblies, of which between 100 and 200 were installed and operated in many countries of the world between 1950 and 1970, did lead to a more widespread availability of neutrons. In many cases, the principle uses were in basic and applied nuclear science and engineering with some radioisotope production for nuclear medicine and tracer studies in other fields. For practical reasons of supply, most radionuclides used off-site were a few longer-lived ones such as C-14, P-32 and I-131. Direct uses of neutrons such as for neutron radioactivation applications, required that interested researchers had to work at the reactor site or to establish effective and reliable means of transporting precious samples to and from the site. Furthermore, research reactors of the period were often in complexity and in scale of operation, beyond the capability of university departments or institutes to manage and to utilize efficiently. As a result relatively few such studies were done by universities or personnel in research institutes or centres and neutron uses in non-nuclear disciplines were far fewer than expected. In retrospect this was both understandable but unfortunate in that many interesting and worthwhile experiments particularly in interdisciplinary fields could have benefitted greatly from neutrons and indeed much of the vigor of present neutron applications lie in such fields. In fact some research reactors particularly those dedicated to neutron physics and engineering, have closed down in the last 15 years often because of gross underutilization whereas those which have found significant uses in other fields have flourished.

An explanation, in part, of the more than exponential increase in neutron applications by other scientific disciplines during the period from 1965 to 1980 is found in the development of safe low-cost nuclear reactors and of other laboratory neutron sources. Neutron flux in the range of 10^{10} to 10^{12} n.cm.$^{-2}$s^{-1} can now be obtained and utilized without having to install and manage a multi-million dollar facility and one which requires from 5 to 20 highly trained staff for operation alone and a high annual financial operating cost. One such development in which this laboratory has participated since 1969 together with the principal inventor and developer: Atomic Energy of Canada Ltd., is the unique Canadian critical pool-type reactor called SLOWPOKE*. SLOWPOKE* is a compact, under-moderated assembly utilizing about 800 g of U-235 in enriched form as aluminum alloy together with a berllyium annular reflector. The basic design, safety features and operating characteristics of the currently-available SLOWPOKE-2 reactor have been described previously by Kay, Jervis et al, 1973. The University of Toronto SLOWPOKE-2 reactor facility is operated regularly with thermal neutron flux in accessible sample ports of from 2.5×10^{11} to 10^{12} and has been shown to operate safely up to 3×10^{12}. In such irradiation sites within a 15 cm thick Be reflector, the fast neutron flux is between 8 and 15% of the thermal flux, and the steady-state sample temperatures are often in the range 30 to 40°C during irradiation such that volatile liquids and less stable biological or organic substances can readily be left in the reactor.

2. Toronto SLOWPOKE Nuclear Reactor Facility:

The facility at the University of Toronto is licensed for continuous operation as necessary and for up to 18 hours unattended operation without the presence of any reactor operator such as for overnight or weekend use. The Toronto reactor is fitted with seven pneumatic samples lines such that up to 10 sample capsules of 7 cm volume and 4 capsules of 28 cm volume can be irradiated simultaneously and independently for periods ranging from a few seconds to many hours. The SLOWPOKE reactor plus a number of associated gamma-ray spectrometers for analysis of short-lived radionuclides has been installed on the main campus in the centre of Toronto and is situated safely within a short distance from principal traffic arteries of the city and a subway line. However this location adjacent to pure and applied science departments and associated institutes, centres and teaching hospitals of a major university have greatly facilitated use of SLOWPOKE neutrons by up to 150 university researchers from 20-30 different disciplines. These range widely from clinical and forensic medicine through the pure and applied sciences to industrial applications and innovation associated with the study of new processes and products. The recent years have proven how very much SLOWPOKE has contributed to research and teaching, and yet how safe, reliable and low cost the installation and operation have been: it has probably been as a consequence of having neutrons so readily available on the main campus of a major research-oriented university such as the University of Toronto that their uses there have grown steadily and strongly since the day of installation with a doubling period of 2-3 years. As an example in the following table details of the use of a ten-year period are summarized and divided among the types of use, whether for teaching laboratory, graduate thesis research, faculty members' research done with postdoctorates and research associates, and uses by outside, non-university groups. (Hancock, Jervis, Hewitt, 1982)

---

* SLOWPOKE is the name of a unique Canadian reactor and is an acronym for: <u>S</u>afe <u>LOW</u> <u>P</u>ower <u>K</u>ritical <u>E</u>xperiment.

Table 1

	1981/82		1980-81		1971-72	
	Uses	Hours	Uses	Hours	Uses	Hours
Undergrad.	1854	1501 (12%)	2456	2976	250	220 (25%)
Postgrad.	6514	6317 (52%)	2789	2620	309	349 (40%)
Faculty	5648	3680 (30%)	4780	3470	257	298 (33%)
Non-Univ.	2796	664 (6%)	3119	1159	61	19 (2%)
	16,812	12,162	13,144	10,225	859	886

Each use typically involved the irradiation of one or more sample specimens for NAA determination of between 10 and 25 trace components each by utilizing both short-lived radionuclides (e.g. V-51, Al-28, Cu-66, etc.) and intermediate half-life nuclides (e.g. Br-82, Sb-124, As-76, Mn-56, etc.) Thus it is seen that between 200,000 and 300,000 individual measurements are completed each year at the university using on-campus neutrons from SLOWPOKE.

3. Uses of Laboratory Neutrons From SLOWPOKE:

The topics of study involved in all these uses vary widely and are very interesting: they mainly lie at present in the earth sciences, environmental studies, materials studies and trace elements in biomedical materials. As indicated, during 1981-82 more than 160 on-going projects were assisted by SLOWPOKE neutrons involving 76 professors and 107 of their graduate students and researchers, as well as about 350 undergraduate students pursuing assigned laboratory experiments and their own research topics. Many of these are interdisciplinary researches in which teams of researchers assembled from different fields blend their knowledge and skills and in pursuing their experiments use a variety of measurement techniques including instrumental NAA.

While many examples could be given of current interdisciplinary research involving neutrons, a few are described here briefly that involve environmental dispersal of inorganic pollutants and some associated public health investigations.

Instrumental neutron activation (INAA) is well-suited by means of the measurement of short-lived nuclides, to accomplish the trace analysis of 25 to 40 trace elements in the kind of sample involved in environmental studies and these include several heavy metals and elements with non-trivial environmental toxicology. A particular advantage of a medium neutron-flux reactor is that a wide variety of sample types can be subjected to INAA under the same conditions and without any particular pre-treatment. For example, in one recent study, trace elements in oil sands, in the host sand and clay, the bitumen and product oils derived from it and in aqueous effluents streams could all be determined (Jervis, 1982). These included highly volatile and combustible fuel oils. In related studies, coal, their liquification slurries and coal-derived liquids are being analyzed by INAA (Pringle, 1982). Fly ash, solid residues and airborne particulate matter are also amenable to these techniques (Paciga and Jervis, 1976; Jervis, 1982). Aerosols collected by the use of filtration or sedimentation/fractionating air samplers can be neutron-activated directly on the collector backing material or filter provided the latter are of sufficient purity.

In table 2 are given results from the study of oil sands and the bitumen and heavy oils extracted from it and of conventional crude oil as pumped from W. Canada oil wells. The comparison indicates that many

Table 2 - Oil Sands and Derived Synthetic Crude

Element	Concentration (ppm)			
	SUNCOR Oil sand	Bitumen	Synthetic Crude	Pumped* Crude
S	8000	33000-44000	800-2100	1000-4000
Cl	35	7-31	5-6	3.8-371
Al	9100	310-1600	3	0.5-20
Mn	12	2.4-14	0.1	0.1-2.5
Na	310	30-120	7	3-5.5
V	34	145-200	0.002-0.007	0.02-6
As	0.9	0.2-0.6	<0.02	1.5
Ti	610	32-130	<0.6	7-19
Fe	550	100-600	<10	0.7
Ba	210	5-12	<0.9	6
Sc	0.5	0.04-0.3	<0.002	0.1
Cr	7.9	1.8-3	<0.2	0.09
Br	0.2	0.1-0.2	0.02-0.03	0.1-0.8
La	6.7	0.5-1.6	<0.02	-
U	0.5	0.08-0.3	<0.1	-

* Adapted from crude oil data on Cretaceous and Devonian formations in Alberta, analyzed by R. Filby and K. Shah; and recent data from C. Davis and R.E. Jervis.

impurities which are concentrated in the bitumen are greatly reduced during coking and up-grading to syncrude oil. However it should be pointed out that, effectively, the syncrude has undergone a first stage of refining compared to the as-mined crude so that the petroleum product streams derived from syncrude do not contain lower concentrations in general compared to those derived from conventional oil refining.

Studies of coals have revealed many trace inorganic impurities in addition to those which are generally considered to make up the incombustible ash component of coal that constitutes up to 25% by mass of some low-grade thermal coals (table 3). Recent studies of major Canadian coal deposits have shown large variation in the trace elements, as seen from the results for Br (viz. a range from 0.1 to 3550 ppm), K (70 ppm to 3%) and As (0.3 to 320 ppm). The value of using a simple, direct instrumental technique such as INAA with SLOWPOKE neutrons is seen when the total number of elemental determinations involved in a national coal survey to obtain representative results, is fully appreciated.

Table 3 - Representative Canadian Coals

Element	No. of Det.	Concentration (ppm)	Median
Al%	60	0.2 - 3.9	1.1
As	56	0.3 - 320	12
Ba	60	13 - 3900	124
Br	50	0.13 - 3550	12
Ca	59	284 - 18180	1754
Ce	55	1.8 - 52	12
Cl	60	16 - 3710	829
Co	56	0.3 - 18	2.6

(Table 3 continued next page)

Element	No. of Det.	Concentration (ppm)	Median
Cr	59	0.16 - 35	10
Cs	56	0.02 - 4.1	0.60
Dy	60	0.2 - 2.6	0.98
Eu	52	0.04 - 0.8	0.28
Fe%	60	0.15 - 9.57	1.06
Hf	55	0.11 - 4.9	0.60
I	56	0.2 - 4.0	1.1
K	58	74 - 29000	1061
La	59	1.3 - 35	6.9
Mg	58	50 - 5000	598
Mn	60	6 - 285	80
Na	60	43 - 8361	457
Rb	31	1.3 - 30	9.4
Sb	59	0.08 - 23	0.78
Sc	60	0.5 - 8.6	2.5
Sm	60	0.3 - 16	1.2
Sn	25	8 - 267	98
Sr	55	8 - 735	89
Ta	41	0.02 - 1.2	0.15
Tb	19	0.02 - 0.5	0.25
Th	55	0.3 - 14	2.1
Ti	60	50 - 1950	542
U	60	0.1 - 4	0.84
V	59	2.4 - 87	17

Current graduate research on coal liquifaction (Pringle, 1982) have included measurements of the mass balances of many of the trace constituents of the coal, several of which are suspected of playing a role as auto-catalyst of a solvent-promoted coal liquifaction process. Preliminary results indicate that much of the mineral matter in coal remains in dispersed solid form throughout; however most of the halogens are co-extracted with the coal liquid. Of some environmental importance when coal is used directly without up-grading, such as liquifaction, is the fate of trace impurities upon combustion. In table 4 is given a summary of findings for Canadian sub-bituminous coals in which it is evident that appreciable fractions (viz. >10%) are volatilized of more than half of the inorganic trace elements measured including the halogens, mercury, selenium and some of those elements thought of as mineral-derived: the alkalis, Ba, Sr, V and Sn.

Table 4 - Effect of Static Coal Ashing

Mostly volatilized (75-100%)	Cl, Br, I, Se, Hg
Partially volatilized (25-75%)	Na, K, Sb, Sc, Sr
Mostly retained in ash (10-25% volatilized)	Ba, Sn, Hf, Eu, U
Retained in ash (<10% volatilized)	Al, Ca, Ce, Co, Cr, Cs, Dy, Fe, Mg, Mn, Ta, Ti, Th, V, Zr

An approach made possible by multi-elemental NAA is the evaluation of inter-element ratios such as V/Al and Mn/Al (Paciga and Jervis, 1976) which can point to particular dominant sources of airborne contaminants. Furthermore, distinction between natural dusts and those arising from combustion, metal refining or abrasion can be made through inter-element ratio normalization (Paciga and Jervis, 1976) in which normalization of the relative concentrations of trace elements in the atmosphere to those of elements such as Al, Sc or Fe in proportion to their abundance in soil and rocks can show substantial contributions from industrial activities. Other published environmental studies from this laboratory of research done by INAA in cooperation with ecologists, botanists, environmental health specialists and chemists have involved studies of aerosols and depositing dusts, vegetation, soils, foodstuffs, water and human tissues such as hair and blood. (Roberts, 1974)

4. Summary:

These few examples of work done at Toronto suggest that the availability of laboratory neutrons has greatly facilitated interdisciplinary applied research there. Large numbers of samples, rather complex in nature, have been amenable to INAA such that more than 200,000 trace analyses are done per year using a small reactor with a very limited number of sample irradiations sites. Estimates of the approximate cost of these analyses lie in the range of $0.50 - 1.00 per result whereas most comparable determinations by other instrumental approaches cost more than $5.00 per analysis and for some less common elements, may range to $25.00, even for large numbers of samples. Thus, laboratory neutrons have not only been made readily available and safely, but they provide for rapid, fairly accurate, low cost instrumental trace analysis.

5. References:

Hancock, R.G.V., Jervis, R.E., Hewitt, J.S., 1982, SLOWPOKE Report, University of Toronto.
Jervis, R.E., Ho, K-L., Tiefenbach, B., 1982 J. Radioanal. Chem., 71 No. 1-2, 225.
Kay, R.E., Stevens-Guille, P.D., Hilborn, J.W., and Jervis, R.E., 1973 Internat. J. Appl. Rad. & Isotopes, 24, 509.
Paciga, J.J., Jervis, R.E., 1976 Environ. Sci. & Technology, 10 1124.
Pringle, T.G. 1982 Redistribution of Trace and Minor Elements During Coal Liquifaction, M.A.Sc. thesis (unpublished).
Roberts, T.M., Hutchinson, T.C., Paciga, J.J., Chattophyay, A., Jervis, R.E., VanLoon, J., Parkinson, D.K., 1974 Science 186 1120; Inst. For Environmental Studies Report EE#1, Univ. of Toronto, Canada, 1975.

Neutron techniques in safeguards

Martin S. Zucker

Brookhaven National Laboratory, Upton, New York, U.S.A.

Abstract. An essential part of Safeguards is the ability to quantitatively and nondestructively assay those materials with special neutron interactive properties involved in nuclear industrial or military technology. Neutron techniques have furnished most of the important ways of assaying such materials, which is no surprise since the neutronic properties are what characterizes them. The techniques employed rely on a wide selection of the many methods of neutron generation, detection, and data analysis that have been developed for neutron physics and nuclear science in general.

The term 'Safeguards'[1] refers to a continuing and developing effort by national and international organizations to prevent by certain control measures the diversion to illicit use of 'special nuclear material' ('SNM'). Usually SNM refers to various fertile or fissile nuclides, especially U or Pu, but is also interpreted to include substances such as D_2O or graphite useful for production of such nuclides. The control measures being developed are an integrated triad of physical security, classical accounting, and the actual assay of the flow of SNM in its various forms into, out of, and within nuclear plants and installations. The goal of assay systems is a material balance whose uncertainty is less than a 'strategic quantity'. No law of nature guarantees that this goal is attainable; indeed, more often it may not be. Thus the importance of considering Safeguards as being composed of the three mutually supporting factors, each insufficient by itself.

The radioactivity of typical SNM makes non-destructive assay ('NDA') attractive since it can be accomplished without breaching containment, and nuclear NDA methods usually the obvious choice. Most of the nuclear NDA involves neutrons in a significant way, not surprising since it is the unusual manner in which neutrons react with them that defines SNM. Though this paper discusses only neutron techniques it should be noted that other techniques, nuclear or not, are often needed to provide auxiliary information needed to interpret neutron technique derived data. For example, interpreting a spontaneous fission rate would require isotopic ratios obtained through gamma or mass spectrometry.

In the ~15 years of Safeguards instrumentation development it seems that virtually every trick known to neutron, nuclear, or reactor physics, and in various combinations, has been tried as a way of assaying the various forms in which SNM appears in the nuclear economy: feed stock or raw material, scrap, waste, finished product (nuclear fuel rods, plates, or assemblies), spent fuel, and processing or recovery solutions. The most manageable classification of the myriad of schemes seems to be in terms of the incoming or interrogation radiation, its energy spectrum and means of production: accelerator, reactor, isotopic or none (relying on spontaneous fission), and the outgoing radiation, and its manner of detection, including the electronic processing of the detector pulse train. We elect here to first consider three techniques of widest applicability and importance as primary examples of neutron safeguards techniques.

(i) Neutron correlation[2] was originally developed to take advantage of the multiplicity of neutrons emitted in the spontaneous fission (s.f.) of the 'even' ('fertile') nuclides such as ^{238}U and ^{238,240,242}Pu, expressed as the probability P_ν that ν prompt neutrons are emitted in a time $\sim 10^{-15}$s. The point of using the number of fissions rather than just the number of neutrons as an assay 'signature' is that these nuclides are α emitters and therefore much or even most of the neutrons from the sample may be due to (α,n) reactions with light elements present perhaps incidentally, e.g. F, O, Al, etc. These reactions will depend on the chemical and physical composition, rather than only on the SNM. In fact, the problem is to separate out the bursts of 'correlated' neutrons, from fissions that are randomly distributed in time, from the 'uncorrelated' (α,n) neutrons (and those from background), which occur singly and also at random. (However when chemical and physical form are rigidly controlled, and background minimal, 'gross n' counting can be satisfactory.)

The basic neutron correlation detector, the 'well counter', is a hollow cylinder of moderator with (gamma insensitive) ^{10}BF$_3$ or ^{3}He proportional counters embedded. The well is used to hold the sample. The purpose of moderator is reducing the energy of the neutrons to where they react efficiently with the ^{10}B or ^{3}He. The fast neutrons will slow down in a few μs, followed by diffusion in the counter until they either escape, are captured by the moderator, or react with the ^{10}B or ^{3}He, giving rise to pulses. The time τ_o for diffusion, $\sim(15-150)$μs depending on the size, material, and geometry of the detector assembly, is the lifetime of neutrons introduced into the well. Thus, in a trade-off, to obtain efficient detection, the original $\sim 10^{-15}$s distribution has been broadened by the stochastic slowing down process to a distribution of the form $\tau_o^{-1} \exp(-t/\tau_o)$. A kind of correlation performed on a pulse train formed from the added outputs of all the detectors now becomes preferable to coincidence, since τ_o is much greater than the pulse width.

An incoming pulse causes the electronics to interrogate what is in effect a delay line of length τ (chosen to be of the order of τ_o) as to how many pulses preceded it in that time. A tally of the number of such preceding pulses is kept and compared with a similar tally of the pulses in an identical delay line which is separated from the first by many times τ_o. Pulses in the first delay can have two origins. Either they originated in the same fission as the incoming pulse (are correlated) or did not and are uncorrelated. Pulses in the second delay however must be uncorrelated. The difference or 'net' count N then depends on the correlated count alone. A proportionality between N and the number of fissions can be derived from first principles in terms of the P_ν and half-life for s.f. of the nuclides, the detection efficiency ε and neutron lifetime τ_o, τ and other parameters of the electronics which can be measured independently. The theory starts to break down for two different reasons as sample size increases. For large enough samples there is an appreciable probability that neutrons from a fission will induce further fission, so that the number of fissions is no longer in one to one correspondence with the amount of SNM. However,[3] often the region where the linear calibration holds can be extended by introducing a 'multiplication' factor M such that the neutron emission multiplicity ν is replaced by $M\nu$; otherwise the calibration must be done empirically. Analogous to the way in which a radioactive source emitting coincident radiation can be calibrated absolutely without knowledge of the detector efficiency, it can be shown that G^2/N, where G, the 'gross' count, is also independent of detector efficiency to a high degree and so this can be used to reduce geometry and matrix effects. The other flaw is that for very high count rates particularly if largely uncorrelated, even if the electronics performs as designed, the relative statistical precision in N suffers. This is affected by ε and τ_o, and there are natural limits to further improvement.

Neutron correlation has been extended to 'odd' ('fissile') nuclides principally ^{235}U by using an external isotopic neutron source to induce fission ('active' mode) as well as the 'passive' mode described above, and the basic well counter has been proliferated into many specialized forms for particular applications such as fuel rods, assemblies, raw material, scrap, and waste of both Pu and U.

(ii) Fast coincidence techniques[4] employing plastic scintillator-photomultiplier detectors offer an interesting contrast with the preceding as a way of getting a signature for fission. The detectors are sensitive to fast neutrons and gammas and so take advantage of the fact that $\sim$2.5n and 7.5γ are produced within 10^{-15}s of each fission. The coincidence resolution time is however limited to (10-50)ns or more by the transit time of the radiation in the sample volume and detectors, whose dimensions are typically in the range (0.2-2.0)m. Since fission gammas tend to be less penetrating than neutrons, some instruments try to rely solely on neutrons by interposing Pb shielding to lessen sensitivity to 'matrix'. This does however limit count rate, hence the statistical precision for a given sample size and assay time, and, since only two-fold coincidence can be used, increases sensitivity to background. The calibrations tend to be more linear and less matrix dependent than instruments that also accept gammas. With the latter three-and even four-fold coincidences can be used and still obtain useable count rates.

Usually these instruments are used in the active mode, typically with a 'subthreashold' isotopic source such as ^{241}Am-Li(α,n), $\bar{E}_n \sim$.4MeV, or ^{124}Sb-Be(γ,n), $\bar{E}_n \sim$.025 MeV, although 'superthreshold' ^{252}Cf(s.f.,n), $\bar{E}_n \sim$2.3MeV, is also used. In principle, by using these alternately, information on both the even and odd nuclides would result, since the superthreshold source would induce fission in both.

In contrast to correlation, where all counters form one superdetector, in coincidence the total detector volume is divided among several channels, with a resulting possible trade off between signal (true coincidences) to background (accidentals) ratio, and statistical precision. Also there is no simple theory or calculation of response. The calibration is purely empirical, using functions of the form $Am(1+Bm)^{-1}$ or $A(1-e^{-Bm})$ with m the mass of SNM, and A and B determined from standards and strongly matrix, geometry, and 'interrogation' source dependent. Large samples cause problems because of multiplication, and also from absorption by the sample of ingoing and outgoing radiation; this latter leads to a lack of sensitivity (decreased slope or 'saturation') in the calibration. An interesting recent development, analagous to using G^2/N in correlation, is to use as a signature various algebraic combinations of the various multiplicities of true coincidences available in a several-fold coincidence circuit to derive quantities proportional to the amount of fission but independent of detector efficiency, alleviating geometry and some matrix effects. Thus, e.g. for a 3-fold arrangement, instead of using just N_{123}, quantities such as $(N_{12}N_{23}N_{31}/N_{123}^2)$ and $(N_1N_2N_3/N_{123})^{1/2}$ are used as indicators.

(iii) Bombardment with neutrons from an isotopic source followed by observation of the delayed gammas from the induced fission is an important technique with nearly all fuel rods assayed this way. The rods are moved at a controlled speed past a collimated neutron source, ^{252}Cf or ^{238}Pu-Li(α,n) is favored, so that only a small length, say $\sim$1 cm, is irradiated. Down stream a collimated gamma sensitive detector picks up the induced activity. The method is sensitive enough to detect individual pellets of incorrect enrichment, as well as assay the total content in minutes per rod, and is an outstanding example of how the manufacturer's interest in quality control often coincides with Safeguards.

The above methods implemented in various ways make up the bulk of applications, but by no means all.[5] In the following several others will be described briefly for general interest because of the physical principles entailed, but with relatively less claim for universality or importance:

(i) Accelerators, and accelerator and isotopic pulsed sources. Pulsed source techniques have the virtue that the detectors can be operated to sense radiation emitted from the sample when the interrogation source is not operating, in principle enhancing the signal to background ratio. Certain accelerators are readily operated in a suitable pulsed mode. 'Sealed beam' devices, miniaturized Cockroft-Walton generators of the type used for oil well logging, have been used to provide $E_n \sim$14 MeV via the T(d,n)^{3}He reaction. Electron linacs developed for industrial or medical radiography have been used in several ways. Electron energies of $\gtrsim$5MeV can be used to make Bremmsstrahlung gammas capable of inducing photofission directly in the SNM sample, or produce from a ^{238}U target neutrons then used to bombard the sample; lesser electron energies can be used to make photoneutrons from D or Be targets. Van de Graaff accelerators are particularly flexible in the wide range of energies of monoenergetic neutron yielding reactions, e.g., T(p,n)^{3}He, ^{7}Li(p,n)^{7}Be, D(d,n)^{3}He, T(d,n)^{4}He, etc. that can be used with them. Isotopic sources can also be pulsed. In the 'Shuffler' arrangement, an intense ($\sim 10^9$ n/s) ^{252}Cf(s.f.,n) source briefly irradiates the sample and is then rapidly withdrawn into a neutron shield by a flexible cable. Another method uses a gamma source passing through a hollow Be (or D containing) cylinder. Most used is ^{124}Sb yielding $\sim$0.025 MeV neutrons with Be. The detectors are of moderator pierced by BF$_3$ or ^{3}He proportional counters surrounding the volume where the sample is introduced.

Whatever the pulsing method, the delayed fission neutrons (and or gammas) can be used as a signature. Since the delayed neutrons are only $\sim$1% of the prompt, the advantage of the interrogation neutrons not being present is only partly realized. In fact, 'room return' and the 'on/off' ratio of neutrons from the source become important considerations. Typically, irradiation is for $\sim$1s, followed by a several second observation of the delayed neutron decay. The initial slopes of the properly normalized decay curves is found to be different for ^{238}U compared to ^{235}U or ^{239}Pu.

It is also possible to use the prompt neutrons as a signature, but record them after the interrogating pulse of neutrons. The 'slab detector', consisting of a flat array of BF_3 or ^3He counters in moderating material, has a characteristic neutron lifetime. The prompt neutrons entering the slab during the irradiation time therefore will give rise to an exponentially decreasing count rate, conveniently viewed on a few hundred μs time scale. This takes advantage of the relative intensity of the prompt group and the fact that the neutron source is off during the counting period, but does require more sophisticated electronics, e.g., to desensitize the counters during the massive initial pulse, so that they will recover soon enough after the pulse to take advantage of the maximum signal-to-background ratio available then.

Up to now, accelerators have not been very successful in industrial settings because of insufficient reliability and lack of qualified operators, while isotopic sources have reliable and predictable behavior and are comparatively inexpensive, and so have been favored. However, recent availability of better medical linacs, and significantly improved sealed beam sources, combined with a growing need to assay large containers (210 ℓ drums or boxes 1-1.5m on a side) which requires the greater penetrability of a more energetic neutrons source, may change this. Van de Graaff accelerators do not seem as amenable to industrial use and so their greater versatility may be confined to laboratories.

(ii) The lead 'slowing-down' spectrometer has been used to assay fuel 'rods'. A pulsed 14 MeV neutron beam from a sealed beam source enters a block of lead ∿2m on a side. The neutrons tend to slow down as a group in such a medium with their average energy a function of time. Fissions induced in the rods inserted into the lead block are detected with proton recoil counters sensitive primarily to the high energy fission neutrons. Cross sections for induced fission are energy dependent, with differences among the nuclides. Thus, the fission neutron production will be a function of time, which, normalized with respect to the slowing-down neutron flux (measured with a BF_3 proportional counter) provides the total fissile content and isotopic ratios.

(iii) An interesting illustration of neutron methods in Safeguards is furnished by recent work to develop NDA techniques for measuring D_2O/H_2O concentration of mixtures in 210 ℓ drums without opening them. Several neutron based nuclear methods were tried and all proved basically workable: (a) transmission of neutrons from an ^{241}Am-Li source through the drum, taking advantage of differences in the scattering cross section of D vs. H, (b) the detection of 2.226 MeV gammas from the $H(n,\gamma)D$ reaction, which clearly indicates the presence of H, again using an isotopic source for the neutrons, (c) neutron detection from $D(\gamma,n)H$ reactions using 2.61 MeV gammas from $^{208}Tℓ$ to bombard the drum (d) measurement of the lifetime of a neutron bunch injected into the drum from a pulsed accelerator, this lifetime being a delicate function of the respective absorption cross sections, transport mean free paths, etc., of D and H.

It is difficult to quote typical accuracies (which involve random and systematic errors) or precisions (stability or reproducibility), or the ultimate sensitivity, of a particular technique. This is because they are usually extremely dependent on the nature of the sample including total amount, concentration, matrix, isotopics, physical or chemical form, possible interferences, and the availability of suitable standards. Establishing such accuracies and precisions is in fact a major part of the Safeguards instrumentation effort, with obvious implications as to how well the system can work. At their best, these methods can achieve accuracies and precisions of a few tenths of a percent, more typically (1-10)%, at their worst, say (20-50)%. Nature is, however, often kind in that the poorer results are usually in those cases where the material is very diffuse and poorly characterized, therefore of low monetary and/or strategic value, and the quantities need not be so well known. Ultimate sensitivities to SNM are often measured in mg to dg, and the largest assayable quantities in a few kg, though the best accuracy doesn't go with such extremes. The minimal assayable amounts depend on the allowable assay time and instrument environment, and the development of specialized versions of the basic instruments will aid assay of both minimum and maximum quantities.

This report covers work done at the U.S. DOE laboratories (principally LANL, but also BNL, ANL, LLNL, and Mound), the members of the ESARDA consortium (principally JRC ISPRA, KfK, and Harwell), under the auspices of the US NRC, and certain U.S. Companies: SAI, IRT, and NNC.[7]

Notes and References

1. As in other fields, a mostly self-explanatory jargon has developed; such words will be introduced with quotation marks ' '.

2. Selected references on neutron correlation counting by M. Zucker, et al. are (a) "Neutron Correlation Counting", Proc. ANS, May 15-17, 1978; (b) "Assay of Low-Enriched Uranium........", Proc. of the 2nd Annual ESARDA Conf., p. 313, May 1981; (c) "Apparatus Characterizations as a Standard for Neutron Correlation Counting" Proc. of the 4th An. ESARDA Conf., p. A1-A10, April 27-29, 1982; (d) R. Sher, "...Characteristics of Neutron Well Counters...." BNL 50332, 1972.

3. Selected references for multiplication are (a) N. Ensslin, et al., "Self-Multiplication Corrections Factors for Neutron Coincidence Counting", Nuclear Materials Management, Vol. VIII, no. 2, p. 60, 1979; (b) M.S. Krick, "Neutron Multiplication Corrections, etc." LANL Report LA-8460-MS, 1980; (c) see also ref. 2(b).

4. The best overall single reference for this section and the following (which contains exhaustive specific references to work at LANL, IRT, etc.) is T. Gozani, "Active Nondestructive Assay of Nuclear Materials" U.S. NRC, Jan. 1981 (NUREG/CR-0602, SAI-MLM-2585). We regard this also as an excellent review of the basic physics and technology involved in Safeguards. See also Sher and Untermyer, "The Detection of Fissionable Materials....", ANS monograph, 1980.

5. The wide range and scope of safeguards methods as revealed by available instrumentation, both neutron based NDA and other methods as well, is illustrated by Fishbone and Keisch "Safeguards Instrumentation A Computer-Based Catalog", BNL Report BNL 51450, August 1981.

6. Fainberg, Zucker, et al. "Assay of Heavy Water..." 3rd ESARDA Symposium May 1981. Though technically and aesthetically satisfying, the nuclear methods proved inferior to a non-nuclear technique (measurement of the acoustic velocity) as far as field applications go, showing the danger in riding one's hobby horse too hard!

7. In preparing this paper we appreciated the efforts and consultation of T. Gozani (SAI), H. Menlove (LANL), and L. Kelly (BNL).

Inst. Phys. Conf. Ser. No. 64: Section 8
Paper presented at Conf. on Neutron and its Applications, Cambridge, 1982

Neutrons in medicine

J.F. Fowler,

Gray Laboratory of the Cancer Research Campaign, Mount Vernon Hospital, Northwood, Middlesex HA6 2RN.

Abstract. Neutrons have had a profound impact on medical science. The following applications can be listed:
(1) Production of radionuclide tracers.
(2) Production of radionuclides for treating cancer (e.g. Co-60, I-125, Cs-137, Ir-192) and earlier P-32, Au-198, I-131.
(3) Activation analysis.
and the direct use of neutrons for treating cancer. It should be remembered that about a quarter of the population of Western countries gets cancer, so that even small improvements in treatment would help large numbers of people.
(4) Neutron-emitting californium, Cf-252.
(5) Boron capture therapy with slow neutrons.
(6) Fast neutron radiotherapy.
I shall concentrate on the last two aspects.

1. Introduction

The discovery of the neutron in 1932 was followed at once by scientific curiosity about its biological effects and the active interest of Rutherford in whether neutrons would be useful in the treatment of cancer. Sir Cuthbert Wallace at Mount Vernon Hospital recruited from the Cavendish Laboratory a postdoctoral physicist with experience of measuring radiation doses, and L.H. Gray came to Mount Vernon in 1933. He was in charge of the physics department and was one of the first Hospital Physicists. He went back to the Cavendish in 1935 to irradiate a rat, some bean roots, some tissue cultures and chick embryos with neutrons produced by a D-T generator under M.L. Oliphant's supervision. No significant results were obtained and it was evident that neutron radiobiology could not be started on a part-time visiting basis.

When John Read, a New Zealand physicist with experience of constructing a million volt X-ray tube, came to Mount Vernon in 1934 he joined forces with Hal Gray to request £500 to build the first neutron generator designed for biological research. The British Empire Cancer Campaign (as the Cancer Research Campaign was then called) granted this sum and Gray and Read set to work to build it with their own hands. The hospital gave them a wooden hut and in order to save money they built their own diffusion pumps and wound their own transformers. They sawed up angle iron and cemented water pipes into the floor. In spite of a supplementary grant of £100 in 1937, they did not finish the neutron generator until 1938. They only had a few years before the war broke up their partnership. There is a sense in which fast neutron radiobiology is still suffering from this lack of continuous

0305-2346/83/0064-0469 $02.25 © 1983 The Institute of Physics

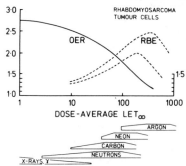

Fig 1 Variation of RBE and OER with LET for cells of an experimental tumour in rats (from Curtis, pers comm). The upper broken curve is for hypoxic cells, showing the higher RBE than for well-oxygenated tumour cells (lower dotted curve)

early development. At the same time, John Lawrence in Berkeley, California, began to do neutron radiobiology using the cyclotron that his brother Ernest had built. Gray and Read invented the "energy unit" for measuring radiation dose which has developed into the international unit of absorbed dose, the Gray, one joule per kilogram of matter.

Gray and Read demonstrated a high radiobiological effectiveness (RBE) of neutrons and alpha particles for mitotic delay in chick embryo cells and in bean roots. This is one of the potential advantages of fast neutrons for the treatment of cancer, and they irradiated tumours transplanted into the flanks of mice. They also showed the fall of RBE between neutrons and alpha particles, when the overkill region of high LET was reached. The other potential advantage, a lower oxygen enhancement ratio (OER) with neutron than with gamma ray photon irradiation, was pointed out by Gray and others in 1953.

2. The Potential Advantages of High LET Radiation (i.e. High Ion Density Tracks)

There are in principle two main biological effects which provide a potential advantage for the treatment of solid human tumours with neutrons or heavy charged particles.

(1) The reduced oxygen enhancement ratio (Fig 1, Curtis pers comm).
(2) The reduced variation in radiosensitivity with age of cell in the cell cycle (Fig 2, Raju et al 1975; Sinclair 1970).

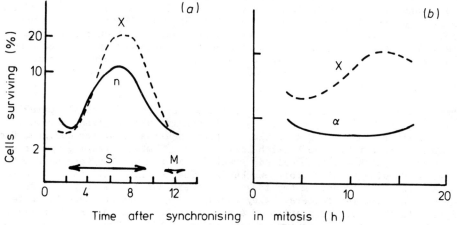

Fig. 2 Cell survival after constant doses of radiation given at different phases of the cell cycle (a) X-rays and neutrons (b) X-rays and α particles.

The first potential advantage depends on the presence in tumours of hypoxic cells because the blood supply in tumours is disorganised. The hypoxic cells are much more resistant to low LET radiation, by a dose ratio of about three, than are well oxygenated cells, i.e. cells in normal tissues. This is a disadvantage in the treatment of cancer with X or gamma rays. It is not possible to increase X-ray doses by a factor of three to kill hypoxic cells, because even an increase of 10% will cause unacceptable damage to normal tissues. This disadvantage is of course smaller for neutrons because the OER is smaller. A "hypoxic gain factor" (HGF) has been defined, being the OER for X-rays divided by the OER for the high LET radiation.

Fig 3 shows the maximum possible values of HGF on the right of the diagram: being 1.4 to 1.8 for neutrons of various energy (Withers and Peters 1979). The values of HGF decrease, of course, as the proportion of hypoxic cells in a tumour decrease. In an extreme case, if all the hypoxic cells were eliminated (by reoxygenation of the surviving cells as the treatment progressed), the HGF would fall to unity, i.e. no gain. We do not yet know anything about rates of reoxygenation in human tumours but they are very variable in mouse tumours. Attempts are now under way to measure the proportion of hypoxic cells in tumour biopsy specimens, using either radio-actively labelled nitroimidazoles or fluorescent dyes which are fixed preferentially in hypoxic cells. Achievement of this aim will be an important step, because then patients can be selected in whom benefit would be expected from the use of neutrons.

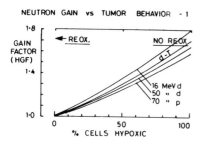

Fig 3 The calculated decrease of gain factor with decrease in the proportion of hypoxic cells in a tumour (after Withers and Peters 1979). The four curves are for different energies of fast neutron beam, as indicated.

The second potential advantage is more difficult to predict quantitatively. It depends on the fact that cells are particularly radio-resistant to low LET radiation at certain phases of the cell cycle. If a tumour contains a high proportion of such cells, it cannot be so well treated with X-rays as with high LET radiations; this much is obvious. A "kinetic gain factor" (KGF) has been defined as the average RBE of the total population of tumour cells, divided by the RBE of normal tissues, with respect only to the relative sensitivities in different phases of the cell cycle. The present evidence points to slowly growing tumours as those with the most resistant cells (Batterman et al 1982) and hence with the highest potential KGF obtainable from the use of neutrons.

However, if the tumours are kinetically active, they might redistribute the resistant cells into a more sensitive phase before the next daily dose; there would then be a reduced gain from the use of high LET radiation.

Fig 4 illustrates possible values for this kinetic gain factor, showing the decrease in gain, and even its change into a disadvantage, if the tumour cells are quickly redistributed into X-ray sensitive phases (Withers and

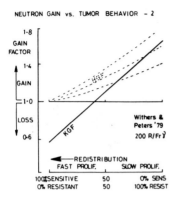

Fig. 4 Kinetics gain factor (full line) as a function of the proportion of low-LET-resistant cells in population. As this proportion falls, due to redistribution through the cycle before the next dose fraction, KGF obviously decreases. It can fall below 1.0, becoming then a disadvantage. Both KGF and HGF (dotted) fall for "kinetically active" tumours.

Peters 1979). It would be expected that the less kinetically active tumours would also reoxygenate less rapidly, as shown by the dotted lines in Fig 4. There is no contradiction here in that kinetically inactive tumours appear to offer the biggest gains for neutron radiotherapy, but we do not yet know how to measure this reliably.

So much for the principles. Let us look at the practice of neutron radiotherapy.

3. Boron Capture Therapy with Slow Neutrons

If a drug containing ^{10}B can be concentrated in malignant tumours, then flooding that region of the body with thermal neutrons will release more radiation energy in the tumour than in normal tissues:

$$^{10}B \ (n, \ \alpha) \ ^{7}Li + 2.4 \ MeV$$

The alpha particles have a short range (a few μm) and a high LET so they are very damaging to the tumour cells.

This method was tried at Brookhaven, USA, in the 1950's and has been revived at Tokyo, Japan, in the last decade. A beam of thermal neutrons can be extracted from a reactor in the usual way. The physical problem is that a thermal neutron beam is rapidly attenuated by elastic collisions with the hydrogen in tissue, so that the flux falls to half in 2 cm. For a tumour 8 cm deep, in the middle of the brain, the beam would be attenuated to about 1/16th. Even applying the beam twice, from opposite sides of the head, would give a depth-surface ratio of only 1/8. The pharmacological problem is to concentrate the boron in the tumour with a better ratio than 8:1. There are no known drugs which achieve a ratio better than 3 or 5:1 and in brain these ratios are at their most favourable because the blood-brain barrier to drug diffusion is intact in normal brain but defective in tumours. The results obtained at Brookhaven, on several dozen incurable patients, were so poor - not surprisingly, in view of the physical and pharmaological factors given above - that the method was discontinued (Farr & Yamamoto 1961).

Recently, however, results from Tokyo have reported longer survival of patients treated by boron capture therapy at the time of brain surgery than by conventional cobalt-60 and surgery and we await further results with interest (Hatanaka et al 1979). Clinical trials take a long time to mature.

The physical problem would be largely overcome if epithermal neutrons were used instead of thermal neutrons in the incident beam. They would penetrate throughout a head before being thermalised (Fowler 1981). However, a reactor with a very high flux is needed for sufficiently short irradiation sessions.

The disadvantages are the difficulty of ensuring a uniform and known high concentration of boron in all parts of the tumour; the scattered neutron and gamma doses to other tissues; and the inconvenience or serious trauma of taking patients out of hospitals to distant reactor sites.

4. Fast-neutron Radiotherapy

Fast neutrons are now in clinical use in about two dozen radiotherapy centres in ten countries, some using D-T machines and some cyclotrons. The emphasis is now on high-energy cyclotrons (40-60 MeV protons on Be targets) which are based in hospitals and have steerable treatment beams.

The first clinical treatments of cancer with fast neutrons were done at Berkeley, California, using the 37-inch and 60-inch cyclotrons in 1938 - 1942. They were based on radiobiological experiments on three types of animal tumour which showed higher RBEs than for normal tissue effects (Lawrence et al 1936; Lawrence 1967). About 250 patients were treated A few remarkable cases of tumour control were achieved but many cases of severe late damage to normal tissues were reported (Stone 1948) and the method was not used for twenty years.

Radiobiological work on the skin of pigs at Hammersmith Hospital, London, and on human cells in tissue culture at Rijswijk, the Netherlands, showed why this late damage was so severe, at least in part, and suggested a way of overcoming it (Barendsen et al 1963, Fowler and Morgan 1963). The RBE was shown to increase with decreasing dose per session, because the dose-response curve is relatively straight for neutrons but has an initial inefficient "shoulder" for X-rays. Stone and Lawrence had obtained clinical values of RBE for skin reactions in patients using large single doses from the 37- inch cyclotron. When the 60-inch Crocker cyclotron became available they used 14 to 18 small doses, spread over several weeks in sometimes irregular schedules. It was inevitable that the RBE values were higher after these small doses per fraction.

What is still controversial is whether the RBEs for late effects are consistently higher than for aute effects. This is where the absence of continuing radiobiological experiments from the early days is missed. A careful review of Stone's clinical records found that the early reactions had also been excessive and no evidence was found for significantly higher RBEs for late injury (Sheline et al 1971). Animal experiments have repeatedly failed to show higher RBEs for late than early damage to normal tissues, except some experiments using a large number of very small fractions. Recent clinical results have also shown discrepant results, some centres reporting worse late damage than expected from the early reactions and some not.

It now appears possible - although we await definitive confirmation - that at unusually low doses per fraction normal tissues have a greater capacity to repair late than early radiation injury after low LET irradiation. The same differential does not apply after high LET radiation, so that neutrons are equally damaging to late and early reactions. Thus it is not that neutrons are surprisingly harmful, but that X-rays are surprisingly ineffective, for late damage particularly. Provided that tumours react like acutely-reacting tissues, and provided that the acute reaction in normal tissue can be minimised, this gives X-rays an advantage which has only just been realised and has as yet only been exploited empirically (Fowler et al 1982, Withers 1982).

As I said, we await definitive confirmation of this difference. It will be ironical if the neutron clinical work thus shows up an advantage of low LET radiation instead. But it would be an advantage none the less, if it turns out to be true. It will require some organizationally difficult changes in radiotherapy procedure to exploit such a difference - to treat each patient 2 or 3 times a day instead of daily.

The way in which neutrons could be used constructively in this context is to use shorter overall times than the usual 4 to 7 weeks, so as to avoid the tumours growing during this rather long treatment time. Non-standard scheduling is indicated: for example twelve sessions in 2 weeks or shorter, instead of in four as at present. Two fractions per day could be used. Acute normal tissue injury would be the limiting factor, but this could be allowed for in the pilot studies leading to the selection of the dose level to be used. Late damage would not be made worse by the shortening of overall treatment time, and the tumours should be treated more effectively.

5. Clinical results from fast neutron radiotherapy

Clinical results to date have been controversial. It is clear that neutrons, in common with any other new modalities, do not provide an instant cure for cancer. Fig. 5 represents the well known results from Hammersmith Hospital of the treatment of advanced tumours of the mouth, nose and throat using neutrons from the MRC cyclotron (16 MeV deuterons on a Be target). The results showed a persisting control of the tumours for 54/71 patients treated with neutrons but only 12/62 patients randomized to the conventional photon treatments. This difference 76% against 19%, is highly significant (Catterall et al., 1975, 1977). The patients all had very advanced cancer, so most of them died soon of distant metastases and a

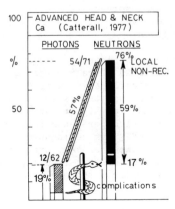

Fig. 5 Results of the clinical trial of neutron against photon therapy for advanced head and neck tumours at Hammersmith Hospital (Catterall et al 1975, 1977). The ladder represents the gain and the medical snake represents the loss due to side-effects (from Fowler 1981).

difference in survival was neither expected nor observed as a result of the two different forms of local treatment. Local control of tumours was unusually low in the photon group; about three quarters of these patients were treated in other radiotherapy centres by conventional methods and some photon-treated patients received biologically lower doses than most of the neutron-treated patients received. This point is sometimes criticised, but the criticism is not valid: when the analysis of the results was repeated without the low-dose patients, the difference was similar: 30/70 (81%) against 3/11 (27%). Neutrons still showed a significant improvement (P = 0.02) in spite of the smaller number of patients analysed.

However, the complication rate (unwanted side effects) was higher after the neutron treatments than after the photon treatments, 17% against 4% when the number of patients entering each arm of the trial was taken as 100%. This is a more important criticism, although if the complication rate was expressed instead as a percentage of the patients with successfully controlled tumours it was similar (12/54 for neutrons vs 3/12 for photons). This way of expressing the results is, however, controversial. Ideally this clinical trial should be repeated with a slightly lower neutron dose to obtain equal normal tissue complications.

A clinical trial with the same aims is nearing completion at Edinburgh, where a cyclotron with the same energy as at Hammersmith (i.e. rather low) is run by the MRC. Here the neutron doses have been reduced somewhat and the side effects are reported to be similar. The neutron results are also similar to those at Hammersmith (68%). But the photon treatments are giving the same good result: 68% (Duncan, 1982). Perhaps it is the shortness of the overall time used - 4 weeks instead of the conventional 6 or 7 weeks - that has helped control the tumours, rather than the neutrons. We are seeing again the influence of neutrons in helping to improve the understanding and practical results of radiotherapy generally, whether by using neutrons themselves or by leading to improved use of conventional photon beams.

No significant differences have been reported from the neutron clinical trials in other countries, although the numbers of patients in any one trial are as yet small. It is clear that differences are likely to be marginal unless some ways are found of identifying those individual tumours which would benefit most from the use of high LET radiation (Section 2). Gliomas - brain tumours - are not well treated with neutrons. Conflicting results have been obtained for soft tissue sarcomas. Encouraging results have been reported for neck glands, to which cancer of the mouth, nose and throat may spread; and for treatment of salivary glands. Centres in the USA have reported better results for mixed schedules than for neutrons alone: two fractions per week of neutrons with photons given on the other three working days.

It must be pointed out that most neutron beams used up to now have provided very poor depth doses, similar to those from old-fashioned X-ray sets that were phased out of common use in the 1950's. In some centres the neutron beams have been fixed in a horizontal position. In others using D.T. generators the beam angle could be adjusted but the dose rates were too low for reliable positioning to be maintained. A recent review of the world-wide neutron results (Dutreix and Tubiana, 1979) conclude that "Due to these handicaps the possible improvement achieved by neutron therapy is possibly underestimated; unfortunately we cannot assess the extent to which it is underestimated."

"Many of these handicaps could now be eliminated through the use of specially designed and hospital-based cyclotrons."

At present, some of these, with proton energies of 40 to 70 MeV, are indeed being installed in a few countries in hospitals. One is being built for the Liverpool Regional Radiotherapy Centre at Clatterbridge on the Wirral.

Fig. 6 shows the penetration obtainable using neutron beams of various energy in comparison with conventional sources of photon beams. The half-value depths in tissue should be 13 - 15 cm but for the Hammersmith and Edinburgh cyclotrons they are only about 8 cm.

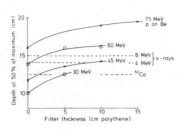

Fig. 6 Depth of 50% doses in water for neutron beams generated by protons of the stated energy on thick Be targets. Full curves - neutron beams 125 cm SSD. Broken curves - X-ray and Co-60 beams 100 cm SSD. The polythene filter increases the mean neutron energy but reduces the dose rate (after Catterall and Bewley 1979).

Radiotherapy with Charged Nuclear Particles

Although the emphasis must be on neutrons for their birthday, this paper would not be complete without a brief mention of charged particle radiotherapy. Accelerated charged particles provide superb distributions of physical dose (Fig. 7). The small scatter provides sharp edges to the beam. The limited range and small straggle provides a sharp cut-off beyond the target volume. The Bragg peak provides higher doses at depth than in the overlying tissues. As well as these obvious physical advantages there are the potential biological advantages of densely ionizing radiation which were described in Section 2 (Fowler 1981). Charged particles - both heavy stripped nuclei and negative pi mesons (P.H. Fowler 1965) - yield high LET tracks in their Bragg peak region, of course. This peak is so narrow that it has to be spread out in depth by varying the energy of the particle beam to scan its range through several cm. The resulting average LET in the spread-out peak is much lower than at the unmodulated Bragg peak. However, for particles as heavy as Si, Ne or Ar the LET is comparable with neutrons (table 1 and Fig. 1). For negative pi mesons the average LET is lower; perhaps pi mesons act in a "ready mixed" way like the American schedules

TABLE 1 Dose-average LET values keV μm^{-1}) for heavy particles in 10 cm spread peak (from Raju 1979).

PARTICLE	IN THE PEAK REGION (10 cm DEEP)		
	PROXIMAL	CENTRAL	DISTAL
Neutrons	75	75	75
Pi mesons	15	30	60
Helium	8	16	30
Carbon	30	40	130
Neon	70	100	300
Argon	200	300	1500

using mixed neutrons and photons. Pi mesons have been used to treat patients in Los Alamos, USA and at Villigen near Zurich in Switzerland. The TRIUMF synchrocyclotron at Vancouver, Canada, is being readied for clinical use and there is a dose rate problem. Heavy charged particles are used at Berkeley, California, to treat deep-seated cancers as well as pituitary disease. For heavy charged particles, the potential biological advantages described in Section 2 are therefore added to the good physical distributions. Proton beams provide an excellent physical dose distribution but no biological advantage. Proton beams are being used to treat cancer at Boston, USA, at Moscow, Dubna and Leningrad in the USSR and they have been used at Uppsala in Sweden. We shall hear more about charged particle radiotherapy in future, as well as about neutron therapy.

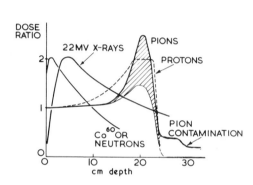

Fig. 7 Doses delivered by radiotherapy beams of various types as a function of depth. Cobalt-60 or neutron doses decrease exponentially with depth after a subcutaneous dose build-up in the first few mm. 22 MV X-rays from betatrons or linear accelerators do the same after a few cm of build-up.

Charged particles yield a peak at depth although the Bragg peak is so narrow that it has to spread out in depth by varying the particle energy. The shaded area represents the high LET dose (Table 1). Pions and heavier charged particles than protons deliver a small dose beyond the range of the primary particles.

It is necessary to compare clinical results from neutrons and proton beams in order to sort out whether it is the biological or the physical advantage which is the more important. Both advantages are present in the more expensive charged particle beams.

Acknowledgements

It is a pleasure to acknowledge the stimulating discussions I have had with many workers in the field, especially Drs. Alper, Barendsen, Bewley, Catterall, Curtis, Denekamp, Duncan, Field, Peter Fowler, Hall, Hornsey, Peters, Raju, Tobias and Withers.

References

Barendsen G W, Walter H M D, Fowler J F and Bewley D K 1963 Radiat. Res. 18 106
Batterman J J, Breur K, Hart G A M and Van Peperzeel H A 1982 Europ. J. Cancer 00 000
Catterall M and Bewley D K 1979 Fast Neutrons in the Treatment of Cancer (London: Academic Press)
Catterall M, Bewley D K and Sutherland I 1977 Br. Med. J. 1 1642
Catterall M, Sutherland I and Bewley D K 1975 Br. Med. J. 2 653
Duncan W 1982 Proc CROS/RTOG Int. Workshop on Particle Accelerators in Radiation Therapy - PART III. Int. J. Rad. Oncol. Biol. Phys. 8 Dec 1982

Dutreix J and Tubiana M 1979 High LET Radiations in Clinical Radiotherapy ed G W Barendsen et al (Oxford: Pergamon) p 243
Gray L H, Conger A D, Ebert M, Hornsey S and Scott O C A 1953 Br. J. Radiol. $\underline{26}$ 638
Farr L E and Yamamoto L Y 1961 J. Nucl. Med. $\underline{2}$ 253
Fowler J F 1981 Nuclear Particles in Cancer Treatment (Bristol: Adam Hilger)
Fowler J F and Morgan R L 1963 Br. J. Radiol. $\underline{36}$ 115
Fowler P H 1965 Proc. Physics Society 85 1051
Fowler J F, Parkins C S, Denekamp J, Terry, N H A and Maughan R L 1982 Proc CROS/RTOG Int. Workshop on Particle Accelerators in Radiation Therapy - PART III. Int. J. Radiat. Oncol. Biol. Phys. $\underline{8}$ Dec 1982
Hatanaka H et al 1979 Abstracts of 6th Int. Congr. Radiat. Res., Tokyo No. B-25-4 p 161
Lawrence J H Aerbersold P C and Lawrence E O 1936 Proc. Natl. Acad. Sci. USA $\underline{22}$ 543
Lawrence J H 1967 Radiat. Res. Suppl. 7 360
Raju M R 1980 Heavy Particle Radiotherapy (New York: Academic Press)
Raju MR 1979 In High LET Radiations in Clinical Radiotherapy ed G W Barendsen et al (Oxford: Pergamon) p 209
Raju M R, Tobey R A, Jett J H and Walters R A 1975 Radiat. Res. $\underline{63}$ 422
Sinclair W K 1970 in Time and Dose Relationships in Radiation Biology as Applied to Radiotherapy BNL Report 50203 (C 57) 932
Stone R S 1948 Am. J. Roentgenol. $\underline{59}$ 771
Sheline G E, Phillips T L, Field S B, Brennan J T and Raventos A 1971 Am. J. Roentgenol. $\underline{111}$ 31
Thomlinson R H and Gray L H 1955 Br. J. Cancer $\underline{9}$ 539
Withers H R and Peters L J 1979 in High LET Radiations in Clinical Radiotherapy (Oxford: Pergamon) p 257
Withers H R, Thames H D and Peters L J 1982 Proc CROS/RTOG Int. Workshop on Particle Accelerators in Radiation Therapy - PART III. Int. J. Rad. Oncol. Biol. Phys. $\underline{8}$ Dec 1982

Neutron radiobiology: dependence of the relative biological effectiveness on neutron energy, dose and biological response

G.W. Barendsen

Radiobiological Institute TNO, Lange Kleiweg 151, Rijswijk, and Laboratory for Radiobiology, University of Amsterdam, Plesmanlaan 121, Amsterdam, the Netherlands.

Abstract. The high Linear Energy Transfer of charged particles produced by interactions of fast neutrons with elements of cells and tissues, results in an increased radiobiological effectiveness in comparison with electrons. In addition, biological effects of neutrons are less repairable and less dependent on the presence of oxygen and on other conditions of the cells. The relative biological effectiveness of fast neutrons is not a unique factor, but depends on the energy spectrum, on the applied dose and on various biological parameters. For the same type of response RBE values are also different for cells of different origin.

1. Introduction

Energy deposition which results from the passage of fast neutrons through living cells and tissues, can cause a spectrum of biological effects, a.o. expressed as cell reproductive death, chromosome aberrations, malignant transformation and altered biophysical or biochemical characteristics. Experimental evidence indicates that these effects are not qualitatively different from the effects induced by electromagnetic radiation, but dose-effect relationships generally show a larger effectiveness per unit dose of neutrons. Thus the same effect, e.g. 50 percent death in a cell population, can be produced by a dose of D_n Gray of neutrons and D_x Gray of X-rays, with $D_x/D_n > 1$. The value of D_x/D_n is called relative biological effectiveness (RBE). For many biological effects, RBE values of fast neutrons are in the range of 1-20; with a few systems the RBE of the most efficient neutrons, at about 0.5 MeV energy, may attain at low doses high values in the range of 50-100.

The large RBE values of neutrons for various responses of mammalian cells are due to the high local concentration of energy deposition along tracks of the charged particles produced by fast neutrons. The interactions of fast neutrons with nuclei of biological material, mostly H, C, N and O, result in the production of various charged particles, mainly protons and recoils of the heavier nuclei due to scattering, and alpha particles from nuclear reactions. In comparison with electrons these particles are characterised by high values of the linear energy transfer (LET). The relative contributions to the kerma or absorbed dose in tissue from the various interactions depend on the neutron energy, whereby the largest contribution is due to recoil protons (ICRU 26, 1977). The energy spectra of the charged particles depend also on the neutron energy. As a consequence the spectrum of LET varies widely with neutron energy and these differences correspond to significant variations in RBE. However, a general characteristic of the spatial distribution of energy deposition by fast neutrons in tissue is the considerably larger average LET compared to corresponding values for the fast electrons produced in tissue by electromagnetic radiation. With thermal neutrons the energy deposition in tissue is different from fast neutrons. Interactions of thermal neutrons, mainly capture processes, result in energy dissipation through secondary processes, whereby the LET spectrum depends on the nature of the emitted particles, in some processes protons, in other cases electrons. Discussions in the present paper will not be concerned with thermal neutrons.

2. Effects of fast neutrons on cells

The effect which has been investigated most extensively with mammalian cells is impairment of the capacity for unlimited proliferation, mostly called reproductive death. This effect is evidently of great importance for the integrity and function of normal tissues, as well as for the application of radiation in tumour therapy. In order to illustrate some of the differences between dose-effect relationships of X-rays and fast neutrons, a number of survival curves are presented in figure 1. (Barendsen and Broerse 1977).

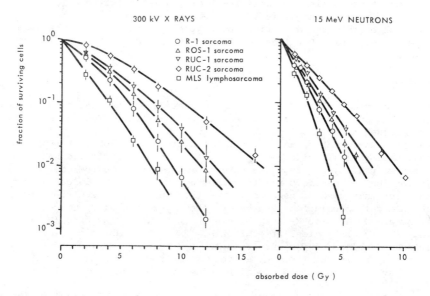

Fig. 1. Survival curves of mammalian cells derived from different types of animal tumours. Survival was measured by the capacity for unlimited proliferation in culture.

The curves obtained with X-rays all show an initial negative slope and a significant downward curvature in the first decade between 100 percent and 10 percent survival. The initial negative slope can be interpreted as a measure of the effectiveness for induction of lethal damage by single ionizing particles. This probability depends for a given cell type on various conditions of the cells, but it is not subject to modification by dose fractionation and dose rate. The increase in slope at larger doses is due to a second mode of damage induction, whereby effective lethal lesions are produced through accumulation of sub-lethal lesions, initiated close together in time by independent particles (Barendsen, 1979). The right panel of figure 1. shows clearly that with 15 MeV neutrons the initial negative slopes are steeper, i.e. the RBE for induction of lethal damage by single ionizing particles is larger as compared to X-rays. Furthermore it is evident that less curvature is obtained, i.e. the contribution of accumulation of damage is relatively less important. As a consequence the influence of dose fractionation and of dose rate, which depends on the occurrence of repair of sub-lethal damage during intervals between fractions or during low dose rate irradiation, is of less importance with fast neutrons.

The differences in shapes of the survival curves obtained with X-rays and 15 MeV neutrons respectively, result in variation of the RBE of neutrons with the level of damage considered, i.e. RBE values are largest for low doses, corresponding to high percentages survival between 100 percent and 80 percent (Barendsen, 1968). For the cell types represented in figure 1., RBE values of 15 MeV neutrons at low doses range between 2 and 4. At 10 percent survival, RBE values of 15 MeV neutrons range between 1.5 and 2.0. For other neutron energies different RBE values are obtained with maxima in the range of 10-20 for neutrons of about 0.5 MeV energy.

Survival curves of mammalian cells in the region of survival between 100 percent and 10 percent can be described by the formula:

$$S(D)/S(0) = \exp -(a_1 D + a_2 D^2) \qquad (1)$$

corresponding to a frequency of induction of randomly distributed effective lethal lesions in cells which is related to the dose D by

$$F(D) = a_1 D + a_2 D^2 \qquad (2)$$

The parameter a_1 represents the effectiveness for induction of those lethal cellular lesions which increase linearly with the dose. In terms of biophysical concepts, a_1 represents the effectiveness with which cellular lethal events are induced by individual particles. This effectiveness depends on the fraction of the total number of lesions which remains unrepaired or is not rendered ineffective by the cells. It is important to point out that this fraction is not a fixed constant for a given cell, but may be modified by many factors, including the repair capacity of the cells, the presence of oxygen and other conditions such as stage of the cell cycle. However, the probability for induction of lethal lesions represented by a_1, is not dependent on interaction of sub-lethal lesions produced by independent particles. The value of a_1 is equal to $1/D_0$ (initial slope) of the survival curve, in which D_0 is called "mean lethal dose".

In order to illustrate the dependence of the RBE on the neutron energy spectrum, values of a_1 derived from survival curves for different cell lines are presented in figure 2. as a function of the $\bar{Y}_F$ (1 μm), the frequency mean of the lineal energy of fast neutrons, which depends on the neutron energy spectrum (Barendsen, 1979). $\bar{Y}_F$ is the microdosimetric quantity equivalent to the frequency average LET. It can be concluded from figure 2 that large differences in absolute sensitivity are observed among different cell types. For the neutron energy spectra of greatest relevance to radiotherapy, which are characterised by $\bar{Y}_F$ values between 10 and 20 keV/μm, the a_1 values differ by at least a factor of 4.

In addition a strong dependence on the neutron energy is observed with maximum a_1 values for the neutron energy range between 0.5 and 1 MeV. In the range of $\bar{Y}_F$ values between 10 and 20 KeV/μm, the lines in figure 2. are closely parallel and the increase of RBE with $\bar{Y}_F$ can be described to a first approximation by:

$$RBE = k \cdot (\bar{Y}_F/10)^{0.75} \qquad (3)$$

whereby k is equal to the RBE at $\bar{Y}_F = 10$ keV/μm. Absolute values of the RBE, which are always calculated relative to low LET photon radiation, depend to a large extent on the a_1 value of the survival curves for these photons. These values show even larger differences than corresponding a_1 values for fast neutrons, as shown by the broken lines in figure 2.

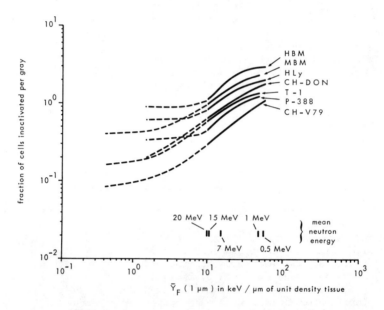

Fig. 2. Effectiveness per unit dose for loss of reproductive capacity of different types of mammalian cells as a function of $\bar{Y}_F$, the frequency mean of the lineal energy of fast neutrons (Barendsen 1979). The solid lines pertain to the range of $\bar{Y}_F$ values for which neutron data are available. The broken lines show the relation to corresponding data for photons. For survival curves which are not exponential the effectiveness values pertain to the initial slopes of survival curves determined from a_1 in formula (1) (Barendsen 1979).

The second parameter represented in the formulae (1) and (2) which contributes to the effectiveness of a given dose of radiation, and which influences the RBE of neutrons, is a_2. The term $a_2 D^2$ describes the influence of accumulation of damage which plays an increasing part with increasing dose. This factor must be evaluated in relation to the corresponding value a_1, and for this purpose the ratio a_1/a_2 is most relevant. This ratio a_1/a_2 represents the dose at which the linear and quadratic terms in (1) and (2) are equal. For X-rays, a_1/a_2 values range for different cell types between 2 and 10 Gy (Barendsen, 1982). If the quadratic term has a negligible influence for effects induced by fast neutrons, a_1/a_2 for X-rays represents the X-ray dose at which the RBE of neutrons has decreased by a factor of 2 from the highest value at infinitely small doses.

For other biological effects induced in cells, e.g. chromosome aberrations and malignant transformation, less experimental information is available as compared to cell reproductive death. Several characteristic differences between dose effect relationships for cell reproductive death induced by neutrons and X-rays are also obtained for chromosome aberrations. The induction of dicentric chromosomes by X-rays can generally be represented by formula 2, while for the induction by fast neutrons the quadratic term is relatively smaller, especially for neutron energies of less than 5 MeV. The frequencies of observed chromosomal structural changes per unit dose are generally smaller than the corresponding frequencies for cell reproductive death, but the RBE values of fast neutrons for chromosomal aberrations frequently do not differ widely from RBE values for cell death. In some experiments, especially at low doses, RBE values of neutrons for induction of dicentrics were observed to be larger by a factor of 2 to 4 as compared to values derived from survival curves. The large RBE values at low doses in these cases were due to relatively small values of a_1 for high energy photons (Barendsen 1979). Considerable differences have been observed for chromosome damage among cells of different origin, RBE values ranging up to 40 for the most effective neutron energy in the range of 0.5 - 1 MeV.

Cell transformation, assessed by altered colony morphology in culture after neutron irradiation, has been studied in only a limited number of experiments (Borek et al. 1978, Hall et al. 1982). From experiments in which the RBE of 0.43 MeV neutrons was determined with hamster embryo cells and the RBE of 35 MeV $d^+ \to$ Be was measured with C3H mouse fibroblast, it can be deduced that the values are for equal doses very similar to RBE values for cell reproductive death. The values increase with decreasing doses and range between 4 and 10 for 0.43 MeV and between 2 and 4 for the 35 MeV $d^+ \to$ Be neutrons.

Many experiments have been performed indicating that cellular effects of fast neutrons depend to a lesser extent on the influence of various factors in the cellular environment and on the stage in the cell cycle, than effects of low LET radiation. A notable example is the influence of the presence of oxygen. Hypoxia renders cells a factor of about 3 less sensitive to X-rays but only a factor of 1.4 - 1.9 less sensitive to fast neutrons. As a consequence the RBE of fast neutrons tends to be larger by a factor of 1.5 to 2.0 for effects produced in severely hypoxic cells (Fowler 1982).

3. Effects of neutrons on tissues and tumours

Similar to low LET radiations, fast neutrons can produce impairment of the integrity and function of normal tissues. Most effects are presumably induced through the responses of individual cells, in particular reproductive death, and it is therefore not surprising to observe that RBE values are similar to those for cellular effects, with a similar range of variability and dependence on the type of tissue, the dose and the neutron energy. Most studies have been performed with high energy neutrons (5 MeV and more). Reviews of these results can be found in recent publications (Barendsen, Broerse and Breur 1979). Similar to effects on normal tissues, studies of effects on tumours have yielded RBE values of neutrons approximately equal to values for effects on cell reproductive death, modified for large doses by the influence of hypoxia as discussed by Fowler (1982).

4. Carcinogenesis by fast neutrons

The relative biological effectiveness of neutrons for cancer induction has been studied with a variety of neutron energies and for tumours arising in various tissues. Proceedings of a symposium with reviews of this subject will appear shortly (Broerse and Gerber 1982). In several studies on the induction of mammary cancer in rats, RBE values for the most effective neutrons in energy range of 0.5 to 1 MeV, have been observed which increase at very los doses of the order of 0.01 Gy to values in the range of 50 to 100. These high values have been obtained for strains of rats with a very high spontaneous frequency of mammary tumours. In other strains values of the RBE of neutrons for induction of mammary cancer have been shown to be lower. Cancer induction and expression involves of course an extremely complex series of processes and the experiments yield complete data only after long intervals. Experimental data on neutron carcinogenesis have been published also for lung cancer and leukemia, indicating a variation of RBE values, mostly in the range observed for chromosome aberrations and cell transformation. However, the analysis of many of these results involves complex corrections for intercurrent death from other causes and accurate data at very low doses are difficult to obtain. Nevertheless this information is very important for the estimation of risks from neutron irradiation in man, in particular since the only epidemiological data on man for survivors of atomic bomb explosions in Japan are now believed to yield little or no data on neutron effectiveness (Dobson and Straume 1982).

References

Barendsen G W 1968 Curr. Top. Radiation Research 4 295-354
Barendsen G W 1979 Int. J. Radiat. Biol. 36 49-63
Barendsen G W 1982 (to be published) Int. J. Radiation Oncology Biol. Phys.
Barendsen G W and Broerse J J 1977 Int. J. Radiation Oncology Biol. Phys. 3 87-91
Barendsen G W, Broerse J J and Breur K 1979 High-LET radiations in Clinical Radiotherapy, Suppl. Eur. J. Cancer
Borek C, Hall E J, Rossi H H 1978 Cancer Research 38 2997-3005
Broerse J J and Gerber G B 1982 (to be published) ed. Neutron Carcinogenesis Europ. Communities
Dobson R L and Straume T 1982 (to be published) in Neutron Carcinogenesis ed Broerse J J and Gerber G B Eur. Communities
Fowler J J 1982 Proc. Conf. The Neutron and its Applications (to be published)
Hall E J, Rossi H H, Zaider M, Miller R C and Borek C 1982 (to be published) in Neutron Carcinogenesis ed. Broerse J J and Gerber G B Eur. Communities
ICRU (International Commission on Radiation Units and Measurements) 1977 Report 26

In vivo activation ananlysis in medicine

Professor Keith Boddy

Regional Medical Physics Department, Newcastle General Hospital, Newcastle upon Tyne, NE4 6BE.

1. Introduction

IN VIVO neutron activation analysis (IVAA) involves the exposure of the living human body, or some relevant part of it, to a small dose of neutrons. The radioactivity induced in the elements of body composition, or the prompt gamma radiation emitted on neutron capture, is measured using external counters. The principle is simple. The practice is more difficult. This review indicates some of the special problems of IN VIVO neutron activation analysis of patients and outlines methods adopted for clinical measurements.

2. Special Features

In conventional activation analysis, considerable attention is given to the preparation of samples with respect to homogeneity and uniformity. However, for total body IVAA, the physical size of the human body, its very irregular shape and the variation in body habitus from one patient to another pose significant complications. Analogously, the measurement of elements in body organs, such as the thyroid, liver and kidneys, by partial-body IVAA involves a "sample" whose size, depth and precise location in the body varies among individuals. Ideally, the irradiation and counting efficiencies should not be influenced by these geometrical factors!

It is not uncommon for IN VITRO samples to be chemically processed post-irradiation. Interfering radioisotopes produced simultaneously are removed before counting. Obviously, patients cannot be subjected to such procedures. The irradiation and counting techniques must endeavour to minimise interferences.

3. Total Body IN VIVO Activation Analysis

3.1 Flux uniformity along the length and across the width of the body

Various geometries, illustrated in Fig.1, have been employed to obtain flux uniformity along the length and across the width of the body.

With a single source, the Kings College Hospital/Harwell group made the first measurements in man using a 1.1m arc geometry (Fig.1.a) and 14 MeV neutrons produced by a Cockcroft-Walton generator. However, a more prolific source of neutrons is necessary if the subject is to be irradiated statically in the vertical or horizontal positions, because the source must be several metres from the subject for inverse square law variations of the flux to be negligible at the head and feet compared with the centre-line. Consequently, major cyclotron facilities, usually established for other purposes, have been pressed into service by groups in Birmingham, Washington, London and Edinburgh. More recently, the Leeds group have used a single sealed source generator at 2.5m from the supine subject with the mid-saggital plane orthogonal to the beam. A bilateral irradiation is achieved by rotation of the subject in the geometries of Fig.1.a,b,&c.

Multiple radioactive neutron sources have been adopted by workers in Brookhaven National Laboratory (U.S.A.) and the University of Toronto (Fig.1.d). A scanning geometry (Fig.1.e), using two sealed-tube 14 MeV neutron generators to give a simultaneous bilateral irradiation was introduced by the East Kilbride (Scotland) group. A variant of this technique is proposed in Newcastle upon Tyne. The Birmingham group has used a scanning method in conjunction with a collimated neutron beam from the cyclotron for prompt gamma examinations. At Brookhaven, for prompt-gamma-ray measurements, the scanning geometry has been adopted with a Pu-Be source beneath the subject and detectors above.

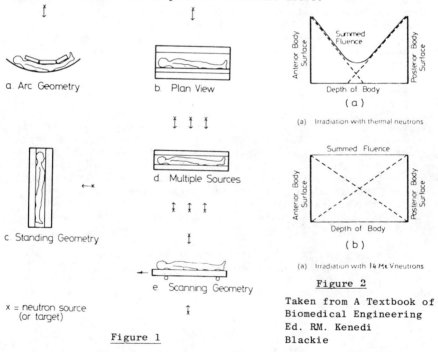

Figure 1

Figure 2

Taken from A Textbook of Biomedical Engineering Ed. RM. Kenedi Blackie

3.2 Flux uniformity through the depth of the body

Although thermal neutrons produce several of the radionuclides induced in the body, they are attenuated rapidly by tissue. As illustrated in Fig.2.a, even a bilateral irradiation results in a very non-uniform summed fluence. However, irradiation with neutrons of higher energy results in partial thermalisation within the body, producing a thermal neutron flux that decreases almost linearly with depth. Bilateral irradiation yields a fairly uniform summed flux, as shown idealistically in Fig.2.b. At the same time, both relevant and interfering radionuclides are induced in the body by the fast neutrons themselves. The best uniformity has been reported using incident 14 MeV neutrons but acceptable uniformity has also been obtained with neutrons from cyclotrons and radioactive sources such as Pu-Be and Am-Be. The reported variations of fluence with depth range from $\pm$ 3.5% to $\pm$ 30% for thermal neutrons and from $\pm$ 3% to $\pm$ 20% for fast neutrons.

3.3 Counting techniques

Most groups use conventional whole-body counters to measure radioactivity induced in the body. An array of sodium iodide detectors, preferred for counting efficiency and consequent lower dose of neutrons, is housed in a shielded room, built from steel or lead, 15cm or 10cm thickness respectively, to reduce the background counting-rate. In contrast, the East Kilbride group uses a "shadow-shield" whole-body counter whose scanning speed matches that of the irradiation facility. This simple procedure automatically corrects for radioactive decay, associated with the scanning geometry, for all of the induced radionuclides.

An ingenious alternative approach for measuring body calcium was introduced by Palmer of the Washington group. Instead of using the conventional $^{48}Ca(n,\gamma)^{49}Ca$ reaction, the $^{40}Ca(n,\alpha)^{37}Ar$ was employed. The potential advantages include the much greater natural abundance ($\sim$97%) of ^{40}Ca, a smaller neutron dose and counting of exhaled ^{37}Ar in a proportional counter. Unfortunately, the slow release of ^{37}Ar from some body compartments in man currently limits its clinical usefulness.

3.4 Total body elements measured and principal interferences

The major elements of total body composition measured to date by IVAA are summarised in Table 1 together with the principal interfering reactions. A few centres measure Ca, P, Na, Cl, K, N and O simultaneously with a single irradiation. Most measure some of these elements, while N, H and C may be determined individually by prompt-gamma-ray analysis. The total body content of aluminium and silver has also been estimated.

Table 1

	Relevant Reaction	Principal Interference	Comments on Interference
1.	$^{48}Ca(n,\gamma)^{49}Ca$	$^{37}Cl(n,p)^{37}S$	Emits 3.10 MeV gamma ray.
2.	$^{23}Na(n,\gamma)^{24}Na$	$^{24}Mg(n,p)^{24}Na$	Produces identical radionuclide.
3.	$^{37}Cl(n,\gamma)^{38}Cl$	$^{39}K(n,2n)^{38m}K$	Emits 2.16 MeV gamma ray.
4.	$^{31}P(n,\alpha)^{28}Al$	$^{28}Si(n,p)^{28}Al$	Produces identical radionuclide.
5.	$^{14}N(n,2n)^{13}N$	$^{16}O(p,\alpha)^{13}N$	Recoil protons produce identical radionuclide.
6.	$^{16}O(n,p)^{16}N$		
7.	$^{40}Ca(n,\alpha)^{37}Ar$		
8.	$^{14}N(n,\gamma)^{15}N$	)	
9.	$^{1}H(n,\gamma)^{2}D$	)	Prompt Capture Gamma
10.	$^{12}C(n,\gamma)^{13}C$	)	

3.5 Calibration procedure

Essentially three approaches have been adopted for calibration so that the body content of the elements measured could be expressed as grammes or millimoles. For obvious reasons, direct assessments of accuracy are difficult.

With remarkable fortitude, the Washington group examined five cadavers, whose weight ranged from 44.5 to 92.3 kg, by TBIVAA and then by chemical analysis after ashing. Most groups have used anthropomorphic phantoms of various sizes, containing known quantities of relevant and interfering elements in solution, for calibration purposes. Some phantoms have contained "pseudo-bone" distributed anthropomorphically. The Brookhaven group has described a calibration procedure for the whole-body counter, which is claimed to give a geometrically invariant response, and has used a single phantom containing a human skeleton for irradiation calibration. In the exceptional case of nitrogen, hydrogen is sometimes used as an "internal standard" in capture gamma-ray analysis. For sequential measurements, the patient acts as his or her own "control" or phantom and accurate absolute quantification of changes is often unnecessary for clinical purposes. Precision of measurements is usually quoted in the range of $\pm 1\%$ to $\pm 5\%$.

4. Partial Body IN VIVO Activation Analysis

4.1 Irradiation techniques and applications

The requirements for PBIVAA are obviously rather different from those of TBIVAA and, perhaps, more demanding. Although the activating neutron flux should be as uniform as possible throughout the relevant organ or region of the body, in other tissues the radiation dose should be minimal. Those organs of interest, such as the thyroid, liver and kidneys, or regions of bone in the spine and extremities are relatively superficial. Consequently, neutrons of lower energy than for total body analysis will often be suitable.

Nuclear reactors have been employed to measure thyroidal iodine and calcium in sections of bone. The East Kilbride group used collimated beams of epithermal and fast neutrons to give a suitable uniformity of fluence. At Orsay, France, where sodium was also measured, thermal and "cold" neutrons were utilised with iodine-129 as an internal standard for the thyroid to correct for substantial fluence inhomogeneity.

Application of the fission spectrum was extended by the East Kilbride group who developed a facility using californium-252 neutron sources for iodine and calcium measurements. The Edinburgh group extended the calcium measurements into clinical practice. These sources have the advantages of being suitably cheap and small in physical size but suffer from a comparatively short half-life. They are used by the Orsay, Lyons and Texas groups for calcium measurements and by the Swansea group for determining cadmium in kidney.

Radioactive (α,n) sources emit neutron spectra of higher mean energy than californium-252 and have been applied to PBIVAA as well as TBIVAA. The Aberdeen group measured calcium in the hand with a source of ^{241}Am-Be and its potential for measuring phosphorus in bone by the $^{31}P(n,\alpha)^{28}Al$ reaction was demonstrated by the East Kilbride group. Recently, the feasbility of measuring cadmium in the liver and mercury in the brain has been demonstrated by workers in Baghdad using sources of this type. More commonly, ^{238}Pu-Be sources have been used for the measurement of calcium and nitrogen in the trunk by the Toronto group, to measure cadmium in the liver or kidneys by the groups at Birmingham, Brookhaven and Brisbane (Australia) and in conjunction with ^{252}Cf by the Orsay group to determine calcium and phosphorus in the hand.

Collimated beams of neutrons generated by a cyclotron have been exploited by the Birmingham group in measuring spinal calcium, cadmium in liver and kidney, to demonstrate the feasibility of measuring iron and copper in the liver and, recently, to estimate the lower detection limits IN VIVO for mercury in kidney.

4.2 Counting techniques

In general terms, two types of counting systems have been employed. For the detection of delayed gamma rays from induced radionuclides, specially shielded sodium iodide detectors are commonly used over the irradiated organ or region of the body. Examples are the measurement of thyroidal iodine and bone calcium. When capture gamma rays, emitted promptly, are to be detected, Ge(Li) detectors have usually been adopted, without exception in determining cadmium, for instance. Crudely, the detectors range from about 13% relative efficiency upwards, one centre using two detectors of 25% relative efficiency. The specialised shielding of the detectors usually incorporates neutron absorbing materials.

4.3 Calibration procedures

Simulated organs, containing known amounts of the relevant elements in solution, are almost invariably incorporated into a phantom (representing the relevant body region) for calibration purposes. Account must be taken of interferences, such as induced sodium-24 and chlorine-38 in thyroid or bone measurements or hydrogen in capture gamma measurements. It is becoming increasingly common to estimate the size, precise location and depth in the body of the organ of interest by ultrasound scanning to reduce uncertainties attributable to these factors. More accurate simulation or correction of results can then be achieved.

5. Clinical Applications

Most clinical examinations using IVAA compare the result in the patient with those from healthy control subjects and/or involve sequential measurements in the same patient to detect changes as the disease or treatment progresses. Calcium and phosphorus are major constituents of bone mineral and reflect the status of bone. Many reports in the literature concern comparative and longitudinal studies in a variety of bone disorders. Sodium, chlorine and potassium are important as body electrolytes and their relationship has been described in relevant conditions such as hypertension. Nitrogen is an excellent indicator of protein in nutritional studies. Oxygen and hydrogen reflect the body content of water. Carbon is expected to provide estimates of body fat. Iodine has been measured in thyroid diseases. Cadmium has been determined as a toxicological burden from industrial and environmental contamination. Silver was studied in argyria associated with anti-smoking tablets and aluminium in dialysed patients. Copper and iron are of interest in Wilson's disease and in haemochromatosis respectively.

Fifty years ago when the neutron was discovered, its contribution to IN VIVO activation analysis was almost certainly unforeseen. Nevertheless, it is perhaps, a fitting tribute that much of the modern work is based on the (α,n) reaction from which its discovery derived.

6. Bibliography

IN VIVO Neutron Activation Analysis 1973 (Vienna:IAEA).
Proc. 2nd East Kilbride Conference on Progress and Problems of IN VIVO Activation Analysis SURRC57/76. ed. K. Boddy (East Kilbride: Scottish Universities Research and Reactor Centre).
Bibliography on IN VIVO Activation Analysis SURRC65/78. ed. K. Boddy (East Kilbride: Scottish Universities Research and Reactor Centre).
Cohn S.H. 1980 Atomic Energy Review 18 599.

Neutron sources, future plans and possibilities

G.A. Bartholomew

Atomic Energy of Canada Limited, Chalk River, Ontario K0J 1J0

1. Introduction

Methods of neutron production is a broad subject covered by many recent reviews (see e.g. Barschall (1978), Bartholomew (1979), and various papers in publications edited by Ullmaier (1977), Okamoto (1980), Bhat and Pearlstein (1980), Cierjacks (1982)) and one cannot do it justice in a short paper. I shall begin with a brief discussion of requirements and limitations plus highlights from various corners of the field to sketch the current status and will end by outlining the accelerator breeder program at CRNL which offers several opportunities for advanced sources.

2. Status/Highlights

2.1 Reactors

Demand for high-flux research reactors is still increasing notably to satisfy the need for high intensities for condensed matter studies (Teal et al. 1977), but also occasionally to meet needs for neutron beams in radically new experiments, as witness the recent intense interest in neutron-antineutron oscillations (Schwarzschild 1982). Steady-state fluxes might eventually reach $\sim 5 \times 10^{15}$ $cm^{-2}s^{-1}$ in upgraded designs of existing reactors (Teal et al. 1977). The limitations are core cooling and costs. More radical concepts, e.g. the liquid-jet reactor (King 1971) might exceed 10^{16} $cm^{-2}s^{-1}$ but have yet to be developed. The pulsed reactor (see following talk) appears limited ultimately by heat dissipation and properties of materials exposed to pulsed thermal stresses and radiation damage. An exhaustive review of pulsed reactor sources is given by Whittemore and West (1982).

2.2 Epithermal Sources

Sources for epithermal energies (<1 eV) include pulsed electron beam devices sometimes coupled to fission boosters and pulsed and continuous reactors and spallation sources. The main application is for research in condensed matter physics. That eight laboratories in F.R. Germany, Japan, Switzerland, U.K., USA and Canada have or plan to acquire, such spallation facilities represents the largest single recent development in neutron sources. The laboratories meet under the International Collaboration on Advanced Neutron Sources (ICANS) (see proceedings Bauer and Filges 1981, Carpenter 1982; also Manning and Stiller, this session). Spallation sources are reviewed by Carpenter (1977) and Fraser and Bartholomew (1982).

The main requirement of sources for condensed-matter studies has been high intensity and the main limitation, costs (see e.g. Brinkman et al. 1980).

However, some concepts also approach the heat dissipation limit. Target heat and neutron yields are plotted vs beam energy for several reactions in Fig. 1. The superiority of the spallation reaction as evident from these curves is the main reason for its adoption by ICANS laboratories (see e.g. Carpenter 1977, Bauer and Vetter 1981).

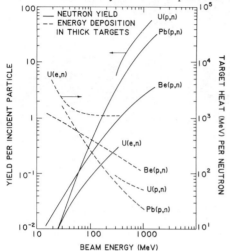

Fig. 1 Comparison of energy depositions and of neutron yields for various reactions in targets of thickness equal to beam penetration (Lone et al. 1982a)

2.3 Resonance Sources

Sources in this range include electron linacs using (e,n) reactions in heavy elements, isochronous cyclotrons using the U(d,n) reaction, and spallation systems. The main application is measurement of microscopic nuclear data. Sharp timing and high intensity are paramount requirements.

Electron linac systems give good performance at relatively low cost and are therefore attractive in spite of unfavourable target heat (Harvey 1982, Lynn 1982). The possibility of extracting individual or bunched micropulses for very high resolution may also be easier with electrons than with heavier particles. This use of micropulses, so far extensively exploited for neutron time-of-flight only with proton accelerators (Cierjacks 1966, 1982), is an anticipated future advance (Lynn 1982).

A most productive early source, the Nevis cyclotron, (Rainwater et al. 1964) was a spallation source as is the current WNR facility, LANL, (Lisowski et al 1980). Following development of the RFQ structure which has improved the beam transmission and reliability of proton linacs, spallation is now being more widely considered for linac plus accumulator ring systems at beam energies of 200-300 MeV (Bartine 1982, Wasson 1982). Proton induction linacs also show great promise (Keefe and Hoyer 1981).

2.4 Monoenergetic Neutron Sources

Sources of this type (see e.g. Smith 1980) are needed over all energies to supply basic nuclear data. The 0° differential cross sections for several of the more useful reactions are shown in Fig. 2. All are monoenergetic over only a limited range. For years experimenters have lamented the 8 to 14 MeV "window" where no convenient monoenergetic sources were available. Actually one reaction spanning the window, viz H(t,n), was well known but little used because of the inconvenience of accelerating tritium. Recently Drosg (1981) has argued that the H(^{7}Li,n) reaction has many attractive features including a high neutron

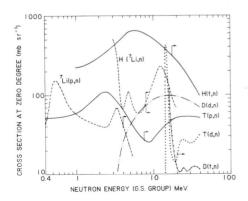

Fig. 2 Zero-degree neutron production from high-energy monoenergetic neutron source reactions. Energies above which neutrons from breakup reactions are present are identified by arrows (adapted from Drosg 1980, 1981).

production cross section. For such inverse (projectile heavier than target nucleus) reactions, neutron production is extremely well collimated around zero degrees, a highly desirable property for cross-section measurements. In future, inverse reactions may play an important role in this field.

For neutrons above about 30 MeV, quasi monoenergetic neutron sources may be produced by (p,n) reactions on thin targets of light nuclei, in particular D, T, ^{7}Li and ^{9}Be (Batty et al. 1969). A facility employing this method for neutron time-of-flight applications to 600 MeV is in use at the SIN accelerator (Fischer et al. 1976, 1978).

2.5 Fast Neutron Sources

Included in this category are fission-spectrum sources, the monoenegetic 14 MeV T(d,n) source and white sources based on (d,n) and (p,n) reactions on Li and Be targets. Such sources are used for materials damage tests, cancer therapy, activation analysis, and other applications.

The T(d,n) rotating target (see Barschall 1978, Barschall in Cierjacks 1982 p57) operating at near the target heat limit, can provide $\sim 2 \times 10^{13}$ neutrons s^{-1} (Heikkinen and Logan 1981). Major further advances would require a gas target possibly with a fission blanket (Battat et al. 1977).

Yields and spectra for (d,n) and (p,n) reactions on Li and Be targets now seem established to good precision (see Lone and Bigham 1982). The planned Fusion Materials Irradiation Test (FMIT) facility based on the Li(d,n) reaction will provide about 3×10^{16} neutrons s^{-1}. The liquid Li FMIT target, dissipating some 3.5 MW, is now well developed (Trego and Miller 1980).

A conceptually simple facility for bathing extended samples with neutrons with a fusion-like spectrum (67% between 12 and 16 MeV) could be provided by an enclosure lined with ^{6}LiD exposed to thermal neutrons in a reactor (Lone et al. 1980). The fast neutron current incident on samples would be about 10^{-4} of the thermal current at the liner.

In final analysis the most relevant data on materials damage by fusion neutrons will come from fusion reactors. However, the earliest operating tokamaks, TFTR and JET, will not match the fluences available from RTNS-II or FMIT while INTOR may just do so (Jassby, 1977, 1979). Fusion neutron sources with geometry suitable for neutron time-of-flight applications will likely be of inertial confinement (point source) variety (Brugger 1971).

Figure 3 shows neutron intensity
and duty factor ranges covered by
all the above sources. Hatched
areas and solid vertical lines
represent sources operating or
committed; unshaded areas and
dotted lines include high
performance design concepts from
the literature. The high intensity
boundaries are determined by
limitations discussed above, particularly heat dissipation and costs.

3. Accelerator Breeder

3.1 Characteristics

In electrical breeding systems, a
neutron source would be surrounded
by a blanket of fertile material
from which fissile material and heat
would be extracted. Electrical

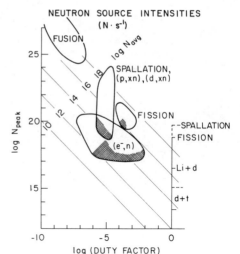

Fig. 3 Neutron Source Intensities (Bartholomew 1979)

breeding is seen as one option to ensure fissile supplies for advanced
CANDU thorium fuel cycles some time beyond 2000 (Fraser et al. 1981). An
output of 1 Mg·a^{-1} per unit, requiring a source of $\sim 4 \times 10^{19}$ s^{-1}, could be
obtained by spallation or fusion. In our view the best approach with
today's technology is the accelerator breeder (AB). A development program
leading to the AB automatically calls up several advanced neutron
sources. Such opportunities are also present with fusion, e.g. FMIT.

An AB with ~ 1 Mg·a^{-1} output requires acceleration of 300 mA of protons to
1 GeV in continuous (cw) mode. Production of 300 mA cw has been
demonstrated (Shubaly and de Jong 1982), quasi-cw acceleration of such
currents has been accomplished in injector linacs for high energy machines
(see Schriber 1978), and acceleration of 0.5 mA quasi-cw beams to 800 MeV
is achieved in LAMPF (Hagerman 1979). Some problems are anticipated, but
with this experience and recent advances, e.g. in RFQ structures (Stokes
et al. 1981), 1 GeV 300 mA cw beams now seem achievable.

Serious problems are anticipated in developing a target-blanket assembly
that can adequately withstand the radiation and thermal stresses in the
AB. However, neither a high-flux nor an exceptionally bright (point-like)
source is needed; indeed the fuel would be in extended assemblies and so a
distributed uniform source would be most suitable. Before designs are
finalized much information will be needed on neutron cross sections, on
the behaviour of fuel, and on radiation damage by spallation neutrons.
The spallation spectrum is peaked at ~ 3 MeV with a low-intensity high-
energy tail extending to the proton bombarding energy.

3.2 Test and Spin-Off Sources

The AB development program at CRNL is envisaged in four steps (Bartholomew
1981, Fraser et al. 1981). Step 1, the Zero Energy Breeder Accelerator
(ZEBRA), will produce protons at full beam (300 mA), but at only 10 MeV,
i.e. 1% of the required energy, primarily for the study of accelerator
problems. With a Li target this beam could produce a cw neutron source of
6×10^{15} s^{-1} with energies $< \sim 8$ MeV. A possible spin-off source from Step 1

would produce neutrons <700 keV using the Li(p,n) reaction with 2.5 MeV protons (Lone et al. 1982b). This "clean" source (no MeV neutrons and few γ-rays) would be attractive for neutron radiography or for keV filtered-beam applications. With currents limited by target heating to $\sim$50 mA, the output would be $\sim 4 \times 10^{13}$ s^{-1}, and thermal neutron intensities on specimen would be $\sim 6 \times 10^6$ cm^{-2}s^{-1} (well-collimated).

Step 2, the Electronuclear Materials Test (EMTF) facility would use a cw 70 mA proton beam accelerated to 200 MeV and a Pb-Bi target such as the hollow-jet (Hoffmann et al. 1981). The neutron spectrum would be almost identical to that from the AB and would therefore be ideal for obtaining basic data needed for target-blanket design. With a D$_2$O moderator, an unperturbed thermal flux of 2×10^{15} cm^{-2}s^{-1} (Lone et al. 1982a) would be available for condensed matter research and other uses.

In Step 3, the EMTF would be extended to a 1 GeV, 70 mA in a PILOT breeder facility. This facility is the minimum needed for AB fuel engineering tests. The beam might also be time-shared with another target providing fluxes of $\sim 10^{16}$ cm^{-2}s^{-1} for research. Hopefully, by cost sharing in this way sources of this scale can be justified.

In Step 4, the PILOT breeder would be uprated to deliver the 1 GeV 300 mA beam required for a full-scale DEMO facility. A possible spin-off might be the breeding of tritium.

ZEBRA would be located at a laboratory but the EMTF-to-DEMO system would be located at a power station to reclaim output heat and take advantage of fuel reprocessing, waste management, etc. We are now launched on the ZEBRA program with operation scheduled for 1990 and EMTF to follow about 2000.

I wish to thank M.A. Lone for much help with this paper.

References

Barschall H H 1978 An. Rev. Nucl. Part. Sci. 28 207
Bartholomew G A 1979 Neutron Capture Gamma-Ray Spectroscopy eds
   R E Chrien and W R Kane (Plenum) p503
Bartholomew G A 1981 ICANS-V Proc. of the 5th Meeting of the International
   Collaboration on Advanced Neutron Sources Jülich eds G S Bauer and
   D Filges KFA-Jülich Report Jül-Conf-45 p29
Bartine D 1982 private communication
Battat M, Dierckx R and Emigh C R 1977 Proc. Symp. on Neutron Cross
   Sections from 10 to 40 MeV, Brookhaven eds M R Bhat and S Pearlstein
   BNL-NCS-50681 p185
Batty C J, Bonner B E, Kilvington A I, Tschalär C and Williams L E 1969
   Nucl. Instr. and Meth 68 273
Bauer G S and Filges D eds 1981 ICANS-V Proc. of the 5th Meeting of the
   International Collaboration on Advanced Neutron Sources Jülich
   KFA-Jülich Report Jül-Conf-45
Bauer G S and Vetter J E 1981 Proc. IIASA Workshop on a Perspective on
   Adaptive Nuclear Energy Evolutions towards a World of Neutron Abundance,
   Laxenburg
Bhat M R and Pearlstein S, eds 1980 Proc. Symp. Neutron Cross-Sections
   from 10 to 50 MeV Brookhaven BNL-NCS-51245
Brinkman W F et al 1980 Report on the Review Panel on Neutron Scattering
   US DOE Report IS 4761 UC25
Brugger R M 1971 Aerojet Nuclear Company ANCR-1034 TID-4500
Carpenter J M 1977 Nucl. Instr. and Meth. 145 91
Carpenter J M ed 1982 ICANS VI Proc. 6th Meeting of the International
   Collaboration on Advanced Neutron Sources Argonne 1982

Cierjacks S, Forti P, Kropp L and Unseld H 1966 Proc. of Seminar on Intense Neutron Sources Santa Fe CONF660925 p589
Cierjacks S 1982 ed Neutron Sources for Applications and Basic Nuclear Research (Pergamon)
Drosg M 1980 Proc. of the IAEA Consultant's Meeting on Neutron Source Properties Debrecen ed K Okamoto INDC(NDS)-114/GT p201
Drosg M 1981 Los Alamos National Laboratory Report LA-8842-MS
Fischer Th et al. 1976 SIN Jahresbericht pE31
Fischer Th et al. 1978 SIN Newsletter No.10 p29
Fraser J S, Hoffmann C R, Schriber S O, Garvey P M and Townes B M 1981 Atomic Energy of Canada Limited Report AECL-7260
Fraser J S and Bartholomew G A 1982 Neutron Sources for Applications and Basic Nuclear Research, ed S. Cierjacks (Pergamon) p217
Hagerman D C 1979 Proceedings of 1979 Linear Accelerator Conference Brookhaven BNL-51134, p78
Harvey J A 1982 private communication
Heikkinen D W and Logan C M 1981 IEEE Trans. Nucl. Sci. NS28 1490
Hoffmann H, Piesche M, Wild E, Martin K and Baumgärtner E 1981 ICANS-V Proc. of the 5th Meeting of the International Collaboration on Advanced Neutron Sources Jülich KFA-Jülich Report Jül-Conf-45 p549
Jassby D L 1977 Nucl. Fus. 17 373
Jassby D L, Caldwell C S and Amherd N 1979 Electric Power Research Institute Report EPRI TPS-79-705
Keefe D and Hoyer E 1981 Proc. Workshop on High Intensity Accelerators and Compression Rings, Karlsruhe ed M Kuntze KfK 3228 p64
King L D P 1971 Los Alamos Scientific Laboratory Report LA-DC-12736
Lisowski P W, Auchampauch G F, Moore M S, Morgan G L and Shamu R E 1980 Symp. Neutron Cross-Sections for 10 to 50 MeV Brookhaven eds M.R. Bhat and S. Pearlstein, BNL-NCS-51245 p301
Lone M A, Santry D C and Inglis W N 1980 Symp. Neutron Cross-Sections for 10 to 50 MeV Brookhaven eds M R Bhat and S Pearlstein, BNL-NCS-51245 p193
Lone M A and Bigham C B 1982 Neutron Sources for Applications and Basic Nuclear Research, ed S. Cierjacks (Pergamon)
Lone M A, Selander W N, Latouf J, Townes B W and Bartholomew G A 1982a Atomic Energy of Canada Limited Report AECL-7839
Lone M A, Ross A M, Fraser J S, Schriber S O, Kushneriuk S A and Selander W N 1982b Atomic Energy of Canada Limited Report AECL-7413
Lynn J E 1982 private communication
Okamoto K 1980 ed Proc. of the IAEA Consultant's Meeting on Neutron Source Properties Debrecen IAEA INDC(NDS)-114/GT
Rainwater J, Havens W W and Garg J B 1964 Rev. Sci. Instr. 35 263
Schriber S O 1978 Atomkernenergie 32 49
Schwarzschild B M 1982 Physics Today 35 19
Shubaly M R and de Jong M 1982 Atomic Energy of Canada Ltd AECL-7605 p79
Smith A B 1980 Proc. of the IAEA Consultant's Meeting on Neutron Source Properties Debrecen ed K Okamoto INDC(NDS)-114/GT p19
Stokes R H, Wangler T P and Crandall K R 1981 IEEE Trans. Nucl. Sci. NS28 1999
Teal G K et al. 1977 Neutron Research on Condensed Matter: A Study of the Facilities and Scientific Opportunities in the United States, Panel Report National Research Council Washington D C
Trego A L and Miller W C 1980 Hanford Engineering Development Laboratory Report HEDLSA-1918FP
Ullmaier H ed 1977 High Energy and High Intensity Neutron Sources Nucl. Instr. and Meth. 145 1-218
Wasson O A 1982 private communication
Whittemore W L and West G B 1982 see Cierjacks 1982 Vol 2 p157

The IBR-2 reactor as a pulsed neutron source for scientific research

V.D.Ananiev, V.A.Arkhipov, B.N.Bunin, Yu.M.Bulkin, N.A.Dollezhal,
A.D.Zhirnov, V.L.Lomidze, V.I.Luschikov, Yu.I.Mitiaev, Yu.M.Ostanevich,
Yu.N.Pepelyshev, V.S.Smirnov, I.M.Frank, N.A.Khriastov, E.P.Shabalin,
Yu.S.Yazvitskij

Laboratory of Neutron Physics, Joint Institute for Nuclear Research,Dubna;
Research and Design Institute of Power Technology, Moscow, USSR

On April 9, 1982 the repetitively pulsed fast reactor IBR-2 of the Laboratory of Neutron Physics (JINR,Dubna, USSR) reached the 2 MW power level. (50% design power). The first stage neutron beam experiments were started since then. The IBR-2 is the only realized project of many those stimulated in 60-70 years by a successfull operation of the first IBR type reactor, 6 kW,(1,2) . The IBR-2 reactor core consists of plutonium dioxide fuel elements performed in the form of an asymmetrical hexahedron enveloped by a double-walled stainless steel jacket (3). It is cooled by sodium. From five sides the core is surrounded by a stationary reflector with tungsten insertions playing the role of control and safety blocks.

The IBR-2 differs from a conventional steady-state reactor by a moving reflector (MR) which is placed at the broadest side of the core (Fig.1). The main rotor of the MR is made of very strong steel and has a long wing. It rotates at a speed of 3000 rpm passing periodically by the core and varying reactivity so that the reactor generates power pulses in a fixed time interval. The auxillary rotor (AR) has the same rotation axis, but rotates slower. The power pulses are generated only when both rotors simultaneously pass by the core. In this case the frequency of pulses is 5 Hz which is favourable for thermal neutron experiments.

The IBR-2 core generates fast neutrons only. In order to obtain the lower energy neutrons the four outer water moderators are used. One of them has a complicate exterior surface form (a grooved type moderator) to increase the thermal neutron flux (4).

Because of the innovation and complicity of the IBR-2 design the procedure of getting the reactor into operation was performed carefully and thus took more time than usual. The IBR-2 physical start-up, i.e. "zero power" experi-

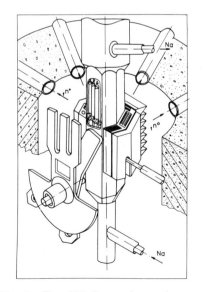

Fig.1. The IBR-2 reactor scheme

ments, was carried out in 1977-1978 without cooling (5) and repeated in 1980 with sodium. The power start-up took place in 1981-1982. During the reactor start-up experiments at power levels from 100 W to 2 MW were studied the static and transient characteristics of the reactor itself as well as of the cooling system, and heat and temperature distributions in the reactor and in the shielding, reactor vessel stresses, random noises of power, the power noise source identification, vibrations of MR, stability and reliability of the control and safety systems, radiation fields, reactor behaviour during the imitation of failures of the reactor equipment. This proved the reactor itself and its equipment to operate in accordance with the project.

The most important characteristic of a pulsed reactor is a pulse duration. The power pulse duration depends on the velocity of MR relative to the core, on its ability to reflect neutrons and on the neutron generation time. The measurements performed during the reactor start-up in 1978 resulted in 200 μs FWHM of the pulse instead of expected 92 μs. The discrepancy was due to the effect of AR on the power pulse shape. So, the previous rotor-- an aluminium disk with the berillium insertion-- was replaced by the steel "trident type" rotor (see Fig. 1). This allowed us to make the pulse shorter almost by a factor of 2. But just before the power start-up it was decided to decrease the main rotor rotation rate to 1500 rpm in order to make operation more reliable under high pulse irradiation. As a result the pulse duration today remains large and equals to 230 μs.

Much attention paid to the operation of MR was not only due to its effect on pulse shape, but also because the pulsed reactor is very sensitive to reactivity excursions. So 1 mm axial displacement of the MR would result in the increasing of the energy of power pulses by a factor of 10. However, the vibrations of the rotor were found to be small : standard deviation 0.01÷0.02 mm, the maximum value registered is equal to 0.1 mm. The reactor power oscillations were about 2% and appeared to be mainly due to oscillations of the sodium flow.

One of the most important operational characteristics of the reactor is the power reactivity coefficient. For the IBR-2 the fast component of this parameter (time constant about 6÷8 sec) is negative and sufficiently great to provide the effective shutdown mechanism which makes the reactor operation safe (Fig.2).

Now a few remarks about the IBR-2 today and future status. After the 2 MW power tests were completed, the neutron beam scientific experiments were started at a 0.4 MW mean power level and at 5 p.p.s. This is 20 times over the power of the IBR-30 reactor which is enough for the first step of neutron beam research. By now the IBR-2 reactor ammounts 2000 hrs of operation time. In the near future the power will be gradually increased to a nominal level of 4 MW. The efforts to decrease the pulse duration are also continued. Last year an

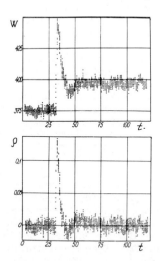

Fig. 2 Transients of power (W) and reactivity (ρ) of 1BR-2 following fast insertion of control block (experiment); W--arbitrary units, t(time)--sec, ρ --in β_p--unit of pulsed reactor reactivity, $\beta_p = 2 \cdot 10^{-4}$ K_{eff}.

extensive experimental work on selection of the optimal configuration of rotors was successfully completed. Now the MR of an optimal shape is being designed. The other opportunity is to find a more suitable material for the main rotor that would allow to reach the rotation speed of 3000 rpm initially planned and to reduce correspondingly the FWHM of the pulse to 110-120 μs.

The pulsed reactor IBR-2 is intended to be used mainly as an extracted neutron beam source for the investigations by the time-of-flight method (TOF). The duration of neutron bursts of the reactor is close to the diffusion time spread of thermal neutrons in water being 200 μs for a grooved type moderator. This allows us to make the best use of thermal neutron flux. The merits of the IBR-2 as a thermal neutron source are the great intensity of the peak neutron flux ($\sim 10^{16}$ n/cm^{2}s) and the large area of the "luminous" surface of moderators (up to 1000 cm^2 each).

Table 1 Performances of intense thermal neutron sources

	Thermal mean power MW	FWHM of neutron pulses μs	Pulse frequency	Thermal neutron flux 10^{15} n/cm^{2}s	
				time averaged	peak
IBR-30 (JINR, Dubna)	0.025	90	4-100	$6 \cdot 10^{-5}$	0.15
GMER (ILL, Grenoble)	60	-	-	1.0	-
IPNS I (ANL, Chicago)	-	50	30	-	0.3
SNS (project) (Rutherford Lab. Chilton)	0.16	50	50	-	4
IBR-2 (JINR, Dubna) today	2	230	25	$5 \cdot 10^{-3}$	-
	0.4	230	5	-	2
in the end of 1982	2	230	5	$5 \cdot 10^{-3}$	10
future	4	100	5	0.02	30

The time structure of the IBR-2 neutron flux measured is plotted in Fig.3. The IBR-2 is compared with other high efficiency thermal neutron sources in Table 1 (6).

There are 14 horizontal neutron beams and 3 inclined ones. The lay-out of neutron beams and physical instruments at IBR-2 is shown in Fig. 4, where one can see two rings of biological shielding with a ring corridor 3 m wide between them. Inside it the collimators and choppers may be placed. Choppers sinchronized with the bursts of the reactor allow the useful part of the neutron spectra to pass through, but shut the beam for the rest of time cutting down the background and satellite neutrons. The neutron beams are transported inside the evacuated tubes shielded by concrete blocks 1 m thick along the two experimental halls (30x60 m^2 each) Therefore, the radiation field in the halls are on safety level and, moreover, each spectrometer is decoupled of others so that the background is sufficiently low. Some of the beams are prolonged out of the building, some are equipped with curved neutron guides.

The program for condensed matter research with the IBR-2 is comprehensive. Four of the physical installations are constructed to investigate elastic processes -- two diffractometers (both on beam No.6), the spectrometer for small-angle scattering (beam No.4) and the spectrometer for the investigation of polycrystal structures (beam No.7). A few words about them. The two are already in operation. The diffractometer is supplied with the special system for the three-dimensional analysis (two space coordinates and a time one). One can observe up to 100 reflections of albuminous crystal simultaneously. The performances of the diffractometer are as follows : thermal neutron flux on the sample (at 4 MW power)-- 2.10^7 n/cm^{2}s; wavelength resolution --0.03 Å,; available interval of λ -- 1.5 ÷ 18 Å; scattering angles -- 0 ÷ 165°; the resolution of Fourier series -- $\leq$ 1Å.

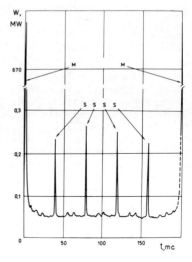

Fig. 3 Time structure of the IBR-2 power for one pulsation period at a mean power of 1 MW; M--main pulse, S--satellite

Unlike other modern instruments of the same purpose the small-angle scattering spectrometer has an axial symmetry for

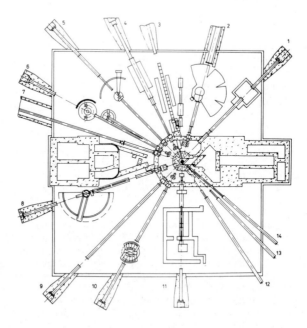

Fig. 4 Arrangement of neutron beams and physical instruments at IBR-2. Concrete shielding of beams is not shown

all the components : collimators, sample, detectors. This fact as well as
the use of the TOF method for the monochromatization of neutrons serve to
a considerable increase of the luminosity of the spectrometer, i.e. by
10÷20 times. Its performances : neutron flux (at 4 MW) -- 6.10^7 n/cm^{2}s; the
wavelength interval used -- 0.7÷7 Å; the angular resolution -- 3 ÷ 20%;
available momentum transfer -- 8.10^{-3} ÷ 2 Å^{-1}. The physical investigations
with a small-angle scattering spectrometer were carried out already at IBR-
30 reactor, and now are being continued at the IBR-2. They include the study of biological macromolecules in solutions (their size and shape), of the
structure of polyelectrolytes and simple electrolytes in solvents as well
as of synthetic polymers. The second diffractometer on beam No.6 is not
completed yet and is meant for the study of the kinetics of magnetic phase
transformations. For this purpose it has a pulse magnet which will create
the magnetic field with a strength up to 500 kOe.

The facilities for a study of neutron inelastic scattering constructed for
the IBR-2 have both direct (beam No.2) and inverse (beam No.10) geometry,
i.e. the TOF analysis may be applied to either scattered or incident neutrons. In addition to them a correlation spectrometer is already operating
for the analysis of the energy of both scattered and incident neutrons with
the TOF method (beam No.5). As a result the whole neutron spectrum is effectively used in the measurements : thermal neutron flux on the sample -- 10^8
n/cm^{2}s, wavelength interval -- 0.7÷4 Å, scattering angles 15÷65°, 75°, 90°;
momentum transfer -- 0.4÷12 Å^{-1}; energy transfer -- 0÷80 meV; energy resolution --$\geq 0.8 \cdot 10^{-2} \sqrt{E}$, E in meV. In the spectrometer on beam No.2 the incident beam of neutrons is monochromized with the help of the mechanical
chopper synchronized to power pulses of the reactor. This instrument is a
combination of two spectrometers. One of them with a first flight path of
100 m ensures very high resolution (10^{-6} eV) while the other provides
high intensity of monoenergetic neutrons on the sample -- up to 10^7 n/cm^{2}s.

Experiments with ultracold neutrons were initiated at Dubna in 1968. The
ultracold neutron facility on beam No.3 is one of the four operating now
physical installations at the IBR-2. Its first version allows one to have
an ultracold neutron intensity of 400 n/cm^{2}s. Later on the pulse shutter
will be installed in order to achieve the density of ultracold neutrons in
the storage tank the same as in the convertor during the reactor burst,
i.e. to use effectively the pulse mode of the IBR-2 operation.

For the first time the spectrometer on polarized thermal neutrons with polarizing mirror neutron guides is being constructed on beam No.8. The polarized neutron flux on the sample at a 30 m distance from the reactor is
expected to be 10^7 n/cm^{2}s.

One of the IBR-2 beams (No.11) meant for the medical and biological investigations is equipped with the special instruments to form a "pure" neutron
beam.

The IBR-2 research program in applied physics includes the neutron activation and capture gamma-ray analysis of the elements, the study of the
radiation stability of materials under pulsed irradiation field, the analysis of implanted atoms in semiconductors, the study of the chemistry of
"hot" atoms in condensed matters. For these purposes there are made some
irradiation holes with a fast neutron flux up to 3.10^{14} n/cm^{2}s (time averaged).

The duration of the power pulse of the IBR-2 is too long to perform successfully the nuclear investigations with resonance neutrons. At present

the nuclear physics research program on the IBR-2 is restricted to the experiments requiring high intensity at a moderate resolution. These are physics of fission and rare nuclear processes in low energy resonances. The two beams are picked out for them (No.1 and No.9).

References

1. G.E.Blokhin et al. "Atomnaja Energija", 1961, v.10, part 5, p.437.
2. I.M.Frank In: "Problemy Fiziki elementarnykh chastitz i atomnogo jadra" (Particles and Nucleus), v.2, part 4, Moscow Atomizdat, 1972, p.806.
3. V.D.Ananiev et al. "Pribory i tekhnika eksperimenta", 1977, No.5, p.17.
4. N.A.Gundorin, V.M.Nazarov JINR Preprint, P3-80-721, Dubna, 1980.
5. V.D.Ananiev et al. "Atomnaja Energija",1979, v.46, part 6, p.393.
6. E.P.Shabalin, "Atomnaja Energija, 1982, v.52, part 2, p.92.
7. Yu.M.Ostanevich, I.M.Frank, E.P.Shabalin Proc. III Intern. School on Neutron Physics, 1978, Dubna, 1978, part 1, p.5.

Neutrons and nuclear power

Sir Walter Marshall, FRS

The discovery of the neutron by Chadwick in 1932 and the subsequent discovery of fission by Hahn and Strassman has led to two enormous technological projects. The first and most urgent, to build nuclear weapons, has been discussed by Dr Garwin earlier today. The second, to use neutrons for the peaceful production of nuclear energy, is the subject of my own talk. I propose to determine as best I can what was in the minds of those early pioneers concerning the use of neutrons to produce civil power. I will then examine to what extent their thinking has been substantiated in the event, to what extent were they successful in anticipating the actual evolution of nuclear power and what points did they fail to anticipate.

The first formal reference I can find to the use of neutrons and fission for civil power is the report by the MAUD Committee on "The use of Uranium as a Source of Power". This report from the MAUD Committee concluded that the "uranium boiler" had considerable possibilities for peacetime development but that the scheme was not worth serious consideration from the point of view of the "present war". They recommended, however, that it was essential that Britain should take an active part in the research work on the use of uranium for power. That early MAUD Committee correctly anticipated the importance of uranium 235 plus the importance of a moderator to slow down the neutrons, and it drew attention to the need for protection against radioactive effects and also the fact that the presence of uranium 238 would automatically produce plutonium (which they referred to as "a new element of mass 239". They correctly anticipated that the possible moderators were heavy water, very pure graphite or light water if enriched uranium can be used. Given the overwhelming pre-occupation with the possibility of producing the atomic bomb – an understandable pre-occupation in the middle of a world war, especially in 1941 – these technical insights into the future nuclear power industry are very impressive.

There are various accounts of thinking on civil nuclear power during the war but all that was, of course, overshadowed by the vast Manhattan Project. Earlier in this Conference we have been given some account of those early days and I shall therefore pass quickly to the "consolidation phase" which took place immediately after the war when scientists throughout the world settled down to decide what to make of these new opportunities.

Looking back to the early archives of Harwell, the thinking seems to have been dominated by the scarcity of uranium. It was argued that because uranium was such a rare element then thermal reactors, whatever their moderator, would have only a short transient role to play. There was, therefore, great emphasis on the need to build fast reactors as soon as possible.

The first reference I can find to fast reactors in the Harwell archives in in an appendix to the Minutes of the Harwell Power Committee held in July 1946 and a quotation from that appendix is shown in Figure 1.

FIRST MENTION OF A FAST REACTOR AT HARWELL

EXTRACT FROM THE APPENDIX TO MINUTES OF THE SECOND MEETING OF THE HARWELL POWER COMMITTEE HELD ON 2nd. JULY 1946

In piles designed for power it is important that δ should be positive in order that each pile may be self-supporting, but very small values of δ are acceptable. In such a pile one may regard the thorium (or uranium 238) as the fuel and the fissile material as a catalyst. Another kind of pile which might sometimes be required is a so called breeder pile, the object of which is to increase the amount of fissile material available. In such a pile obviously one wants δ to be significantly greater than zero, otherwise the whole operation would be extravagant.

Figure 1   First reference to Fast Reactors in Harwell archives

This same theme occurs also in the Romanes Lecture which Cockcroft gave in 1950 where he suggested that the primary development task facing the nuclear power industry was how to find ways of using uranium more efficiently, preferably by developing, "breeder piles". Thermal reactors might be used in the interim but this was definitely second best.

We see, therefore, that even in those early days the development of nuclear power was influenced by uncertainty about the world uranium supplies. It is a sobering thought that that remains true today although sometimes it is argued that uranium is so plentiful, that work on fast reactors can be set aside indefinitely, and sometimes it is argued that uranium is so scarce that we should give fast reactors the priority they clearly had in the minds of those early pioneers. Of course there is no meaningful answer to the question "what uranium resources does the world have?" Instead, we must ask the more refined question "what resources are available at each particular cost level?". The early pioneers did not look at the matter in that way – except by implication – and the only conclusion we can draw is that they were neither worse nor better than ourselves in looking into the future and anticipating the results of a market place reflecting the law of supply and demand.

We are on surer grounds when we examine the thinking of the early pioneers on purely technical questions. So far as I can tell, they correctly anticipated all the main points which would be important for the operation of a civil nuclear industry in later years. I have already referred to the early thinking on breeding potential. Figure 2, which is taken from Glasstone and Sesonke shows the breeding potential as a function of neutron energy from the main isotopes uranium 235, uranium 233 and plutonium 239. The curve shown in this figure gives the general trend, but not the fine structure in the resonance region, and leads immediately to

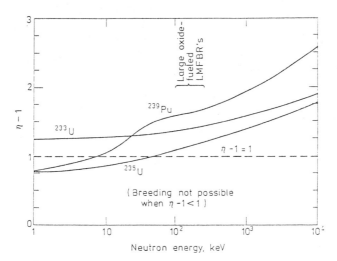

Figure 2 Breeding potential as a function of neutron energy. (The curves show general trends but not the fine structure in the response regions.) - Figure 8.16 from Nuclear Reactor Engineering, Third Edition, S. Glassstone and A. Sesonske, Van Nostrand

the conclusion that nuclear power must start by relying upon the fission of uranium 235 at low energies; raises the possibility of breeding uranium 233 in a very efficiently moderated (ie heavy water) reactor and raises also the larger opportunity of breeding plutonium 239 in reactors where the neutrons were not modified. All the early discussions concerning these possibilities were technically correct but commercially unrealistic. Thermal reactors using uranium 235 have become well established but the fast reactor using plutonium 239 and using the technology exactly as envisaged by the early pioneers remains a promise for the future and the prospect of using uranium 233 remains just that; a prospect.

It is worthwhile examining also the early thinking on the control and safety problems of a reactor. The control of a fission reactor depends upon the existence of delayed neutrons. The characteristics of delayed thermal neutrons in thermal fission are given in Figure 3. It is this small fraction of delayed neutrons which makes the safe control of the reaction feasible and this was correctly anticipated by the early pioneers.

The importance of temperature coefficients was recognised quite clearly in those early days, for example at the start of reactor studies in the UK it was realised that a natural uranium/graphite moderated water cooled pile (the type initially built at Hanford during the war) had a positive coolant temperature coefficient. This was unacceptable in the UK and was one of the reasons why the air cooled piles which led to the magnoxz reactors and later the AGR's. The Hanford piles were re-designed using slightly enriched fuel so that the temperature coefficient was negative.

The present PWR has a very large negative temperature coefficient of reactivity and although there is a positive temperature coefficient in the CANDU reactor, it is small enough to be controllable.

The existence of a "Doppler effect" was realised early on. In a thermal reactor the neutrons can be captured in the resonances of the ^{238}U. Because of the increased kinetic energy of the target (^{238}U) nuclei with

Approximate Half-life (seconds)	Number of Fission Neutrons Delayed per Fission			Energy (MeV)
	U-233	U-235	Pu-239	
55	5.7×10^{-4}	5.2×10^{-4}	2.1×10^{-4}	0.25
23	19.7	34.6	13.2	0.46
5.2	16.6	31.0	12.9	0.41
2.3	18.4	62.4	19.9	0.45
0.61	3.4	18.2	5.2	0.41
0.23	2.2	6.6	2.7	
Total delayed	0.0066	0.0158	0.0061	
Total fission neutrons	2.49	2.42	2.93	
Fraction delayed	0.0026	0.0065	0.0020	

Figure 3 Characteristics of delayed fission neutrons in thermal fission – Table 2.10 from Nuclear Reactor Engineering, S. Glasstone and A. Sesonske, Van Nostrand

increasing temperature, the width of a resonance is increased, but the height of the peak is decreased, the total area beneath the resonance curve remaining constant. If the resonance absorption cross sections are large, so that essentially all neutrons with energies in the resonance region are captured, the widening of the region will result in a decrease in the resonance escape probability as the reactor operates and the temperature rises.

These are all the technical examples I need to give you today. Altogether we can conclude that the early pioneers foresaw accurately all the fundamental technical points which would dominate the civil nuclear power programme for decades to come. But they underestimated the timescale for the commercial development of that technology and they misread the future supply and demand market for uranium ore. Nevertheless, on all the aspects we have discussed so far, we must conclude that they did very well. I would be proud to have done so well myself.

It is also pertinent to ask the following question. It is clear that the early pioneers had a golden vision that civil nuclear power would supply the energy needs of mankind cheaply and economically for the indefinite future, and certainly long after the world's resources of fossil fuels had been exhausted. That was a golden vision; perhaps it was rather premature and optimistic but is there anything fundamentally wrong with it? In my opinion there is nothing wrong with that fundamental vision; it remains as correct today as it was then. Even given the benefit of hindsight over the best part of 50 years, we can still conclude that the use of neutrons and fission power offers the only real promise of an abundant supply of energy to mankind for the indefinite future. Furthermore, we are now seeing the first signs of the exhaustion of fossil fuels. There would be no OPEC cartel if the supply of oil grossly exceeded demand, the quadrupling of oil prices came about because oil is an exhaustible resource and its exhaustion, while still distant in time, is apparent to all the oil producing states. Here at home, some of the first North Sea oil fields have already passed their peak production and we are planning the use of gas from the far north because of the foreseeable exhaustion of gas from the south part of the North Sea.

To my mind the proper vision of the role of nuclear power in the future is not significantly different from that of the early pioneers. There is no long term alternative to the use of neutrons and the fission of uranium and plutonium to provide mankind's energy needs in the future. The coal resources of the world, have a vital role to play in replacing oil and gas as suppliers to our chemical industries and coal also has a vital role to play for many decades to come in electricity production. Nevertheless, the long term vision of a world dependent upon neutrons and nuclear power for its energy requirements remains as true today as it was in the days of those early pioneers.

However, if indeed that is the case and if indeed that is obvious to me and, I believe, obvious to all thinking, informed scientists, why is it not apparent to everyone?

Many newspaper correspondents and TV commentators refer to the death of nuclear power. They point to the fears the public have of this new technology. They point to the difficulties of constructing large nuclear power stations to time and to cost and they point to the stagnation of the nuclear industry in America, Germany and elsewhere. They point to the hesitations and uncertainties in the British nuclear power programme and argue that the confidence and vigour of the admirable French programme is dependent on a disregard of French public opinion.

Here then is a paradox! It the vision of the early pioneers was correct, if that vision is still correct, why is the nuclear power business in such disarray throughout the world? I would like to devote the remainder of this lecture to try and understand this paradox and to give you my thoughts about it. The sense of my argument will be that the early pioneers and those of us who have followed in their footsteps grossly underestimated the problems of communication with the general public concerning this vast, new and apparently frightening technology.

This failure in communication seems to have been foreseen first by the novelist C P Snow, later Lord Snow, who wrote of the danger that the western world was developing two separate cultures; the scientific and technical culture upon which our civilisation was becoming dependent, and the general mass of non-scientific, generally non-numerate remainder of the population. I can remember listening to his early lectures with some complacency because I was clearly a member of the scientific and technical community of which he approved and I interpreted his lectures as an exhortation to the general mass of the population to try and understand what we were doing. Rather belatedly I have realised I can interpret his lectures in the reverse sense, that we in the scientific culture must not run so fast that we lose the confidence of the rest of the population. As I now see it, C P Snow was giving a message to us that we should take care to explain ourselves to the public. In some ways he wished to give more emphasis to that than to the message I first perceived; namely that he was urging the general population to try and understand science.

It is my firm opinion that the present day impasse which faces nuclear power is due overwhelmingly to this lack of communication betwee4n scientists and the general population. It certainly is not due to fundamental weaknesses in the concept of nuclear power. For two decades now I have looked at almost all the allegations of risk and danger supposedly posed by nuclear power and I have concluded that none of them have real content. The case for nuclear power is truly overwhelming. The general public may have a different perception of nuclear power; that perception in itself does not determine that nuclear power has failed although I have heard some TV commentators argue just that, presumably because all media people are more interested in the appearance of the facts than the facts themselves. However, it does mean that we scientists have to work harder and with more care at explaining ourselves. We must take care to use a vocabulary that permits us to do that both accurately and effectively and, as an illustration of the case in presentation which I believe is necessary, I would like to devote the rest of this lecture to a discussion of the phenomena which frightens the general public more than any other; namely the risk of large uncontrolled accidents in a nuclear reactor.

None of the facts and figures I shall quote would be a surprise to the early pioneers of neutron physics. The facts then were the same as they are now and the same as they will be in the distant future. Our scientific appreciation of them has remained constant but the public's awareness of them and perception of them has changed alarmingly. That must be our fault, it must be an error of communication and, quite possibly, these difficulties would not have arisen if the early neutron pioneers had been psychologists as well as good physicists.

Let me start my discussion on reactor accidents by suggesting a typical newspaper headline which might read "Reactor Accident at Sizewell could kill 1250 people in London". It is an invented headline but it is not artificial. Television programmes have said something similar but multiplied the number of people by a factor of ten. We use the word 'could' in our everyday language in very ambiguous ways. A jumbo jet 'could' crash killing 300 people; this does not happen very often and the aircraft industry does its best to keep it a rare event, nevertheless it does happen. This is therefore an accident that could be described as 'expected'. Two jumbo jets 'could' collide killing 600 people; this is much rarer but it is possible. In thinking about the safety of jumbo jets we concentrate on the 'expected' not the 'possible' accidents because we know intuitively that 'expected' accidents and their frequency determine the safety of jumbo jets. One jumbo jet 'could' fall on a crowded Wembley Stadium and kill 10000 people. This is not actually impossible, therefore we have to say that it is 'imaginable'. Here are three sentences where we have used the world 'could' and we have used it in three different ways. We could have meant it to be 'expected', 'possible' or 'imaginable'.

In the nuclear industry the accidents that we expect, will kill no one. The accidents which are possible will not harm the general public. The imaginable accidents are not actually impossible, they could happen but they are very unlikely, say once in 10 million years. The public would perhaps more easily understand the vocabulary of 'expected', 'possible' and 'imaginable' than the probability numbers which we might guess at something like 10^{-2}, 10^{-4}, 10^{-6}. I think, therefore, that the first step is to put over to the public that the word 'could' has a vast range of meaning; simply because it could 'happen' does not mean it will happen.

The second step is to avoid comparing accidents from nuclear reactors with accidents from earthquakes; dams bursting or major fires. We can use these comparisons in our technical discussions. But, while trying to explain to the general public the rareness of the event and arguing that a nuclear accident is less frequent than a severe earthquake, we put into public mind the image of a large disaster with a large number of casualties. This brings us to another problem of vocabulary. To go back to our first heading "Reactor accident at Sizewell could kill 1250 people in London". What does the word 'kill' mean? At first sight that seems a silly question to ask because we all think we know what it means. In fact in everyday language and in our technical discussions, we use it in an ambiguous way, or at least a way that is not sharply defined. Does the word 'kill' mean sudden death, or does it mean an adverse health effect leading to reduced life expectancy? We use the word in both ways in our everyday language - we determine the meaning solely by context. Let us look at possible newspaper headlines using the word 'kill'.

"Earthquake kills 10000 people":

we all know that here kill means sudden death. We have seen pictures of earthquakes and their consequences.

"Cigarette smoking could kill 10000 people":

here we know immediately that kill does not mean that if someone smokes a cigarette he will drop down dead. We know it means that somebody has calculated the adverse long-term effects of cigarette smoking which may cause death earlier than for non-smokers. Indeed the Government health warning about cigarette smoking talks about adverse health effects. It is only for brevity that such headlines are used. This is perfectly fair because it describes effects within our knowledge and does not, therefore, mislead the public.

"Atom bomb could kill 10000 people":

this is, of course, unambiguous. We know that an atom bomb could kill 10000 people, and that kill means sudden death, because the knowledge of atomic weapons is impressed on the public mind.

"Reactor accident could kill 10000 people":

this is the ambiguous headline. Here kill means an adverse health effect leading to reduced life expectancy. That is a simplification but it is, overwhelmingly, the interpretation which should be put on it. But it is not the interpretation the public puts on it. They think kill means not an adverse health effect but sudden death. We have now reached a headline which, because it is outside their area of knowledge, misleads the public; although it is perfectly fair and does not mislead us.

It is legitimate to ask what these words mean, and it is legitimate to ask how the risks from large nuclear accidents should be discussed. A large accident, although a remote possibility, would, if it should occur, cause some deaths in the short-term. If it does that we should say so. It is a proper explanation to make and it is entirely appropriate to compare the risks from a reactor accident causing short-term immediate deaths with those from earthquakes or dam burst. But it is not helpful to talk about long-term deaths in the same way, it is not a valid way of explaining the risks of a big accident. I think they are better described in terms of adverse health effects leading to a reduced life expectancy.

I think it is fair to describe it as an increased chance of cancer death. But the explanation I favour is to compare the risks from a reactor accident to the equivalent pattern of compulsory cigarette smoking, because the risks are similar in type and both carry a risk of death from cancer. Because Governments, certainly the UK Government, have invested such an enormous effort in educating the general public about the nature of risks from cigarette smoking, people do understand that if they smoke they have an increased chance of dying from cancer. Let us discuss these ways of describing the consequences of an 'imaginable' accident using one example.

A gigantic "imaginable" accident could deliver one rem to the whole population of London, 10 million people. The total dose is 10^7 rem. This is an imaginable accident, the chances of it occurring are very, very remote. The adverse health effect of this accident can be calculated. The simple method is to say that if a person receives one rem at one point in time then his chances of death at a later time are slightly increased over a 30-year interval ranging from 10 to 40 years later (see figure 4). This

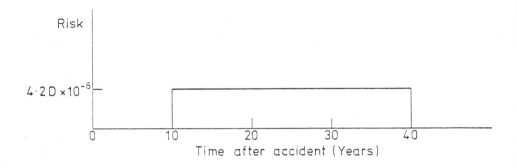

Figure 4   The additional risk of radiation induced fatal cancer following an imaginable nuclear accident

is a mathematical simplification of the best medical evidence we have. If this risk is used to determine the consequences of the accident, short-term deaths are zero. Calculations also show that 1250 people will die earlier than they otherwise would - in other words the accident has 'killed' 1250 people. Alternatively we can calculate that these long-term effects give a loss of life expectancy for the population of London of 19 hours. It is, therefore, fair to say that this accident, which is so remote that we can only imagine it, with a probability within the range of 1 in 10 million to 1 in 100 million, possibly even lower, would produce a loss of life expectancy of only 19 hours, less than a day. For some people that is a satisfactory and understandable way of explaining the magnitude of the effect. It does, however, have a weakness. The loss of life expectancy for the population as a whole is an average figure made up of negligible effects on most people and a substantial loss of life expectancy for a very small number of people. It is possible to argue, therefore, that an average figure is misleading because it does not allow for the stochastic nature of the risk.

Talking about increasing the chances of death from cancer again leads to the difficulty that, to be meaningful, we are driven to quoting numbers, and I fear that the public is not sufficiently numerate to understand the slight increase in risk which any hypothetical accident would produce. So while it is a valid way to describe the risk it has its limitations for presentation to the public. You will have gathered that I have lost faith in the public's ability to understand small numbers and I have also lost faith in the type of presentation which Lord Rothschild used on television some years ago which depended heavily on explaining to people the smallness of probabilities.

Finally we look at a pattern of compulsory cigarette smoking. It is not satisfactory to compare reactor accidents directly with the risks of smoking because smoking cigarettes is a voluntary matter. Neither is it sensible to compare the risks of cigarette smoking with the risks from nuclear power in normal operation. It is much better to simulate the risk by proposing a smoking pattern which reproduces the risk from a very large accident. The equivalent description is to ask all Londoners to smoke 1/20th of a cigarette each week over that 30 year period (figure 5). I

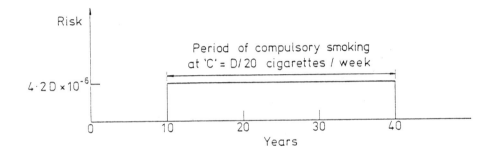

Figure 5   Smoking pattern with a risk of death equivalent to that following a big nuclear accident as in Figure 3

have some hope that this presentation of risk is one that the public have a chance of understanding because they do understand the nature of the risk they run from cigarette smoking. Most important, they understand the stochastic nature of that risk. They will not use the word "stochastic", they will simply say that perhaps they will be lucky or perhaps they will be unlucky as a result of their smoking pattern. Nowadays the public has such an instinctive understanding of the risk of smoking cigarettes that it must surely be sensible to explain the risks from radiation in terms of the equivalent compulsory smoking pattern. We must stress that we are not trying to compare voluntary and involuntary risks, we are trying to ask them to imagine that this is a smoking pattern which is imposed on them. That, in fact, one rem to each of 10 million people is equivalent to a compulsory smoking pattern of 1/20th of a cigarette a week. This level of smoking is so low, it must be less than the involuntary level of cigarette smoking forced on them by other people smoking alongside them. If Londoners smoke 1/20th of a cigarette a week 1250 of them would be killed, using the word in the sense I used it before. When we say "it will kill 1250 people" we do not mean in the sense of an atomic bomb, we mean that it is equivalent to 1/20th of a cigarette each week.

This must, of course, be supplemented with an explanation that in practice there will be a distribution of doses and, therefore, there will be a distribution of effects. I have over-simplified an imaginable accident but I think I have given a better appreciation of what the consequences might be than many of the discussions which I have seen in the literature and certainly in the newspapers over the last few years. I am not talking about the risks from nuclear power averaged with the possibility that the accident will occur; I am describing the consequences given that the accident has occurred and I have explained that the probability of this event happening is very low - less than one in a million.

Conclusion

I have in this lecture only discussed one problem of communication; that concerning large accidents. We need to give a similar assessment of the risks from nuclear waste and the risks of the so-called "plutonium economy". To return to the opening theme of this lecture, how would the early neutron pioneers react today? They would, I think, be pleased at the way the technology had developed and that development would not surprise them at all. They would be disappointed by the slowness of commercialising the breeder fast reactor. They would be critical about the fact that we take much longer than they did to accomplish anything but, most of all, they would be astonished and dismayed by the hostility to nuclear power expressed by a number of opposition groups. They would be puzzled by that. They saw neutrons leading to an energy source which would be a servant and a boon to civilisation and the human race. They would recognise in us, their successors, the same ideals. They would be shocked to see how this great adventure to serve mankind with limitless energy could bring upon itself such hostility and they would, I think, agree with my diagnosis that a vast communication gap has opened up between the scientific and non-scientific cultures. I believe they would share my astonishment of how it has come about. They would then, I think, recall the many attempts they made collectively to explain to Governments the implications of this new technology and they would chide us for not doing as much as we ought to have done in this direction.

I hope that this lecture, which links their triumphs and insights to the present state of nuclear power in the world is a first step in re-acquainting the public with the promise of nuclear power and the certainty that it is the lowest risk source of energy available to mankind.

The neutron and nuclear weapons

Richard L. Garwin

IBM Thomas J. Watson Research Center, Yorktown Heights, NY 10598
(also Department of Physics, Columbia University, New York, NY 10025)

Abstract. The evolution of nuclear weapons is traced in outline from 1943 and their effects recalled. Technical enthusiasm for nuclear weapons development impedes efforts to understand, control, and avoid the devastating use of nuclear weapons, which have little military utility. Modern non-nuclear battlefield weapons are more effective and more usable. I recommend skepticism about the need and utility of modernization programs and against the massive programs to build MX missiles, B-1 bombers, Trident submarines, and to undertake to deploy space weapons. Nuclear deterrence of war can better be achieved by a greater regard for the nuclear weapons we have and by minimal-cost improvements toward objective needs.

1. Introduction

The fiftieth year of the neutron is the thirty-seventh of the fission bomb and the thirtieth of the hydrogen bomb. Certainly it was extraordinary in thirteen years to move from the discovery of the neutron to the construction and use of nuclear weapons of energy release 20,000 tons of high explosive ("20 kilotons" yield). Addressing a group of physicists, I am tempted to rehearse the history of nuclear weapons-- from the fancy of H.G. Wells to the insight of Leo Szilard in 1934, to the discovery of fission and the successful effort to provide the fissile isotopes U-235 and Pu-239 used in the Hiroshima and Nagasaki bombs. And there is much physics and personal interest in the realization of fission and thermonuclear weapons. But to accede to this temptation would be to miss the larger opportunity to review the effects and significance of nuclear weapons, setting the stage for the recommendations and the panel discussion which follow, emphasizing what to do about nuclear weapons.

My own involvement with this subject began in 1950 at the Los Alamos Scientific Laboratory and occupied three to five months of each year throughout that decade and much of the following. I devised improved means for detailed diagnosis during nuclear weapons test explosions, contributed to new initiator and explosive systems and did the first work on fratricide-- the impairment of function of one nuclear explosion by another. I was involved also in the design and realization of the first few iterations of the hydrogen bomb-- the liquid deuterium 8-megaton explosion of November 1952 and the higher-yield solid-fueled tests of 1954. Since that time I have worked also on the missiles and aircraft which carry nuclear weapons, the permissive action links (PAL) which prevent nuclear explosions except on the insertion of a valid code, on defensive systems, on basing modes for strategic weapons, on space weapons, and on arms control, including the limitation of nuclear weapon tests. I served eight years on the President's Science Advisory Committee (Kennedy, Johnson, Nixon) and three years on the Defense Science Board, advisory to the Secretary of Defense.

In view of the well-warranted restrictions on providing information on nuclear weapons, I shall limit my talk precisely to materials declassified by the US government or those I have previously published, emphasizing the "Los Alamos Primer (LA-1 by Robert Serber of April 1943)-- notes written up by E.U. Condon based on a set of five Serber lectures providing an indoctrination course on the plans and status of the fission weapon."

2. The Physics of the Fission Bomb

The idea of the "atomic bomb" (so-called) is to store energy in nuclei metastable against fission because of the Coulomb repulsion of assumed fragments:

170 MeV is 2.8×10^{-4} erg, or 7×10^{20} erg/kg;

1 g high explosive is 1000 cal or 4×10^{10} erg;

So 1 kg (4 moles) of fission is 17 kg of H.E. and leaves 4-6 moles of neutrons-- 56 e-foldings or 81 doublings. The neutron absorption time is some $(13 \text{ cm})/(1.4 \times 10^9 \text{ cm/s})$ or 10^{-8} s, or 1 "shake" (in modern terms, 10 ns).

Fission is the (n,2n) process sought by Szilard such that a neutron provokes the release not only of energy but of more than one neutron, so that one can have a prompt chain reaction. In fact, the number of neutrons released per fission (ν) exceeds two, so that even a finite mass can have a reproduction factor greater than unity by retaining half the generated neutrons against leakage from the mass.

In 1982 we can replace 1943 estimates of critical masses of U-235 and (alpha-phase) Pu-239 by published values of 56 kg and 11 kg for bare spheres, and less for these fissile materials surrounded by "tampers"-- at once neutron reflectors and inertial impediments to expansion and termination of the chain reaction.

For a mass supporting a chain reaction, Serber writes for the neutron population

$$n = n_0 e^{\alpha t}; \quad \alpha = (\nu-1)\tau^{-1} \{1 - R^2/R^2(t)\}$$

To have an explosion, we must release the energy rapidly-- fast compared with sound travel time to the targets of destruction (seconds-- easy) but fast also compared with the quenching time of the bomb by disassembly.

Calculating the efficiency, determined by assuming that the internal pressure corresponding to the instantaneous energy density produces expansion of the bomb materials opposed only by the inertia of the bomb, Serber suggests for the energy density, $\varepsilon = \varepsilon_0 e^{\alpha t}$; $P \sim \varepsilon$

$$\rho R \, dv/dt = P = \rho R \, d^2R/dt^2 = \rho R^2 d^2\Delta/dt^2$$
$$(\Delta \equiv \Delta R/R). \text{ So } d^2\Delta/dt^2 = P/\rho R^2 = \varepsilon_0 e^{\alpha t}/\rho R^2$$

Integrating twice, $\Delta_0 = Ef/\alpha^2 R^2$, or $f = \alpha^2 R^2 \Delta_0/E$, with E the fission energy per gram at 100% efficiency. Thus Serber writes $f = K\Delta_0^3 R^2 (\nu-1)^2/E\tau^2$ and estimates $K = 2/3$. For $x = 2$ (a mass twice critical), $\Delta_0 = 0.25$. Then $f = 1.5\%$, giving a yield of 20 KT for a mass of 60 kg. This is the yield for proper timing of initiation, after the assembly of several (x) critical masses. How to keep the stored energy from being released prematurely by a stray neutron (cosmic ray, spontaneous fission, α-n reaction) and how to provide a neutron at the time desired to initiate the chain reaction? The initiating source suggested by Serber is a radium-beryllium (or Po-Be) α-n source of multi-Curie strength <u>suddenly</u> mixed within 10 microseconds or less.

At worst, there will be a neutron at the moment the assembly passes through critical. Because the reproduction factor is (by definition) low at that time, the neutron population will not build so high before the bomb disassembles, and there will be a "fizzle." There is no hope to assemble plutonium by means of a gun-type assembly as used for the Hiroshima bomb with U-235, not so much because of the enormous α-n background provided by the 20,000-year half-life of Pu-239 (after all, 10 kg of Pu is then 1000 Curies), but because of the spontaneous fission of Pu-240 present to 1-5% in weapons-grade Pu of few-month exposure and 10-30% in power-reactor spent fuel. Nevertheless, the more rapid assembly provided by the high-explosive implosion method suggested by Neddermeyer (high velocity and shorter distance than the gun) both allow one to assemble Pu with modest probability of predetonation and <u>also</u> provide a substantial fizzle yield of about one kiloton, as stated in an Oppenheimer letter written before the July 1945 test. In fact, around 1978 the US government announced that it had successfully detonated a fission bomb made with power-reactor plutonium.

3. Thermonuclear Weapons

As described by Herbert York in his book "The Advisors-- Oppenheimer, Teller, and the Superbomb," in the 1950s fission bombs were augmented by providing more-or-less intimate contact between the fissile material and a small charge containing deuterium and tritium, so that the kilovolt temperatures reached in the fission reaction induced the D+T reaction, and the 14-MeV neutron then provoked more fission, thus accounting for the term "boosted fission bomb."

In a thermonuclear weapon, a fission explosion is used to provide initial conditions for a D-D reaction in some solid compound of deuterium. The largest thermonuclear weapon yield to date is some 60 megatons from a Soviet explosion in 1961, and the highest yield weapon in active inventory is said to be a 20-MT warhead on the large Soviet ICBM SS-18 (or RS-20 in Soviet terminology). The US had formerly deployed 20-MT free-fall bombs and still has some 10-MT-class warheads on its 52 Titan-II missiles.

According to US defense department testimony, Soviet strategic weapons derive about 50% of their energy release from fission-- perhaps from normal uranium exposed to the neutrons from the D-D (and D-T) reactions. Since only a few percent of the thermonuclear neutrons need cause fission to reach a 50% energy fraction from fission. However, weapons of any yield can be made with only a few KT of fission yield, as was disclosed in conjunction with the now-abandoned "Plowshare" program of peaceful nuclear explosions (PNE).

Although not news when it was first achieved, the <u>neutron</u> bomb was a matter of some dissension in NATO about five years ago. The ultimate neutron bomb would provide 1 KT of energy from the burning of 12 grams of D and T, yielding the prompt radiation of a 10-KT fission bomb, but with blast effects only to about half the distance. Dubbed ERW (for Enhanced Radiation Weapon), it should instead be called "suppressed-blast weapon," and in my judgment is an irrelevant detail.

Because of considerably smaller size and required investment in fissile material, and because of the capability to produce higher efficiency even with U-235, implosion weapons dominate current inventories.

Of course a nuclear weapon needs devices to fire the implosion symmetrically and synchronously, reliable batteries or the equivalent, barometric and radar fuzes, an adequate initiator, permissive action links, and the like. However, the assembly without the U-235 or Pu is fully susceptible to nondestructive or destructive component tests, and means exist for testing with normal uranium substituted for the U-235 or an appropriate material for the Pu-- as by flash radiography with linear accelerator sources.

4. Weapons Effects

An integral view of a nuclear weapon may consider the mass transformed into energy in the explosion, so that 1 megaton corresponds to the disappearance of about 40 grams of mass. Another integral view observes that a megaton explosion (4×10^{22} erg) will produce one calorie per square centimeter at 100 km distance (in the absence of air with its unpredictable absorption). For a ground-burst weapon, the area exposed to 4 lbs per square inch (4 psi = 0.3 atmosphere) overpressure extends to 5.4 km, and this distance is proportional to yield to the one-third power, (so 540 m for 1-KT ground burst, etc).

Enough about the theory and practice of nuclear weapons. They may, however, cost 10% the cost of the vehicle which delivers them, which in turn may have a cost 10% that of the overall weapons system (including basing, manning, maintenance, and the like).

G.I. Taylor calculated the evolution of an initial state in which a prescribed yield Y of pure energy is deposited in air. At high enough temperature, the energy density in the thermal radiation field has a long mean free path in the ionized atoms and eats faster than sound (in the hot medium) into the undisturbed air, producing an isothermal sphere whose radius increases and temperature decreases with time. The hot surface of this fireball radiates even in the visible until the speed of the shockwave in the undisturbed air exceeds the rate of growth of the fireball, at which time the fireball continues to cool by adiabatic expansion. The shocked air is sufficiently hot to form molecular species (nitrogen oxide) which absorb the fireball radiation, masking it until later in the explosion. For various yields Y, the aerodynamic phase of the explosion is similar with a length and time scale each proportional to the cube root of the yield, so that energy densities and velocities at scaled distances and times are identical.

The fireball at pressure equilibrium with the ambient air then rises at acceleration 2 g and cools, freezing about 1% of the contained oxygen as nitrogen oxide and forming the familiar mushroom cloud. In the ozonosphere, this nitrogen oxide will survive for a year or two and destroy ozone catalytically. The exchange of ten thousand MT is calculated in the National Academy of Science's "Long-term Worldwide Effects of Multiple Nuclear Weapons Detonations" (1978) to result in a 50% decrease in ozone column and substantial incidence of crop failure, blindness in animals and man, and the like. In fact, this effect of nuclear weapons was analyzed but ten years ago in the pioneering calculations of Malvin Ruderman and the late Henry Foley. This first-order and perhaps most important worldwide effect of nuclear explosions was observed in the course of work performed by them to investigate the acceptability of the supersonic transport aircraft!

Bomb-produced nitrogen oxides is one of the bigger influences on "the fate of the earth" following nuclear war-- i.e., global non-economic, non-target-country effects of large-scale nuclear war. Others are possible climate change and zonal species extinction from the gigatons of particulates raised by groundbursts and from forest fires. Fallout of radioactive material, lethal in the target countries and probably their down-wind neighbors, does not loom large on a global scale. At present, I do not believe there is a likelihood of extinction of the human species in nuclear war, although I believe there is a high probability that such a war would destroy our nations, ourselves, and our civilization. My recommendations would be no different if extinction were a fact.

Now you are ready to read Samuel Glasstone's book "The Effects of Nuclear Weapons-- 1977" for detailed graphs. This is a very useful book published by the United States government.

In a 1954 paper at Los Alamos, I calculated the radiated electromagnetic wave from a nuclear explosion in the atmosphere or in space, driven by the asymmetry of the Compton recoil electron current, and measurements were in general agreement until the big space explosions of 1962 which showed that I and others had not anticipated the biggest effect. This was then explained by Conrad Longmire as the local generation at the top of the atmosphere of radiating asymmetry as the sheet of bomb gamma-rays converts to Compton recoils, predominantly forward, which electrons are then deflected by the earth's magnetic field, giving rise to electromagnetic fields out to line of sight of magnitude as great as 50 kV/m. These high-frequency fields are readily screened by conductive shields (and restated as 500 V/cm don't seem so impressive). In any case, the 1962 series of space explosions did essentially nothing to electronics and electrical systems in Hawaii except perhaps for operating the switch on a string of street lights.

This is a synopsis of nuclear weapons as seen by the developer. Unfortunately, the 4-psi overpressure is significant in providing a rule-of-thumb contour at which about half the inhabitants of the city will be killed by blast from the nuclear explosion. Many more can die from burns and firestorm; and from disease and disruption of society if not one but hundreds of the tens of thousands of existing nuclear weapons are used. Just multiply the population density per square kilometer by the area within the 4 psi contour-- 92 square kilometers times the yield in megatons to the two-thirds power. More concretely, the 34 largest Soviet cities have a population density averaging 4200 per sq km, so that 92 sq km would contain 380,000 people-- the expected prompt fatalities from a (first) single megaton weapon.

5. Weapons Systems

You all know the nuclear weapon delivery modes-- ballistic missiles in free fall toward their target and based on submarines (SLBM) at intercontinental range (ICBM) or at intermediate range (IRBM); free-fall gravity bombs; long-range air-launched cruise missiles (ALCM), and the like. There are now about ten thousand strategic nuclear weapons in the Soviet inventory and about the same number in the US. There are probably more than an equal number of nuclear weapons for carriage on short-range aircraft, short-range missiles, surface-to-air missiles, air-to-air missiles, antisubmarine torpedoes, atomic demolition munitions (nuclear landmines), neutron bombs, and all kinds of weapons intended to destroy an opponent's military forces. They are more useful however against an enemy without nuclear weapons. Why?

Because a single delivered megaton weapon will kill half a million people, but a battlefield weapon delivered against an advancing tank column may kill five tanks. Similarly, in the long history of development of nuclear-armed ballistic-missile defense, the problem has always been the necessity of a 99%+ perfect defense of cities, and so long as weapons are deliverable against one's population centers, the escalation from conventional munitions to nuclear weapons is neither militarily credible nor useful. Incidentally, it is my opinion that modern conventional munitions are more effective as well as more useful than nuclear weapons against forces in combat:

- homing non-nuclear torpedoes for ASW,
- homing air-to-air or surface-to-air missiles,
- anti-tank guided weapons,
- ABM defense of silos (to the extent needed).

The use of nuclear weapons against troops on their bases or in rear areas provides effectiveness beyond that of conventional weapons. Still, the vulnerability of cities is so great that even destroying almost all of the strategic weapons on the other side does not preclude nuclear retaliation and the loss of tens of millions or hundreds of millions of lives. As the threat of strategic defenses emerged in the 1950s and 1960s, the US and the Soviet Union built larger forces, hardened their basing, provided penetration aids in the form of multiple and now multiple-independently-targeted reentry vehicles (MIRVs).

6. The Control of Nuclear Weapons

The nuclear weapon forces of the US and the Soviet Union have probably kept an uneasy peace. But the increase in numbers threatens destruction through accident or miscalculation, especially because of the very large numbers of battlefield weapons with the troops, even though we have provided ours with permissive action links (PALs) in the attempt to guarantee that they can't be detonated without specific authorization of the President (and of NATO command, in many cases). but will this authorization come in time to be useful? Lord Montgomery answered that question by stating that he would have used those nuclear weapons first and asked permission later; he went on to characterize battlefield nuclear weapons as an intolerable threat for this reason.

We have built nuclear weapons first in recent decades, confident that there would be some productive military use for them, and ignoring the possibility realized in so many instances that our initiative and our opponent's response have actually impaired our security. It did not have to be so; we could have had MIRVs for defeating possible ABM systems and (if we had abjured accuracy) still not have threatened hardened silos. I preached for many years the inadvisability of ICBM accuracy capable of killing silos hardened to 100-atm overpressure or more, and the greater security of a world in which even if one side had that accuracy, the other did not. But both sides are now becoming able to imperil the survival of the land-based ballistic missile forces of the other; which will clearly (but not disastrously) drive the participants to be ready to launch these ICBMs while they are under attack but before most of the attacking RVs land.

In the US, "Minuteman vulnerability" has been used as the excuse for developing the MX missile, whose hundred-ton mass and ten warheads pose great difficulty (both practical and logical) in finding some land-basing mode which would not be equally suitable for the Minuteman itself.

The passion for technical development has overwhelmed the modest urge for negotiation (arms control) for slowing the advance of the opponent's military capability. In regard to nuclear weapons, I believe it has also imperiled the fight against proliferation of nuclear weapons to additional nations, for instance by arguing against the acceptability of a total (comprehensive) ban on nuclear tests, going beyond the 1963 limited nuclear test ban in which the US, Soviet Union, and United Kingdom agreed not to detonate nuclear weapons in the atmosphere, in space, or in the waters of the world. In effect, the nuclear weapons laboratories in the United States have argued that a comprehensive test ban is unacceptable because nuclear weapons in stockpile would degrade in ways which could be detected only by nuclear testing and for which solutions could be found only by nuclear testing. In fact, the argument comes down to the weapons laboratories' assertion that they could not resist the temptation (no matter what the rules) to improve small aspects of the nuclear weaponry! The then-Director of the Los Alamos Scientific Laboratory, Harold Agnew, testified several times that given money for a non-nuclear testing program and when-necessary exemption from the occupational safety and health act, he definitely did not need nuclear weapons testing to ensure the continued viability of stockpile weapons. He later reversed this position, probably on the political judgment that without a nuclear test program he would not receive these funds and exemptions.

7. Recommendations

I believe that the current technical, qualitative, and quantitative competition in superpower nuclear weaponry and weapon delivery systems, and especially the exaggerated and provably false arguments advanced by proponents of these systems, seriously threaten our survival. I simply restate what I have argued extensively in print:

1. A comprehensive test ban would be an important measure against the spread of nuclear weapons to other nations and should be concluded immediately, banning nuclear explosions except in above-ground, permanently occupied buildings.
2. The US and NATO should on their own initiative withdraw and recycle battlefield nuclear weapons, without renouncing battlefield nuclear explosions from SLBMs, ALCMs, or ICBMs.
3. The US should promptly perfect a reliable and safe capability to launch its ICBMs before they are destroyed by already-launched Soviet ICBMs, said capability being necessary to continue to deter attack if ever the SSBN forces are determined to be vulnerable and are attacked.
4. There should be no program to build MX missiles, B-1 bombers, or a Stealth intercontinental bomber aircraft. They are unnecessary and detract from more urgent defense needs. The MX has demonstrated by five years of desperate search that there is no land basing mode which will provide enduring survival for a large missile.
5. If silo killers are demanded (contrary to this view of the national security), the Trident-I missile in the Poseidon submarines (as well as the Minuteman) should be upgraded in accuracy by the use of the global positioning system (GPS). Minuteman could also profit from ground beacons in the Minuteman fields, which would guard against the possibility of Soviet attack on the Navstar GPS satellites. NATO should

not deploy the 108 Pershing-II and 464 ground-launched cruise missiles authorized in this so-called "double decision" of December 1979, but should depend on Trident-I SLBM and Minuteman ICBMs dedicated to NATO.
6. The UK should not use the Trident-II missile to modernize its deterrent force. Arguments of hull corrosion, metal fatigue and short operating life of the Polaris subs are false (from the facts as set forth in a letter to me of 03/28/78 from the Assistant Secretary of the Navy, David E. Mann) and real problems of the UK "independent deterrent" are not solved by the proposed program.
7. The US and NATO should not renounce first use of nuclear weapons. To do so will impair deterrence, reassurance, and the credibility of western governments.
8. The US and NATO should <u>finally</u> take stock of what needs to be done in standardization, competitive sole-source procurement, of effective-technology non-nuclear weapons. Theater-range weapons are usable in concentration against an advancing salient and are therefore far more valuable than artillery shells; they can also be based with lesser vulnerability.
9. The only deterrent to attack we really have in view is the "countervailing strategy," as advanced before hyperbole by the Carter White House. We thus make calmly and amply clear to the Warsaw Pact that we will use our weapons (if necessary, our nuclear weapons) to enforce losses on the Soviet Union at any time substantially greater (according to our understanding of their values) than what they have gained, and we will persist in this strategy no matter what threats they make or damage they cause. For modest damages we will clearly injure the Soviet Union more than the West, and in case of destruction of the West, we'll destroy the Soviet Union about as well. The purpose of course is to <u>deter</u> the Soviet Union by showing that they have nothing to gain from attack on the West, and a lot to lose.
10. With this goal for nuclear weapons, we should immediately but tentatively reduce the ready strategic force by 50% because the last thousands of warheads contribute cost and hazard but not effective destruction if the strategic nuclear force is applied at all economically-- with its first warheads causing the most destruction per warhead. We should therefore aim toward a 96% reduction-- to 1000 warheads in the United States and in the Soviet Union, and to a lesser proportional reduction in UK, French, and Chinese forces. The Chinese have already offered to reduce their strategic force if the Soviet Union does.

8. Conclusion

In this half hour, I have recalled the state of knowledge of nuclear weapons in early 1943, reminded you of the various effects of nuclear weapons and of their current numbers, delivery means, and basing. Because of their vast potential for destruction I have concluded that nuclear weapons have little effective military utility, less than appropriate conventional weapons, and I have given you my recommendations which will allow nuclear weapons to continue to keep an uneasy peace while reducing the likelihood of war by accident or miscalculation.

Author Index

Ackroyd R T, *377*
Ananiev V D, *497*
Arkhipov V A, *497*
Armbruster P, *105*
Audouze J, *89*
Axe J D, *255*

Bacon G E, *187*
Bahcall J N, *71*
Baker L J, *395*
Barendsen G W, *479*
Bartholomew G A, *491*
Barton J P, *447*
Baym G, *45*
Beynon T D, *395*
Bishop J F W, *417*
Boddy K, *485*
Bohr J, *289*
Brissenden R J, *383*
Brockhouse B N, *193*
Bulkin Yu M, *497*
Bullough R, *429*
Bunin B N, *497*
Buyers W J L, *221*
Byrne J, *15*

Clayton C G, *451*
Clayton D D, *33*
Clough S, *295*
Corngold N R, *371*
Cowley R A, *245*
Crocker V S, *395*
Cusack S, *351*

Dollezhal N A, *497*

Edwards S F, *329*
Egelstaff P A, *267*
Ellis J, *57*
Enderby J E, *271*

Fowler J F, *469*
Fowler W A, *83*
Frank I M, *497*
Friedman E, *143*

Garwin R L, *513*
Goldhaber M, *29*

Hellens R L, *365*
Hendry J, *1*
Higgins J S, *311*
Hodgson P E, *137*

Ishikawa Y, *227*

Jervis R E, *457*

Keller A, *317*
Khriastov N A, *497*
Kjaer K, *289*
Koehler W C, *215*

Lander G H, *233*
Lane A M, *125*
Leadbetter A J, *277*
Llewellyn-Smith C H, *21*
Lomidze V L, *497*
Luschikov V I, *497*
Lynn J E, *119*

Mackintosh A R, *199*
McTague J P, *289*
Marshall W, *503*
Matzke Hj, *423*
May R P, *347*
Mezei F, *181*
Michaudon A, *389*
Miller A, *357*
Mitiaev Yu I, *497*

Nelson R S, *411*
Nicklow R M, *251*
Nielsen M, *289*

Oberthür R, *321*
Ostanevich Yu M, *497*

Pepelyshev Yu N, *497*

Raghaven V, *361*
Ramsey N F, *5*
Rauch H, *169*
Reines F, *75*
Ruderman M, *51*

Schmatz W, *301*
Schoenborn B P, *361*

Seaborg G T, *101*
Shabalin E P, *497*
Shull C G, *157*
Smirnov V S, *497*
Steigman G, *65*
Steyerl A, *177*
Stuhrmann H B, *335*
Svensson E C, *261*

Tayler R J, *61*
Thompson M W, *403*
Träger F, *149*

Truran J W, *95*

Villain J, *239*

Walker J, *435*
White J W, *283*
Wright A C, *305*

Yazvitskij Yu S, *497*

Zhirnov A D, *497*
Zucker M S, *463*